Feedstock-based Bioethanol Fuels. I. Non-Waste Feedstocks

This book aims to inform readers about the recent developments in the production, evaluation, and utilization of bioethanol fuels from non-waste feedstocks. It covers the production of bioethanol fuels from first-generation starch feedstocks and sugar feedstocks, grass biomass, wood biomass, cellulose, biosyngas, and third-generation algae. In this context, there are nine key sections where the first four chapters cover the production of bioethanol fuels from feedstocks at large and non-waste feedstocks.

This book shows that pretreatments and hydrolysis of the non-waste feedstocks, fermentation of hydrolysates, and separation and distillation of bioethanol fuels are the fundamental processes for bioethanol fuel production from these non-waste feedstocks with the exception of the biosyngas feedstocks.

This book is a valuable resource for the stakeholders primarily in the research fields of energy and fuels, chemical engineering, environmental science and engineering, biotechnology, microbiology, chemistry, physics, mechanical engineering, agricultural sciences, food science and engineering, materials science, biochemistry, genetics, molecular biology, plant sciences, water resources, economics, business and management, transportation science and technology, ecology, public, environmental, and occupational health, social sciences, toxicology, multidisciplinary sciences, and humanities among others.

Feedstock-based Bioethanol Fuels. I. Non-Waste Feedstocks

Starch, Sugar, Grass, Wood, Cellulose, Algae, and Biosyngas-based Bioethanol Fuels

Edited by
Ozcan Konur

CRC Press
Taylor & Francis Group
Boca Raton London New York

CRC Press is an imprint of the
Taylor & Francis Group, an **informa** business

Designed cover image: © Shutterstock

First edition published 2024
by CRC Press
2385 NW Executive Center Drive, Suite 320, Boca Raton FL 33431

and by CRC Press
4 Park Square, Milton Park, Abingdon, Oxon, OX14 4RN

CRC Press is an imprint of Taylor & Francis Group, LLC

© 2024 selection and editorial matter, Ozcan Konur; individual chapters, the contributors

ISBN: 978-1-032-12752-1 (hbk)
ISBN: 978-1-032-12849-8 (pbk)
ISBN: 978-1-003-22645-1 (ebk)

DOI: 10.1201/9781003226451

Typeset in Times
by codeMantra

Contents

PART 7　Introduction to Feedstock-based Bioethanol Fuels

PART 8 *Introduction to Non-Waste Feedstock-based Bioethanol Fuels*

PART 9   *First Generation Starch-based Bioethanol Fuels*

Contents

PART 10 First Generation Sugar Feedstock-based Bioethanol Fuels

Ozcan Konur

PART 11 Grass-based Bioethanol Fuels

PART 12   *Wood-based Bioethanol Fuels*

PART 13   Cellulose-based Bioethanol Fuels

PART 14   *Third Generation Algal Bioethanol Fuels*

PART 15 *Biosyngas-based Bioethanol Fuels*

Contents

Preface

The recent supply shocks caused first by the COVID-19 pandemics and later by the Ukrainian war have shown that biofuels such as bioethanol, biohydrogen, biogas, biosyngas, and biodiesel fuels could play a vital role to maintain the energy security and indirectly food security at the global scale. These shocks have also resulted in the need for further setup of incentive structures for the production and consumption of bioethanol fuels in blends with crude oil-based gasoline, petrodiesel, or liquefied natural gas (LNG) in the gasoline and diesel engines, in direct ethanol fuel cells (DEFC), and in the production of biohydrogen fuels for fuel cells and valuable biochemicals from bioethanol fuels.

Thus, it is essential to assess the research on the production, evaluation, and utilization of bioethanol fuels from a wide range of biomass including the first generation starch and sugar feedstocks, wood, and grass, second generation lignocellulosic biomass including waste biomass and agricultural residues such as starch feedstock residues and sugar feedstock residues, and the third generation algal biomass.

Thus, this six-volume *Handbook of Bioethanol Fuels* assesses the research on the production, evaluation, and utilization of bioethanol fuels and presents a representative sample of this interdisciplinary research population with a collection of 110 chapters (Table 1.1).

The first two volumes provide an overview of research on the fundamental processes for bioethanol fuel production with a collection of 39 chapters: Pretreatments of the biomass, hydrolysis of the pretreated biomass, microbial fermentation of the hydrolysates with yeasts, and separation and distillation of bioethanol fuels from the fermentation broth. They also provide an overview of the research on bioethanol fuels and production processes for bioethanol fuels (Tables 1.2 and 1.3).

The third and fourth volumes provide an overview of research on the production of bioethanol fuels from the non-waste and waste biomass, respectively, with a collection of 36 chapters. In this context, the third volume covers the production of bioethanol fuels from first generation starch feedstocks and sugar feedstocks, grass biomass, wood biomass, cellulose, biosyngas, and third generation algae (Table 1.4) while the fourth volume covers the production of second generation bioethanol fuels from residual sugar feedstocks, residual starch feedstocks, food waste, industrial waste, urban waste, forestry waste, and lignocellulosic biomass at large (Table 1.5). They also provide an overview of research on feedstock-based bioethanol fuels, non-waste feedstock-based bioethanol fuels, and second generation waste biomass-based bioethanol fuels (Tables 1.4 and 1.5).

Finally, the fifth and sixth volumes provide an overview of research on the evaluation and utilization of bioethanol fuels with a collection of 37 chapters. In this context, the fifth volume covers the evaluation and utilization of bioethanol fuels in general, gasoline fuels, nanotechnology applications in bioethanol fuels, utilization of bioethanol fuels in transport engines, evaluation of bioethanol fuels, utilization of bioethanol fuels, and development and utilization of bioethanol fuel sensors (Table 1.6). Further, the sixth volume of this handbook provides an overview of research on the country-based experience of bioethanol fuels at large, Chinese, US, and European experience of bioethanol fuels, production of bioethanol fuel-based biohydrogen fuels for fuel cells, bioethanol fuel cells, and bioethanol fuel-based biochemicals with a collection of 19 chapters (Table 1.7).

Thus, the third volume of this handbook provides an overview of research on the production of bioethanol fuels from the non-waste feedstocks such as first generation starch and sugar feedstocks, grass, wood, cellulose, third generation algal feedstocks, and biosyngas with 19 chapters (Table 1.4). In this context, there are nine key sections where the first four chapters cover the production of bioethanol fuels from feedstocks at large and non-waste feedstocks.

Hence, the third volume indicates that research on the production of bioethanol fuels from these non-waste feedstocks has intensified in recent years to become a major part of the bioenergy and biofuels research together primarily with biodiesel, biohydrogen, and biogas research as a sustainable alternative to crude oil-based gasoline and petrodiesel fuels as well as LNG.

The third volume also indicates that a wide range of pretreatments alone or in combination with each other fractionate the biomass to its constituents of cellulose, lignin, and hemicellulose and improve both sugar and bioethanol fuel yield, making the bioethanol fuels from these non-waste feedstocks, with the exception of the biosyngas feedstock, more competitive in relation to crude oil- and natural gas-based fuels.

The third volume also indicates that hydrolysis of the biomass, microbial hydrolysate fermentation, and separation and distillation of bioethanol fuels from fermentation broth together with biomass pretreatments are the fundamental production processes for bioethanol fuel production from these non-waste feedstocks with the exception of the biosyngas feedstock, making bioethanol fuels more competitive in relation to crude oil- and natural gas-based fuels.

The third volume also indicates that a small number of documents, authors, institutions, publication years, source titles, countries, Scopus subject categories, Scopus keywords, and research fronts have shaped the research on bioethanol fuels from these non-waste feedstocks.

The third volume also indicates that the level of funding for the research on both bioethanol fuels from these non-waste feedstocks has not been sufficient with the resulting loss of momentum in the research output in recent years. Thus, there is a crucial need to improve the incentive structures for the major stakeholders such as researchers and their institutions as well source titles and academic databases to improve the volume and quality of the research output in these fields. This is a crucial need to maintain the energy security and indirectly food security at a global scale in the light of the recent supply shocks caused by the COVID-19 pandemics and the Ukrainian war.

The third volume also indicates that the contribution of the social sciences and humanities to the research in these fields has been minimal, due in part to the restrictive editorial policies of the source titles in these fields toward social science- and humanities-based interdisciplinary studies. Thus, there is ample room to improve incentive structures for the inclusion of social sciences and humanities into these fields.

The third volume also indicates that China, Europe as a whole, and the USA have been major producers of the research in these research fields, and there has been heavy competition among them in terms of both volume and citation impact of the research output. The USA and Europe as a whole have had a higher citation impact in relation to China, benefiting from their first-mover advantage starting their research in these fields in the 1970s. China as a late mover has had more intensive research funding initiatives in relation to the USA and Europe, improving both its research output and citation impact through the provision of the efficient incentive structures for its major stakeholders in the last two decades. In this way, China might also overtake both the USA and Europe in terms of citation impact of the research output in addition to the volume of the research output in the future.

This handbook at large and its third volume are a valuable resource for the stakeholders primarily in the research fields of energy and fuels, chemical engineering, environmental science and engineering, biotechnology, microbiology, chemistry, physics, mechanical engineering, agricultural sciences, food science and engineering, materials science, biochemistry, genetics, molecular biology, plant sciences, water resources, economics, business and management, transportation science and technology, ecology, public, environmental, and occupational health, social sciences, toxicology, multidisciplinary sciences, and humanities among others.

Ozcan Konur

Acknowledgments

This handbook has been a multi-stakeholder project from its conception to its publication. CRC Press and Taylor and Francis Group have been the major stakeholders in financing and executing it. Marc Gutierrez has been the executive director of the project. A large number of teams from the Publisher have contributed immensely to the production of the handbook. Only a limited number of authors have participated in this project due to the low level of incentives, compared to journals. A small number of highly cited scholars have shaped the research on bioethanol fuels. The contribution of all these and other stakeholders to this handbook is greatly acknowledged.

Editor

The Editor has interdisciplinary research interests and has published primarily in the areas of bioenergy and biofuels, algal bioenergy and biofuels, nanoenergy and nanofuels, nanobiomedicine, algal biomedicine, disability studies, higher education, biodiesel fuels, algal biomass, lignocellulosic biomass, scientometrics, and bioethanol fuels. He has edited a book titled *Bioenergy and Biofuels* (CRC Press, 2018), a handbook titled *Handbook of Algal Science, Technology, and Medicine* (Elsevier, 2020), and a handbook titled *Handbook of Biodiesel and Petrodiesel Fuels: Science, Technology, Health, and Environment* (CRC Press, 2021) in three volumes.

Contributors

Hongzhang Chen
State Key Laboratory of Biochemical
    Engineering, Beijing Key Laboratory of
    Biomass Refining Engineering, Institute of
    Process Engineering
Chinese Academy of Sciences,
    Beijing, China

Mst. Husne Ara Khatun
State Key Laboratory of Biochemical
    Engineering, Beijing Key Laboratory of
    Biomass Refining Engineering, Institute of
    Process Engineering
Chinese Academy of Sciences,
    Beijing, China

Ozcan Konur
Department of Materials Engineering
(Formerly) Ankara Yildirim Beyazit University,
    Ankara, Turkey

Bahareh Nowruzi
Department of Biotechnology, Science and
    Research Branch
Islamic Azad University,
    Tehran, Iran

Lan Wang
State Key Laboratory of Biochemical
    Engineering, Beijing Key Laboratory of
    Biomass Refining Engineering, Institute of
    Process Engineering
Chinese Academy of Sciences,
    Beijing, China

Part 7

Introduction to Feedstock-based
Bioethanol Fuels

40 Feedstock-based Bioethanol Fuels
Scientometric Study

Ozcan Konur
(Formerly) Ankara Yildirim Beyazit University

40.1 INTRODUCTION

The crude oil-based gasoline fuels (Ma et al., 2002; Newman and Kenworthy, 1989) have been widely used in the transportation sector since the 1920s. However, there have been great public concerns over the adverse environmental and human impact of these fuels (Hill et al., 2006, 2009). Hence, biomass-based bioethanol fuels (Hill et al., 2006; Konur, 2012e, 2015, 2019, 2020a) have increasingly been used in blending gasoline fuels (Hsieh et al., 2002; Najafi et al., 2009), in the fuel cells (Antolini, 2007, 2009), and in the biochemical production (Angelici et al., 2013; Morschbacker, 2009) in a biorefinery context (Fernando et al., 2006; Huang et al., 2008).

Bioethanol fuels also play a critical role in maintaining the energy security (Kruyt et al., 2009; Winzer, 2012) in the supply shocks (Kilian, 2008, 2009) related to oil price shocks (Hamilton, 1983, 2003), COVID-19 pandemics (Fauci et al., 2020; Li et al., 2020), or wars (Hamilton, 1983; Jones, 2012) in the aftermath of the Russian invasion of Ukraine (Reeves, 2014).

However, it is necessary to pretreat the biomass (Taherzadeh and Karimi, 2008; Yang and Wyman, 2008) to enhance the yield of the bioethanol (Hahn-Hagerdal et al., 2006; Sanchez and Cardona, 2008) prior to the bioethanol production through the hydrolysis (Sun and Cheng, 2002; Taherzadeh and Karimi, 2007) and fermentation (Lin and Tanaka, 2006; Olsson and Hahn-Hagerdal, 1996) of the biomass and hydrolysates, respectively.

The research in the field of the bioethanol fuels has intensified in this context in the key research fronts of the pretreatment (Hendriks and Zeeman, 2009; Mosier et al., 2005) and hydrolysis (Sun and Cheng, 2002; Zhang and Lynd, 2004) of the feedstocks, fermentation (Jonsson and Martin, 2016; Palmqvist and Hahn-Hagerdal, 2000) of the hydrolysates, and production (Balat, 2011; Limayem and Ricke, 2012) and evaluation (Hamelinck et al., 2005; Pimentel and Patzek, 2005) of the bioethanol fuels.

The research in this field has also intensified for the feedstocks of wood biomass (Galbe and Zacchi, 2002; Zhu and Pan, 2010), algal biomass (Ho et al., 2013; John et al., 2011), grass biomass (Keshwani and Cheng, 2009; Pimentel and Patzek, 2005), sugar feedstock biomass (Bai et al., 2008; Canilha et al., 2012), and starch feedstock biomass (Bai et al., 2008; Bothast and Schlicher, 2005), lignocellulosic biomass at large (Mosier et al., 2005; Sun and Cheng, 2002), cellulose (Pinkert et al., 2009; Zhang and Lynd, 2004), sugar (Cardona et al., 2010; Laser et al., 2002) and starch feedstock residues (Binod et al., 2010; Talebnia et al., 2010), food waste (Guimaraes et al., 2010; Ravindran and Jaiswal, 2016), industrial waste (Cardona et al., 2010; Prasad et al., 2007), urban waste (Prasad et al., 2007; Ravindran and Jaiswal, 2016), and forestry waste (Duff and Murray, 1996). Thus, it complements the research on the processes for the production, evaluation, and utilization of bioethanol fuels at large.

However, it is essential to develop efficient incentive structures (North, 1991) for the primary stakeholders to enhance the research in this field (Konur, 2000, 2002a,b,c, 2006a,b, 2007a,b).

The scientometric analysis has been used in this context to inform the primary stakeholders about the current state of the research in a selected research field (Garfield, 1955; Konur, 2011, 2012a,b,c,d,e,f,g,h,i, 2015, 2018b, 2019, 2020a).

As there have been no published scientometric studies in this field, this book chapter presents a scientometric study of the research in the bioethanol fuels. It examines the scientometric characteristics of both the sample and population data presenting scientometric characteristics of these both datasets in the order of documents, authors, publication years, institutions, funding bodies, source titles, countries, Scopus subject categories, Scopus keywords, and research fronts.

40.2   MATERIALS AND METHODS

The search for this study was carried out using Scopus database (Burnham, 2006) in October 2022.

As a first step for the search of the relevant literature, the keywords were selected using the most-cited first 300 population papers for each feedstock. The selected keyword list was then optimized to obtain a representative sample of papers for each research field. These five keyword lists were then integrated to obtain the keyword list for this research field (Konur, 2023a,b).

As a second step, two sets of data were used for this study. First, a population sample of 26,850 papers was used to examine the scientometric characteristics of the population data. Secondly, a sample of 269 most-cited papers, corresponding to 1% of the population papers, was used to examine the scientometric characteristics of these citation classics.

The scientometric characteristics of these both sample and population datasets were presented in the order of documents, authors, publication years, institutions, funding bodies, source titles, countries, Scopus subject categories, Scopus keywords, and research fronts.

Lastly, the key scientometric findings for both datasets were discussed to highlight the research landscape for the bioethanol fuels. Additionally, a number of brief conclusions were drawn and a number of relevant recommendations were made to enhance the future research landscape.

40.3   RESULTS

40.3.1   The Most Prolific Documents in the Bioethanol Fuels

The information on the types of documents for both datasets is given in Table 40.1. The articles and conference papers, published in journals, dominate both the sample (71%) and population (93%) papers with 22% deficit. Further, review papers and short surveys have a 25% surplus as they are

TABLE 40.1
Documents in the Bioethanol Fuels

Documents	Sample Dataset (%)	Population Dataset (%)	Surplus (%)
Article	68.0	90.8	−22.8
Review	26.8	3.4	23.4
Conference paper	2.6	2.4	0.2
Short Survey	2.6	0.2	2.4
Book chapter	0.0	2.0	−2.0
Letter	0.0	0.5	−0.5
Note	0.0	0.4	−0.4
Book	0.0	0.2	−0.2
Editorial	0.0	0.1	−0.1
Sample size	269	26,850	

Population dataset, The number of papers (%) in the set of the 26,850 population papers; Sample dataset, The number of papers (%) in the set of 269 highly cited papers.

over-represented in the sample papers as they constitute 29% and 4% of the sample and population papers, respectively.

It is further notable that 97%, 2%, and 1% of the population papers were published in journals, books, and book series, respectively. Similarly, 97% and 3% of the sample papers were published in the journals and book series, respectively.

40.3.2 THE MOST PROLIFIC AUTHORS IN THE BIOETHANOL FUELS

The information about the most prolific 24 authors with at least 1.5% of sample papers each is given in Table 40.2.

The most prolific author is John N. Saddler with 4.1% of the sample papers, followed by Bruce E. Dale, Charles E. Wyman, and Guido Zacchi with 3.7%, 3.7%, and 3.0% of the sample papers, respectively. The other prolific authors are Blake A. Simmons, Michael E. Himmel, and Barbel Hahn-Hagerdal with 2.6% of the sample papers each.

On the other hand, the most influential author is John N. Saddler with 3.5% surplus, followed by Bruce E. Dale and Charles E. Wyman with 3.3% surplus each. The other influential authors are Guido Zacchi, Barbel Hahn-Hagerdal, Michael E. Himmel, Blake A. Simmons, and Robin D. Rogers with 2.1%–2.6% surplus each.

The most prolific institution for the sample dataset is the Lund University with four authors, followed by University of British Columbia with two authors. On the other hand, the most prolific country for the sample dataset is the USA with 12 authors, followed by Sweden and Canada with six and two authors, respectively. In total, only seven countries house these top authors.

The most prolific research front for these top authors is the pretreatments of the feedstocks with 23 authors followed by the hydrolysis of the feedstocks with 21 authors. The other prolific research fronts are the fermentation of the hydrolysates and the bioethanol production in general with 12 and 11 authors, respectively.

On the other hand, there is significant gender deficit (Beaudry and Lariviere, 2016) for the sample dataset as surprisingly only three of these top researchers are female with a representation rate of 13%.

Additionally, there are other authors with the relatively low citation impact and with 0.1%–0.4% of the population papers each: Run Cang Sun, Juan C. Parajo, Venkatesh Balan, Akihiko Kondo, Mercedes Ballesteros, Qiang Yong, Feng Xu, Junyong Zhu, Jie Bao, Jun Zhang, Bruce S. Dien, Hasan Jameel, Yunqiao Pu, Zhenhong Yuan, Yanqun Xu, Eulogio Castro, Ignacio Ballesteros, Xianzhi Meng, Lisbeth Olsson, Paul Christakopoulos, Vijay Singh, Herbert Sixta, Anuj K. Chandel, Tae Hyun Kim, Xinshu Zhuang, Verawat Champreda, Caoxing Huang, Kyoung Heon Kim, Navadol Laosiripojana, Liangcai Peng, Hongyan Chen, Gunnar Liden, Chang G. Yoo, Paloma Manzanares, Solange I. Mussatto, Maria J. Negro, Carlos R. Soccol, Yongcan Jin, Carlos Martin, Xuebing Zhao, Michael A. Cotta, Caoxing Lai, Lee R. Lynd, Anne S. Meyer, Luiz P. Ramos, Wen Wang, Johnn F. Jorgens, Rajaev Kumar, Shiro Saka, Silvio S. Da Silva, Andre Ferraz, Gwi-Taek Jeong, Adriana M. F. Milagres, Jose M Oliva, Qiang Yu, Dehua Liu, Wei Qi, Aloia Romani, Donghai Wang, and Zhanrong Zhang.

40.3.3 THE MOST PROLIFIC RESEARCH OUTPUT BY YEARS IN THE BIOETHANOL FUELS

Information about papers published between 1970 and 2022 is given in Figure 40.1. This figure clearly shows that the bulk of the research papers in the population dataset were published primarily in the 2010s and the early 2020s with 52% and 20% of the population dataset, respectively. Similarly, the publication rates for the 2000s, 1990s, 1980s, and 1970s were 12%, 6%, 5%, and 2%, respectively. Further, the rate for the pre-1970s was 2%.

Similarly, the bulk of the research papers in the sample dataset were published in the 2000s and 2010s with 52% and 28% of the sample dataset, respectively. Similarly, the publication rates for the 1990s, 1980s, and 1970s were 10%, 2%, and 1% of the sample papers, respectively. Further, the rate for the pre-1970s was 0%.

TABLE 40.2

Most Prolific Authors in the Bioethanol Fuels

No.	Author Name	Author Code	Sample Papers (%)	Population Papers (%)	Surplus	Institution	Country	HI	N	Res. Front
1	Saddler, John N.	7005297559 57202481615	4.1	0.6	3.5	Univ. British Columbia	Canada	99	420	P, H, F, R
2	Dale, Bruce E.	7201511969	3.7	0.4	3.3	Michigan State Univ.	USA	92	430	P, H, F, R
3	Wyman, Charles E.	7004396809	3.7	0.4	3.3	Univ. Calf. Riverside	USA	80	287	P, H, F, R
4	Zacchi, Guido	7006727748	3.0	0.4	2.6	Lund Univ.	Sweden	68	204	P, H, F, R
5	Simmons, Blake A.	7102183263	2.6	0.3	2.3	Lawrence Berkeley Natl. Lab.	USA	76	446	P, H
6	Himmel, Michael E.	7007125552	2.6	0.2	2.4	Natl. Renew. Ener. Lab.	USA	74	423	P, H
7	Hahn-Hagerdal, Barbel*	7005389381	2.6	0.1	2.5	Lund Univ.	Sweden	76	258	P, H, F, R
8	Rogers, Robin D.	35474829200	2.2	0.1	2.1	Univ. Alabama	USA	118	891	P
9	Ragauskas, Arthur J.	7006265204	1.9	0.4	1.5	Univ. Tennessee	USA	93	762	P, H
10	Galbe, Mats	7003788758	1.9	0.4	1.5	Lund Univ.	Sweden	51	131	P, H, F, R
11	Pandey, Ashok	7201771319	1.9	0.2	1.7	CSIR	India	93	917	P, H, F, R
12	Holtzapple, Mark T.	7004167004	1.9	0.2	1.7	Texas A&M Univ.	USA	47	199	P
13	Pan, Xuejun	57203296000	1.9	0.1	1.8	Univ. Wisconsin Madison	USA	45	118	P, H
14	Karimi, Keikhosro	10046195700	1.5	0.3	1.2	Vrije Univ.	Belgium	57	224	P, H, F, R
15	Taherzadeh, Mohammad J.	6701407496	1.5	0.3	1.2	Univ. Boras	Sweden	67	426	P, H, F, R
16	Singh, Seema*	35264950300	1.5	0.2	1.3	Sandia Natl. lab.	USA	59	186	P, H
17	Ladisch, Michael R.	7005670397	1.5	0.2	1.3	Purdue Univ.	USA	75	334	P, R
18	Jonsson, Leif J.	7102349315	1.5	0.2	1.3	Umea Univ.	Sweden	41	148	P, H, F
19	Jorgensen, Henning	7202554496	1.5	0.1	1.4	Univ. Copenhagen	Denmark	35	84	P, H
20	Chundawat, Shishir P. S.	12803763300	1.5	0.1	1.4	Rutgers	USA	32	91	P, H
21	Gilkes, Neil	35493889000	1.5	0.1	1.4	Univ. British Columbia	Canada	51	88	P, H
22	Nilvebrant, Nils-Olof	57209815309	1.5	0.1	1.4	Borregaard	Norway	22	43	H, F
23	Palmqvist, Eva A.*	6603821896	1.5	0.1	1.4	Lund Univ.	Sweden	19	31	P, H, F, R
24	Vogel, Kenneth P.	57204256821 7102498441	1.5	0.1	1.4	Univ. Nebraska-Lincoln	USA	63	146	P, H

*, Female; Author code, the unique code given by Scopus to the authors; F, Fermentation of the hydrolysates; H, Hydrolysis of the feedstock; P, Pretreatment of the feedstock; Population papers, the number of papers authored in the population dataset; R, Bioethanol fuel production; Sample papers, the number of papers authored in the sample dataset.

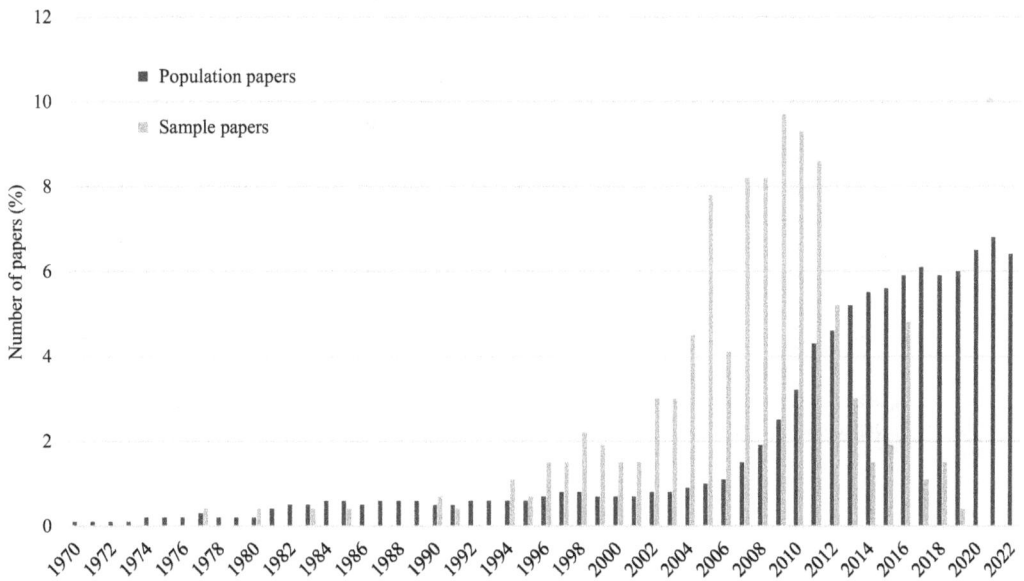

FIGURE 40.1 The research output by years regarding the bioethanol fuels.

The most prolific publication years for the population dataset were 2021, 2020, and 2022 with 6.8, 6.5%, and 6.4% of the dataset, respectively, whilst 76% of the population papers were published between 2008 and 2022. Similarly, 83% of the sample papers were published between 2002 and 2016, whilst the most prolific publication years were 2009, 2010, and 2011 with 9.7%, 9.3%, and 8.6% of the sample papers, respectively.

It is notable that there is a rising trend for the population papers between 2005 and 2011. However, it appears that it lost its momentum after 2011.

40.3.4 The Most Prolific Institutions in the Bioethanol Fuels

Information about the most prolific 32 institutions publishing papers on the bioethanol fuels with at least 1.5% of the sample papers each is given in Table 40.3.

The most prolific institutions are the Lund University and National Renewable Energy Laboratory (NREL) with 6.7 and 6.3% of the sample papers, respectively, followed by the University of British Columbia, Michigan State University, Dartmouth College, and Technical University Denmark with 3%–5.2% of the sample papers each. The other prolific institutions are University Sao Paulo, Oak Ridge National Laboratory, NC State University, USDA Forest Service, Joint Bioenergy Institute, and Georgia Institute of Technology with 2.6% of the sample papers each.

Similarly, the top country for these most prolific institutions is the USA with 19 institutions. The other prolific countries are Sweden and Denmark with three and two institutions, respectively. In total, ten countries house these top institutions.

On the other hand, the institutions with the most citation impact are the Lund University and NREL with 5.7% and 5.2% surplus, respectively, followed by the University of British Columbia, Michigan State University, and Dartmouth College with 4.2%, 3.7%, and 3.4% surplus, respectively. The other influential institutions are Technical University of Denmark, Georgia Institute of Technology, and Joint Bioenergy Institute with 2.2% surplus each.

Additionally, there are other institutions with the relatively low citation impact and with 0.4%–1.6% of the population papers each: South China University of Technology, Nanjing Forestry University, Beijing Forestry University, State University of Campinas, United States Department

TABLE 40.3
The Most Prolific Institutions in the Bioethanol Fuels

No.	Institutions	Country	Sample Papers (%)	Population Papers (%)	Surplus (%)
1	Lund Univ.	Sweden	6.7	1.0	5.7
2	Natl. Renew. Ener. Lab.	USA	6.3	1.1	5.2
3	Univ. British Columbia	Canada	5.2	1.0	4.2
4	Michigan State Univ.	USA	4.5	0.8	3.7
5	Dartmouth Coll.	USA	3.7	0.3	3.4
6	Tech. Univ. Denmark	Denmark	3.0	0.8	2.2
7	Univ. Sao Paulo	Brazil	2.6	2.1	0.5
8	Oak Ridge Natl. Lab.	USA	2.6	0.8	1.8
9	NC State Univ.	USA	2.6	0.8	1.8
10	USDA Forest Serv.	USA	2.6	0.7	1.9
11	Jnt. Bioener. Inst.	USA	2.6	0.4	2.2
12	Georgia Inst. Technol.	USA	2.6	0.4	2.2
14	Chinese Acad. Sci.	China	2.2	2.4	-0.2
15	Univ. Wisconsin Madison	USA	2.2	0.7	1.5
16	Purdue Univ.	USA	2.2	0.6	1.6
17	Univ. Calif. Berkeley	USA	2.2	0.4	1.8
18	Sandia Natl. Lab.	USA	2.2	0.3	1.9
19	Univ. Alabama	USA	2.2	0.2	2.0
20	USDA Agr. Res. Serv.	USA	1.9	0.9	1.0
21	Univ. Copenhagen	Denmark	1.9	0.4	1.5
22	Auburn Univ.	USA	1.9	0.4	1.5
23	Lawrence Berkeley Natl. Lab.	USA	1.9	0.3	1.6
24	Natl. Inst. Intdis. Sci. Technol.	India	1.9	0.3	1.6
25	VTT Tech. Res. Ctr.	Finland	1.5	0.4	1.1
26	Texas A&M Univ.	USA	1.5	0.4	1.1
27	Isfahan Univ. Technol.	Iran	1.5	0.3	1.2
28	Umea Univ.	Sweden	1.5	0.3	1.2
29	Virginia Polytech. Inst. State Univ.	USA	1.5	0.3	1.2
30	Imperial Coll.	UK	1.5	0.3	1.2
31	Univ. Boras	Sweden	1.5	0.3	1.2
32	Novozymes Biotech Inc.	USA	1.5	0.1	1.4

of Agriculture, State Key Laboratory of Pulp and Paper Engineering, Kyoto University, State University of Paulista, National Scientific Research center (CNRS), University of Tennessee, Federal University of Rio de Janeiro, Tianjin University of Science & Technology, Aalto University, Tianjin University, Russian Academy of Sciences, University of Vigo, Chalmers University of Technology, Korea University, National Biorenewables Laboratory, Qilu University of Technology, Beijing University of Chemical Technology, Tsinghua University, Iowa State University, University of Putra Malaysia, University of Illinois Urbana-Champaign, Chinese Academy of Forestry, DOE Bioenergy Research Centers, Guangxi University, Northwest A&F University, China Agricultural University, Federal University of Parana, University of Tokyo, Ministry of Agriculture of China, Wageningen University & Research, Helsinki University, Jiangnan University, Abo Akademi University, East China University of Science and Technology, King Mongkut's University of Technology, University of California Riverside, University of Georgia, and CIEMAT.

40.3.5 THE MOST PROLIFIC FUNDING BODIES IN THE BIOETHANOL FUELS

Information about the most prolific 16 funding bodies funding at least 1.1% of the sample papers each is given in Table 40.4. Further, only 33% and 46% of the sample and population papers each were funded, respectively

The most prolific funding body is the U.S. Department of Energy with 3.6% of the sample papers, followed by the Natural Sciences and Engineering Research Council of Canada and the European Commission with 2.2% of the sample papers each. The other funding bodies are the National Science Foundation, Swedish National Board for Industrial and Technical Development, National Natural Science Foundation of China, U.S. Department of Agriculture, and NREL with 1.5%–1.9% of the sample papers each.

On the other hand, the most prolific countries for these top funding bodies are the USA and Sweden with five funding bodies each, followed by Canada with two funding bodies. In total, six countries and the EU house these top funding bodies.

The funding body with the most citation impact is the U.S. Department of Energy with 2.6% surplus, followed by the Swedish National Board for Industrial and Technical Development with 1.8% surplus. The other influential funding bodies are the European Commission, Natural Sciences and Engineering Research Council of Canada, and NREL with 1.2%–1.3% surplus each. Further, the funding body with the least citation impact is the National Natural Science Foundation of China

TABLE 40.4
The Most Prolific Funding Bodies in the Bioethanol Fuels

No.	Funding Bodies	Country	Sample Paper No. (%)	Population Paper No. (%)	Surplus (%)
1	U.S. Department of Energy	USA	4.5	1.9	2.6
2	Natural Sciences and Engineering Research Council of Canada	Canada	2.2	0.9	1.3
3	European Commission	EU	2.2	0.9	1.3
4	National Science Foundation	USA	1.9	1.2	0.7
5	Swedish National Board for Industrial and Technical Development	Sweden	1.9	0.1	1.8
6	National Natural Science Foundation of China	China	1.5	9.0	−7.5
7	U.S. Department of Agriculture	USA	1.5	0.7	0.8
8	National Renewable Energy Laboratory	USA	1.5	0.3	1.2
9	Research Support Foundation of the State of Sao Paulo	Brazil	1.1	1.9	−0.8
10	Office of Science	USA	1.1	0.9	0.2
11	Energy Agency	Sweden	1.1	0.4	0.7
12	Swedish Research Council	Sweden	1.1	0.2	0.9
13	Natural Resources Canada	Canada	1.1	0.1	1.0
14	Knut and Alice Wallenberg Foundation	Sweden	1.1	0.1	1.0
15	Carl Tryggers Foundation for Scientific Research	Sweden	1.1	0.1	1.0
16	Technology Information, Forecasting and Assessment Council	India	1.1	0.1	1.0

with 7.5% deficit, followed by Research Support Foundation of the State of Sao Paulo with 0.8% deficit. The National Natural Science Foundation of China is the largest funder of the population papers with a funding rate of 9%.

The other funding bodies with the relatively low citation impact and with 0.4%–3.0% of the population papers each are the National Council for Scientific and Technological Development, Higher Education Personnel Improvement Coordination, National Key Research and Development Program of China, Fundamental Research Funds for the Central Universities, Japan Society for the Promotion of Science, National Research Foundation of Korea, European Regional Development Fund, Chinese Academy of Sciences, National Basic Research Program of China (973 Program), China Postdoctoral Science Foundation, Biological and Environmental Research, Priority Academic Program Development of Jiangsu Higher Education Institutions, National Institute of Food and Agriculture, Natural Science Foundation of Jiangsu Province, China Scholarship Council, Seventh Framework Program, Ministry of Education, Culture, Sports, Science and Technology, National Council of Science and Technology, Ministry of Science and Technology India, Foundation for Science and Technology, National High-tech Research and Development Program, Ministry of Education, Ministry of Education, Science and Technology, Biotechnology and Biological Sciences Research Council, Thailand Research Fund, Ministry of Economy and Competitiveness, Ministry of Higher Education Malaysia, Ministry of Education of China, and Ministry of Science, ICT and Future Planning.

40.3.6   The Most Prolific Source Titles in the Bioethanol Fuels

Information about the most prolific 14 source titles publishing at least 1.5% of the sample papers each in the bioethanol fuels is given in Table 40.5.

The most prolific source title is the Bioresource Technology with 21.9% of the sample papers, followed by Biotechnology and Bioengineering and Green Chemistry with 7.1% of the sample papers each. The other prolific titles are the Biomass and Bioenergy, Biotechnology for Biofuels,

TABLE 40.5
The Most Prolific Source Titles in the Bioethanol Fuels

No.	Source Titles	Sample Papers (%)	Population Papers (%)	Surplus (%)
1	Bioresource Technology	21.9	9.1	12.8
2	Biotechnology and Bioengineering	7.1	1.7	5.4
3	Green Chemistry	7.1	0.9	6.2
4	Biomass and Bioenergy	3.3	2.1	1.2
5	Biotechnology For Biofuels	3.0	2.0	1.0
6	Applied Microbiology and Biotechnology	2.6	0.9	1.7
7	Proceedings of the National Academy of Sciences of the United States of America	2.6	0.1	2.5
8	Biotechnology Advances	2.6	0.1	2.5
9	Enzyme and Microbial Technology	2.2	0.8	1.4
10	Applied Biochemistry and Biotechnology Part A	1.9	0.8	1.1
11	Journal of Biotechnology	1.9	0.3	1.6
12	Biotechnology Progress	1.5	0.5	1.0
13	Renewable and Sustainable Energy Reviews	1.5	0.4	1.1
14	Chemical Communications	1.5	0.1	1.4

Applied Microbiology and Biotechnology, Proceedings of the National Academy of Sciences of the United States of America, and Biotechnology Advances with 2.6%–3.3% of the sample papers each.

On the other hand, the source title with the most citation impact is the Bioresource Technology with 12.8% surplus, followed by Green Chemistry and Biotechnology and Bioengineering with 6.2% and 5.4% surplus, respectively. The other influential titles are the Biotechnology Advances, Proceedings of the National Academy of Sciences of the United States of America, Applied Microbiology and Biotechnology, and Journal of Biotechnology with 1.6%–2.5% surplus each.

The other source titles with the relatively low citation impact with 0.4%–2.8% of the population papers each are Bioresources, Industrial Crops and Products, Applied Biochemistry and Biotechnology, Cellulose, Biomass Conversion and Biorefinery, Renewable Energy, ACS Sustainable Chemistry and Engineering, Process Biochemistry, Biotechnology Letters, Holzforschung, Bioenergy Research, Fuel, RSC Advances, Industrial And Engineering Chemistry Research, Journal of Cleaner Production, Journal of Chemical Technology and Biotechnology, Waste and Biomass Valorization, Cellulose Chemistry and Technology, Chemical Engineering Transactions, Applied and Environmental Microbiology, Energy and Fuels, Biochemical Engineering Journal, Wood Science and Technology, Energies, Bioprocess and Biosystems Engineering, Scientific Reports, Applied Energy, Chemical Engineering Journal, Journal of Wood Chemistry and Technology, Energy, Journal of Wood Science, Plos One, and Journal of the American Chemical Society.

40.3.7 The Most Prolific Countries in the Bioethanol Fuels

Information about the most prolific 19 countries publishing at least 1.5% of sample papers each in the bioethanol fuels is given in Table 40.6.

The most prolific country is the USA with 39% of the sample papers, followed by Sweden, China, Canada, Japan, and Denmark with 7%–11% of the sample papers each. The other prolific countries are India, Brazil, and the UK with 4%–6% of the sample papers each.

TABLE 40.6
The Most Prolific Countries in the Bioethanol Fuels

No.	Countries	Sample Papers (%)	Population Papers (%)	Surplus (%)
1	USA	39.4	17.2	22.2
2	Sweden	11.5	3.3	8.2
3	China	8.2	19.6	−11.4
4	Canada	7.4	4.2	3.2
5	Japan	7.1	6.4	0.7
6	Denmark	6.7	1.5	5.2
7	India	6.3	7.4	−1.1
8	Brazil	3.7	7.6	−3.9
9	UK	3.7	3.1	0.6
10	Spain	3.0	3.3	-0.3
11	Germany	3.0	2.7	0.3
12	France	3.0	2.6	0.4
13	Finland	2.6	2.1	0.5
14	Netherlands	2.6	1.1	1.5
15	S. Korea	2.2	3.7	−1.5
16	Iran	1.9	1.4	0.5
17	Australia	1.5	1.6	−0.1
18	S. Africa	1.5	1.0	0.5
19	Ireland	1.5	0.3	1.2

It is notable that ten countries listed in Table 40.6 produce 45% and 24% of the sample and population papers, respectively, with 21% surplus and China is the largest producer of the population papers with a publication rate of 19.6%.

On the other hand, the country with the most citation impact is the USA with 22% surplus, followed by Sweden and Denmark with 8% and 5% surplus, respectively. The other prolific countries are Canada and Netherlands with 3% and 2% surplus, respectively. Similarly, the country with the least citation impact is China with 11% deficit, followed by Brazil, S. Korea, and India with 1%–4% deficit each.

Additionally, there are other countries with relatively low citation impact and with 0.4%–2.4% of the sample papers each: Malaysia, Thailand, Italy, Indonesia, Turkey, Russia, Mexico, Portugal, Taiwan, Poland, Egypt, Nigeria, Pakistan, Austria, Belgium, Greece, Colombia, Argentina, Chile, Czech Republic, Switzerland, Slovakia, New Zealand, Saudi Arabia, Vietnam, Cuba, Romania, Hungary, Norway, and Israel.

40.3.8 THE MOST PROLIFIC SCOPUS SUBJECT CATEGORIES IN THE BIOETHANOL FUELS

Information about the most prolific ten Scopus subject categories indexing at least 3% of the sample papers each is given in Table 40.7.

The most prolific Scopus subject category in the bioethanol fuels is the Chemical Engineering with 57% of the sample papers, followed by Environmental Science, Biochemistry, Genetics and Molecular Biology, Energy, and Immunology and Microbiology with 31%–45% of the sample papers each. It is notable that Social Sciences including Economics and Business account for only 1% and 3% of the sample and population studies, respectively.

On the other hand, the Scopus subject category with the most citation impact is the Chemical Engineering with 17% surplus. The other influential subject areas are Biochemistry, Genetics and Molecular Biology, Environmental Science, and Immunology and Microbiology with 12%–13% of the sample papers each. Similarly, the least influential subject categories are Agricultural and Biological Sciences, and to a lesser extent Engineering, Materials Science, and Chemistry with 6%–7% deficit each.

40.3.9 THE MOST PROLIFIC KEYWORDS IN THE BIOETHANOL FUELS

Information about the Scopus keywords used with at least 6.7% or 3.6% of the sample or population papers, respectively, is given in Table 40.8. For this purpose, keywords related to the keyword set given in the appendix of the related papers are selected from a list of the most prolific keyword set provided by Scopus database.

TABLE 40.7
The Most Prolific Scopus Subject Categories in the Bioethanol Fuels

No.	Scopus Subject Categories	Sample Papers (%)	Population Papers (%)	Surplus (%)
1	Chemical Engineering	56.5	39.6	16.9
2	Environmental Science	44.6	31.8	12.8
3	Biochemistry, Genetics and Molecular Biology	40.9	27.5	13.4
4	Energy	36.8	28.2	8.6
5	Immunology and Microbiology	30.9	18.5	12.4
6	Chemistry	12.6	18.3	−5.7
7	Agricultural and Biological Sciences	8.9	19.4	−10.5
8	Materials Science	5.6	11.7	−6.1
9	Engineering	4.5	11.4	−6.9
10	Multidisciplinary	3.0	2.4	0.6

TABLE 40.8
The Most Prolific Keywords in the Bioethanol Fuels

No.	Keywords	Sample Papers (%)	Population Papers (%)	Surplus (%)
1	Feedstocks			
	Cellulose	65.1	32.5	32.6
	Lignin	54.6	23.9	30.7
	Biomass	44.2	19.5	24.7
	Lignocellulose	34.2	11.3	22.9
	Hemicellulose	20.8	5.8	15.0
	Lignocellulosic biomass	17.5	8.0	9.5
	Zea	15.2	5.5	9.7
	Carbohydrate	12.6	4.0	8.6
	Wood	11.2	7.4	3.8
	Corn	8.9	1.6	7.3
	Straw	7.4	4.9	2.5
	Corn stover	6.7	2.4	4.3
	Bagasse	4.5	6.2	−1.7
	Maize	3.7	3.6	0.1
2	Pretreatments			
	Enzymes	24.9	11.6	13.3
	Pretreatment	19.0	8.5	10.5
	ILs	16.0	7.6	8.4
	Pre-treatment	13.8	10.2	3.6
	Solvent	13.0	5.1	7.9
	Temperature	11.9	6.1	5.8
	Sulfuric acid	8.2	3.6	4.6
	Dissolution	8.2	2.5	5.7
	Degradation	7.4	4.1	3.3
	ammonia	7.4	2.0	5.4
	Trichoderma	7.1		7.1
	Hypocrea	6.7	1.6	5.1
	pH	4.8	5.4	−0.6
	Water	4.8	3.9	0.9
	Delignification	0.0	5.0	−5.0
3	Fermentation	0.0	0.0	0.0
	Fermentation	30.5	22.2	8.3
	Fungi	12.3	7.4	4.9
	Saccharomyces	7.8	6.4	1.4
	Yeast	7.4	7.7	−0.3
4	Hydrolysis and hydrolysates			
	Hydrolysis	55.0	24.5	30.5
	Sugar	32.0	14.4	17.6
	Cellulases	29.7	11.7	18.0
	Enzyme activity	29.4	12.8	16.6
	Enzymatic hydrolysis	19.7	15.4	4.3
	Glucose	17.1	13.2	3.9

(Continued)

TABLE 40.8 (*Continued*)
The Most Prolific Keywords in the Bioethanol Fuels

No.	Keywords	Sample Papers (%)	Population Papers (%)	Surplus (%)
	Saccharification	16.4	11.9	4.5
	Xylose	10.8	5.0	5.8
	Enzymolysis	7.1	3.6	3.5
	Polysaccharides	7.1	2.5	4.6
5	Products			
	Ethanol	38.3	25.1	13.2
	Biofuels	26.0	14.5	11.5
	Bioethanol	12.3	11.9	0.4
	Ethanol production	8.2	5.1	3.1
	Biofuel production	7.1	2.4	4.7
	Bio-ethanol production	3.7	4.8	−1.1

These keywords are grouped under the five headings: feedstocks, pretreatments, fermentation, hydrolysis and hydrolysates, and products.

The most prolific keyword related to the biomass and biomass constituents is cellulose with 65% of the sample papers, followed by lignin, biomass, lignocellulose, and hemicellulose with 21%–55% of the sample papers, respectively. The other prolific keywords are lignocellulosic biomass, zea, carbohydrate, and wood with 11%–18% of the sample papers each.

Further, the most prolific keyword related to the pretreatments is enzymes with 25% of the sample papers, followed by pretreatment, ILs, pre-treatment, solvent, and temperature with 12%–19% of the sample papers each. The other prolific keywords are the sulfuric acid, dissolution, degradation, ammonia, trichoderma, and hypocrea with 7%–8% of the sample papers each.

The most prolific keyword related to the fermentation is fermentation with 31% of the sample papers. The other prolific keyword is fungi with 12% of the sample papers. Further, the most prolific keyword related to the hydrolysis and hydrolysates is hydrolysis with 55% of the sample papers. The other prolific keywords are sugar, cellulases, enzyme activity, enzymatic hydrolysis, glucose, and saccharification with 16%–32% of the sample papers each. Finally, the most prolific keyword related to the products is ethanol with 38% of the sample papers, followed by biofuels with 30% of the sample papers. The other prolific keywords are biofuels and bioethanol with 26 and 12% of the sample papers, respectively.

On the other hand, the most prolific keywords across all of the research fronts are cellulose, hydrolysis, lignin, biomass, ethanol, lignocellulose, sugar, fermentation, cellulases, enzyme activity, biofuels, and enzymes with 25%–65% of the sample papers each. The other prolific keywords are hemicellulose, enzymatic hydrolysis, pretreatment, lignocellulosic biomass, glucose, saccharification, and ILs with 16%–21% of the sample papers each.

Similarly, the most influential keywords are cellulose, lignin, hydrolysis, biomass, and lignocellulose with 23%–33% surplus each. The other influential keywords are cellulases, sugar, enzyme activity, hemicellulose, enzymes, ethanol, biofuels, and pretreatment with 11%–18% surplus each.

40.3.10 THE MOST PROLIFIC RESEARCH FRONTS IN THE BIOETHANOL FUELS

Information about the research fronts for the sample papers in the bioethanol fuels with respect to the feedstocks is given in Table 40.9. As this table shows, the most prolific research front for this field is the lignocellulosic biomass with 36% of the sample papers. The other prolific research fronts are the cellulose and wood biomass with 21% and 10% of these HCPs, respectively. The other research

TABLE 40.9
The Most Prolific Research Fronts for the Bioethanol Fuels

No.	Research Fronts	N Paper (%) Sample	N Paper Population (%)	Surplus (%)
1	Lignocellulosic biomass-based bioethanol fuels	36.1	13.6	22.5
2	Cellulose-based bioethanol fuels	20.8	14.7	6.1
3	Wood-based bioethanol fuels	10.4	20.8	−10.4
4	Grass-based bioethanol fuels	5.6	3.8	1.8
5	Sugar bioethanol fuels	3.7	4.2	−0.5
6	Starch feedstock residues-based bioethanol fuels	3.0	15.3	−12.3
7	Starch bioethanol fuels	2.6	5.2	−2.6
8	Algal biomass-based bioethanol fuels	1.9	2.9	−1.0
9	Sugar feedstock residues-based bioethanol fuels	1.1	7.7	−6.6
10	Urban waste-based bioethanol fuels	0.7	6.1	−5.4
11	Food waste-based bioethanol fuels	0.7	5.3	−4.6
12	Industrial waste-based bioethanol fuels	0.4	18.5	−18.1
13	Forestry waste-based bioethanol fuels	0.4	2.0	−1.6
	Sample size	269	26,852	

N paper population, The number of papers in the population sample of 26,852 papers; N paper (%) sample, The number of papers in the population sample of 269 papers.

TABLE 40.10
The Most Prolific Thematic Research Fronts for the Bioethanol Fuels

No.	Research Fronts	N Paper (%) Sample
1	Biomass pretreatments	78.1
2	Biomass hydrolysis	46.8
3	Bioethanol production	17.8
4	Hydrolysate fermentation	14.9
5	Bioethanol fuel evaluation	7.8
6	Bioethanol fuel utilization	0.7

N paper (%) sample, The number of papers in the population sample of 269 papers.

fronts are the grass, sugars, starch feedstock residues, starch feedstocks, algal biomass, sugar feedstock residues, urban waste, food waste, industrial waste, and forestry waste with 0.4%–5.6% of the sample papers each.

Further, the most influential research fronts are the lignocellulosic biomass and cellulose with 23% and 6% surplus, respectively, whilst industrial waste, starch feedstock residues, and wood are the least influential research fronts with 18%, 12%, and 10% deficits, respectively, followed by sugar feedstock residues, urban waste, food waste, starch feedstocks, and forestry waste with 2%–7% deficit each.

Information about the thematic research fronts for the sample papers in the bioethanol fuels is given in Table 40.10. As this table shows, the most prolific research front is the pretreatments of the non-waste feedstocks with 78% of the sample papers, followed by the hydrolysis of the feedstock with 47% of the sample papers. The other prolific research fronts are the bioethanol production, hydrolysate fermentation, and bioethanol fuel evaluation with 18%, 15%, and 8% of the sample papers, respectively. Further, bioethanol fuel utilization relates to only 1% of the sample papers.

40.4 DISCUSSION

40.4.1 INTRODUCTION

The crude oil-based gasoline fuels have been widely used in the transportation sector since the 1920s. However, there have been great public concerns over the adverse environmental and human impact of these fuels. Hence, biomass-based bioethanol fuels have increasingly been used in blending gasoline fuels, in the fuel cells, and in the biochemical production in a biorefinery context.

However, it is necessary to pretreat the biomass to enhance the yield of the bioethanol prior to the bioethanol production through the hydrolysis and fermentation. The research in the field of the bioethanol fuels has intensified in this context in the key research fronts of the pretreatment and hydrolysis of the feedstocks, fermentation of the hydrolysates, and production, and evaluation of the bioethanol fuels. The research in this field has also intensified for the feedstocks of wood biomass, algal biomass, grass biomass, sugar feedstock biomass, starch feedstock biomass, lignocellulosic biomass at large, cellulose, sugar and starch feedstock residues, food waste, industrial waste, urban waste, and forestry waste. Thus, it complements the research on the processes for the production, evaluation, and utilization of bioethanol fuels at large.

However, it is essential to develop efficient incentive structures for the primary stakeholders to enhance the research in this field. This is especially important to maintain energy security in the cases of supply shocks such as oil shocks, war-related chocks as in the case of Russian invasion of Ukraine, or COVID-19 shocks.

The scientometric analysis has been used in this context to inform the primary stakeholders about the current state of the research in a selected research field. As there has been no scientometric study in this field, this book chapter presents a scientometric study of the research in the bioethanol fuels. It examines the scientometric characteristics of both the sample and population data presenting scientometric characteristics of these both datasets in the order of documents, authors, publication years, institutions, funding bodies, source titles, countries, Scopus subject categories, Scopus keywords, and research fronts.

As a first step for the search of the relevant literature, the keywords were selected using the most-cited first 300 population papers for each feedstock. The selected keyword list was then optimized to obtain a representative sample of papers for each research field. These seven keyword lists were then integrated to obtain the keyword list for this research field (Konur, 2023a,b).

As a second step, two sets of data were used for this study. First, a population sample of 26,850 papers was used to examine the scientometric characteristics of the population data. Secondly, a sample of 269 most-cited papers, corresponding to 1% of the population papers, was used to examine the scientometric characteristics of these citation classics.

The scientometric characteristics of these sample and population datasets were presented in the order of documents, authors, publication years, institutions, funding bodies, source titles, countries, Scopus subject categories, Scopus keywords, and research fronts.

Lastly, the key scientometric findings for both datasets were discussed to highlight the research landscape for bioethanol fuels. Additionally, a number of brief conclusions were drawn and a number of relevant recommendations were made to enhance the future research landscape.

40.4.2 THE MOST PROLIFIC DOCUMENTS IN THE BIOETHANOL FUELS

Articles (together with conference papers) dominate both the sample (71%) and population (93%) papers with 22% deficit (Table 40.1). Further, review papers have a surplus (25%) and the representation of the reviews in the sample papers is quite extraordinary (29%).

Scopus differs from the Web of Science database in differentiating and showing articles (68%) and conference papers (3%) published in the journals separately. However, it should be noted that these conference papers are also published in journals as articles, compared to those published only in the conference proceedings. Hence, the total number of articles and review papers in the sample dataset are 71% and 29%, respectively.

It is observed during the search process that there has been inconsistency in the classification of the documents in Scopus as well as in other databases such as Web of Science. This is especially relevant for the classification of papers as reviews or articles as the papers not involving a literature review may be erroneously classified as a review paper. There is also a case of review papers being classified as articles. For example, the total number of the reviews in the sample dataset was manually found as nearly 38% compared to 29% as indexed by Scopus, decreasing the number of articles and conference papers to 62% for the sample dataset. It is notable that many technoeconomic and life cycle studies were often indexed as reviews by the Scopus database.

In this context, it would be helpful to provide a classification note for the published papers in the books and journals at the first instance. It would also be helpful to use the document types listed in Table 40.1 for this purpose. Book chapters may also be classified as articles or reviews as an additional classification to differentiate review chapters from the experimental chapters as it is done by the Web of Science. It would be further helpful to additionally classify the conference papers as articles or review papers as well as it is done in the Web of Science database.

40.4.3 THE MOST PROLIFIC AUTHORS IN THE BIOETHANOL FUELS

There have been most prolific 24 authors with at least 1.5% of the sample papers each as given in Table 40.2. These authors have shaped the development of the research in this field.

The most prolific authors are John N. Saddler, Bruce E. Dale, Charles E. Wyman, Guido Zacchi, and to a lesser extent Blake A. Simmons, Michael E. Himmel, and Barbel Hahn-Hagerdal. Further, the most influential authors are John N. Saddler, Bruce E. Dale, Charles E. Wyman, and to a lesser extent Guido Zacchi, Barbel Hahn-Hagerdal, Michael E. Himmel, Blake A. Simmons, and Robin D. Rogers. It is notable that all of these top authors are male.

It is important to note the inconsistencies in indexing of the author names in Scopus and other databases. It is especially an issue for the names with more than two components such as 'Blake Sam de Hyun Hahn-Hagerdal'. The probable outcomes are 'Hahn-Hagerdal, B.S.D.H.', 'de Hyun Hahn-Hagerdal, B.S.', or 'Hyun Hahn-Hagerdal, B.S.D.'. The first choice is the gold standard of the publishing sector as the last word in the name is taken as the last name. In most of the academic databases such as PUBMED and EBSCO databases, this version is used predominantly. The second choice is a strong alternative, whilst the last choice is an undesired outcome as two last words are taken as the last name. It is good practice to combine the words of the last name by a hyphen: 'Hyun-Hahn-Hagerdal, B.S.D.'. It is notable that inconsistent indexing of the author names may cause substantial inefficiencies in the search process for the papers as well as allocating credit to the authors as there are different author entries for each outcome in the databases.

There are also inconsistencies in the shortening Chinese names. For example. 'Runcang Zhu' is often shortened as 'Zhu, R.', 'Zhu, R.-C.', and 'Zhu, R.C.' as it is done in the Web of Science database as well. However, the gold standard in this case is 'Zhu, R' where the last word is taken as the last name and the first word is taken as a single forename. In most of the academic databases such as PUBMED and EBSCO, this first version is used predominantly. Nevertheless it makes sense to use the third option to differentiate Chinese names efficiently: 'Zhu, R.C.'. Therefore, there have been difficulties in locating papers for the Chinese authors. In such cases, the use of the unique author codes provided for each author by the Scopus database has been helpful.

There is also a difficulty in allowing credit for the authors especially for the authors with common names such as 'Zhu, R.' in conducting scientometric studies. These difficulties strongly influence the efficiency of the scientometric studies as well as allocating credit to the authors as there are the same author entries for different authors with the same name, for example, 'Zhu, R.' in the databases.

In this context, the coding of authors in Scopus database is a welcome innovation compared to the other databases such as Web of Science. In this process, Scopus allocates a unique number to each author in the database (Aman, 2018). However, there might still be substantial inefficiencies in this coding system especially for common names. For example, some of the papers for a certain author maybe allocated to another researcher with a different author code. It is possible that Scopus uses a number of software programs to differentiate the author names and the program may not be false-proof (Kim, 2018).

In this context, it does not help that author names are not given in full in some journals and books. This makes difficult to differentiate authors with common names and makes the scientometric studies further difficult in the author domain. Therefore, the author names should be given in full in all books and journals at the first instance. There is also a cultural issue where some authors do not use their full names in their papers. Instead, they use initials for their forenames: 'Himmel, H.J.', 'Himmel, H.', or 'Himmel, J.' instead of 'Himmel, Hyun Jae'.

There are also inconsistencies in naming of the authors with more than two components by the authors themselves in journal papers and book chapters. For example. 'Himmel, A.P.C.' might be given as 'Himmel, A.' or 'Himmel, A.C.' or 'Himmel, A.P.' or 'Himmel, C' in the journals and books. This also makes the scientometric studies difficult in the author domain. Hence, contributing authors should use their name consistently in their publications.

The other critical issue regarding the author names is the inconsistencies in the spelling of the author names in the national spellings (e.g., Özgümüş, Şütçöl) rather than in the English spellings (e.g., Ozgumus, Sutcol) in Scopus database. Scopus differs from the Web of Science database and many other databases in this respect where the author names are given only in the English spellings. It is observed that national spellings of the author names do not help much in conducting scientometric studies as well in allocating credits to the authors as sometimes there are often the different author entries for the English and National spellings in the Scopus database.

The most prolific institutions for the sample dataset are Lund University, and to a lesser extent University of British Columbia. Further, the most prolific countries for the sample dataset are the USA and to a lesser extent Sweden and Canada. These findings confirm the dominance of the USA and to a lesser extent Europe and Canada in this field. On the other hand, pretreatments and hydrolysis of the feedstocks and to a lesser extent the fermentation of the hydrolysates and the bioethanol fuels are the key research fronts studied by these top authors.

It is also notable that there is significant gender deficit for the sample dataset as surprisingly with a representation rate of 13%. This finding is the most thought-provoking with strong public policy implications. Hence, institutions, funding bodies, and policymakers should take efficient measures to reduce the gender deficit in this field as well as other scientific fields with strong gender deficit. In this context, it is worth to note the level of representation of the researchers from the minority groups in science on the basis of race, sexuality, age, and disability, besides the gender (Blankenship, 1993; Dirth and Branscombe, 2017; Konur, 2000, 2002a,b,c, 2006a,b, 2007a,b).

40.4.4 THE MOST PROLIFIC RESEARCH OUTPUT BY YEARS IN THE BIOETHANOL FUELS

The research output observed between 1970 and 2022 is illustrated in Figure 40.1. This figure clearly shows that the bulk of the research papers in the population dataset were published primarily in the 2010s and early 2020s. Similarly, the bulk of the research papers in the sample dataset were published in the 2000s and 2010s.

These findings suggest that the most prolific sample and population papers were primarily published in the 2010s. Further, a significant portion of the sample and population papers were published in the early 2020s and 2000s, respectively.

These are the thought-provoking findings as there has been significant research boom since 2009 and 2002 for the population and sample papers, respectively. In this context, the increasing public concerns about climate change (Change, 2007), greenhouse gas emissions (Carlson et al., 2017), and global warming (Kerr, 2007) have been certainly behind the boom in the research in this field since 2007. Furthermore, the recent supply shocks experienced due to the COVID-19 pandemics and the Ukrainian war might also be behind the research boom in this field since 2019.

Based on these findings, the size of the population papers is likely to more than double in the current decade, provided that the public concerns about climate change, greenhouse gas emissions, and global warming, as well as the supply shocks are translated efficiently to the research funding in this field.

40.4.5 The Most Prolific Institutions in the Bioethanol Fuels

The most prolific 32 institutions publishing papers on the bioethanol fuels with at least 1.5% of the sample papers each given in Table 40.3 have shaped the development of the research in this field.

The most prolific institutions are the Lund University, NREL, and to a lesser University of British Columbia, Michigan State University, Dartmouth College, Technical University Denmark, University Sao Paulo, Oak Ridge National Laboratory, NC State University, USDA Forest Service, Joint Bioenergy Institute, and Georgia Institute of Technology. Similarly, the top countries for these most prolific institutions are the USA, and to a lesser extent Sweden and Denmark. In total, ten countries house these top institutions.

On the other hand, the institutions with the most citation impact are the Lund University, NREL, and to a lesser extent the University of British Columbia, Michigan State University, Dartmouth College, Technical University of Denmark, Georgia Institute of Technology, and Joint Bioenergy Institute. These findings confirm the dominance of the institutions from the USA, Europe, and to a lesser extent Brazil and Canada.

40.4.6 The Most Prolific Funding Bodies in the Bioethanol Fuels

The most prolific 16 funding bodies funding at least 1.1% of the sample papers each is given in Table 40.4. It is notable that only 33% and 46% of the sample and population papers were funded, respectively.

The most prolific funding bodies are the U.S. Department of Energy, and to a lesser extent the Natural Sciences and Engineering Research Council of Canada, European Commission, National Science Foundation, Swedish National Board for Industrial and Technical Development, National Natural Science Foundation of China, U.S. Department of Agriculture, and NREL. On the other hand, the most prolific countries for these top funding bodies are the USA, Sweden, and to a lesser extent Canada. In total, six countries and the EU house these top funding bodies.

The funding bodies with the most citation impact are the U.S. Department of Energy, and to a lesser extent the Swedish National Board for Industrial and Technical Development, European Commission, Natural Sciences and Engineering Research Council of Canada, and NREL. Further, the funding bodies with the least citation impact are the National Natural Science Foundation of China and to a lesser extent Research Support Foundation of the State of Sao Paulo.

These findings on the funding of the research in this field suggest that the level of the funding, mostly since 2009, is moderately intensive and it has been largely instrumental in enhancing the research in this field (Ebadi and Schiffauerova, 2016) in light of North's institutional framework (North, 1991). It is also notable that the funding rate in this field is relatively modest compared to those in the other research fronts of the bioethanol fuels such as algal bioethanol fuels. Further, it is expected that this funding rate would improve in light of the recent supply shocks. Further, it emerges that the USA, Sweden, and China have heavily funded the research on the bioethanol fuels.

40.4.7 THE MOST PROLIFIC SOURCE TITLES IN THE BIOETHANOL FUELS

The most prolific 14 source titles publishing at least 1.5% of the sample papers each in the bioethanol fuels have shaped the development of the research in this field (Table 40.5).

The most prolific source titles are the Bioresource Technology and to a lesser extent Biotechnology and Bioengineering, Green Chemistry, Biomass and Bioenergy, Biotechnology for Biofuels, Applied Microbiology and Biotechnology, Proceedings of the National Academy of Sciences of the United States of America, and Biotechnology Advances.

On the other hand, the source titles with the most impact are the Bioresource Technology and to a lesser extent by Green Chemistry, Biotechnology and Bioengineering, Biotechnology Advances, Proceedings of the National Academy of Sciences of the United States of America, Applied Microbiology and Biotechnology, and Journal of Biotechnology.

It is notable that these top source titles are primarily related to the bioresources, biotechnology, and to a lesser extent energy and microbiology. This finding suggests that Bioresource Technology and the other prolific journals in these fields have significantly shaped the development of the research in this field as they focus primarily on the bioethanol fuels with a high yield. In this context, the influence of the top journal is quite extraordinary.

40.4.8 THE MOST PROLIFIC COUNTRIES IN THE BIOETHANOL FUELS

The most prolific 19 countries publishing at least 1.5% of the sample papers each have significantly shaped the development of the research in this field (Table 40.6).

The most prolific countries are the USA and to a lesser extent Sweden, China, Canada, Japan, Denmark, India, Brazil, and the UK. It is notable that ten countries listed in Table 40.6 produce 45% and 24% of the sample and population papers, respectively, with 21% surplus, and China is the largest producer of the population papers with a publication rate of 19.6%.

On the other hand, the countries with the most citation impact are the USA and to a lesser extent Sweden, Denmark, Canada, and Netherlands. Similarly, the countries with the least impact are China and to a lesser extent Brazil, S. Korea, and India.

The close examination of these findings suggests that the USA, Europe, and to a lesser extent China, Canada, India, Japan, and Brazil are the major producers of the research in this field. It is a fact that the USA has been a major player in science (Leydesdorff and Wagner, 2009). The USA has further developed a strong research infrastructure to support its corn- and grass-based bioethanol industry (Gillon, 2010).

However, China has been a rising mega star in scientific research in competition with the USA and Europe (Leydesdorff and Zhou, 2005). China is also a major player in this field as a major producer of bioethanol (Fang et al., 2010).

Next, Europe has been a persistent player in the scientific research in competition with both the USA and China (Leydesdorff, 2000). Europe has also been a persistent producer of bioethanol along with the USA and Brazil (Gnansounou, 2010).

Further, Canada (Tahmooresnejad et al., 2015), Japan (Negishi et al., 2004), India (Karpagam et al., 2011), and Brazil (Glanzel et al., 2006) are the other countries with substantial research activities in bioethanol fuels.

40.4.9 THE MOST PROLIFIC SCOPUS SUBJECT CATEGORIES IN THE BIOETHANOL FUELS

The most prolific ten Scopus subject categories indexing at least 3% of the sample papers each, given in Table 40.7, have shaped the development of the research in this field.

The most prolific Scopus subject categories in the bioethanol fuels are Chemical Engineering, and to a lesser extent Environmental Science, Biochemistry, Genetics and Molecular Biology, Energy, and Immunology and Microbiology. It is also notable that Social Sciences including Economics and Business have a minimal presence in both sample and population studies.

On the other hand, the Scopus subject categories with the most citation impact are Chemical Engineering, and to a lesser extent Biochemistry, Genetics and Molecular Biology, Environmental Science, and Immunology and Microbiology. Similarly, the least influential subject categories are Agricultural and Biological Sciences and to a lesser extent Engineering, Materials Science, and Chemistry.

These findings are thought-provoking, suggesting that the primary subject categories are related to chemical engineering, environmental sciences, and to a lesser extent biochemistry, energy, and microbiology as the core of the research in this field concerns with the production and utilization of the bioethanol fuels. The other finding is that social sciences are not well represented in both the sample and population papers as in line with the most fields in bioethanol fuels. The social, environmental, and economics studies account for the field of social sciences.

40.4.10 The Most Prolific Keywords in the Bioethanol Fuels

A limited number of keywords have shaped the development of the research in this field as shown in Table 40.8. These keywords are grouped under the five headings: feedstock, pretreatments, fermentation, hydrolysis and hydrolysates, and products.

The most prolific keywords across all of the research fronts are cellulose, hydrolysis, lignin, biomass, ethanol, lignocellulose, sugar, fermentation, cellulases, enzyme activity, biofuels, enzymes, and to a lesser extent hemicellulose, enzymatic hydrolysis, pretreatment, lignocellulosic biomass, glucose, saccharification, and ILs.

Similarly, the most influential keywords are cellulose, lignin, hydrolysis, biomass, lignocellulose, and to a lesser extent cellulases, sugar, enzyme activity, hemicellulose, enzymes, ethanol, biofuels, and pretreatment.

These findings suggest that it is necessary to determine the keyword set carefully to locate the relevant research in each of these research fronts. Additionally, the size of the samples for each keyword highlights the intensity of the research in the relevant research areas for both sample and population datasets. These findings also highlight different spelling of some strategic keywords such as pretreatment v. pre-treatment v. treatment and bioethanol v. ethanol v. bio-ethanol, etc. However, there is tendency toward the use of the connected keywords without using a hyphen: Bioethanol or pretreatment.

40.4.11 The Most Prolific Research Fronts in the Bioethanol Fuels

Information about the research fronts for the sample papers in the bioethanol fuels is given in Table 40.9. As this table shows, the most prolific research fronts for this field are the lignocellulosic biomass-based bioethanol fuels, cellulose and wood biomass-based bioethanol fuels, and to a lesser extent the grass-based bioethanol fuels, sugar bioethanol fuels, starch feedstock residues-based bioethanol fuels, starch bioethanol fuels, algal biomass-based bioethanol fuels, sugar feedstock residues-based bioethanol fuels, urban waste-based bioethanol fuels, food waste-based bioethanol fuels, industrial waste-based bioethanol fuels, and forestry waste-based bioethanol fuels.

Further, the most influential research fronts are the lignocellulosic biomass-based bioethanol fuels, and to a lesser extent cellulose-based bioethanol fuels whilst industrial waste-based bioethanol fuels, starch feedstock residues-based bioethanol fuels, and wood-based bioethanol fuels, and to a lesser extent sugar feedstock residues-based bioethanol fuels, urban waste-based bioethanol fuels, food waste-based bioethanol fuels, starch bioethanol fuels, and forestry waste-based bioethanol fuels are the least influential research fronts.

Thus, the first three research fields have substantial importance, complementing the remaining bioethanol fuel research fields. It is important to note that the research on the production of the first generation bioethanol fuels from sugar and starch feedstocks comprises only 6.3% of the sample papers in total. These first generation bioethanol fuels are not much desirable as they undermine the food security (Makenete et al., 2008; Wu et al., 2012).

Table 40.10 shows that the most prolific thematic research fronts are the pretreatments and hydrolysis of the feedstocks, and to a lesser extent bioethanol production, hydrolysate fermentation, bioethanol fuel evaluation, and bioethanol fuel utilization.

These findings are thought-provoking in seeking ways to increase feedstock-based bioethanol yield at the global scale. It is clear that all of these research fronts have public importance and merit substantial funding and other incentives. Further, it is notable that bioethanol fuels have become a core unit of the bioethanol research to make it more competitive with the crude oil-based gasoline and petrodiesel fuels, especially for the USA, Europe, and China.

It is notable that the pretreatment and hydrolysis of the feedstocks emerge as primary research fronts for this field. These processes are required to improve the ethanol yield. However, the research fronts of the fermentation of the hydrolysates and the bioethanol production from the hydrolysates are also important.

Further, the field of the evaluation and utilization of bioethanol fuels is a neglected area. This suggests that the primary stakeholders have been primarily interested in these key processes of the bioethanol production. It is also notable that evaluation of the bioethanol fuels such as technoeconomics, life cycle, economics, social, land use, labor, and environment-related studies emerges as a case study for the bioethanol fuels. Similarly, the utilization of these biofuels in the gasoline or diesel engines is also an important research field from a societal perspective. In this context, the USA and Brazil have been the global leaders in the production and use of the corn- and sugarcane-based bioethanol fuels since the 1970s in the aftermath of the global crude oil crisis in the early 1970s.

In the end, these most-cited papers in this field hint that the production of bioethanol fuels could be optimized using the structure, processing, and property relationships of these feedstocks in the fronts of the feedstock pretreatment and hydrolysis, and hydrolysate fermentation (Formela et al., 2016; Konur, 2018a, 2020b, 2021a,b,c,d; Konur and Matthews, 1989).

40.5   CONCLUSION AND FUTURE RESEARCH

The research on the feedstock-based bioethanol fuels has been mapped through a scientometric study of both sample (269 papers) and population (26,850 papers) datasets.

The critical issue in this study has been to obtain a representative sample of the research as in any other scientometric study. Therefore, the keyword set has been carefully devised and optimized after a number of runs in the Scopus database. It is a representative sample of the wider population studies. This keyword set was provided in the appendix of the related studies, and the relevant keywords are presented in Table 40.8. However, it should be noted that it has been very difficult to compile a representative keyword set since this research field has been connected closely with many other fields. Therefore, it has been necessary to compile a keyword list to exclude papers concerned with the other research fields.

The other issue has been the selection of a multidisciplinary database to carry out the scientometric study of the research in this field. For this purpose. Scopus database has been selected. The journal coverage of this database has been notably wider than that of Web of Science and other multisubject databases.

The key scientometric properties of the research in this field have been determined and discussed in this book chapter. It is evident that a limited number of documents, authors, institutions, publication years, institutions, funding bodies, source titles, countries, Scopus subject categories, Scopus keywords, and research fronts have shaped the development of the research in this field.

There is ample scope to increase the efficiency of the scientometric studies in this field in the author and document domains by developing consistent policies and practices in both domains across all the academic databases. In this respect, it seems that authors, journals, and academic databases have a lot to do. Furthermore, the significant gender deficit as in most scientific fields emerges as a public policy issue. The potential deficits on the basis of age, race, disability, and sexuality need also to be explored in this field as in other scientific fields.

The research in this field has boomed since 2009 and 2002 for the population and sample papers, respectively, possibly promoted by the public concerns on global warming, greenhouse gas emissions, and climate change. Furthermore, the recent COVID-19 pandemics and Russian invasion of Ukraine have resulted in global supply shocks shifting the recent focus of the stakeholders from the crude oil-based fuels to biomass-based fuels such as bioethanol fuels. It is expected that there would be further incentives for the key stakeholders to carry out the research for the bioethanol fuels to increase the ethanol yield and to make it more competitive with the crude oil-based gasoline and diesel fuels. This might be truer for the crude oil- and foreign exchange-deficient countries to maintain the energy and food security at the face of the global supply shocks.

The relatively modest funding rate of 33% and 46% for the sample and population papers, respectively, suggests that funding in this field significantly enhanced the research in this field primarily since 2009, possibly more than doubling in the current decade. However, it is evident that there is ample room for more funding and other incentives to enhance the research in this field further.

The institutions from the USA, and to a lesser extent Sweden and Denmark have mostly shaped the research in this field. Further, USA, Europe, and to a lesser extent China, Canada, Japan, India, and Brazil have been the major producers of the research in this field as the major producers and users of bioethanol fuels. It is evident that these countries have well-developed research infrastructure in bioethanol fuels and their derivatives.

It emerges that ethanol is more popular than bioethanol as a keyword with strong implications for the search strategy. In other words, the search strategy using only bioethanol keyword would not be much helpful. On the other hand, the Scopus keywords are grouped under the five headings: biomass, pretreatments, fermentation, hydrolysis and hydrolysates, and products.

Table 40.9 shows that the most prolific research fronts for this field are the lignocellulosic biomass-based bioethanol fuels, cellulose-based bioethanol fuels, wood biomass-based bioethanol fuels, and to a lesser extent the grass-based bioethanol fuels, sugar bioethanol fuels, starch feedstock residues-based bioethanol fuels, starch bioethanol fuels, algal biomass-based bioethanol fuels, sugar feedstock residues-based bioethanol fuels, urban waste-based bioethanol fuels, food waste-based bioethanol fuels, industrial waste-based bioethanol fuels, and forestry waste-based bioethanol fuels. It is important to note that the research on the production of the first generation bioethanol fuels from sugar and starch bioethanol fuels comprises only a small part of the sample papers in total. These first generation bioethanol fuels are not much desirable as they undermine the food security.

Further, Table 40.10 shows that the most prolific research fronts are the pretreatments and hydrolysis of the feedstocks, and to a lesser extent bioethanol production, hydrolysate fermentation, bioethanol fuel evaluation, and bioethanol fuel utilization.

The first four research fronts dominate the research in this field whilst the field of the utilization and evaluation of the bioethanol fuels is relatively a neglected research field. In this context, it is notable that there is ample room for the improvement of the research on social and humanitarian aspects of the research on the bioethanol fuels from the feedstock such as scientometric and user studies.

These findings are thought-provoking in seeking ways to increase bioethanol yield at the global scale. It is clear that all of these research fronts have public importance and merit substantial funding and other incentives. Further, it is notable that feedstock-based bioethanol fuels have become a core unit of the bioethanol research to make it more competitive with the crude oil-based gasoline and diesel fuels, especially for the USA, Europe, Brazil, and China.

Thus, the scientometric analysis has a great potential to gain valuable insights into the evolution of the research in this field as in other scientific fields especially in the aftermath of the significant global supply shocks such as COVID-19 pandemics and the Russian invasion of Ukraine.

It is recommended that further scientometric studies are carried out for the primary research fronts. It is further recommended that reviews of the most-cited papers are carried out for each primary research front to complement these scientometric studies. Next, the scientometric studies of the hot papers in these primary fields are carried out.

ACKNOWLEDGMENTS

The contribution of the highly cited researchers in the field of the feedstock-based bioethanol fuels has been gratefully acknowledged.

REFERENCES

Aman, V. 2018. Does the Scopus author ID suffice to track scientific international mobility? A case study based on Leibniz laureates. *Scientometrics* 117:705–720.

Angelici, C., B. M. Weckhuysen and P. C. A. Bruijnincx. 2013. Chemocatalytic conversion of ethanol into butadiene and other bulk chemicals. *ChemSusChem* 6:1595–1614.

Antolini, E. 2007. Catalysts for direct ethanol fuel cells. *Journal of Power Sources* 170:1–12.

Antolini, E. 2009. Palladium in fuel cell catalysis. *Energy and Environmental Science* 2:915–931.

Bai, F. W., W. A. Anderson and M. Moo-Young. 2008. Ethanol fermentation technologies from sugar and starch feedstocks. *Biotechnology Advances* 26:89–105.

Balat, M. 2011. Production of bioethanol from lignocellulosic materials via the biochemical pathway: A review. *Energy Conversion and Management* 52:858–875.

Beaudry, C. and V. Lariviere. 2016. Which gender gap? Factors affecting researchers' scientific impact in science and medicine. *Research Policy* 45:1790–1817.

Binod, P., R. Sindhu and R. R. Singhania, et al. 2010. Bioethanol production from rice straw: An overview. *Bioresource Technology* 101:4767–4774.

Blankenship, K. M. 1993. Bringing gender and race in: US employment discrimination policy. *Gender & Society* 7:204–226.

Bothast, R. J. and M. A. Schlicher. 2005. Biotechnological processes for conversion of corn into ethanol. *Applied Microbiology and Biotechnology* 67:19–25.

Burnham, J. F. 2006. Scopus database: A review. *Biomedical Digital Libraries* 3:1–8.

Canilha, L., A. K. Chandel and T. S. dos Santos Milessi, et al. 2012. Bioconversion of sugarcane biomass into ethanol: An overview about composition, pretreatment methods, detoxification of hydrolysates, enzymatic saccharification, and ethanol fermentation. *Journal of Biomedicine and Biotechnology* 2012:989572.

Cardona, C. A., J. A. Quintero and I. C. Paz. 2010. Production of bioethanol from sugarcane bagasse: Status and perspectives. *Bioresource Technology* 101:4754–4766.

Carlson, K. M., J. S. Gerber and D. Mueller, et al. 2017. Greenhouse gas emissions intensity of global croplands. *Nature Climate Change* 7:63–68.

Change, C. 2007. Climate change impacts, adaptation and vulnerability. *Science of the Total Environment* 326:95–112.

Dirth, T. P. and N. R. Branscombe. 2017. Disability models affect disability policy support through awareness of structural discrimination. *Journal of Social Issues* 73:413–442.

Duff, S. J. B. and W. D. Murray. 1996. Bioconversion of forest products industry waste cellulosics to fuel ethanol: A review. *Bioresource Technology* 55:1–33.

Ebadi, A. and A. Schiffauerova. 2016. How to boost scientific production? A statistical analysis of research funding and other influencing factors. *Scientometrics* 106:1093–1116.

Fang, X., Y. Shen, J. Zhao, X. Bao and Y. Qu. 2010. Status and prospect of lignocellulosic bioethanol production in China. *Bioresource Technology* 101:4814–4819.

Fauci, A. S., H. C. Lane and R. R. Redfield. 2020. Covid-19-navigating the uncharted. *New England Journal of Medicine* 382:1268–1269.

Fernando, S., S. Adhikari, C. Chandrapal and M. Murali. 2006. Biorefineries: Current status, challenges, and future direction. *Energy & Fuels* 20:1727–1737.

Formela, K., A. Hejna, L. Piszczyk, M. R. Saeb and X. Colom. 2016. Processing and structure-property relationships of natural rubber/wheat bran biocomposites. *Cellulose* 23:3157–3175.

Galbe, M. and G. Zacchi. 2002. A review of the production of ethanol from softwood. *Applied Microbiology and Biotechnology* 59:618–628.

Garfield, E. 1955. Citation indexes for science. *Science* 122:108–111.

Gillon, S. 2010. Fields of dreams: Negotiating an ethanol agenda in the Midwest United States. *Journal of Peasant Studies* 37:723–748.

Glanzel, W., J. Leta and B. Thijs. 2006. Science in Brazil. Part 1: A macro-level comparative study. *Scientometrics* 67:67–86.

Gnansounou, E. 2010. Production and use of lignocellulosic bioethanol in Europe: Current situation and perspectives. *Bioresource Technology* 101:4842–4850.

Guimaraes, P. M. R., J. A. Teixeira and L. Domingues. 2010. Fermentation of lactose to bio-ethanol by yeasts as part of integrated solutions for the valorisation of cheese whey. *Biotechnology Advances* 28:375–384.

Hahn-Hagerdal, B., M. Galbe, M. F. Gorwa-Grauslund, G. Liden and G. Zacchi. 2006. Bio-ethanol - The fuel of tomorrow from the residues of today. *Trends in Biotechnology* 24:549–556.

Hamelinck, C. N., G. van Hooijdonk and A. P. C. Faaij. 2005. Ethanol from lignocellulosic biomass: Techno-economic performance in short-, middle- and long-term. *Biomass and Bioenergy*, 28:384–410.

Hamilton, J. D. 1983. Oil and the macroeconomy since World War II. *Journal of Political Economy* 91:228–248.

Hamilton, J. D. 2003. What is an oil shock? *Journal of Econometrics* 113:363–398.

Hendriks, A. T. W. M and G. Zeeman. 2009. Pretreatments to enhance the digestibility of lignocellulosic biomass. *Bioresource Technology* 100:10–18.

Hill, J., E. Nelson, D. Tilman, S. Polasky and D. Tiffany. 2006. Environmental, economic, and energetic costs and benefits of biodiesel and ethanol biofuels. *Proceedings of the National Academy of Sciences of the United States of America* 103:11206–11210.

Hill, J., S. Polasky and E. Nelson, et al. 2009. Climate change and health costs of air emissions from biofuels and gasoline. *Proceedings of the National Academy of Sciences of the United States of America* 106:2077–2082.

Ho, S. H., S. W. Huang and C. Y. Chen, et al. 2013. Bioethanol production using carbohydrate-rich microalgae biomass as feedstock. *Bioresource Technology* 135:191–198.

Hsieh, W. D., R. H. Chen, T. L. Wu and T. H. Lin. 2002. Engine performance and pollutant emission of an SI engine using ethanol-gasoline blended fuels. *Atmospheric Environment* 36:403–410.

Huang, H. J., S. Ramaswamy, U. W. Tschirner and B. V. Ramarao. 2008. A review of separation technologies in current and future biorefineries. *Separation and Purification Technology* 62:1–21.

John, R. P., G. S. Anisha, K. M. Nampoothiri and A. Pandey. 2011. Micro and macroalgal biomass: A renewable source for bioethanol. *Bioresource Technology* 102:186–193.

Jones, T. C. 2012. America, oil, and war in the Middle East. *Journal of American History* 99:208–218.

Jonsson, L. J. and C. Martin. 2016. Pretreatment of lignocellulose: Formation of inhibitory by-products and strategies for minimizing their effects. *Bioresource Technology* 199:103–112.

Karpagam, R., S. Gopalakrishnan, M. Natarajan and B. R. Babu. 2011. Mapping of nanoscience and nanotechnology research in India: A scientometric analysis, 1990-2009. *Scientometrics* 89:501–522.

Kerr, R. A. 2007. Global warming is changing the world. *Science* 316:188–190.

Keshwani, D. R. and J. J. Cheng. 2009. Switchgrass for bioethanol and other value-added applications: A review. *Bioresource Technology* 100:1515–1523.

Kilian, L. 2008. Exogenous oil supply shocks: How big are they and how much do they matter for the US economy? *Review of Economics and Statistics* 90:216–240.

Kilian, L. 2009. Not all oil price shocks are alike: Disentangling demand and supply shocks in the crude oil market. *American Economic Review*, 99:1053–69.

Kim, J. 2018. Evaluating author name disambiguation for digital libraries: A case of DBLP. *Scientometrics* 116:1867–1886.

Konur, O. 2000. Creating enforceable civil rights for disabled students in higher education: An institutional theory perspective. *Disability & Society* 15:1041–1063.

Konur, O. 2002a. Access to nursing education by disabled students: Rights and duties of nursing programs. *Nurse Education Today* 22:364–374.

Konur, O. 2002b. Assessment of disabled students in higher education: Current public policy issues. *Assessment and Evaluation in Higher Education* 27:131–152.

Konur, O. 2002c. Access to employment by disabled people in the UK: Is the Disability Discrimination Act working? *International Journal of Discrimination and the Law* 5:247–279.

Konur, O. 2006a. Participation of children with dyslexia in compulsory education: Current public policy issues. *Dyslexia* 12:51–67.

Konur, O. 2006b. Teaching disabled students in higher education. *Teaching in Higher Education* 11:351–363.

Konur, O. 2007a. A judicial outcome analysis of the *Disability Discrimination Act*: A windfall for the employers? *Disability & Society* 22:187–204.

Konur, O. 2007b. Computer-assisted teaching and assessment of disabled students in higher education: The interface between academic standards and disability rights. *Journal of Computer Assisted Learning* 23:207–219.

Konur, O. 2011. The scientometric evaluation of the research on the algae and bio-energy. *Applied Energy* 88:3532–3540.

Konur, O. 2012a. Prof. Dr. Ayhan Demirbas' scientometric biography. *Energy Education Science and Technology Part A: Energy Science and Research* 28:727–738.

Konur, O. 2012b. The evaluation of the biogas research: A scientometric approach. *Energy Education Science and Technology Part A: Energy Science and Research* 29:1277–1292.

Konur, O. 2012c. The evaluation of the global energy and fuels research: A scientometric approach. *Energy Education Science and Technology Part A: Energy Science and Research* 30:613–628.

Konur, O. 2012d. The evaluation of the research on the biodiesel: A scientometric approach. *Energy Education Science and Technology Part A: Energy Science and Research* 28:1003–1014.

Konur, O. 2012e. The evaluation of the research on the bioethanol: A scientometric approach. *Energy Education Science and Technology Part A: Energy Science and Research* 28:1051–1064.

Konur, O. 2012f. The evaluation of the research on the biofuels: A scientometric approach. *Energy Education Science and Technology Part A: Energy Science and Research* 28:903–916.

Konur, O. 2012g. The evaluation of the research on the biohydrogen: A scientometric approach. *Energy Education Science and Technology Part A: Energy Science and Research* 29:323–338.

Konur, O. 2012h. The evaluation of the research on the microbial fuel cells: A scientometric approach. *Energy Education Science and Technology Part A: Energy Science and Research* 29:309–322.

Konur, O. 2012i. The scientometric evaluation of the research on the production of bioenergy from biomass. *Biomass and Bioenergy* 47:504–515.

Konur, O. 2015. Current state of research on algal bioethanol. In *Marine Bioenergy: Trends and Developments*, Ed. S. K. Kim and C. G. Lee, pp. 217–244. Boca Raton, FL: CRC Press.

Konur, O., Ed. 2018a. *Bioenergy and Biofuels*. Boca Raton, FL: CRC Press.

Konur, O. 2018b. Bioenergy and biofuels science and technology: Scientometric overview and citation classics. In *Bioenergy and Biofuels*, Ed. O. Konur, pp. 3–63. Boca Raton: CRC Press.

Konur, O. 2019. Cyanobacterial bioenergy and biofuels science and technology: A scientometric overview. In *Cyanobacteria: From Basic Science to Applications*, Ed. A. K. Mishra, D. N. Tiwari and A. N. Rai, pp. 419–442. Amsterdam: Elsevier.

Konur, O. 2020a. The scientometric analysis of the research on the bioethanol production from green macroalgae. In *Handbook of Algal Science, Technology and Medicine*, Ed. O. Konur, pp. 385–401. London: Academic Press.

Konur, O., Ed. 2020b. *Handbook of Algal Science, Technology and Medicine*. London: Academic Press.

Konur, O., Ed. 2021a. *Handbook of Biodiesel and Petrodiesel Fuels: Science, Technology, Health, and Environment*. Boca Raton, FL: CRC Press.

Konur, O., Ed. 2021b. *Handbook of Biodiesel and Petrodiesel Fuels: Science, Technology, Health, and Environment. Volume 1. Biodiesel Fuels: Science, Technology, Health, and Environment*. Boca Raton, FL: CRC Press.

Konur, O., Ed. 2021c. *Handbook of Biodiesel and Petrodiesel Fuels: Science, Technology, Health, and Environment. Volume 2. Biodiesel Fuels based on the Edible and Nonedible Feedstocks, Wastes, and Algae: Science, Technology, Health, and Environment*. Boca Raton, FL: CRC Press.

Konur, O., Ed. 2021d. *Handbook of Biodiesel and Petrodiesel Fuels: Science, Technology, Health, and Environment. Volume 3. Petrodiesel Fuels: Science, Technology, Health, and Environment*. Boca Raton, FL: CRC Press.

Konur, O. 2023a. Non-waste feedstock-based bioethanol fuel production: Scientometric study. In *Feedstock-based Bioethanol Fuels. I. Non-Waste Feedstocks: Starch, Sugar, Grass, Wood, Cellulose, Algae, and Biosyngas-based Bioethanol Fuels. Handbook of Bioethanol Fuels Volume 3*, Ed. O. Konur. Boca Raton, FL: CRC Press.

Konur, O. 2023b. Second generation waste biomass-based bioethanol fuels: Scientometric study. In *Feedstock-based Bioethanol Fuels. II. Waste Feedstocks: Agricultural, Food, Industrial, Urban, Forestry, and Lignocellulosic Waste-based Bioethanol Fuels. Handbook of Bioethanol Fuels Volume 4*, Ed. O. Konur. Boca Raton, FL: CRC Press.

Konur, O. and F. L. Matthews. 1989. Effect of the properties of the constituents on the fatigue performance of composites: A review. *Composites* 20:317–328.

Kruyt, B., D. P. van Vuuren, H. J. de Vries and H. Groenenberg. 2009. Indicators for energy security. *Energy Policy* 37:2166–2181.

Laser, M., D. Schulman and S. G. Allen, et al. 2002. A comparison of liquid hot water and steam pretreatments of sugar cane bagasse for bioconversion to ethanol. *Bioresource Technology* 81:33–44.

Leydesdorff, L. 2000. Is the European Union becoming a single publication system? *Scientometrics* 47:265–280.

Leydesdorff, L. and C. Wagner. 2009. Is the United States losing ground in science? A global perspective on the world science system. *Scientometrics* 78:23–36.

Leydesdorff, L. and P. Zhou. 2005. Are the contributions of China and Korea upsetting the world system of science? *Scientometrics* 63:617–630.

Li, H., S. M. Liu, X. H. Yu, S. L. Tang and C. K. Tang. 2020. Coronavirus disease 2019 (COVID-19): Current status and future perspectives. *International Journal of Antimicrobial Agents* 55:105951.

Limayem, A. and S. C. Ricke. 2012. Lignocellulosic biomass for bioethanol production: Current perspectives, potential issues and future prospects. *Progress in Energy and Combustion Science*, 38:449–467.

Lin, Y. and S. Tanaka. 2006. Ethanol fermentation from biomass resources: Current state and prospects. *Applied Microbiology and Biotechnology* 69:627–642.

Ma, X., L. Sun and C. Song. 2002. A new approach to deep desulfurization of gasoline, diesel fuel and jet fuel by selective adsorption for ultra-clean fuels and for fuel cell applications. *Catalysis Today* 77:107–116.

Makenete, A. L., W. J. Lemmer and J. Kupka. 2008. The impact of biofuel production on food security: A briefing paper with a particular emphasis on maize-to-ethanol production. *International Food and Agribusiness Management Review* 11:101–110.

Morschbacker, A. 2009. Bio-ethanol based ethylene. *Polymer Reviews* 49:79–84.

Mosier, N., C. Wyman and B. Dale, et al. 2005. Features of promising technologies for pretreatment of lignocellulosic biomass. *Bioresource Technology* 96:673–686.

Najafi, G., B. Ghobadian and T. Tavakoli, et al. 2009. Performance and exhaust emissions of a gasoline engine with ethanol blended gasoline fuels using artificial neural network. *Applied Energy* 86:630–639.

Negishi, M., Y. Sun and K. Shigi. 2004. Citation database for Japanese papers: A new bibliometric tool for Japanese academic society. *Scientometrics* 60:333–351.

Newman, P. W. G. and J. R. Kenworthy. 1989. Gasoline consumption and cities: A comparison of U.S. cities with a global survey. *Journal of the American Planning Association* 55:24–37.

North, D. C. 1991. Institutions. *Journal of Economic Perspectives* 5:97–112.

Olsson, L. and B. Hahn-Hagerdal. 1996. Fermentation of lignocellulosic hydrolysates for ethanol production. *Enzyme and Microbial Technology* 18:312–331.

Palmqvist, E. and B. Hahn-Hagerdal. 2000. Fermentation of lignocellulosic hydrolysates. II: Inhibitors and mechanisms of inhibition. *Bioresource Technology* 74:25–33.

Pimentel, D. and T. W. Patzek. 2005. Ethanol production using corn, switchgrass, and wood; Biodiesel production using soybean and sunflower. *Natural Resources Research* 14:65–76.

Pinkert, A., K. N. Marsh, S. Pang and M. P. Staiger. 2009. Ionic liquids and their interaction with cellulose. *Chemical Reviews* 109:6712–6728.

Prasad, S., A. Singh and H. C. Joshi. 2007. Ethanol as an alternative fuel from agricultural, industrial and urban residues. *Resources, Conservation and Recycling* 50:1–39.

Ravindran, R. and A. K. Jaiswal. 2016. A comprehensive review on pre-treatment strategy for lignocellulosic food industry waste: Challenges and opportunities. *Bioresource Technology* 199:92–102.

Reeves, S. 2014. To Russia with love: How moral arguments for a humanitarian intervention in Syria opened the door for an invasion of the Ukraine. *Michigan State University International Law Review* 23:199.

Sanchez, O. J. and C. A. Cardona. 2008. Trends in biotechnological production of fuel ethanol from different feedstocks. *Bioresource Technology* 99:5270–5295.

Sun, Y. and J. Cheng. 2002. Hydrolysis of lignocellulosic materials for ethanol production: A review. *Bioresource Technology* 83:1–11.

Taherzadeh, M. J. and K. Karimi. 2007. Enzyme-based hydrolysis processes for ethanol from lignocellulosic materials: A review. *Bioresources* 2:707–738.

Taherzadeh, M. J. and K. Karimi. 2008. Pretreatment of lignocellulosic wastes to improve ethanol and biogas production: A review. *International Journal of Molecular Sciences* 9:1621–1651.

Tahmooresnejad, L., C. Beaudry, C and A. Schiffauerova. 2015. The role of public funding in nanotechnology scientific production: Where Canada stands in comparison to the United States. *Scientometrics* 102:753–787.

Talebnia, F., D. Karakashev and I. Angelidaki. 2010. Production of bioethanol from wheat straw: An overview on pretreatment, hydrolysis and fermentation. *Bioresource Technology* 101:4744–4753.

Winzer, C. 2012. Conceptualizing energy security. *Energy Policy* 46:36–48.

Wu, F., D. Zhang and J. Zhang. 2012. Will the development of bioenergy in China create a food security problem? Modeling with fuel ethanol as an example. *Renewable Energy* 47:127–134.

Yang, B. and C. E. Wyman. 2008. Pretreatment: The key to unlocking low-cost cellulosic ethanol. *Biofuels, Bioproducts and Biorefining* 2:26–40.

Zhang, Y. H. P. and L. R. Lynd. 2004. Toward an aggregated understanding of enzymatic hydrolysis of cellulose: Noncomplexed cellulase systems. *Biotechnology and Bioengineering* 88:797–824.

Zhu, J. Y. and X. J. Pan. 2010. Woody biomass pretreatment for cellulosic ethanol production: Technology and energy consumption evaluation. *Bioresource Technology* 101:4992–5002.

41 Feedstock-based Bioethanol Fuels

Review

Ozcan Konur
(Formerly) Ankara Yildirim Beyazit University

41.1 INTRODUCTION

The crude oil-based gasoline fuels (Ma et al., 2002; Newman and Kenworthy, 1989) have been widely used in the transportation sector since the 1920s. However, there have been great public concerns over the adverse environmental and human impact of these fuels (Hill et al., 2006, 2009). Hence, biomass-based bioethanol fuels (Hill et al., 2006; Konur, 2012, 2015, 2019, 2020) have increasingly been used in blending gasoline fuels (Hsieh et al., 2002; Najafi et al., 2009), in the fuel cells (Antolini, 2007, 2009), and in the biochemical production (Angelici et al., 2013; Morschbacker, 2009) in a biorefinery context (Fernando et al., 2006; Huang et al., 2008).

However, it is necessary to pretreat the biomass (Alvira et al., 2010; Taherzadeh and Karimi, 2008) to enhance the yield of the bioethanol (Hahn-Hagerdal et al., 2006; Sanchez and Cardona, 2008) prior to the bioethanol fuel production from the feedstocks through the hydrolysis (Sun and Cheng, 2002; Taherzadeh and Karimi, 2007) and fermentation (Lin and Tanaka, 2006; Olsson and Hahn-Hagerdal, 1996) of the biomass and hydrolysates, respectively.

The research in the field of the feedstock-based bioethanol fuels has intensified in this context in the key research fronts of the pretreatment (Hendriks and Zeeman, 2009; Mosier et al., 2005) and hydrolysis (Sun and Cheng, 2002; Zhang and Lynd, 2004) of the feedstock, fermentation (Jonsson and Martin, 2016; Palmqvist and Hahn-Hagerdal, 2000) of the feedstock-based hydrolysates, and bioethanol fuel production (Balat, 2011; Limayem and Ricke, 2012), and bioethanol fuel evaluation (Hamelinck et al., 2005; Pimentel and Patzek, 2005).

The research in this field has also intensified for the feedstocks of wood biomass (Galbe and Zacchi, 2002; Zhu and Pan, 2010), algal biomass (Ho et al., 2013; John et al., 2011), grass biomass (Keshwani and Cheng, 2009; Pimentel and Patzek, 2005), sugar feedstock biomass (Bai et al., 2008; Canilha et al., 2012), and starch feedstock biomass (Bai et al., 2008; Bothast and Schlicher, 2005), lignocellulosic biomass at large (Mosier et al., 2005; Sun and Cheng, 2002), cellulose (Pinkert et al., 2009; Zhang and Lynd, 2004), sugar (Cardona et al., 2010; Laser et al., 2002) and starch feedstock residues (Binod et al., 2010; Talebnia et al., 2010), food waste (Guimaraes et al., 2010; Ravindran and Jaiswal, 2016), industrial waste (Cardona et al., 2010; Prasad et al., 2007), urban waste (Prasad et al., 2007; Ravindran and Jaiswal, 2016), and forestry waste (Duff and Murray, 1996). Thus, it complements the research on the processes for the production, evaluation, and utilization of bioethanol fuels at large.

However, it is essential to develop efficient incentive structures (North, 1991) for the primary stakeholders to enhance the research in this field (Konur, 2000, 2002a,b,c, 2006a,b, 2007a,b). Although there has been a large number of review papers on the feedstock-based bioethanol fuels (Hendriks and Zeeman, 2009; Mosier et al., 2005; Sun and Cheng, 2002), there has been no review of the most-cited 25 papers in this field.

DOI: 10.1201/9781003226451-52

Thus, this book chapter presents a review of the most-cited 25 articles in the field of the feedstock-based bioethanol fuels. Then, it discusses the key findings of these highly influential papers and comments on the future research priorities in this field.

41.2 MATERIALS AND METHODS

The search for this study was carried out using Scopus database (Burnham, 2006) in October 2022.

As a first step for the search of the relevant literature, the keywords were selected using the most-cited first 300 population papers for each feedstock. The selected keyword list was then optimized to obtain a representative sample of papers for each research field. These keyword lists were then integrated to obtain the keyword list for this research field (Konur, 2023a,b).

As a second step, a sample dataset was used for this study. The first 25 articles with at least 608 citations each were selected for the review study. Key findings from each paper were taken from the abstracts of these papers and were discussed. Additionally, a number of brief conclusions were drawn and a number of relevant recommendations were made to enhance the future research landscape.

41.3 RESULTS

The brief information about 25 most-cited papers with at least 608 citations each on the feedstock-based bioethanol fuels is given below. The primary research fronts are the pretreatments and hydrolysis of the feedstocks and production and evaluation of the feedstock-based bioethanol fuels with nine, nine, and seven highly cited papers (HCPs), respectively.

41.3.1 FEEDSTOCK PRETREATMENTS

The brief information about nine most-cited papers on the pretreatments of feedstocks with at least 618 citations each is given below (Table 41.1).

Schwanninger et al. (2004) studied the effects of short-time vibratory ball milling on the shape of FTIR spectra of wood and cellulose in a paper with 940 citations. They observed that this milling process had a strong influence on the shape of FTIR spectra of wood and cellulose, even if the samples were milled for only a short time. Further, the mechanical treatment itself, rather than temperature or particle size or oxidation processes, was the main influencing factor for the observed changes in the structure of the FTIR spectra. They observed the most visible changes in the spectra of cellulose and wood at wave numbers 1,034, 1,059, 1,110, 1,162, 1,318, 1,335, 2,902 cm^{-1}, and in the OH-stretching vibration region from 3,200 to 3,500 cm^{-1}. These changes were mainly associated with a decrease in the degree of crystallinity and/or a decrease in the degree of polymerization of the cellulose.

Sun et al. (2009) dissolved and partially delignified yellow pine and red oak in 1-ethyl-3-methyl-imidazolium acetate ([C$_2$mim]OAc) after mild grinding in a paper with 833 citations. They showed [C$_2$mim]OAc was a better solvent for wood than 1-butyl-3-methylimidazolium chloride ([C$_4$mim]Cl) and that type of wood, initial wood load, particle size, etc. affected dissolution and dissolution rates. For example, red oak dissolved better and faster than yellow pine. They obtained carbohydrate-free lignin and cellulose-rich materials by using the proper reconstitution solvents (e.g., acetone/water 1:1 v/v) and they achieved approximately 26.1% and 34.9% reductions of lignin content in the reconstituted cellulose-rich materials from pine and oak, respectively, in one dissolution/reconstitution cycle. For pine, they recovered 59% of the holocellulose in the original wood in the cellulose-rich reconstituted material, while they recovered 31% and 38% of the original lignin, respectively, as carbohydrate-free lignin and as carbohydrate-bonded lignin in the cellulose-rich materials.

Cai and Zhang (2005) dissolved cellulose in LiOH/urea and NaOH/urea aqueous solutions in a paper with 751 citations. They evaluated the dissolution behavior and solubility of cellulose. They found that cellulose having viscosity-average molecular weight of 11.4×10^4 and 37.2×10^4 could be dissolved, respectively, in 7% NaOH/12% urea and 4.2% LiOH/12% urea aqueous solutions

TABLE 41.1
The Pretreatment of Feedstocks

No.	Papers	Biomass	Prts.	Parameters	Keywords	Lead Authors	Affil.	Cits
4	Schwanninger et al. (2004)	Wood	Milling	Pretreatment, temperature, particle size, oxidation processes, FT-IR spectra, wave numbers, crystallinity	Wood, milling	Schwanninger, Manfred 6602877236	Univ. Natr. Res. Life Sci. Austria	943
8	Sun et al. (2009)	Wood Pine, oak	IL	IL types, dissolution rates, wood types, delignification, cellulose recovery	Wood, delignification, IL	Rogers, Robin D. 35474829200	Univ. Alabama USA	833
13	Cai and Zhang (2005)	Cellulose	IL	Cellulose dissolution, alkaline solvents, cellulose dissolution behavior, solubility, hydrogen bonding	Cellulose, dissolution	Zhang, Lina* 55917992100	Wuhan Univ. China	751
15	Fort et al. (2007)	Wood	IL	Wood dissolution, IL, cellulose reconstitution, delignification	Wood, IL	Rogers, Robin D. 35474829200	Univ. Alabama USA	725
17	Martinez et al. (2004)	Lignocellulosic biomass Wood	Enzymes	Fungal genome, wood degradation, hydrolytic enzymes, lignocellulose degradation	Lignocellulose, degrading, fungus	Cullen, Dan 7202109135	USDA Forest Serv. USA	685
18	Fukaya et al. (2008)	Cellulose	ILs	Cellulose dissolution, ILs, anions, cellulose solubilization	Cellulose dissolution	Ohno, Hiroyuki 7403244652	Tokyo Univ. Agr. Technol. Japan	672
19	Quinlan et al. (2011)	Cellulose	Enzymes	Cellulose degradation, GH61 enzymes, cellodextrin, methylated histidine	Cellulose, degradation	Johansen, Katja S.* 36473579400	Univ. Copenhagen Denmark	670
20	Li et al. (2007)	Wood Aspen	Steam	Lignin depolymerization and repolymerization, delignification	Wood, delignification	Li, Jiebing 7410069979	Rise Res. Inst. Sweden	665
22	Remsing et al. (2006)	Cellulose	IL	Cellulose dissolution mechanism, IL, hydrogen bonding, carbohydrate hydroxyl protons	Cellulose, IL	Rogers, Robin D. 35474829200	Univ. Alabama USA	657

*, Female; Cits., Number of citations received for each paper; Na, non-available; Prt, Biomass pretreatments.

pre-cooled to −10°C within 2 min, whereas all of them could not be dissolved in KOH/urea aqueous solution. Further, the dissolution power of the solvent systems was in the order of LiOH/urea > NaOH/urea ≫ KOH/urea aqueous solution. LiOH/urea and NaOH/urea aqueous solutions as non-derivatizing solvents broke the intra- and intermolecular hydrogen bonding of cellulose and prevented the approach toward each other of the cellulose molecules, leading to the good dispersion of cellulose to form an actual solution.

Fort et al. (2007) dissolved wood in the IL in a paper with 725 citations. They used 1-n-butyl-3-methylimidazolium chloride ([C_4mim]Cl). They presented the dissolution profiles for woods of different hardness. They also showed that cellulose could be readily reconstituted from the IL-based wood liquors in fair yields by the addition of a variety of precipitating solvents. Further, the polysaccharide obtained in this manner was virtually free of lignin and hemicellulose and had characteristics that were comparable to those of pure cellulose samples subjected to similar processing conditions.

Martinez et al. (2004) sequenced the 30-million base pair genome of *Phanerochaete chrysosporium* strain RP78 using a whole-genome shotgun approach in a paper with 685 citations. They found that the *P. chrysosporium* genome had an impressive array of genes encoding secreted oxidases, peroxidases, and hydrolytic enzymes that cooperated in wood decay. Analysis of the genome data would enhance the understanding of lignocellulose degradation and provide a framework for further development of bioprocesses for biomass utilization.

Fukaya et al. (2008) dissolved cellulose with polar ionic liquids (ILs) under mild conditions in a paper with 672 citations. They obtained a series of alkylimidazolium salts containing dimethyl phosphate, methyl methylphosphonate, or methyl phosphonate prepared by a facile, one-pot procedure as room temperature ILs, which had the potential to solubilize cellulose under mild conditions. Especially, they found that N-ethyl-N′-methylimidazolium methylphosphonate ([C_2mim]MeO) enabled the preparation of 10 wt% cellulose solution by keeping it at 45°C for 30 min with stirring and rendered soluble 2–4 wt% cellulose without pretreatments and heating.

Quinlan et al. (2011) performed the oxidative degradation of cellulose by a copper metalloenzyme that exploited biomass components in a paper with 670 citations. They showed that glycoside hydrolase (CAZy) GH61 enzymes were a unique family of copper-dependent oxidases and copper was needed for GH61 maximal activity and that the formation of cellodextrin and oxidized cellodextrin products by GH61 is enhanced in the presence of small molecule redox-active cofactors such as ascorbate and gallate. Further, the active site of GH61 contained a type II copper and, uniquely, a methylated histidine in the copper's coordination sphere.

Li et al. (2007) studied the lignin depolymerization and repolymerization and its critical role for delignification of aspen wood by steam explosion pretreatment in a paper with 665 citations. They analyzed the lignin portion and observed the competition between lignin depolymerization and repolymerization and identified the conditions required for these two types of reaction. Further, the addition of 2-naphthol inhibited the repolymerization reaction strongly, resulting in a highly improved delignification by subsequent solvent extraction and an extracted lignin of uniform structure.

Remsing et al. (2006) explored the mechanism of cellulose dissolution in 1-n-butyl-3- methylimidazolium chloride ([C_4mim]Cl) in a paper with 657 citations. Through ^{13}C and [35/37]Cl NMR relaxation measurements on several model systems, they found that the solvation of cellulose by this IL involved hydrogen bonding between the carbohydrate hydroxyl protons and the IL chloride ions in a 1:1 stoichiometry.

41.3.2 HYDROLYSIS OF THE FEEDSTOCKS

There are nine HCPs for the pretreatments of feedstock with at least 608 citations (Table 41.2).

Park et al. (2010) compared four different techniques incorporating X-ray diffraction (XRD) and solid-state ^{13}C nuclear magnetic resonance (NMR) for the measurement of the cellulose crystallinity index (CI) using eight different cellulose preparations in a paper with 2,026 citations. They found that the simplest method which involved measurement of just two heights in the X-ray

TABLE 41.2
The Hydrolysis of Feedstocks

No.	Papers	Biomass	Prts.	Parameters	Keywords	Lead Authors	Affil.	Cits
1	Park et al. (2010)	Cellulose	Enzymes	Cellulose crystallinity index, measurements, XRD, NMR, cellulose digestibility and accessibility, cellulases	Cellulose, cellulase	Johnson, David K. 24550868900	Natl. Renew. Ener. Lab. USA	2026
6	Suganuma et al. (2008)	Cellulose	C catalysts	Hydrolysis, solid carbon catalysts, crystalline cellulose, activation energy, catalyst recycling	Cellulose, hydrolysis	Hara, Michkazu 7403345875	Tokyo Inst. Technol. Japan	885
7	Li et al. (2010)	Switchgrass	Acids, IL, enzymes	Pretreatment types, acid hydrolysis, delignification, saccharification, sugar yields, crystallinity, surface area, lignin content, enzymatic hydrolysis	Switchgrass, pretreatment, saccharification, recalcitrance, IL	Singh, Seema* 35264950300	Sandia Natl. Lab. USA	860
9	Kilpelainen et al. (2007)	Wood Spruce, pine	IL, enzymes	IL pretreatment, enzymatic hydrolysis, IL types, wood types	Wood, IL, dissolution	Kilpelainen, Ilkka 7006830888	Univ. Helsinki Finland	832
11	Lee et al. (2009)	Wood	IL, enzymes	Enzymatic hydrolysis, IL pretreatment, delignification, cellulose crystallinity index, enzyme recycling	Wood, IL, hydrolysis	Dordick, Jonathan S. 7102545507	Rensselaer Polytech Inst. USA	816
12	Eriksson et al. (2002)	Lignocellulose Spruce	Enzymes, surfactants	Enzymatic hydrolysis, pretreatment, mechanisms, enzyme adsorption, cellulase stability, surfactant–lignin interactions	Lignocellulose, hydrolysis	Tjerneld, Folke 7006446969	Lund Univ. Sweden	754
14	Kumar et al. (2009)	Wood Poplar	Ammonia, acids, alkali, SO_2, enzymes	Pretreatment types, enzymatic hydrolysis, cellulase adsorption capacity, biomass crystallinity, cellulose degree of polymerization	Poplar, pretreatment	Wyman, Charles E. 7004396809	Univ. Calif. Riverside USA	726
24	Lloyd and Wyman (2005)	Corn stover	Acids, enzymes	Enzymatic hydrolysis, xylose and glucose yields, total potential sugar yield	Corn stover, pretreatment, hydrolysis	Wyman, Charles E. 7004396809	Univ. Calif. Riverside USA	626
25	Sasaki et al. (2000)	Cellulose	Acids, enzymes	Cellulose dissolution and hydrolysis, cellulose, glucose and cellobiose decomposition rates, hydrogen linkages	Cellulose, dissolution, hydrolysis	Sasaki, Mitsuru 54962464300	Tohoku Univ. Japan	608

*, Female; Cits., Number of citations received for each paper; Na, non-available; Prt, Biomass pretreatments.

diffractogram produced significantly higher CI values than did the other methods. However, the alternative XRD and NMR methods, which consider the contributions from amorphous and crystalline cellulose to the entire XRD and NMR spectra, provided a more accurate measure of the CI of cellulose. Although celluloses having a high amorphous content were usually more easily digested by enzymes, it was unclear whether CI actually provides a clear indication of the digestibility of a cellulose sample. Cellulose accessibility should be affected by crystallinity, but was also likely to be affected by several other parameters, such as lignin/hemicellulose contents and distribution, porosity, and particle size. Given the methodological dependency of cellulose CI values and the complex nature of cellulase interactions with amorphous and crystalline celluloses, they cautioned against trying to correlate relatively small changes in CI with changes in cellulose digestibility.

Suganuma et al. (2008) hydrolyzed cellulose by amorphous carbon bearing SO_3H, COOH, and OH groups in a paper with 885 citations. They found that crystalline pure cellulose was not hydrolyzed by conventional strong solid Brønsted acid catalysts such as niobic acid, H-mordenite, Nafion, and Amberlyst-15, while amorphous carbon bearing solid catalysts functioned as an efficient catalyst for the reaction. The apparent activation energy for the hydrolysis of cellulose into glucose using the carbon catalyst was 110 kJ/mol, smaller than that for sulfuric acid under optimal conditions (170 kJ/mol). Further, this carbon catalyst could be readily separated from the saccharide solution after reaction for reuse in the reaction without loss of activity. They attributed the catalytic performance of the carbon catalyst to the ability of the material to adsorb β-1,4 glucan, which did not adsorb to other solid acids.

Li et al. (2010) compared the efficiency of dilute acid hydrolysis and dissolution in an IL in terms of delignification, saccharification efficiency, and saccharide yields with switchgrass in a paper with 860 citations. When subject to IL pretreatment, they observed that switchgrass exhibited reduced cellulose crystallinity, increased surface area, and decreased lignin content compared to dilute acid pretreatment. Further, the IL pretreatment enabled a significant enhancement in the rate of enzymatic hydrolysis of the cellulose component of switchgrass, with a rate increase of 16.7-fold, and a glucan yield of 96.0% obtained in 24 h. In conclusion, the IL pretreatment offered unique advantages compared to the dilute acid pretreatment process for switchgrass.

Kilpelainen et al. (2007) dissolved wood in ILs in a paper with 832 citations. They found that 1-butyl-3-methylimidazolium chloride ([Bmim]Cl) and 1-allyl-3-methylimidazolium chloride ([Amim]Cl) had good solvating power for Norway spruce sawdust and Norway spruce and Southern pine thermomechanical pulp fibers as these ILs provided solutions which permitted the complete acetylation of the wood. Alternatively, they obtained transparent amber solutions of wood when the dissolution of the same lignocellulosic samples was attempted in 1-benzyl-3-methylimidazolium chloride ([BzMim]Cl). They then digested the cellulose of the regenerated wood to glucose by a cellulase enzymatic hydrolysis treatment. Furthermore, completely acetylated wood was readily soluble in chloroform.

Lee et al. (2009) used 1-ethyl-3-methylimidazolium acetate ([Emim]CH_3COO) to extract lignin from wood flour in a paper with 816 citations. They observed that the cellulose in the pretreated wood flour became far less crystalline without undergoing solubilization. When 40% of the lignin was removed, the cellulose CI dropped below 45, resulting in >90% of the cellulose in wood flour to be hydrolyzed by *Trichoderma viride* cellulase. They then reused this IL, thereby resulting in a highly concentrated solution of chemically unmodified lignin.

Eriksson et al. (2002) explored the mechanism of surfactant effect in enzymatic hydrolysis of lignocellulose in a paper with 754 citations. They screened a number of surfactants for their ability to improve enzymatic hydrolysis of steam-pretreated spruce (SPS). They found that the non-ionic surfactants were the most effective. Studies of adsorption of the dominating cellulase of *Trichoderma reesei*, Cel7A (CBHI), during hydrolysis showed that the anionic and non-ionic surfactants reduced enzyme adsorption to the lignocellulose substrate. The approximate reduction of enzyme adsorption was from 90% adsorbed enzyme to 80% with surfactant addition. Surfactants had only a weak effect on cellulase temperature stability. They explained the improved conversion of lignocellulose with surfactant by the reduction of the unproductive enzyme adsorption to the lignin part of the substrate. This was due to hydrophobic interaction of surfactant with lignin on the lignocellulose surface, which released unspecifically bound enzyme.

Kumar et al. (2009) characterized corn stover and poplar solids resulting from leading pretreatment technologies in a paper with 726 citations. They performed ammonia fiber expansion (AFEX), ammonia recycled percolation (ARP), controlled pH, dilute acid, flowthrough, lime, and SO_2 pretreatments. They found that lime pretreatment removed the most acetyl groups from both corn stover and poplar, while AFEX removed the least and the low pH pretreatments depolymerized cellulose and enhanced biomass crystallinity much more than higher pH approaches. Further, the lime pretreated corn stover solids and flowthrough pretreated poplar solids had the highest cellulase adsorption capacity, while dilute acid pretreated corn stover solids and controlled pH pretreated poplar solids had the least. Furthermore, enzymatically extracted AFEX lignin preparations for both corn stover and poplar had the lowest cellulase adsorption capacity. SO_2 pretreated solids had the highest surface O/C ratio for poplar, while for corn stover, they observed the highest value for dilute acid pretreatment with a Parr reactor. Although dependent on pretreatment and substrate, along with changes in cross-linking and chemical changes, they reasoned that these pretreatments might also decrystallize cellulose and change the ratio of crystalline cellulose polymorphs.

Lloyd and Wyman (2005) performed the dilute sulfuric acid pretreatment of corn stover followed by enzymatic hydrolysis of the remaining solids in a paper with 625 citations. They reported the individual xylose and glucose yields as a percentage of the total potential yield of both sugars over a range of sulfuric acid concentrations of 0.22%, 0.49%, and 0.98% w/w at 140°C, 160°C, 180°C and 200°C. They found that up to 15% of the total potential sugar in the substrate could be released as glucose during pretreatment and between 15% and 90+% of the xylose remaining in the solid residue could be recovered in subsequent enzymatic hydrolysis, depending on the enzyme loading. Glucose yields increased from as high as 56% of total maximum potential glucose plus xylose for just enzymatic digestion to 60% when glucose released in pretreatment was included. Xylose yields similarly increased from as high as 34% of total potential sugars for pretreatment alone to between 35% and 37% when credit was taken for xylose released in digestion. Yields were much lower if no acid was used. Conditions that maximized individual sugar yields were often not the same as those that maximized total sugar yields. Overall, they obtained up to about 92.5% of the total sugars originally available in the corn stover used or coupled dilute acid pretreatment and enzymatic hydrolysis.

Sasaki et al. (2000) dissolved and hydrolyzed cellulose in subcritical and supercritical water in a paper with 608 citations. They performed the decomposition experiments of microcrystalline cellulose in subcritical and supercritical water (25 MPa, 320°C–400°C, and 0.05–10.0 s). At 400°C, they obtained mainly hydrolysis products, while in 320°C–350°C water, aqueous decomposition products of glucose were the main products. Further, below 350°C the cellulose decomposition rate was slower than the glucose and cellobiose decomposition rates, while above 350°C, the cellulose hydrolysis rate drastically increased and became higher than the glucose and cellobiose decomposition rates. On the other hand, below 280°C, cellulose particles became gradually smaller with increasing reaction time but, at high temperatures (300°C–320°C), cellulose particles disappeared with increasing transparency and much more rapidly than expected from the lower temperature results. In conclusion, cellulose hydrolysis at high temperature took place with dissolution in water. This was probably because of the cleavage of intra- and intermolecular hydrogen linkages in the cellulose crystal. Thus, a homogeneous atmosphere was formed in supercritical water, and this resulted in the drastic increase of the cellulose decomposition rate above 350°C.

41.3.3   Feedstock-based Bioethanol Production and Evaluation

There are seven HCPs for the production and evaluation of the feedstock-based bioethanol fuels with at least 638 citations each (Table 41.3). Further, there are two and five HCPs for the production and evaluation of the bioethanol fuels, respectively. As the pretreatment and hydrolysis of the feedstock are the fundamental parts of the bioethanol production, these narrated papers often cover these processes too.

41.3.3.1 Feedstock-based Bioethanol Production

There are two HCPs for the production of the feedstock-based bioethanol fuels with at least 664 citations each (Table 41.3). These papers also cover the fermentation of the hydrolysates of the feedstock. As the pretreatment and hydrolysis are the fundamental parts of the bioethanol production, these narrated papers often cover these processes too.

Larsson et al. (1999) studied the effect of the combined severity (SC) of dilute sulfuric acid hydrolysis of spruce on sugar yield and the fermentability of the hydrolysate by *Saccharomyces cerevisiae* in a paper with 889 citations. When the CS of the hydrolysis conditions increased, they observed that the yield of fermentable sugars increased to a maximum between CS 2.0 and 2.7 for mannose, and 3.0 and 3.4 for glucose above which it decreased. Further, the decrease in the yield of monosaccharides coincided with the maximum concentrations of furfural and 5-HMF. With the further increase in CS, the concentrations of furfural and 5-HMF decreased while the formation of formic acid and levulinic acid increased. The yield of ethanol decreased at approximately CS 3, while the volumetric productivity decreased at lower CS. They then assayed the effect of acetic acid, formic acid, levulinic acid, furfural, and 5-HMF on fermentability in model fermentations. Ethanol yield and volumetric productivity decreased with increasing concentrations of acetic acid, formic acid, and levulinic acid. However, furfural and 5-HMF decreased the volumetric productivity but did not influence the final yield of ethanol. The decrease in volumetric productivity was more pronounced when 5-HMF was added to the fermentation, and this compound was depleted at a lower rate than furfural. Further, the inhibition observed in hydrolysates produced in higher CS could not be fully explained by the effect of the furfural, 5-HMF, acetic acid, formic acid, and levulinic acid.

Saha et al. (2005) performed the dilute acid pretreatment, enzymatic saccharification, and fermentation of wheat straw to produce ethanol in a paper with 664 citations. They found that the maximum yield of monomeric sugars from wheat straw (7.83%, w/v, DS) by dilute H_2SO_4 (0.75%, v/v) pretreatment and enzymatic saccharification (45°C, pH 5.0, 72 h) using cellulase, β-glucosidase, xylanase, and esterase was 565 mg/g. Under this condition, no measurable quantities of furfural and hydroxymethylfurfural (HMF) were produced. The yield of ethanol (per liter) from acid-pretreated and enzyme-saccharified wheat straw (78.3 g) hydrolysate by recombinant *Escherichia coli* strain FBR5 was 19 g with a yield of 0.24 g/g DS. Detoxification of the acid- and enzyme-treated wheat straw hydrolysate by overliming reduced the fermentation time from 118 to 39 h in the case of separate hydrolysis and fermentation (SHF) (35°C, pH 6.5), and increased the ethanol yield from 13 to 17 g/L and decreased the fermentation time from 136 to 112 h in the case of simultaneous saccharification and fermentation (SSF) (35°C, pH 6.0).

41.3.3.2 Feedstock-based Bioethanol Evaluation

There are five HCPs for the evaluation of the feedstock-based bioethanol fuels with at least 638 citations each (Table 41.3).

Hamelinck et al. (2005) evaluated ethanol production costs from lignocellulosic biomass in a paper with 1,234 citations. They found that the technology available as of the early 2000s which was based on dilute acid hydrolysis had about 35% efficiency (higher heating value, HHV) from biomass to ethanol. The overall efficiency, with electricity co-produced from the not-fermentable lignin, was about 60%. They foresaw that the improvements in pretreatment and advances in biotechnology, especially through process combinations, could bring the ethanol efficiency to 48% and the overall process efficiency to 68%. They estimated investment costs as of the early 2000s at 2.1 k€/kWHHV (at 400 MWHHV input, i.e., a nominal 2,000 tonne dry/day input). However, a future technology in a five times larger plant (2 GWHHV) could have investments of 900 k€/kWHHV. They further found that a combined effect of higher hydrolysis and fermentation efficiency, lower specific capital investments, increase of scale, and cheaper biomass feedstock costs (from 3 to 2 €/GJHHV) could bring the ethanol production costs from 22 €/GJHHV in the next 5 years, to 13 €/GJ over the 10–15 years timescale, and down to 8.7 €/GJ in 20 or more years.

Pimentel and Patzek (2005) evaluated the energy balance of ethanol fuels from corn, switchgrass, and wood in comparison with biodiesel fuels from soybean and sunflower in a paper with 1,020

TABLE 41.3
The Production and Evaluation of Feedstock-Based Bioethanol Fuels

No.	Papers	Biomass	Res. Fronts	Prts.	Yeasts	Parameters	Keywords	Lead Authors	Affil.	Cits
2	Hamelinck et al. (2005)	Lignocellulosic biomass	Evaluation	Na	Na	Ethanol techno-economics, hydrolysis, fermentation, production costs, short-, middle- and long-term, investment costs, ethanol efficiency	Lignocellulosic, ethanol	Hamelinck, Carlo N. 6603008025	Ecorys Netherlands	1234
3	Pimentel and Patzek (2005)	Corn, switchgrass, wood	Evaluation	Na	Na	Bioethanol energy balance, biodiesel energy balance	Ethanol, corn, switchgrass, wood	Pimentel, David P. 7005471319	Cornell Univ. USA	1020
5	Larsson et al. (1999)	Softwood Spruce	Production	Acids	S. cerevisiae	Acid hydrolysis severity, fermentation, sugar and ethanol yield, fermentation inhibitors, volumetric productivity	Softwood, fermentation, hydrolysis	Hahn-Hagerdal, Barbel* 7005389381	Lund Univ. Sweden	889
10	Schmer et al. (2008)	Switchgrass	Evaluation	Na	Na	Energy balance, GHG emissions, biomass and ethanol yield, energy yield	Switchgrass, ethanol	Vogel, Kenneth P. 7102498441	Univ. Nebraska-Lincoln USA	831
16	Klein-Marcuschamer et al. (2012)	Lignocellulosic biomass Corn stover	Evaluation	Enzymes	Na	Cellulase techno-economics, cellulase production cost, feedstock prices and fermentation times	Lignocellulosic, biofuels, enzymes	Blanch, Harvey W. 7006259341	Univ. Calif. Riverside USA	703
21	Saha et al. (2005)	Wheat straw	Production	Acids, enzymes	E. coli	Ethanol production, pretreatments, hydrolysis, fermentation, ethanol yield, detoxification, SHF, SSF	Straw, saccharification, fermentation, pretreatment, ethanol	Saha, Badal C. 7202946302	USDA Agr. Res. Serv. USA	664
23	Macedo et al. (2008)	Sugarcane	Evaluation	Na	Na	Energy balance and GHG emissions, sugarcane productivity, ethanol yield, GHG emissions avoidance	Sugarcane, ethanol	Macedo, Isaias C. 6603142465	State Univ. Campinas Brazil	638

*, Female; Cits., Number of citations received for each paper; Na, non-available; Prt, Biomass pretreatments.

citations. They found that the energy outputs from ethanol produced using these feedstocks were each less than the respective fossil energy inputs. Ethanol production using corn grain, switchgrass, and wood required 29%, 50%, and 57% more fossil energy than the ethanol fuel produced, respectively.

Schmer et al. (2008) evaluated the net energy of ethanol fuels from switchgrass in a paper with 831 citations. They managed switchgrass in field trials of 3–9 ha on marginal cropland on ten farms across a wide precipitation and temperature gradient in the midcontinental USA to determine net energy and economic costs based on known farm inputs and harvested yields. Annual biomass yields of established fields averaged 5.2–11.1 Mg/ha with a resulting average estimated net energy yield (NEY) of 60 GJ/ha/y. Switchgrass produced 540% more renewable than non-renewable energy consumed. Switchgrass monocultures managed for high yield produced 93% more biomass yield and an equivalent estimated NEY than previous estimates from human-made prairies that received low agricultural inputs. Estimated average greenhouse gas (GHG) emissions from cellulosic ethanol derived from switchgrass were 94% lower than estimated GHG from gasoline. This was a baseline study that represented the genetic material and agronomic technology available for switchgrass production in 2000 and 2001.

Klein-Marcuschamer et al. (2012) constructed a technoeconomic model for the production of fungal cellulase production of lignocellulosic ethanol fuels in a paper with 703 citations. They constructed a technoeconomic model for the production of fungal cellulases. They found that the cost of producing enzymes was much higher than that commonly assumed in the literature. For example, the cost contribution of enzymes to ethanol produced by the conversion of corn stover was $0.68/gal if the sugars in the biomass could be converted at maximum theoretical yields, and $1.47/gal if the yields were based on saccharification and fermentation yields that have been previously reported in the scientific literature. They performed a sensitivity analysis to study the effect of feedstock prices and fermentation times on the cost contribution of enzymes to ethanol price.

Macedo et al. (2008) evaluated the energy balance and GHG emissions in the production and use of ethanol fuels from sugarcane in Brazil for 2005/2006 for a sample of mills processing up to 100 million tons of sugarcane per year in a paper with 638 citations. They proposed a conservative scenario proposed for 2020. They found that the fossil energy ratio was 9.3 for 2005/2006 and might reach 11.6 in 2020 with technologies already commercial. For anhydrous ethanol production, the total GHG emission was 436 kg CO_2 eq/m^3 ethanol for 2005/2006, decreasing to 345 kg CO_2 eq/m^3 in the 2020 scenario. Avoided emissions depended on the final use: for E100 use in Brazil, they were (in 2005/2006) 2,181 kg CO_2 eq/m^3 ethanol, and for E25, they were 2,323 kg CO_2 eq/m^3 ethanol (anhydrous). Both values would increase about 26% for the conditions assumed for 2020 mostly due to the large increase in sales of electricity surpluses. They found the high impact of cane productivity and ethanol yield variation on these balances (and the impacts of average cane transportation distances, level of soil cultivation, and some others) and of bagasse and electricity surpluses on GHG emissions avoidance.

41.4 DISCUSSION

41.4.1 Introduction

The crude oil-based gasoline fuels have been widely used in the transportation sector since the 1920s. However, there have been great public concerns over the adverse environmental and human impact of these fuels. Hence, biomass-based bioethanol fuels have increasingly been used in blending gasoline and petrodiesel fuels, in the fuel cells, and in the biochemical production in a biorefinery context.

However, it is necessary to pretreat the biomass to enhance the yield of the bioethanol prior to the bioethanol fuel production from the feedstocks through the hydrolysis and fermentation of the biomass, respectively.

The research in the field of the feedstock-based bioethanol fuels has intensified in this context in the key research fronts of the pretreatment and hydrolysis of the feedstocks, fermentation of the feedstock-based hydrolysates, and production and evaluation of the feedstock-based bioethanol fuels.

The research in this field has also intensified for the feedstocks of wood biomass, algal biomass, grass biomass, sugar feedstock biomass, starch feedstock biomass, lignocellulosic biomass at large, cellulose, sugar and starch feedstock residues, food waste, industrial waste, urban waste, and for-estry waste. Thus, it complements the research on the processes for the production, evaluation, and utilization of bioethanol fuels at large.

However, it is essential to develop efficient incentive structures for the primary stakeholders to enhance the research in this field. Although there has been a limited number of review papers for this field, there has been no review of the most-cited 25 articles in this field.

Thus, this book chapter presents a review of the most-cited 25 articles on the bioethanol fuel production and evaluation from the feedstocks. Then, it discusses the key findings of these highly influential papers and comments on the future research priorities in this field.

As a first step for the search of the relevant literature, the keywords were selected using the most-cited first 300 population papers for each feedstock. The selected keyword list was then optimized to obtain a representative sample of papers for each research field. These keyword lists were then integrated to obtain the keyword list for this research field (Konur, 2023a,b).

As a second step, a sample dataset was used for this study. The first 25 articles with at least 608 citations each were selected for the review study. Key findings from each paper were taken from the abstracts of these papers and were discussed. Additionally, a number of brief conclusions were drawn and a number of relevant recommendations were made to enhance the future research landscape.

Information about the thematic research fronts for the sample papers in the feedstock-based bioethanol fuels is given in Table 41.4. As this table shows, the most prolific research front for this field is the wood biomass feedstocks such as poplar and pine with 40% of the HCPs, followed by the cellulose with 28% of these HCPs. The other prolific feedstocks are the lignocellulosic biomass at large, grass biomass, and starch feedstock residues with 16%, 12%, and 12% of the HCPs, respec-tively. Further, the other minor feedstocks are the sugar and starch feedstocks with 4% of the HCPs each. On the other hand, it is notable that there are no HCPs for the food waste, algal biomass, sugar feedstock residues, urban waste, forestry waste, and industrial waste. Thus, the first five feedstocks

TABLE 41.4
The Most Prolific Research Fronts for the Feedstock-Based Bioethanol Fuels

No.	Research Fronts	N Paper (%) Review	N Paper (%) Sample	Surplus (%)	N Paper Population (%)
1	Wood-based bioethanol fuels	40.0	10.4	29.6	20.8
2	Cellulose-based bioethanol fuels	28.0	20.8	7.2	14.7
3	Lignocellulosic biomass-based bioethanol fuels	16.0	36.1	−20.1	13.6
4	Grass-based bioethanol fuels	12.0	5.6	6.4	3.8
5	Starch feedstock residues-based bioethanol fuels	12.0	3.0	9.0	15.3
6	Sugar feedstock-based bioethanol fuels	4.0	3.7	0.3	4.2
7	Starch feedstock-based bioethanol fuels	4.0	2.6	1.4	5.2
8	Other feedstock-based bioethanol fuels	0.0	5.2	−5.2	42.5
	Food waste-based bioethanol fuels	0.0	0.7	−0.7	5.3
	Algal biomass-based bioethanol fuels	0.0	1.9	−1.9	2.9
	Sugar feedstock residues-based bioethanol fuels	0.0	1.1	−1.1	7.7
	Urban waste-based bioethanol fuels	0.0	0.7	−0.7	6.1
	Forestry waste-based bioethanol fuels	0.0	0.4	−0.4	2.0
	Industrial waste-based bioethanol fuels	0.0	0.4	−0.4	18.5
	Sample size	25	269		26,852

N paper population, The number of papers in the population sample of 26,852 papers; N Paper (%) review, The number of papers in the sample of 25 reviewed papers; N paper (%) sample, The number of papers in the population sample of 269 papers.

TABLE 41.5
The Most Prolific Thematic Research Fronts for the Feedstock-based Bioethanol Fuels

No.	Research Fronts	N Paper (%) Review	N Paper (%) Sample	Surplus (%)
1	Biomass pretreatments	92.0	78.1	13.9
2	Biomass hydrolysis	48.0	46.8	1.2
3	Bioethanol fuel evaluation	20.0	7.8	12.2
4	Hydrolysate fermentation	16.0	14.9	1.1
5	Bioethanol production	8.0	17.8	−9.8
6	Bioethanol fuel utilization	0.0	0.7	−0.7

N Paper (%) review, The number of papers in the sample of 25 reviewed papers; N paper (%) sample, The number of papers in the population sample of 269 papers.

dominate the HCPs in this field where only waste biomass is starch feedstock residues such as corn stover or wheat straw and cellulose is the key constituent of the lignocellulosic biomass. It is also notable that the research on the sugar and starch feedstock-based bioethanol fuels is not substantial for all of the samples studied, reflecting the adverse effect of undermining the food security (Bentivoglio et al., 2016; Elobeid and Hart, 2007).

Further, the most influential research fronts are the wood biomass with 30% surplus, followed by starch feedstock residues and cellulose with 15% surplus each. Further, the lignocellulosic biomass is the least influential feedstock with 20% deficit since a significant part of the papers for the lignocellulosic biomass-based bioethanol fuels are the reviews and short surveys.

Information about the thematic research fronts for the sample papers in the feedstock-based bioethanol fuels is given in Table 41.5. As this table shows, the most prolific research fronts for this field are the pretreatment and hydrolysis of the feedstock with 92% and 48% of the HCPs, respectively. The other prolific research fronts are the bioethanol evaluation and hydrolysate fermentation with 20% and 16% of the HCPs, respectively. Further, the other minor research front is bioethanol production with 8% of the HCPs and there is no HCP for the bioethanol fuel utilization.

On the other hand, biomass pretreatments and bioethanol fuel evaluation are the most influential research fronts with 14% and 12% surplus, respectively, while bioethanol production is the least influential research front with 10% deficit.

41.4.2 FEEDSTOCK PRETREATMENTS

The brief information about nine most-cited papers on the pretreatments of feedstocks with at least 618 citations each is given below (Table 41.1). On the other hand, it is notable that as the Table 41.5 shows, 92% of these HCPs are related to the pretreatments of the feedstock, respectively.

These findings show that both pretreatments and hydrolysis of the feedstock are the fundamental processes for the bioethanol production from the feedstocks. These narrated studies highlight the importance of the pretreatment and hydrolysis processes for the production of the bioethanol fuels from the feedstock with a high ethanol yield. These pretreatments, primarily enzymatic and chemical pretreatments, fractionate the feedstock and enhance the enzymatic digestibility of the biomass.

Schwanninger et al. (2004) studied the effects of short-time vibratory ball milling on the shape of FTIR spectra of wood and cellulose and observed that this milling process had a strong influence on the shape of FTIR spectra of wood and cellulose. Further, Sun et al. (2009) dissolved and partially delignified yellow pine and red oak in [C$_2$mim]OAc and showed this IL was a better solvent for wood than [C$_4$mim]Cl and that type of wood, initial wood load, particle size, etc. affected dissolution and dissolution rates.

Cai and Zhang (2005) dissolved cellulose in LiOH/urea and NaOH/urea aqueous solutions and found that these non-derivatizing solvents broke the intra- and intermolecular hydrogen bonding of cellulose and prevented the approach toward each other of the cellulose molecules, leading to the good dispersion of cellulose to form an actual solution. Further, Fort et al. (2007) dissolved wood in the IL and showed that cellulose could be readily reconstituted from the IL-based wood liquors in fair yields by the addition of a variety of precipitating solvents.

Martinez et al. (2004) sequenced the 30-million base pair genome of *P. chrysosporium* strain RP78 using a whole-genome shotgun approach and found that the *P. chrysosporium* genome had an impressive array of genes encoding secreted oxidases, peroxidases, and hydrolytic enzymes that cooperated in wood decay. Further, Fukaya et al. (2008) dissolved cellulose with polar ILs under mild conditions and found that [C$_2$mim]MeO enabled the preparation of 10 wt% cellulose solution.

Quinlan et al. (2011) performed the oxidative degradation of cellulose by a copper metalloenzyme that exploited biomass components and showed that GH61 enzymes were a unique family of copper-dependent oxidases. Further, Li et al. (2007) studied the lignin depolymerization and repolymerization and its critical role for delignification of aspen wood by steam explosion pretreatment and observed the competition between lignin depolymerization and repolymerization and identified the conditions required for these two types of reactions. Finally, Remsing et al. (2006) explored the mechanism of cellulose dissolution in [C$_4$mim]Cl and found that the solvation of cellulose by this IL involved hydrogen bonding between the carbohydrate hydroxyl protons and the IL chloride ions in a 1:1 stoichiometry.

41.4.3 HYDROLYSIS OF THE FEEDSTOCKS

There are nine HCPs for the pretreatments of feedstock with at least 608 citations (Table 41.2). On the other hand, it is notable that as the Table 41.5 shows, 48% of these HCPs are related to the hydrolysis of the feedstocks, respectively.

These findings show that both pretreatments and hydrolysis of the feedstock are the fundamental processes for the bioethanol production from the feedstock. These narrated studies highlight the importance of the pretreatment and hydrolysis processes for the production of the bioethanol fuels from the feedstock with a high ethanol yield. These pretreatments, primarily enzymatic and chemical pretreatments, fractionate the feedstock and enhance the enzymatic digestibility of the biomass.

Park et al. (2010) compared four different techniques incorporating XRD and solid-state^{13}C NMR for the measurement of the cellulose crystallinity index (CI) using eight different cellulose preparations and found that the simplest method produced significantly higher CI values than did the other methods. Further, Suganuma et al. (2008) hydrolyzed cellulose by amorphous carbon catalysts and found that these amorphous carbon bearing solid catalysts functioned as an efficient catalyst for the reaction.

Li et al. (2010) compared the efficiency of dilute acid hydrolysis and dissolution in an IL with switchgrass and observed that switchgrass exhibited reduced cellulose crystallinity, increased surface area, and decreased lignin content compared to dilute acid pretreatment. Further, Kilpelainen et al. (2007) dissolved wood in ILs and found that [Bmim]Cl and [Amim]Cl had good solvating power for Norway spruce sawdust and Norway spruce and Southern pine thermomechanical pulp fibers.

Lee et al. (2009) used [Emim]CH$_3$COO to extract lignin from wood flour and found that cellulose in the pretreated wood flour became far less crystalline without undergoing solubilization. Further, Eriksson et al. (2002) explored the mechanism of surfactant effect in enzymatic hydrolysis of lignocellulose and found that the non-ionic surfactants were the most effective.

Kumar et al. (2009) characterized corn stover and poplar solids resulting from leading pretreatment technologies and found that lime pretreatment removed the most acetyl groups from both corn stover and poplar, while AFEX removed the least and the low pH pretreatments depolymerized cellulose and enhanced biomass crystallinity much more than higher pH approaches. Further, Lloyd and Wyman (2005) performed the dilute sulfuric acid pretreatment of corn stover followed by enzymatic hydrolysis and found that up to 15% of the total potential sugar in the substrate could be released as glucose during pretreatment and between 15 and 90+% of the xylose could be recovered in subsequent enzymatic hydrolysis, depending on the enzyme loading. Finally, Sasaki et al. (2000) dissolved and hydrolyzed cellulose in subcritical and supercritical water and found that cellulose hydrolysis at high temperature took place with dissolution in water.

41.4.4 Feedstock-based Bioethanol Production and Evaluation

There are seven HCPs for the production and evaluation of the feedstock-based bioethanol fuels with at least 638 citations each (Table 41.3). Further, there are two and five HCPs for the production and evaluation of the bioethanol fuels, respectively. As the pretreatment and hydrolysis of the feedstock are the fundamental parts of the bioethanol production, these narrated papers often cover these processes too.

It is notable that as the Table 41.5 shows, 8% and 20% of these HCPs are related to the production and evaluation of the bioethanol fuels from the feedstocks, respectively. However, there is no HCP on the utilization of these biofuels in diesel or gasoline engines, partially displacing diesel or gasoline fuels.

41.4.4.1 Feedstock-based Bioethanol Production

There are two HCPs for the production of the feedstock-based bioethanol fuels with at least 664 citations each (Table 41.3). These papers also cover the fermentation of the hydrolysates of the feedstock. As the pretreatment and hydrolysis are the fundamental parts of the bioethanol production, these narrated papers often cover these processes too. It is notable that as the Table 41.5 shows, 8% of these HCPs are related to the production of the bioethanol fuels from the feedstocks.

These narrated studies highlight the importance of the pretreatment (primarily chemical, enzymatic or hydrothermal) and hydrolysis (primarily enzymatic or acid) processes as well as of the fermentation processes (SSF or SHF) on the production of the bioethanol fuels from the feedstocks with a high ethanol yield. Further, some fermentation studies focus on the detoxification of the lignocellulosic hydrolysates to improve the ethanol yield.

Larsson et al. (1999) studied the effect of the combined severity (SC) of dilute sulfuric acid hydrolysis of spruce on sugar yield and the fermentability of the hydrolysate and when the CS of the hydrolysis conditions increased, they observed that the yield of fermentable sugars increased to a maximum between CS 2.0 and 2.7 for mannose, and 3.0 and 3.4 for glucose above which it decreased. Further, Saha et al. (2005) performed the dilute acid pretreatment, enzymatic saccharification, and fermentation of wheat straw to produce ethanol and found that the yield of ethanol (per liter) from this hydrolysate was 19 g with a yield of 0.24 g/g DS.

41.4.4.2 Feedstock-based Bioethanol Evaluation

There are five HCPs for the evaluation of the feedstock-based bioethanol fuels with at least 638 citations each (Table 41.3). It is notable that as the Table 41.5 shows, 20% of these HCPs are related to the evaluation of the bioethanol fuels from the feedstocks.

These narrated studies often focus the technoeconomics and environmental impact of the bioethanol fuels from the feedstocks. These technoeconomic studies show that the feedstock-based ethanol fuels are cost-competitive in relation to the crude oil-based gasoline and petrodiesel fuels, which ethanol fuels partially replace as the ethanol price reached $150 per barrel in 2022 following the invasion of Ukraine by Russia.

Hamelinck et al. (2005) performed the technoeconomic performance of lignocellulosic biomass-based ethanol fuels in short-, middle- and long term and found that a combined effect of higher hydrolysis and fermentation efficiency, lower specific capital investments, increase of scale, and cheaper biomass feedstock costs (from 3 to 2 €/GJHHV) could bring the ethanol production costs from 22 €/GJHHV in the next 5 years, to 13 €/GJ over the 10–15 years timescale, and down to 8.7

€/GJ in 20 or more years. Further, Pimentel and Patzek (2005) evaluated the energy balance of ethanol fuels from corn, switchgrass, and wood in comparison with biodiesel fuels from soybean and sunflower in a paper with 1,020 citations and found that energy outputs from ethanol produced using these biomasses were each less than the respective fossil energy inputs.

Schmer et al. (2008) evaluated the net energy of bioethanol fuels from switchgrass and found that switchgrass produced 540% more renewable than non-renewable energy consumed. Further, Klein-Marcuschamer et al. (2012) constructed a technoeconomic model for the production of fungal cellulase production of lignocellulosic ethanol fuels and found that the cost of producing enzymes was much higher than that commonly assumed in the literature.

Macedo et al. (2008) evaluated the energy balance and GHG emissions in the production and use of ethanol fuels from sugarcane in Brazil for 2005/2006 for a sample of mills processing up to 100 million tons of sugarcane per year, and proposed a conservative scenario for 2020.

41.5   CONCLUSION AND FUTURE RESEARCH

The brief information about the key research fronts covered by the 25 most-cited papers with at least 618 citations each is given under two primary headings: The pretreatments and hydrolysis of the feedstocks and production and evaluation of the bioethanol fuels.

The usual characteristics of these HCPs are that the pretreatments and hydrolysis of the feedstock and fermentation of the resulting hydrolysates are the primary processes for the bioethanol fuel production from feedstock to improve the ethanol yield.

The key findings on these research fronts should be read in light of the increasing public concerns about climate change, GHG emissions, and global warming as these concerns have been certainly behind the boom in the research on the feedstock-based bioethanol fuels as an alternative to crude oil-based gasoline and petrodiesel fuels in the last decades. It is also a sustainable alternative to first generation food crop-based bioethanol fuels such as corn grain-based bioethanol fuels. The recent supply shocks caused by the COVID-19 pandemics and the Russian invasion of Ukraine also highlight the importance of the production and utilization of the bioethanol fuels as an alternative to the crude oil-based gasoline and petrodiesel fuels.

As Table 41.4 shows, the most prolific research fronts for this field are the wood biomass, cellulose, and to a lesser extent lignocellulosic biomass at large, grass biomass, starch feedstock residues, and sugar and starch feedstocks. On the other hand, it is notable that there are no HCPs for the food waste, algal biomass, sugar feedstock residues, urban waste, forestry waste, and industrial waste. Thus, the first five feedstocks dominate the HCPs in this field where only waste biomass is starch feedstock residues such as corn stover or wheat straw and cellulose is the key constituent of the lignocellulosic biomass. Further, the most influential research fronts are the wood biomass, starch feedstock residues, and cellulose, while the lignocellulosic biomass is the least influential feedstock.

As Table 41.5 shows the most prolific research fronts for this field are the pretreatment and hydrolysis of the feedstocks, and to a lesser extent bioethanol evaluation, hydrolysate fermentation, and bioethanol production and there is no HCP for the bioethanol fuel utilization. On the other hand, biomass pretreatments and bioethanol fuel evaluation are the most influential research fronts, while bioethanol production is the least influential research front.

These studies emphasize the importance of proper incentive structures for the efficient production of feedstock-based bioethanol fuels in light of North's institutional framework (North, 1991). In this context, the major producers and users of bioethanol fuels such as the USA and Brazil with vast forests and farmlands have developed strong incentive structures for the efficient feedstock-based bioethanol fuels. In light of the recent supply shocks caused primarily by the COVID-19 pandemics and Russian invasion of Ukraine, it is expected that the incentive structures such as public funding would be enhanced to increase the share of bioethanol fuels in the global fuel portfolio as a strong alternative to crude oil-based gasoline and petrodiesel fuels.

In this context, it is expected that the most prolific researchers, institutions, countries, funding bodies, and journals in this field would have a first-mover advantage to benefit from such potential incentives. This is especially true for the US stakeholders as the USA has become the global leader in both the production and utilization of second generation bioethanol fuels from the feedstocks. It is expected the research would focus more on the algal, wood, and grass biomass-based bioethanol as well as the agricultural residue and waste biomass-based bioethanol fuels at the expense of the starch and sugar feedstock-based bioethanol fuels due to the large societal concerns about the food security in the future.

It is recommended that further review studies are performed for the primary research fronts of the feedstock-based bioethanol fuels.

ACKNOWLEDGMENTS

The contribution of the highly cited researchers in the field of the feedstock-based bioethanol fuels has been gratefully acknowledged.

REFERENCES

Alvira, P., E. Tomas-Pejo, M. Ballesteros and M. J. Negro. 2010. Pretreatment technologies for an efficient bioethanol production process based on enzymatic hydrolysis: A review. *Bioresource Technology* 101:4851–4861.

Angelici, C., B. M. Weckhuysen and P. C. A. Bruijnincx. 2013. Chemocatalytic conversion of ethanol into butadiene and other bulk chemicals. *ChemSusChem* 6:1595–1614.

Antolini, E. 2007. Catalysts for direct ethanol fuel cells. *Journal of Power Sources* 170:1–12.

Antolini, E. 2009. Palladium in fuel cell catalysis. *Energy and Environmental Science* 2:915–931.

Bai, F. W., W. A. Anderson and M. Moo-Young. 2008. Ethanol fermentation technologies from sugar and starch feedstocks. *Biotechnology Advances* 26:89–105.

Balat, M. 2011. Production of bioethanol from lignocellulosic materials via the biochemical pathway: A review: *Energy Conversion and Management* 52:858–875.

Bentivoglio, D., A. Finco and M. R. P. Bacchi. 2016. Interdependencies between biofuel, fuel and food prices: The case of the Brazilian ethanol market. *Energies* 9:464.

Binod, P., R. Sindhu and R. R. Singhania, et al. 2010. Bioethanol production from rice straw: An overview. *Bioresource Technology* 101:4767–4774.

Bothast, R. J. and M. A. Schlicher. 2005. Biotechnological processes for conversion of corn into ethanol. *Applied Microbiology and Biotechnology* 67:19–25.

Burnham, J. F. 2006. Scopus database: A review. *Biomedical Digital Libraries* 3:1–8.

Cai, J. and L. Zhang. 2005. Rapid dissolution of cellulose in LiOH/urea and NaOH/urea aqueous solutions. *Macromolecular Bioscience* 5:539–548.

Canilha, L., A. K. Chandel and T. S. dos Santos Milessi, et al. 2012. Bioconversion of sugarcane biomass into ethanol: An overview about composition, pretreatment methods, detoxification of hydrolysates, enzymatic saccharification, and ethanol fermentation. *Journal of Biomedicine and Biotechnology* 2012:989572.

Cardona, C. A., J. A. Quintero and I. C. Paz. 2010. Production of bioethanol from sugarcane bagasse: Status and perspectives. *Bioresource Technology* 101:4754–4766.

Duff, S. J. B. and W. D. Murray. 1996. Bioconversion of forest products industry waste cellulosics to fuel ethanol: A review. *Bioresource Technology* 55:1–33.

Elobeid, A. and C. Hart. 2007. Ethanol expansion in the food versus fuel debate: How will developing countries fare? *Journal of Agricultural & Food Industrial Organization* 5:7.

Eriksson, T., J. Borjesson and F. Tjerneld. 2002. Mechanism of surfactant effect in enzymatic hydrolysis of lignocellulose. *Enzyme and Microbial Technology* 31:353–364.

Fernando, S., S. Adhikari, C. Chandrapal and M. Murali. 2006. Biorefineries: Current status, challenges, and future direction. *Energy & Fuels* 20:1727–1737.

Fort, D. A., R. C. Remsing and R. P. Swatloski, et al. 2007. Can ionic liquids dissolve wood? Processing and analysis of lignocellulosic materials with 1-n-butyl-3-methylimidazolium chloride. *Green Chemistry* 9:63–69.

Fukaya, Y., K. Hayashi, M. Wada and H. Ohno. 2008. Cellulose dissolution with polar ionic liquids under mild conditions: Required factors for anions. *Green Chemistry* 10:44–46.

Galbe, M. and G. Zacchi. 2002. A review of the production of ethanol from softwood. *Applied Microbiology and Biotechnology* 59:618–628.

Guimaraes, P. M. R., J. A. Teixeira and L. Domingues. 2010. Fermentation of lactose to bio-ethanol by yeasts as part of integrated solutions for the valorisation of cheese whey. *Biotechnology Advances* 28:375–384.

Hahn-Hagerdal, B., M. Galbe, M. F. Gorwa-Grauslund, G. Liden and G. Zacchi. 2006. Bio-ethanol - The fuel of tomorrow from the residues of today. *Trends in Biotechnology* 24:549–556.

Hamelinck, C. N., G. van Hooijdonk and A. P. C. Faaij. 2005. Ethanol from lignocellulosic biomass: Techno-economic performance in short-, middle- and long-term. *Biomass and Bioenergy* 28:384–410.

Hendriks, A. T. W. M and G. Zeeman. 2009. Pretreatments to enhance the digestibility of lignocellulosic biomass. *Bioresource Technology* 100:10–18.

Hill, J., E. Nelson, D. Tilman, S. Polasky and D. Tiffany. 2006. Environmental, economic, and energetic costs and benefits of biodiesel and ethanol biofuels. *Proceedings of the National Academy of Sciences of the United States of America* 103:11206–11210.

Hill, J., S. Polasky and E. Nelson, et al. 2009. Climate change and health costs of air emissions from biofuels and gasoline. *Proceedings of the National Academy of Sciences of the United States of America* 106:2077–2082.

Ho, S. H., S. W. Huang and C. Y. Chen, et al. 2013. Bioethanol production using carbohydrate-rich microalgae biomass as feedstock. *Bioresource Technology* 135:191–198.

Hsieh, W. D., R. H. Chen, T. L. Wu and T. H. Lin. 2002. Engine performance and pollutant emission of an SI engine using ethanol-gasoline blended fuels. *Atmospheric Environment* 36:403–410.

Huang, H. J., S. Ramaswamy, U. W. Tschirner and B. V. Ramarao. 2008. A review of separation technologies in current and future biorefineries. *Separation and Purification Technology* 62:1–21.

John, R. P., G. S. Anisha, K. M. Nampoothiri and A. Pandey. 2011. Micro and macroalgal biomass: A renewable source for bioethanol. *Bioresource Technology* 102:186–193.

Jonsson, L. J. and C. Martin. 2016. Pretreatment of lignocellulose: Formation of inhibitory by-products and strategies for minimizing their effects. *Bioresource Technology* 199:103–112.

Keshwani, D. R. and J. J. Cheng. 2009. Switchgrass for bioethanol and other value-added applications: A review. *Bioresource Technology* 100:1515–1523.

Kilpelainen, I., X. Xie and A. King, et al. 2007. Dissolution of wood in ionic liquids. *Journal of Agricultural and Food Chemistry* 55:9142–9148.

Klein-Marcuschamer, D., P. Oleskowicz-Popiel, B. A. Simmons and H. W. Blanch. 2012. The challenge of enzyme cost in the production of lignocellulosic biofuels. *Biotechnology and Bioengineering* 109:1083–1087.

Konur, O. 2000. Creating enforceable civil rights for disabled students in higher education: An institutional theory perspective. *Disability & Society* 15:1041–1063.

Konur, O. 2002a. Access to nursing education by disabled students: Rights and duties of nursing programs. *Nurse Education Today* 22:364–374.

Konur, O. 2002b. Assessment of disabled students in higher education: Current public policy issues. *Assessment and Evaluation in Higher Education* 27:131–152.

Konur, O. 2002c. Access to employment by disabled people in the UK: Is the Disability Discrimination Act working? *International Journal of Discrimination and the Law* 5:247–279.

Konur, O. 2006a. Participation of children with dyslexia in compulsory education: Current public policy issues. *Dyslexia* 12:51–67.

Konur, O. 2006b. Teaching disabled students in higher education. *Teaching in Higher Education* 11:351–363.

Konur, O. 2007a. A judicial outcome analysis of the *Disability Discrimination Act*: A windfall for the employers? *Disability & Society* 22:187–204.

Konur, O. 2007b. Computer-assisted teaching and assessment of disabled students in higher education: The interface between academic standards and disability rights. *Journal of Computer Assisted Learning* 23:207–219.

Konur, O. 2012. The evaluation of the research on the bioethanol: A scientometric approach. *Energy Education Science and Technology Part A: Energy Science and Research* 28:1051–1064.

Konur, O. 2015. Current state of research on algal bioethanol. In *Marine Bioenergy: Trends and Developments*, Ed. S. K. Kim and C. G. Lee, pp. 217–244. Boca Raton, FL: CRC Press.

Konur, O. 2019. Cyanobacterial bioenergy and biofuels science and technology: A scientometric overview. In *Cyanobacteria: From Basic Science to Applications*, Ed. A. K. Mishra, D. N. Tiwari and A. N. Rai, pp. 419–442. Amsterdam: Elsevier.

Konur, O. 2020. The scientometric analysis of the research on the bioethanol production from green macroalgae. In *Handbook of Algal Science, Technology and Medicine*, Ed. O. Konur, pp. 385–401. London: Academic Press.

Konur, O. 2023a. Non-waste feedstock-based bioethanol fuel production: Scientometric study. In *Feedstock-based Bioethanol Fuels. I. Non-Waste Feedstocks: Starch, Sugar, Grass, Wood, Cellulose, Algae, and Biosyngas-based Bioethanol Fuels. Handbook of Bioethanol Fuels Volume 3*, Ed. O. Konur. Boca Raton, FL: CRC Press.

Konur, O. 2023b. Second generation waste biomass-based bioethanol fuels: Scientometric study. In *Feedstock-based Bioethanol Fuels. II. Waste Feedstocks: Agricultural, Food, Industrial, Urban, Forestry, and Lignocellulosic Waste-based Bioethanol Fuels. Handbook of Bioethanol Fuels Volume 4*, Ed. O. Konur. Boca Raton, FL: CRC Press.

Kumar, R., G. Mago, V. Balan and C. E. Wyman. 2009. Physical and chemical characterizations of corn stover and poplar solids resulting from leading pretreatment technologies. *Bioresource Technology* 100:3948–3962.

Larsson, S., E. Palmqvist and B. Hahn-Hagerdal, et al. 1999. The generation of fermentation inhibitors during dilute acid hydrolysis of softwood. *Enzyme and Microbial Technology* 24:151–159.

Laser, M., D. Schulman and S. G. Allen, et al. 2002. A comparison of liquid hot water and steam pretreatments of sugar cane bagasse for bioconversion to ethanol. *Bioresource Technology* 81:33–44.

Lee, S. H., T. V. Doherty, R. J. Linhardt and J. S. Dordick. 2009. Ionic liquid-mediated selective extraction of lignin from wood leading to enhanced enzymatic cellulose hydrolysis. *Biotechnology and Bioengineering* 102:1368–1376.

Li, C., B. Knierim and C. Manisseri, et al. 2010. Comparison of dilute acid and ionic liquid pretreatment of switchgrass: Biomass recalcitrance, delignification and enzymatic saccharification. *Bioresource Technology* 101:4900–4906.

Li, J., G. Henriksson and G. Gellerstedt. 2007. Lignin depolymerization/repolymerization and its critical role for delignification of aspen wood by steam explosion. *Bioresource Technology* 98:3061–3068.

Limayem, A. and S. C. Ricke. 2012. Lignocellulosic biomass for bioethanol production: Current perspectives, potential issues and future prospects. *Progress in Energy and Combustion Science*, 38:449–467.

Lin, Y. and S. Tanaka. 2006. Ethanol fermentation from biomass resources: Current state and prospects. *Applied Microbiology and Biotechnology* 69:627–642.

Lloyd, T. A. and C. E. Wyman. 2005. Combined sugar yields for dilute sulfuric acid pretreatment of corn stover followed by enzymatic hydrolysis of the remaining solids. *Bioresource Technology* 96:1967–1977.

Ma, X., L. Sun and C. Song. 2002. A new approach to deep desulfurization of gasoline, diesel fuel and jet fuel by selective adsorption for ultra-clean fuels and for fuel cell applications. *Catalysis Today* 77:107–116.

Macedo, I. C., J. E. A. Seabra and J. E. A. R. Silva. 2008. Green house gases emissions in the production and use of ethanol from sugarcane in Brazil: The 2005/2006 averages and a prediction for 2020. *Biomass and Bioenergy* 32:582–595.

Martinez, D., L. F. Larrondo and N. Putnam, et al. 2004. Genome sequence of the lignocellulose degrading fungus *Phanerochaete chrysosporium* strain RP78. *Nature Biotechnology* 22:695–700.

Morschbacker, A. 2009. Bio-ethanol based ethylene. *Polymer Reviews* 49:79–84.

Mosier, N., C. Wyman and B. Dale, et al. 2005. Features of promising technologies for pretreatment of lignocellulosic biomass. *Bioresource Technology* 96:673–686.

Najafi, G., B. Ghobadian and T. Tavakoli, et al. 2009. Performance and exhaust emissions of a gasoline engine with ethanol blended gasoline fuels using artificial neural network. *Applied Energy* 86:630–639.

Newman, P. W. G. and J. R. Kenworthy. 1989. Gasoline consumption and cities: A comparison of U.S. cities with a global survey. *Journal of the American Planning Association* 55:24–37.

North, D. C. 1991. Institutions. *Journal of Economic Perspectives* 5:97–112.

Olsson, L. and B. Hahn-Hagerdal. 1996. Fermentation of lignocellulosic hydrolysates for ethanol production. *Enzyme and Microbial Technology* 18:312–331.

Palmqvist, E. and B. Hahn-Hagerdal. 2000. Fermentation of lignocellulosic hydrolysates. II: Inhibitors and mechanisms of inhibition. *Bioresource Technology* 74:25–33.

Park, S., J. O. Baker and M. E. Himmel, et al. 2010. Cellulose crystallinity index: Measurement techniques and their impact on interpreting cellulase performance. *Biotechnology for Biofuels* 3:10.

Pimentel, D. and T. W. Patzek. 2005. Ethanol production using corn, switchgrass, and wood; biodiesel production using soybean and sunflower. *Natural Resources Research* 14:65–76.

Pinkert, A., K. N. Marsh, S. Pang and M. P. Staiger. 2009. Ionic liquids and their interaction with cellulose. *Chemical Reviews* 109:6712–6728.

Prasad, S., A. Singh and H. C. Joshi. 2007. Ethanol as an alternative fuel from agricultural, industrial and urban residues. *Resources, Conservation and Recycling* 50:1–39.

Quinlan, R. J., M. D. Sweeney and L. L. Leggio, et al. 2011. Insights into the oxidative degradation of cellulose by a copper metalloenzyme that exploits biomass components. *Proceedings of the National Academy of Sciences of the United States of America* 108:15079–15084.

Ravindran, R. and A. K. Jaiswal. 2016. A comprehensive review on pre-treatment strategy for lignocellulosic food industry waste: Challenges and opportunities. *Bioresource Technology* 199:92–102.

Remsing, R.C., R. P. Swatloski, R. D. Rogers and G. Moyna. 2006. Mechanism of cellulose dissolution in the ionic liquid 1-n-butyl-3- methylimidazolium chloride: A^{13}C and$^{35/37}$Cl NMR relaxation study on model systems. *Chemical Communications* 2006:1271–1273.

Saha, B. C., L. B. Iten, M. A. Cotta and Y. V. Wu. 2005. Dilute acid pretreatment, enzymatic saccharification and fermentation of wheat straw to ethanol. *Process Biochemistry* 40:3693–3700.

Sanchez, O. J. and C. A. Cardona. 2008. Trends in biotechnological production of fuel ethanol from different feedstocks. *Bioresource Technology* 99:5270–5295.

Sasaki, M., Z. Fang, Y. Fukushima, T. Adschiri and K. Arai. 2000. Dissolution and hydrolysis of cellulose in subcritical and supercritical water. *Industrial and Engineering Chemistry Research* 39:2883–2890.

Schmer, M. R., K. P. Vogel, R. B. Mitchell and R. K. Perrin. 2008. Net energy of cellulosic ethanol from switch-grass. *Proceedings of the National Academy of Sciences of the United States of America* 105:464–469.

Schwanninger, M., J. C. Rodrigues, H. Pereira and P. Hinterstoisser. 2004. Effects of short-time vibratory ball milling on the shape of FT-IR spectra of wood and cellulose. *Vibrational Spectroscopy* 36:23–40.

Suganuma, S., K. Nakajima and M. Kitano, et al. 2008. Hydrolysis of cellulose by amorphous carbon bearing SO$_3$H, COOH, and OH groups. *Journal of the American Chemical Society* 130:12787–12793.

Sun, N., M. Rahman and Y. Qin, et al. 2009. Complete dissolution and partial delignification of wood in the ionic liquid 1-ethyl-3-methylimidazolium acetate. *Green Chemistry* 11:646–655.

Sun, Y. and J. Cheng. 2002. Hydrolysis of lignocellulosic materials for ethanol production: A review. *Bioresource Technology* 83:1–11.

Taherzadeh, M. J. and K. Karimi. 2007. Enzyme-based hydrolysis processes for ethanol from lignocellulosic materials: A review. *Bioresources* 2:707–738.

Taherzadeh, M. J. and K. Karimi. 2008. Pretreatment of lignocellulosic wastes to improve ethanol and biogas production: A review. *International Journal of Molecular Sciences* 9:1621–1651.

Talebnia, F., D. Karakashev and I. Angelidaki. 2010. Production of bioethanol from wheat straw: An overview on pretreatment, hydrolysis and fermentation. *Bioresource Technology* 101:4744–4753.

Zhang, Y. H. P. and L. R. Lynd. 2004. Toward an aggregated understanding of enzymatic hydrolysis of cellulose: Noncomplexed cellulase systems. *Biotechnology and Bioengineering* 88:797–824.

Zhu, J. Y. and X. J. Pan. 2010. Woody biomass pretreatment for cellulosic ethanol production: Technology and energy consumption evaluation. *Bioresource Technology* 101:4992–5002.

Part 8

Introduction to Non-Waste
Feedstock-based Bioethanol Fuels

42 Non-Waste Biomass-based Bioethanol Fuels
Scientometric Study

Ozcan Konur
(Formerly) Ankara Yildirim Beyazit University

42.1 INTRODUCTION

The crude oil-based gasoline fuels (Ma et al., 2002; Newman and Kenworthy, 1989) have been widely used in the transportation sector since the 1920s. However, there have been great public concerns over the adverse environmental and human impact of these fuels (Hill et al., 2006, 2009). Hence, biomass-based bioethanol fuels (Hill et al., 2006; Konur, 2012e, 2015, 2019, 2020a) have increasingly been used in blending gasoline fuels (Hsieh et al., 2002; Najafi et al., 2009), in the fuel cells (Antolini, 2007, 2009), and in the biochemical production (Angelici et al., 2013; Morschbacker, 2009) in a biorefinery context (Fernando et al., 2006; Huang et al., 2008).

Bioethanol fuels also play a critical role in maintaining the energy security (Kruyt et al., 2009; Winzer, 2012) in the supply shocks (Kilian, 2008, 2009) related to oil price shocks (Hamilton, 1983, 2003), COVID-19 pandemics (Fauci et al., 2020; Li et al., 2020), or wars (Hamilton, 1983; Jones, 2012) in the aftermath of the Russian invasion of Ukraine (Reeves, 2014).

However, it is necessary to pretreat the biomass (Taherzadeh and Karimi, 2008; Yang and Wyman, 2008) to enhance the yield of the bioethanol (Hahn-Hagerdal et al., 2006; Sanchez and Cardona, 2008) prior to the bioethanol production through the hydrolysis (Sun and Cheng, 2002; Taherzadeh and Karimi, 2007) and fermentation (Lin and Tanaka, 2006; Olsson and Hahn-Hagerdal, 1996) of the biomass and hydrolysates, respectively.

One of the most-studied feedstocks for the bioethanol fuels has been the non-waste biomass. The research in the field of the non-waste biomass-based bioethanol fuels has intensified in this context in the key research fronts of the pretreatment (Zhu and Pan, 2010; Zhu et al., 2010) and hydrolysis (Dien et al., 2006; Li et al., 2010) of the non-waste biomass, fermentation (Bai et al., 2008; Wingren et al., 2003) of the non-waste biomass-based hydrolysates, and bioethanol fuel production (Galbe and Zacchi, 2002; John et al., 2011) and bioethanol fuel evaluation (Macedo et al., 2008; Schmer et al., 2008).

The research in this field has also intensified for the feedstocks of wood biomass (Galbe and Zacchi, 2002; Zhu and Pan, 2010), algal biomass (Goh and Lee, 2010; John et al., 2011), grass biomass (Keshwani and Cheng, 2009; Li et al., 2010), sugar feedstock-biomass (Bai et al., 2008; Zabed et al., 2014), and starch feedstock-biomass (Bai et al., 2008; Bothast and Schlicher, 2005). Thus, it emerges as a distinctive research field, complementing the research on the second generation waste biomass-based bioethanol fuels from the agricultural residues and other wastes.

However, it is essential to develop efficient incentive structures (North, 1991) for the primary stakeholders to enhance the research in this field (Konur, 2000, 2002a,b,c, 2006a,b, 2007a,b). The scientometric analysis has been used in this context to inform the primary stakeholders about the current state of the research in a selected research field (Garfield, 1955; Konur, 2011, 2012a,b,c,d,e,f,g,h,i, 2015, 2018b, 2019, 2020a).

As there have been no published scientometric studies in this field, this book chapter presents a scientometric study of the research in the non-waste biomass-based bioethanol fuels. It examines the scientometric characteristics of both the sample and population data presenting scientometric characteristics of both these datasets in the order of documents, authors, publication years, institutions, funding bodies, source titles, countries, Scopus subject categories, Scopus keywords, and research fronts.

42.2   MATERIALS AND METHODS

The search for this study was carried out using Scopus database (Burnham, 2006) in October 2022.

As a first step for the search of the relevant literature, the keywords were selected using the most-cited first 300 population papers for each non-waste biomass. The selected keyword list was then optimized to obtain a representative sample of papers for the each research field. These five keyword lists were then integrated to obtain the keyword list for this research field (Konur, 2023a,b,c,d,e,f,g).

As a second step, two sets of data were used for this study. First, a population sample of 9,820 papers was used to examine the scientometric characteristics of the population data. Second, a sample of 196 most-cited papers, corresponding to 2% of the population papers, was used to examine the scientometric characteristics of these citation classics.

The scientometric characteristics of both these sample and population datasets were presented in the order of documents, authors, publication years, institutions, funding bodies, source titles, countries, Scopus subject categories, Scopus keywords, and research fronts.

Lastly, the key scientometric findings for both datasets were discussed to highlight the research landscape for the non-waste biomass-based bioethanol fuels. Additionally, a number of brief conclusions were drawn, and a number of relevant recommendations were made to enhance the future research landscape.

42.3   RESULTS

42.3.1   THE MOST-PROLIFIC DOCUMENTS IN THE NON-WASTE BIOMASS-BASED BIOETHANOL FUELS

The information on the types of documents for both datasets is given in Table 42.1. The articles and conference papers, published in journals, dominate both the sample (92%) and population (95%) papers with 3% deficit. Further, review papers and short surveys have a 6% surplus as they are over-represented in the sample papers as they constitute 8% and 2% of the sample and population papers, respectively.

It is further notable that 97%, 2%, and 1% of the population papers were published in journals, books, and book series, respectively. Similarly, 100% of the sample papers were published in the journals.

42.3.2   THE MOST-PROLIFIC AUTHORS IN THE NON-WASTE BIOMASS-BASED BIOETHANOL FUELS

The information about the most-prolific 22 authors with at least 2% of sample papers each is given in Table 42.2.

The most-prolific author is John N. Saddler with 9.2% of the sample papers, followed by Guido Zacchi and Xuejun Pan with 6.1% of the sample papers each. The other prolific authors are Mats Galbe, Charles E. Wyman, Junyong Zhu, David J. Gregg, Arthur J. Ragauskas, and Bruce E. Dale with 3.1%–4.6% of the sample papers each.

TABLE 42.1
Documents in the Non-Waste Biomass-based Bioethanol Fuels

Documents	Sample Dataset (%)	Population Dataset (%)	Surplus (%)
Article	85.7	92.2	−6.5
Conference paper	6.6	2.5	4.1
Review	6.1	2.0	4.1
Short Survey	1.5	0.2	1.3
Book chapter	0.0	1.8	−1.8
Note	0.0	0.6	−0.6
Letter	0.0	0.6	−0.6
Book	0.0	0.1	−0.1
Editorial	0.0	0.1	−0.1
Sample size	196	9,820	

Population dataset, the number of papers (%) in the set of the 9,820 population papers; sample dataset, the number of papers (%) in the set of 196 highly cited papers.

TABLE 42.2
Most-Prolific Authors in the Non-Waste Biomass-based Bioethanol Fuels

No.	Author Name	Author Code	Sample Papers (%)	Population Papers (%)	Surplus	Institution	Country	HI	N	Res. Front
1	Saddler, John N.	7005297559 57202481615	9.2	1.0	8.2	Univ. British Columbia	Canada	99	420	P, H, F, R
2	Zacchi, Guido	7006727748	6.1	0.7	5.4	Lund Univ.	Sweden	68	204	P, H, F, R
3	Pan, Xuejun	57203296000	6.1	0.3	5.8	Univ. Wisconsin Madison	USA	45	118	P, H
4	Galbe, Mats	7003788758	4.6	0.5	4.1	Lund Univ.	Sweden	51	131	P, H, F, R
5	Wyman, Charles E.	7004396809	4.1	0.4	3.7	Univ. Calf. Riverside	USA	80	287	P, H, F, R
6	Zhu, Junyong	7405692678	4.1	0.4	3.7	USDA Forest Serv.	USA	64	311	P, H
7	Gregg, David J.	7005324246	3.6	0.2	3.4	Univ. British Columbia	Canada	21	25	P, H
8	Ragauskas, Arthur J.	7006265204	3.1	0.7	2.4	Univ. Tennessee	USA	93	762	P, H
9	Dale, Bruce E.	7201511969	3.1	0.3	2.8	Michigan State Univ.	USA	92	430	P, H, F, R
10	Gilkes, Neil	35493889000	2.6	0.1	2.5	Univ. British Columbia	Canada	51	88	P, H
11	Negro, Maria J.*	6701512649	2.6	0.1	2.5	CIEMAT	Spain	38	74	P, H, F, R
12	Simmons, Blake A.	7102183263	2.0	0.4	1.6	Lawrence Berkeley Natl. Lab.	USA	76	445	P, H
13	Holtzapple, Mark T.	7004167004	2.0	0.2	1.8	Texas A&M Univ.	USA	47	199	P
14	Ladisch, Michael R.	7005670397	2.0	0.2	1.8	Purdue Univ.	USA	75	334	P, R
15	Ruiz, Encarnacion*	25646493300	2.0	0.2	1.8	Univ. Jaen	Spain	34	75	P,H, F, R

(Continued)

TABLE 42.2 (*Continued*)
Most-Prolific Authors in the Non-Waste Biomass-based Bioethanol Fuels

No.	Author Name	Author Code	Sample Papers (%)	Population Papers (%)	Surplus	Institution	Country	HI	N	Res. Front
16	Ballesteros, Ignacio	6602732963	2.0	0.2	1.8	CIEMAT	Spain	38	70	P, H, F, R
17	Cara, Cristobal	22949914200	2.0	0.1	1.9	Univ. Jaen	Spain	33	54	P, H
18	Hahn-Hagerdal, Barbel*	7005389381	2.0	0.1	1.9	Lund Univ.	Sweden	76	258	P, H, F, R
19	Harun, Razif	35315707300	2.0	0.1	1.9	Univ. Putra Malaysia	Malaysia	22	88	P, H, F, R
20	Palmqvist, Eva A.*	6603821896	2.0	0.1	1.9	Lund Univ.	Sweden	19	31	P, H, F, R
21	Vogel, Kenneth P.	7102498441	2.0	0.1	1.9	Univ. Nebraska Lincoln	USA	63	146	P, H, F, R
22	Danquah, Michael K.	12803940900	2.0	0.1	1.9	Univ. Tennessee	USA	39	225	P, H, F, R

*, Female; Author code, the unique code given by Scopus to the authors; E, bioethanol fuel evaluation; F, fermentation of the non-waste biomass-based hydrolysates; H: hydrolysis of the non-waste biomass; P, pretreatment of the non-waste biomass; population papers, the number of papers authored in the population dataset; R, bioethanol fuel production; sample papers, the number of papers authored in the sample dataset.

On the other hand, the most influential author is John N. Saddler with 8.2% surplus, followed by Xuejun Pan and Guido Zacchi with 5.8% and 5.4% surplus, respectively. The other influential authors are Mats Galbe, Charles E. Wyman, Junyong Zhu, David J. Gregg, Bruce E. Dale, Neil Gilkes, and Maria J. Negro with 2.5%–4.1% surplus each.

The most-prolific institution for the sample dataset is the Lund University with four authors, followed by University of British Columbia with three authors. The other prolific institutions are the Center for Energy, Environmental, and Technological Research (CIEMAT), University of Jaen, and University of Tennessee with two authors each. On the other hand, the most prolific country for the sample dataset is the USA with 10 authors, followed by Sweden and Spain with four authors each. The other prolific country is Canada with three authors. In total, only five countries house these top authors.

The most-prolific research front for these top authors is the pretreatment of the non-waste biomass with 22 authors followed by the hydrolysis of the non-waste biomass with 21 authors. The other prolific research fronts are the fermentation of the non-waste biomass-based hydrolysates and the bioethanol production with 13 and 14 authors, respectively.

On the other hand, there is a significant gender deficit (Beaudry and Lariviere, 2016) for the sample dataset as surprisingly only four of these top researchers are female with a representation rate of 18%.

Additionally, there are other authors with the relatively low citation impact and with 0.2%–0.7% of the population papers each: Run Cang Sun, Juan C. Parajo, Feng Xu, Valentin Santos, Gwi-Taek Jeong, Yunqiao Pu, Jun Zhang, Vijay Singh, Qiang Yong, Lakkana Laopaiboon, Pattana Laopaiboon, Bruce S. Dien, Shiro Saka, Sung-Koo Kim, Hasan Jameel, Xianzhi Meng, Herbert Sixta, Caoxing Huang, Hisashi Miyafuji, Jia-Long Wen, Eulogio Castro, Juanita Freer, Akihiko Kondo, Seema Singh, Jose L. Alonso, Chang G. Yoo, Chenhuan Lai, Aloia Romani, Shao-Ni Sun, Qingxi Hou, Leif J Jonsson, B. N. Kuznetsov, Stefan M. Willfor, S. Yi, Gil Garrote, Gunnar Liden, Shouxin Liu, Gan Yang, Tong-Qi Yuan, Venkatesh Balan, In-Gyu Choi, Seung-Hwan Lee, Andrey Pranovich, Shahab Sokhansanj, and Yanqun Xu.

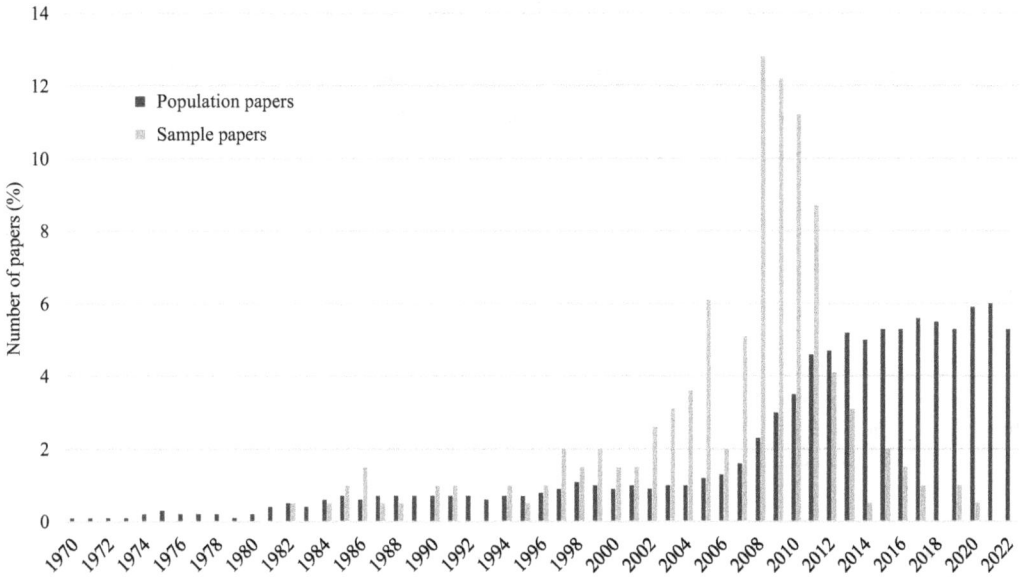

FIGURE 42.1 The research output by years regarding the non-waste biomass-based bioethanol fuels.

42.3.3 THE MOST-PROLIFIC RESEARCH OUTPUT BY YEARS IN THE NON-WASTE BIOMASS-BASED BIOETHANOL FUELS

Information about papers published between 1970 and 2022 is given in Figure 42.1. This figure clearly shows that the bulk of the research papers in the population dataset were published primarily in the 2010s and the early 2020s with 50% and 17% of the population dataset, respectively. Similarly, the publication rates for the 2000s, 1990s, 1980s, and 1970s were 14%, 8%, 6%, and 2% respectively. Further, the rate for the pre-1970s was 4%.

Similarly, the bulk of the research papers in the sample dataset were published in the 2000s and 2010s with 51% and 22% of the sample dataset, respectively. Similarly, the publication rates for the early 2020s, 1990s, 1980s, and 1970s were 1%, 10%, 6%, and 0% of the sample papers, respectively. Further, the rate for the pre-1970s was 1%.

The most-prolific publication years for the population dataset were 2021, 2020, and 2017 with 6%, 5.9%, and 5.6% of the dataset, respectively, while 73% of the population papers were published between 2008 and 2022. Similarly, 75% of the sample papers were published between 2002 and 2013, while the most prolific publication years were 2008, 2009, and 2010 with 12.8%, 12.2%, and 11.2% of the sample papers, respectively.

Further, there is a rising trend for the population papers between 2006 and 2011. However, it lost its momentum after 2011.

42.3.4 THE MOST-PROLIFIC INSTITUTIONS IN THE NON-WASTE BIOMASS-BASED BIOETHANOL FUELS

Information about the most-prolific 29 institutions publishing papers on the non-waste biomass-based bioethanol fuels with at least 2% of the sample papers each is given in Table 42.3.

The most-prolific institution is the University of British Columbia with 9.2% of the sample papers, followed by the Lund University, University of Wisconsin-Madison, USDA Forest Service, and NC State University with 7.1%, 6.1%, 5.6%, and 5.1% of the sample papers, respectively. The other prolific institutions are Dartmouth College, National Renewable Energy Laboratory (NREL),

TABLE 42.3

The Most-Prolific Institutions in the Non-Waste Biomass-based Bioethanol Fuels

No.	Institutions	Country	Sample Papers (%)	Population Papers (%)	Surplus (%)
1	Univ. British Columbia	Canada	9.2	1.4	7.8
2	Lund Univ.	Sweden	7.1	1.1	6.0
3	Univ. Wisconsin Madison	USA	6.1	0.9	5.2
4	USDA Forest Serv.	USA	5.6	1.2	4.4
5	NC State Univ.	USA	5.1	1.3	3.8
6	Dartmouth Coll.	USA	3.6	0.2	3.4
7	NREL	USA	3.6	0.8	2.8
8	Michigan State Univ.	USA	3.6	0.8	2.8
9	Univ. Calif. Riverside	USA	3.6	0.5	3.1
10	Georgia Inst. Technol.	USA	3.1	0.6	2.5
11	Univ. Nebraska-Lincoln	USA	3.1	0.3	2.8
12	CIEMAT	Spain	3.1	0.3	2.8
14	Univ. Helsinki	Finland	2.6	0.5	2.1
15	Univ. Calif. Berkeley	USA	2.6	0.5	2.1
16	Texas A&M Univ.	USA	2.6	0.5	2.1
17	Purdue Univ.	USA	2.6	0.5	2.1
18	Univ. Sao Paulo	Brazil	2.6	2.0	0.6
19	USDA Agr. Res. Serv.	USA	2.6	1.3	1.3
20	Oak Ridge Natl. lab.	USA	2.6	1.2	1.4
21	Monash Univ.	Australia	2.0	0.2	1.8
22	USDA Forest Serv.	USA	2.0	0.8	1.2
23	Auburn Univ.	USA	2.0	0.5	1.5
24	VTT Tech. Res. Ctr.	Finland	2.0	0.5	1.5
25	Jnt. Bioener. Inst.	USA	2.0	0.4	1.6
26	Univ. Putra Malaysia	Malaysia	2.0	0.4	1.6
27	Univ. Jaen	Spain	2.0	0.4	1.6
28	S. China Univ. Technol.	China	2.0	1.8	0.2
29	State Key Lab. Pulp Paper Eng.	China	2.0	1.3	0.7

Michigan State University, and University of California Riverside with 3.6% of the sample papers each. Similarly, the top country for these most prolific institutions is the USA with 17 institutions. The other prolific countries are China, Finland, and Spain with two institutions each. In total, nine countries house these top institutions.

On the other hand, the institutions with the most citation impact are the University of British Columbia with 7.8% surplus, followed by the Lund University, University of Wisconsin-Madison, and USDA Forest Service with 6%, 5.2%, and 4.4% surplus, respectively. The other influential institutions are NC State University, Dartmouth College, University of California Riverside, Michigan State University, NREL, CIEMAT, and University of Nebraska-Lincoln with 2.8%–3.8% surplus each.

Additionally, there are other institutions with the relatively low citation impact and with 4.5%–2.3% of the population papers each: Beijing Forestry University, Nanjing Forestry University, Chinese Academy of Sciences, Kyoto University, State University of Campinas, University of Vigo, Chinese Academy of Forestry, University of Tennessee, University of Illinois Urbana-Champaign, Abo Akademi University, SUNY, Aalto University, Northwest A&F University Technical University in Zvolen, Qilu University of Technology, Khon Kaen University, Iowa State University, Royal Institute of Technology, Chalmers University of Technology, University of Laval, Federal University of Rio de Janeiro, Lulea Technical University, State University of Paulista, Tianjin University of Science &

Technology, Northeast Forestry University, Technical University of Denmark, University of Tokyo, Pukyong National University, Guangxi University, Wageningen University & Research, University of Minnesota, National Scientific Research Center (CNRS), Forestry and Forest Products Research Institute, University of Saskatchewan, University of Georgia, Chonnam National University, Russian Academy of Sciences, SCION, University of Concepcion, Korea University, and University of Toronto.

42.3.5 The Most-Prolific Funding Bodies in the Non-Waste Biomass-based Bioethanol Fuels

Information about the most-prolific 19 funding bodies funding at least 1.0% of the sample papers each is given in Table 42.4. Further, only 37% and 42% of the sample and population papers each were funded, respectively

The most-prolific funding body is the Natural Sciences and Engineering Research Council of Canada with 3.6% of the sample papers, followed by the U.S. Department of Energy and National Natural Science Foundation of China with 3.1% and 2.6% of the sample papers, respectively.

TABLE 42.4
The Most-Prolific Funding Bodies in the Non-Waste Biomass-based Bioethanol Fuels

No.	Funding Bodies	Country	Sample Paper No. (%)	Population Paper No. (%)	Surplus (%)
1	Natural Sciences and Engineering Research Council of Canada	Canada	3.6	1.4	2.2
2	U.S. Department of Energy	USA	3.1	2.2	0.9
3	National Natural Science Foundation of China	China	2.6	7.2	−4.6
4	European Regional Development Fund	EU	2.0	1.1	0.9
5	Natural Resources Canada	Canada	2.0	0.2	1.8
6	Swedish National Board for Industrial and Technical Development	Sweden	2.0	0.1	1.9
7	National Science Foundation	USA	1.5	1.2	0.3
8	Ministry of Higher Education, Malaysia	Malaysia	1.5	0.4	1.1
9	Monash University	Australia	1.5	0.1	1.4
10	National Council for Scientific and Technological Development	Brazil	1.0	2.2	−1.2
11	Office of Science	USA	1.0	1.1	−0.1
12	U.S. Department of Agriculture	USA	1.0	0.9	0.1
13	European Commission	EU	1.0	0.9	0.1
14	Biological and Environmental Research	USA	1.0	0.8	0.2
15	Natural Science Foundation of Jiangsu Province	China	1.0	0.4	0.6
16	Biotechnology and Biological Sciences Research Council	UK	1.0	0.4	0.6
17	Ministry of Science and Technology	Argentina	1.0	0.1	0.9
18	Administrative Department of Science, Technology and Innovation (COLCIENCIAS)	Colombia	1.0	0.1	0.9
19	Ministry of Higher Education, Science and Technology, Dominican Republic	Dominica	1.0	0.1	0.9

The other funding bodies are the European Regional Development Fund, Natural Resources Canada, and Swedish National Board for Industrial and Technical Development with 2% of the sample papers each.

On the other hand, the most-prolific country for these top funding bodies is the USA with five funding bodies, followed by Canada, China, and the EU with two funding bodies each. In total, 11 countries and the EU house these top funding bodies.

The funding body with the most citation impact is the Natural Sciences and Engineering Research Council of Canada with 2.2% surplus. The other influential funding bodies are the Swedish National Board for Industrial and Technical Development and Natural Resources Canada with 1.9 and 1.8% surplus, respectively. Further, the funding body with the least citation impact is the National Natural Science Foundation of China with 4.6% deficit, followed by National Council for Scientific and Technological Development with 1.1% deficit. It is notable that the National Natural Science Foundation of China is the largest funder of the population papers with 7.1% funding rate.

The other funding bodies with the relatively low citation impact and with 0.3%–1.5% of the population papers each are the Higher Education Personnel Improvement Coordination, National Research Foundation of Korea, National Key Research and Development Program of China, Japan Society for the Promotion of Science, Research Support Foundation of the State of Sao Paulo, Fundamental Research Funds for the Central Universities, National Institute of Food and Agriculture, Ministry of Education, China Scholarship Council, Ministry of Economy and Competitiveness, Ministry of Education, Culture, Sports, Science and Technology, Foundation for Science and Technology, Ministry of Education of China, China Postdoctoral Science Foundation, Ministry of Education, Science and Technology, Priority Academic Program Development of Jiangsu Higher Education Institutions, Ministry of Science, ICT and Future Planning, Chinese Academy of Sciences, Energy Agency, National Basic Research Program of China (973 Program), Nanjing Forestry University, Seventh Framework Program, Ministry of Science and Technology of China, State Key Laboratory of Pulp and Paper Engineering, Thailand Research Fund, Oak Ridge National Laboratory, National Council of Science and Technology, Engineering and Physical Sciences Research Council, Horizon 2020 Framework Program, Khon Kaen University, and Academy of Finland.

42.3.6 THE MOST-PROLIFIC SOURCE TITLES IN THE NON-WASTE BIOMASS-BASED BIOETHANOL FUELS

Information about the most-prolific 17 source titles publishing at least 1.5% of the sample papers each in the non-waste biomass-based bioethanol fuels is given in Table 42.5.

The most-prolific source title is the Bioresource Technology with 20.9% of the sample papers, followed by Biotechnology and Bioengineering, Biomass and Bioenergy, and Applied Biochemistry and Biotechnology Part A with 7.7%, 6.1%, and 5.6% of the sample papers, respectively. The other prolific titles are the Enzyme and Microbial Technology and Biotechnology Progress with 4.6 and 4.1% of the sample papers, respectively.

On the other hand, the source title with the most citation impact is the Bioresource Technology with 13.2% surplus, followed by Biotechnology and Bioengineering and Applied Biochemistry and Biotechnology Part A with 6.3% and 4.7% surplus, respectively. The other influential titles are the Enzyme and Microbial Technology, Biomass and Bioenergy, Biotechnology Progress, and Proceedings of the National Academy of Sciences of the United States of America with 3%–4% surplus each.

The other source titles with the relatively low citation impact with 0.4%–3.6% of the population papers each are Bioresources, Industrial Crops and Products, Holzforschung, Biotechnology for Biofuels, Wood Science and Technology, Applied Biochemistry and Biotechnology, Journal of Wood Chemistry and Technology, Journal of Wood Science, ACS Sustainable Chemistry and Engineering, Bioenergy Research, Renewable Energy, Biotechnology Letters, Cellulose, Biomass Conversion and Biorefinery, Fuel, Journal of Cleaner Production, European Journal of Wood and

TABLE 42.5
The Most-Prolific Source Titles in the Non-Waste Biomass-based Bioethanol Fuels

No.	Source titles	Sample Papers (%)	Population Papers (%)	Surplus (%)
1	Bioresource Technology	20.9	7.7	13.2
2	Biotechnology and Bioengineering	7.7	1.4	6.3
3	Biomass and Bioenergy	6.1	2.2	3.9
4	Applied Biochemistry and Biotechnology Part A	5.6	0.9	4.7
5	Enzyme and Microbial Technology	4.6	0.6	4.0
6	Biotechnology Progress	4.1	0.5	3.6
7	Green Chemistry	3.1	0.6	2.5
8	Applied Microbiology and Biotechnology	3.1	0.5	2.6
9	Proceedings of the National Academy of Sciences of the United States of America	3.1	0.1	3.0
10	Applied and Environmental Microbiology	2.6	0.4	2.2
11	Process Biochemistry	2.0	0.7	1.3
12	Industrial and Engineering Chemistry Research	1.5	0.8	0.7
13	Journal of Agricultural and Food Chemistry	1.5	0.6	0.9
14	Energy	1.5	0.4	1.1
15	Renewable and Sustainable Energy Reviews	1.5	0.4	1.1
16	Journal of Biotechnology	1.5	0.2	1.3
17	Energy and Environmental Science	1.5	0.1	1.4

Wood Products, Journal of Chemical Technology and Biotechnology, Energy and Fuels, Energies, Nature, Cellulose Chemistry and Technology, Wood Research, Journal of the American Chemical Society, Carbohydrate Polymers, Scientific Reports, RSC Advances, Sugar Tech, Biofuels Bioproducts and Biorefining, Chemistry of Natural Compounds, Industrial and Engineering Chemistry, Science of the Total Environment, Applied Energy, International Sugar Journal, Algal Research, and Plos One.

42.3.7 THE MOST-PROLIFIC COUNTRIES IN THE NON-WASTE BIOMASS-BASED BIOETHANOL FUELS

Information about the most prolific 18 countries publishing at least 1.5% of sample papers each in the non-waste biomass-based bioethanol fuels is given in Table 42.6.

The most-prolific country is the USA with 41% of the sample papers, followed by Canada, Sweden, and China with 14%, 9%, and 8% of the sample papers, respectively. The other prolific countries are Brazil, Japan, Spain, and Finland with 4%–5% of the sample papers each. Further, eight European countries listed in Table 42.6 produce 26% and 20% of the sample and population papers, respectively, and China is the second largest producer of the population papers with a publication rate of 19.6% after Europe and the USA.

On the other hand, the country with the most citation impact is the USA with 20% surplus, followed by Canada and Sweden with 8% and 6% surplus, respectively. Similarly, the country with the least citation impact is China with 8% deficit, followed by Brazil, Japan, India, S. Korea, the UK, France, and Germany with 1%–2% deficit each.

Additionally, there are other countries with relatively low citation impact and with 0.4 to 2% of the sample papers each: Italy, Thailand, Russia, Turkey, Poland, Indonesia, Portugal, Taiwan, Slovakia, Denmark, Iran, Chile, Mexico, Nigeria, Argentina, Belgium, Czech Republic, Switzerland, Romania, Egypt, Norway, South Africa, Colombia, Serbia, Pakistan, Slovenia, Hungary, and Greece.

TABLE 42.6
The Most-Prolific Countries in the Non-Waste Biomass-based Bioethanol Fuels

No.	Countries	Sample Papers (%)	Population Papers (%)	Surplus (%)
1	USA	40.8	21.0	19.8
2	Canada	13.8	5.9	7.9
3	Sweden	9.2	3.7	5.5
4	China	8.2	16.4	−8.2
5	Brazil	5.1	7.2	−2.1
6	Japan	5.1	7.1	−2.0
7	Spain	4.6	3.6	1.0
8	Finland	4.1	2.9	1.2
9	Malaysia	3.1	1.6	1.5
10	India	2.6	4.3	−1.7
11	S. Korea	2.6	4.3	−1.7
12	Australia	2.6	1.6	1.0
13	Austria	2.0	0.7	1.3
14	UK	1.5	2.8	−1.3
15	France	1.5	2.6	−1.1
16	Germany	1.5	2.3	−0.8
17	Netherlands	1.5	1.1	0.4
18	New Zealand	1.5	0.7	0.8

42.3.8 THE MOST-PROLIFIC SCOPUS SUBJECT CATEGORIES IN THE NON-WASTE BIOMASS-BASED BIOETHANOL FUELS

Information about the most-prolific ten Scopus subject categories indexing at least 3.6% of the sample papers each is given in Table 42.7.

The most-prolific Scopus subject category in the non-waste biomass-based bioethanol fuels is the Chemical Engineering with 55% of the sample papers, followed by Environmental Science, Biochemistry, Genetics and Molecular Biology, Energy, and Immunology and Microbiology with 46%, 39%, 37%, and 33% of the sample papers, respectively. It is notable that Social Sciences including Economics and Business account for 1% and 5% of the sample and population studies, respectively.

On the other hand, the Scopus subject category with the most citation impact is the Chemical Engineering with 19% and 15% surplus. The other influential subject areas are Immunology and Microbiology, Biochemistry, Genetics and Molecular Biology, Environmental Science, and Energy with 19%, 16%, 16%, and 12% surplus, respectively. Similarly, the least influential subject categories are Agricultural and Biological Sciences, Materials Science, and Chemistry with 11% deficit each, followed by Engineering with 4% of the sample papers.

42.3.9 THE MOST-PROLIFIC KEYWORDS IN THE NON-WASTE BIOMASS-BASED BIOETHANOL FUELS

Information about the Scopus keywords used with at least 6.1% or 2.7% of the sample or population papers, respectively, is given in Table 42.8. For this purpose, keywords related to the keyword set given in the Appendix of the related papers are selected from a list of the most prolific keyword set provided by Scopus database.

These keywords are grouped under the five headings: non-waste biomass, pretreatments, fermentation, hydrolysis and hydrolysates, and products.

The most-prolific keyword related to the biomass and biomass constituents is lignin with 45% of the sample papers, followed by biomass, cellulose, wood, carbohydrate, and lignocellulose with

TABLE 42.7
The Most-Prolific Scopus Subject Categories in the Non-Waste Biomass-based Bioethanol Fuels

No.	Scopus Subject Categories	Sample Papers (%)	Population Papers (%)	Surplus (%)
1	Chemical Engineering	54.6	35.3	19.3
2	Environmental Science	45.9	30.2	15.7
3	Biochemistry, Genetics and Molecular Biology	39.3	23.0	16.3
4	Energy	37.2	25.7	11.5
5	Immunology and Microbiology	33.2	14.4	18.8
6	Agricultural and Biological Sciences	13.3	24.7	−11.4
7	Chemistry	8.2	19.0	−10.8
8	Engineering	8.2	12.1	−3.9
9	Multidisciplinary	4.6	2.6	2.0
10	Materials Science	3.6	14.6	−11.0

TABLE 42.8
The Most-Prolific Keywords in the Non-Waste Biomass-based Bioethanol Fuels

No.	Keywords	Sample Papers (%)	Population Papers (%)	Surplus (%)
1	Non-waste biomass			
	Lignin	45.4	21.1	24.3
	Biomass	40.8	16.8	24.0
	Cellulose	38.8	17.7	21.1
	Wood	28.1	16.7	11.4
	Carbohydrate	19.4	8.0	11.4
	Lignocellulose	19.4	4.7	14.7
	Softwood	14.8	2.5	12.3
	Hemicellulose	14.3	4.7	9.6
	Panicum	10.7	2.1	8.6
	Zea	9.7	3.5	6.2
	Grass	9.2	3.3	5.9
	Algae	8.2	2.5	5.7
	Switchgrass	7.7	2.0	5.7
	Populus	6.1	3.2	2.9
	Starch	4.6	3.8	0.8
	Lignocellulosic biomass	3.6	3.5	0.1
	Bamboo		3.4	−3.4
	Eucalyptus		2.8	−2.8
2	Pretreatments			
	Enzymes	25.5	7.1	18.4
	Pretreatment	16.8	8.9	7.9
	Pre-treatment	16.3	7.2	9.1
	Sulfuric acid	14.3	3.6	10.7
	Steam	12.8	4.7	8.1

(Continued)

TABLE 42.8 (*Continued*)
The Most-Prolific Keywords in the Non-Waste Biomass-based Bioethanol Fuels

No.	Keywords	Sample Papers (%)	Population Papers (%)	Surplus (%)
	Temperature	12.2	4.9	7.3
	Acids	9.2		9.2
	Water vapor	6.6	1.7	4.9
	Delignification	6.1	5.3	0.8
	Water	6.1	4.2	1.9
	pH	5.6	4.2	1.4
	Ionic liquids	3.6	4.5	−0.9
	Sodium hydroxide		2.7	−2.7
3	Fermentation			
	Fermentation	34.2	18.7	15.5
	Saccharomyces	13.8	6.0	7.8
	Yeast	8.7	7.1	1.6
	Acetic acids	5.1	3.4	1.7
4	Hydrolysis and hydrolysates			
	Hydrolysis	50.0	19.2	30.8
	Sugar	45.4	13.6	31.8
	Enzyme activity	32.7	10.3	22.4
	Enzymatic hydrolysis	28.1	12.0	16.1
	Glucose	21.9	10.6	11.3
	Cellulases	21.9	4.9	17.0
	Saccharification	17.3	9.7	7.6
	Xylose	10.2	3.7	6.5
	Enzymolysis	5.6	3.2	2.4
	Enzymatic saccharification		3.0	−3.0
5	Products			
	Ethanol	53.1	25.9	27.2
	Biofuels	29.6	15.3	14.3
	Bioethanol	15.8	11.8	4.0
	Ethanol production	12.8	5.5	7.3
	Bioethanol production	7.1	4.8	2.3
	Fuels	6.6		6.6

19%–41% of the sample papers, respectively. The other prolific keywords are softwood, hemicellulose, panicum, and zea with 10%–15% of the sample papers each.

Further, the most-prolific keyword related to the pretreatments is enzymes with 26% of the sample papers, followed by pretreatment, pretreatment, sulfuric acid, steam, and temperature with 12%–17% of the sample papers each. The other prolific keywords are the acids and water vapor with 9% and 7% of the sample papers, respectively.

The most-prolific keyword related to the fermentation is fermentation with 34% of the sample papers. The other prolific keywords are saccharomyces and yeast with 14% and 9% of the sample papers, respectively.

Further, the most-prolific keyword related to the hydrolysis and hydrolysates is hydrolysis with 50% of the sample papers. The other prolific keywords are sugar, enzyme activity, enzymatic hydrolysis, glucose, cellulases, and saccharification with 17%–45% of the sample papers each.

Finally, the most-prolific keyword related to the products is ethanol with 53% of the sample papers, followed by biofuels with 30% of the sample papers. The other prolific keywords are bioethanol and ethanol production with 16% and 13% of the sample papers, respectively.

On the other hand, the most-prolific keywords across all of the research fronts are ethanol, hydrolysis, lignin, sugar, biomass, cellulose, fermentation, enzyme activity, biofuels, wood, enzymatic hydrolysis, enzymes, glucose, and cellulases with 22%–53% of the sample papers each. The other prolific keywords are carbohydrate, lignocellulose, saccharification, pretreatment, pretreatment, bioethanol, and softwood with 15%–19% of the sample papers each.

Similarly, the most influential keywords are sugar, hydrolysis, ethanol, lignin, biomass, enzyme activity, and cellulose with 21%–32% surplus each. The other influential keywords are enzymes, cellulases, enzymatic hydrolysis, fermentation, lignocellulose, and biofuels with 14%–18% surplus each.

42.3.10   THE MOST-PROLIFIC RESEARCH FRONTS IN THE NON-WASTE BIOMASS-BASED BIOETHANOL FUELS

Information about the research fronts for the sample papers in the non-waste biomass-based bioethanol fuels is given in Table 42.9. As this table shows, the most-prolific research front for this field is the wood-based bioethanol fuels with 55% of the sample papers. The other research fronts are the grass-based bioethanol fuels, algal biomass-based bioethanol fuels, sugar feedstock-based bioethanol fuels, and starch feedstock-based bioethanol fuels with 16%, 13%, 13%, and 8% of these HCPs, respectively.

Further, the most influential research fronts are the grass-based bioethanol fuels and algal biomass-based bioethanol fuels with 6% and 5% surplus, respectively, while the starch feedstock-based bioethanol fuels are the least influential research front with 6% deficit, followed by wood-based bioethanol fuels with 2% deficit.

Information about the thematic research fronts for the sample papers in the non-waste biomass-based bioethanol fuels is given in Table 42.10. As this table shows, the most-prolific research front is the pretreatments of the non-waste feedstocks with 74% of the sample papers, followed by the hydrolysis of the non-waste biomass with 58% of the sample papers. The other prolific research

TABLE 42.9
The Most-Prolific Thematic Research Fronts For The Non-Waste Biomass-based Bioethanol Fuels

No.	Research Fronts	N Paper (%) Sample	N Paper Population (%)	Surplus (%)
1	Wood-based bioethanol fuels	54.6	57.0	−2.4
2	Grass-based bioethanol fuels	16.3	10.5	5.8
3	Algal biomass-based bioethanol fuels	13.3	8.0	5.3
4	Sugar feedstock-based bioethanol fuels	12.8	11.4	1.4
5	Starch feedstock-based bioethanol fuels	8.2	14.1	−5.9
	Sample size	196	9,820	

N paper population, the number of papers in the population sample of 9,820 papers; N paper (%) sample, the number of papers in the population sample of 196 papers.

TABLE 42.10

The Most-Prolific Thematic Research Fronts for the Non-Waste Biomass-based Bioethanol Fuels

No.	Research Fronts	N Paper (%) Sample
1	Biomass pretreatments	73.5
2	Biomass hydrolysis	58.2
3	Bioethanol production	37.2
4	Hydrolysate fermentation	27.6
5	Bioethanol fuel evaluation	13.8
6	Bioethanol fuel utilization	1.0

N paper (%) sample, the number of papers in the population sample of 196 papers.

fronts are the bioethanol production, hydrolysate fermentation, and bioethanol fuel evaluation with 37%, 28%, and 14% of the sample papers, respectively. Further, bioethanol fuel utilization relates to 1% of the sample papers.

42.4 DISCUSSION

42.4.1 INTRODUCTION

The crude oil-based gasoline fuels have been widely used in the transportation sector since the 1920s. However, there have been great public concerns over the adverse environmental and human impact of these fuels. Hence, biomass-based bioethanol fuels have increasingly been used in blending gasoline fuels, in the fuel cells, and in the biochemical production in a biorefinery context.

However, it is necessary to pretreat the biomass to enhance the yield of the bioethanol prior to the bioethanol production through the hydrolysis and fermentation. One of the most-studied feedstocks for the bioethanol fuels has been the non-waste biomass. The research in the field of the non-waste biomass-based bioethanol fuels has intensified in this context in the key research fronts of the pretreatment and hydrolysis of the non-waste biomass, fermentation of the non-waste biomass-based hydrolysates, and production and evaluation of the non-waste biomass-based bioethanol fuels. The research in this field has also intensified for the feedstocks of wood biomass, algal biomass, grass biomass, sugar feedstock-biomass, and starch feedstock-biomass. Thus, it emerges as a distinctive research field, complementing the research on the second generation waste biomass-based bioethanol fuels from the agricultural residues and other wastes.

However, it is essential to develop efficient incentive structures for the primary stakeholders to enhance the research in this field. This is especially important to maintain energy security in the cases of supply shocks such as oil price shocks and war-related chocks as in the case of Russian invasion of Ukraine or COVID-19 shocks.

The scientometric analysis has been used in this context to inform the primary stakeholders about the current state of the research in a selected research field. As there has been no scientometric study in this field, this book chapter presents a scientometric study of the research in the non-waste biomass-based bioethanol fuels. It examines the scientometric characteristics of both the sample and population data presenting scientometric characteristics of these both datasets in the order of documents, authors, publication years, institutions, funding bodies, source titles, countries, Scopus subject categories, Scopus keywords, and research fronts.

As a first step for the search of the relevant literature, the keywords were selected using the most-cited first 300 population papers for each non-waste biomass. The selected keyword list was then optimized to obtain a representative sample of papers for the each research field. These keyword lists were then integrated to obtain the keyword list for this research field (Konur, 2023a,b,c,d,e,f,g).

As a second step, two sets of data were used for this study. First, a population sample of 9,820 papers was used to examine the scientometric characteristics of the population data. Second, a sample of 196 most-cited papers, corresponding to 2% of the population papers, was used to examine the scientometric characteristics of these citation classics.

The scientometric characteristics of these sample and population datasets were presented in the order of documents, authors, publication years, institutions, funding bodies, source titles, countries, Scopus subject categories, Scopus keywords, and research fronts.

Lastly, the key scientometric findings for both datasets were discussed to highlight the research landscape for non-waste biomass-based bioethanol fuels. Additionally, a number of brief conclusions were drawn, and a number of relevant recommendations were made to enhance the future research landscape.

42.4.2 THE MOST-PROLIFIC DOCUMENTS IN THE NON-WASTE BIOMASS-BASED BIOETHANOL FUELS

Articles (together with conference papers) dominate both the sample (92%) and population (95%) papers with 3% deficit (Table 42.1). Further, review papers have a surplus (6%), and the representation of the reviews in the sample papers is quite ordinary (8%).

Scopus differs from the Web of Science database in differentiating and showing articles (86%) and conference papers (7%) published in the journals separately. However, it should be noted that these conference papers are also published in journals as articles, compared to those published only in the conference proceedings. Hence, the total number of articles and review papers in the sample dataset are 92% and 8%, respectively.

It is observed during the search process that there has been inconsistency in the classification of the documents in Scopus as well as in other databases such as Web of Science. This is especially relevant for the classification of papers as reviews or articles as the papers not involving a literature review may be erroneously classified as a review paper. There is also a case of review papers being classified as articles. For example, the total number of the reviews in the sample dataset was manually found as nearly 7% compared to 8% as indexed by Scopus, decreasing the number of articles and conference papers to 93% for the sample dataset. It is notable that many technoeconomic and life cycle studies were indexed as reviews by the Scopus database.

In this context, it would be helpful to provide a classification note for the published papers in the books and journals at the first instance. It would also be helpful to use the document types listed in Table 42.1 for this purpose. Book chapters may also be classified as articles or reviews as an additional classification to differentiate review chapters from the experimental chapters as it is done by the Web of Science. It would be further helpful to additionally classify the conference papers as articles or review papers as well as it is done in the Web of Science database.

42.4.3 THE MOST-PROLIFIC AUTHORS IN THE NON-WASTE BIOMASS-BASED BIOETHANOL FUELS

There have been most-prolific 22 authors with at least 2% of the sample papers each as given in Table 42.2. These authors have shaped the development of the research in this field.

The most-prolific authors are John N. Saddler, Guido Zacchi, and Xuejun Pan and to a lesser extent Mats Galbe, Charles E. Wyman, Junyong Zhu, David J. Gregg, Arthur J. Ragauskas, and Bruce E. Dale. Further, the most influential authors are John N. Saddler, Xuejun Pan, Guido Zacchi, and to a lesser extent Mats Galbe, Charles E. Wyman, Junyong Zhu, David J. Gregg, Bruce E. Dale, Neil Gilkes, and Maria J. Negro.

It is important to note the inconsistencies in indexing of the author names in Scopus and other databases. It is especially an issue for the names with more than two components such as 'Blake Sam de Hyun Ladisch'. The probable outcomes are 'Ladisch, B.S.D.H.', 'de Hyun Ladisch, B.S.', or 'Hyun Ladisch, B.S.D.'. The first choice is the gold standard of the publishing sector as the last word in the name is taken as the last name. In most of the academic databases such as PUBMED and EBSCO databases, this version is used predominantly. The second choice is a strong alternative, while the last choice is an undesired outcome as the last two words are taken as the last name. It is a good practice to combine the words of the last name by a hyphen: 'Hyun-Ladisch, B.S.D.'. It is notable that inconsistent indexing of the author names may cause substantial inefficiencies in the search process for the papers as well as allocating credit to the authors as there are different author entries for each outcome in the databases.

There are also inconsistencies in the shortening Chinese names. For example, 'YangYing Pan' is often shortened as 'Pan, Y.', 'Pan, Y.-Y.', and 'Pan, Y.Y.', as it is done in the Web of Science database as well. However, the gold standard in this case is 'Pan, Y', where the last word is taken as the last name and the first word is taken as a single forename. In most of the academic databases such as PUBMED and EBSCO, this first version is used predominantly. Nevertheless, it makes sense to sue the third option to differentiate Chinese names efficiently: 'Pan, Y.Y.'. Therefore, there have been difficulties in locating papers for the Chinese authors. In such cases, the use of the unique author codes provided for each author by the Scopus database has been helpful.

There is also a difficulty in allowing credit for the authors especially for the authors with common names such as 'Pan, X.' in conducting scientometric studies. These difficulties strongly influence the efficiency of the scientometric studies as well as allocating credit to the authors as there are the same author entries for different authors with the same name, e.g., 'Pan, X.' in the databases.

In this context, the coding of authors in Scopus database is a welcome innovation compared to the other databases such as Web of Science. In this process, Scopus allocates a unique number to each author in the database (Aman, 2018). However, there might still be substantial inefficiencies in this coding system especially for common names. For example, some of the papers for a certain author maybe allocated to another researcher with a different author code. It is possible that Scopus uses a number of software programs to differentiate the author names and the program may not be false-proof (Kim, 2018).

In this context, it does not help that author names are not given in full in some journals and books. This makes it difficult to differentiate authors with common names and makes the scientometric studies further difficult in the author domain. Therefore, the author names should be given in all books and journals at the first instance. There is also a cultural issue where some authors do not use their full names in their papers. Instead, they use initials for their forenames: 'Galbe, H.J.', 'Galbe, H.', or 'Galbe, J.' instead of 'Galbe, Hyun Jae'.

There are also inconsistencies in naming of the authors with more than two components by the authors themselves in journal papers and book chapters. For example, 'Galbe, A.P.C.' might be given as 'Galbe, A.' or 'Galbe, A.C.' or 'Galbe, A.P.' or 'Galbe, C' in the journals and books. This also makes the scientometric studies difficult in the author domain. Hence, contributing authors should use their names consistently in their publications.

The other critical issue regarding the author names is the inconsistencies in the spelling of the author names in the national spellings (e.g., Özgümüş, Şençöl) rather than in the English spellings (e.g., Ozgumus, Sencol) in Scopus database. Scopus differs from the Web of Science database and many other databases in this respect where the author names are given only in the English spellings. It is observed that national spellings of the author names do not help much in conducting scientometric studies as well in allocating credits to the authors as sometimes there are the different author entries for the English and National spellings in the Scopus database.

The most-prolific institutions for the sample dataset are Lund University and to a lesser extent University of British Columbia, CIEMAT, University of Jaen, and University of Tennessee. Further, the most-prolific countries for the sample dataset are the USA and to a lesser extent Sweden, Spain, and Canada. These findings confirm the dominance of the USA, Europe, and to a lesser extent

Canada in this field. On the other hand, pretreatments and hydrolysis of the non-waste biomass and to a lesser extent the fermentation of the non-waste biomass-based hydrolysates and the bioethanol fuel production are the key research fronts studied by these top authors.

It is also notable that there is a significant gender deficit for the sample dataset as surprisingly with a representation rate of 18%. This finding is the most thought-provoking with strong public policy implications. Hence, institutions, funding bodies, and policymakers should take efficient measures to reduce the gender deficit in this field as well as other scientific fields with strong gender deficit. In this context, it is worth to note the level of representation of the researchers from the minority groups in science on the basis of race, sexuality, age, and disability, besides the gender (Blankenship, 1993; Dirth and Branscombe, 2017; Konur, 2000, 2002a,b,c, 2006a,b, 2007a,b).

42.4.4 THE MOST-PROLIFIC RESEARCH OUTPUT BY YEARS IN THE NON-WASTE BIOMASS-BASED BIOETHANOL FUELS

The research output observed between 1970 and 2022 is illustrated in Figure 42.1. This figure clearly shows that the bulk of the research papers in the population dataset were published primarily in the 2010s and early 2020s. Similarly, the bulk of the research papers in the sample dataset were published in the 2010s and 2000s. Further, there is a rising trend for the population papers between 2006 and 2011. However, it lost its momentum after 2011.

These findings suggest that the most-prolific sample and population papers were primarily published in the 2010s. Further, a significant portion of the sample and population papers were published in the early 2020s and 2000s, respectively.

These are the thought-provoking findings as there has been significant research boom in since 2009 and 2005 for the population and sample papers, respectively. In this context, the increasing public concerns about climate change (Change, 2007), greenhouse gas emissions (Carlson et al., 2017), and global warming (Kerr, 2007) have been certainly behind the boom in the research in this field since 2007. Furthermore, the recent supply shocks experiences due to the COVID-19 pandemics and the Ukrainian war might also be behind the research boom in this field since 2019.

Based on these findings, the size of the population papers likely to more than double in the current decade, provided that the public concerns about climate change, greenhouse gas emissions, and global warming, as well as the supply shocks are translated efficiently to the research funding in this field.

42.4.5 THE MOST-PROLIFIC INSTITUTIONS IN THE NON-WASTE BIOMASS-BASED BIOETHANOL FUELS

The most-prolific 29 institutions publishing papers on the non-waste biomass-based bioethanol fuels with at least 2% of the sample papers each given in Table 42.3 have shaped the development of the research in this field.

The most-prolific institutions are the University of British Columbia, Lund University, and to a lesser extent University of Wisconsin-Madison, USDA Forest Service, NC State University, Dartmouth College, NREL, Michigan State University, and University of California Riverside. Similarly, the top countries for these most-prolific institutions are the USA and to a lesser extent China, Finland, and Spain. In total, nine countries house these top institutions.

On the other hand, the institutions with the most citation impact are the University of British Columbia, Lund University, and to a lesser extent University of Wisconsin-Madison, USDA Forest Service, NC State University, Dartmouth College, University of California Riverside, Michigan State University, NREL, CIEMAT, and University of Nebraska-Lincoln. These findings confirm the dominance of the institutions from the USA, Europe, and to a lesser extent Canada.

42.4.6   THE MOST-PROLIFIC FUNDING BODIES IN THE NON-WASTE BIOMASS-BASED BIOETHANOL FUELS

The most-prolific 19 funding bodies funding at least 1% of the sample papers each are given in Table 42.4. It is notable that only 37% and 42% of the sample and population papers were funded, respectively.

The most-prolific funding bodies are the Natural Sciences and Engineering Research Council of Canada, the U.S. Department of Energy, and to a lesser extent National Natural Science Foundation of China, European Regional Development Fund, Natural Resources Canada, and Swedish National Board for Industrial and Technical Development. On the other hand, the most-prolific countries for these top funding bodies are the USA and to a lesser extent Canada, China, and the EU. In total, 11 countries and the EU house these top funding bodies.

The funding bodies with the most citation impact are the Natural Sciences and Engineering Research Council of Canada, and to a lesser extent Swedish National Board for Industrial and Technical Development, and Natural Resources Canada. Further, the funding bodies with the least citation impact are the National Natural Science Foundation of China and to a lesser extent National Council for Scientific and Technological Development.

These findings on the funding of the research in this field suggest that the level of the funding, mostly since 2009, is modest and it has been largely instrumental in enhancing the research in this field (Ebadi and Schiffauerova, 2016) in the light of North's institutional framework (North, 1991). It is also notable that the funding rate in this field is relatively modest compared to those in the other research fronts of the bioethanol fuels such as algal bioethanol fuels. Further, it is expected that this funding rate would improve in the light of the recent supply shocks. Further, it emerges that the USA, Canada, China, and Europe have heavily funded the research on the non-waste biomass-based bioethanol fuels.

42.4.7   THE MOST-PROLIFIC SOURCE TITLES IN THE NON-WASTE BIOMASS-BASED BIOETHANOL FUELS

The most-prolific 17 source titles publishing at least 1.5% of the sample papers each in the non-waste biomass-based bioethanol fuels have shaped the development of the research in this field (Table 42.5).

The most-prolific source titles are the Bioresource Technology and to a lesser extent Biotechnology and Bioengineering, Biomass and Bioenergy, Applied Biochemistry and Biotechnology Part A, Enzyme and Microbial Technology, and Biotechnology Progress. On the other hand, the source titles with the most impact are the Bioresource Technology and to a lesser extent by Biotechnology and Bioengineering, Applied Biochemistry and Biotechnology Part A, Enzyme and Microbial Technology, Biomass and Bioenergy, Biotechnology Progress, and Proceedings of the National Academy of Sciences of the United States of America.

It is notable that these top source titles are primarily related to the bioresources, biotechnology, and to a lesser extent energy and microbial technology. This finding suggests that Bioresource Technology and the other prolific journals in these fields have significantly shaped the development of the research in this field as they focus primarily on the non-waste biomass-based bioethanol fuels with a high yield. In this context, the influence of the top journal is quite extraordinary.

42.4.8   THE MOST-PROLIFIC COUNTRIES IN THE NON-WASTE BIOMASS-BASED BIOETHANOL FUELS

The most-prolific 18 countries publishing at least 1.5% of the sample papers each have significantly shaped the development of the research in this field (Table 42.6).

The most-prolific countries are the USA and to a lesser extent Canada, Sweden, China, Brazil, Japan, Spain, and Finland. Further, eight European countries listed in Table 42.6 produce 26% and 20% of the sample and population papers, respectively, and China is the second largest producer of the population papers with a publication rate of 19.6% after Europe and the USA.

On the other hand, the countries with the most citation impact are the USA, Europe, and to a lesser extent Canada. Similarly, the countries with the least impact are China and to a lesser extent Brazil, Japan, India, S. Korea, the UK, France, and Germany.

The close examination of these findings suggests that the USA, Europe, and to a lesser extent China, Canada, Japan, and Brazil are the major producers of the research in this field. It is a fact that the USA has been a major player in science (Leydesdorff and Wagner, 2009). The USA has further developed a strong research infrastructure to support its corn- and grass-based bioethanol industry (Gillon, 2010).

However, China has been a rising mega star in the scientific research in competition with the USA and Europe (Leydesdorff and Zhou, 2005). China is also a major player in this field as a major producer of bioethanol (Fang et al., 2010).

Next, Europe has been a persistent player in the scientific research in competition with both the USA and China (Leydesdorff, 2000). Europe has also been a persistent producer of bioethanol along with the USA and Brazil (Gnansounou, 2010).

Further, Canada (Tahmooresnejad et al., 2015), Japan (Negishi et al., 2004), and Brazil (Glanzel et al., 2006) are the other countries with substantial research activities in bioethanol fuels.

42.4.9 THE MOST-PROLIFIC SCOPUS SUBJECT CATEGORIES IN THE NON-WASTE BIOMASS-BASED BIOETHANOL FUELS

The most-prolific ten Scopus subject categories indexing at least 3.6% of the sample papers each, given in Table 42.7, have shaped the development of the research in this field.

The most-prolific Scopus subject categories in the non-waste biomass-based bioethanol fuels are Chemical Engineering, and to a lesser extent Environmental Science, Biochemistry, Genetics, and Molecular Biology, Energy, and Immunology and Microbiology. It is also notable that Social Sciences including Economics and Business have a minimal presence in both sample and population studies.

On the other hand, the Scopus subject categories with the most citation impact are Chemical Engineering, and to a lesser extent Immunology and Microbiology, Biochemistry, Genetics, and Molecular Biology, Environmental Science, and Energy. Similarly, the least influential subject categories are Agricultural and Biological Sciences, Materials Science, and Chemistry, and to a lesser extent Engineering.

These findings are thought-provoking suggesting that the primary subject categories are related to chemical engineering, environmental sciences, and to a lesser extent energy, genetics, and microbiology as the core of the research in this field concerns with the production and utilization of the non-waste biomass-based bioethanol fuels. The other finding is that social sciences are not well represented in both the sample and population papers as in line with the most fields in bioethanol fuels. The social, environmental, and economics studies account for the field of social sciences.

42.4.10 THE MOST-PROLIFIC KEYWORDS IN THE NON-WASTE BIOMASS-BASED BIOETHANOL FUELS

A limited number of keywords have shaped the development of the research in this field as shown in Table 42.8. These keywords are grouped under five headings: non-waste biomass, pretreatments, fermentation, hydrolysis and hydrolysates, and products.

The most-prolific keywords across all of the research fronts are ethanol, hydrolysis, lignin, sugar, biomass, cellulose, fermentation, enzyme activity, biofuels, wood, enzymatic hydrolysis, enzymes, glucose, cellulases, and to a lesser extent carbohydrate, lignocellulose, saccharification, pre-treatment, pretreatment, bioethanol, and softwood.

Similarly, the most-influential keywords are sugar, hydrolysis, ethanol, lignin, biomass, enzyme activity, cellulose, and to a lesser extent enzymes, cellulases, enzymatic hydrolysis, fermentation, lignocellulose, and biofuels.

These findings suggest that it is necessary to determine the keyword set carefully to locate the relevant research in each of these research fronts. Additionally, the size of the samples for each keyword highlights the intensity of the research in the relevant research areas for both sample and population datasets. These findings also highlight different spelling of some strategic keywords such as pretreatment v. pretreatment and bioethanol v. ethanol v. bio-ethanol, etc. However, there is a tendency toward the use of the connected keywords without using a hyphen.

42.4.11 THE MOST-PROLIFIC RESEARCH FRONTS IN THE NON-WASTE BIOMASS-BASED BIOETHANOL FUELS

Information about the research fronts for the sample papers in the non-waste biomass-based bio-ethanol fuels is given in Table 42.9. As this table shows, the most-prolific research fronts for this field are the wood-based bioethanol fuels and to a lesser extent grass-based bioethanol fuels, algal biomass-based bioethanol fuels, sugar feedstock-based bioethanol fuels, and starch feedstock-based bioethanol fuels. Thus, these research fields have substantial importance as the first or third generation bioethanol fuels, complementing the second generation bioethanol fuel research. It is important to note that the production of these first generation bioethanol fuels from sugar and starch feedstock-based bioethanol fuels is not much desirable as it undermines the food security (Makenete et al., 2008; Wu et al., 2012).

Table 42.10 shows that the most-prolific research fronts are the pretreatments of the non-waste feedstocks, followed by the hydrolysis of the non-waste biomass, bioethanol production, hydrolysate fermentation, bioethanol fuel evaluation, and bioethanol fuel utilization.

These findings are thought-provoking in seeking ways to increase non-waste biomass-based bioethanol yield at the global scale. It is clear that all of these research fronts have public importance and merit substantial funding and other incentives. Further, it is notable that non-waste biomass-based bioethanol fuels have become a core unit of the bioethanol research to make it more competitive with the crude oil-based gasoline and diesel fuels, especially for the USA, Europe, and China.

In comparison to the other feedstock-based research fronts, it is notable that the pretreatment and hydrolysis of the non-waste biomass emerge as primary research fronts for this field. However, the research fronts of the fermentation of the non-waste biomass-based hydrolysates and the bioethanol production from the non-waste biomass-based hydrolysates are also important.

Further, the field of the evaluation and utilization of bioethanol fuels is a neglected area. This suggests that the primary stakeholders have been primarily interested in these key processes of the bioethanol production. It is also notable that evaluation of the non-waste biomass-based bioethanol fuels such as technoeconomics, life cycle, economics, social, land use, labor, and environment-related studies emerges as a case study for the bioethanol fuels. Similarly, the utilization of these biofuels in the gasoline or diesel engines is also an important research field from a societal perspective. In this context, the USA and Brazil have been the global leaders in the production and use of the corn- and sugarcane-based bioethanol fuels since the 1970s in the aftermath of the global crude oil crisis in the early 1970s.

It is further notable that the research on the non-waste biomass-based bioethanol fuels complements the research on the second generation bioethanol fuel research from agricultural residues, urban, food, industrial, forestry wastes, and lignocellulosic and cellulosic biomass at large among others. Thus, it emerges as a distinctive research field complementing the research on the second generation bioethanol fuels.

In the end, these most-cited papers in this field hint that the production of non-waste biomass-based bioethanol fuels could be optimized using the structure, processing, and property relationships of these non-waste biomass in the fronts of the feedstock pretreatment and hydrolysis, and hydrolysate fermentation (Formela et al., 2016; Konur, 2018a, 2020b, 2021a,b,c,d; Konur and Matthews, 1989).

42.5 CONCLUSION AND FUTURE RESEARCH

The research on the non-waste biomass-based bioethanol fuels has been mapped through a scientometric study of both sample (196 papers) and population (9,820 papers) datasets.

The critical issue in this study has been to obtain a representative sample of the research as in any other scientometric study. Therefore, the keyword set has been carefully devised and optimized after a number of runs in the Scopus database. It is a representative sample of the wider population studies. This keyword set was provided in the appendix of the related studies, and the relevant keywords are presented in Table 42.8. However, it should be noted that it has been very difficult to compile a representative keyword set since this research field has been connected closely with many other fields. Therefore, it has been necessary to compile a keyword list to exclude papers concerned with the other research fields.

The other issue has been the selection of a multidisciplinary database to carry out the scientometric study of the research in this field. For this purpose, Scopus database has been selected. The journal coverage of this database has been notably wider than that of Web of Science and other multisubject databases.

The key scientometric properties of the research in this field have been determined and discussed in this book chapter. It is evident that a limited number of documents, authors, institutions, publication years, institutions, funding bodies, source titles, countries, Scopus subject categories, Scopus keywords, and research fronts have shaped the development of the research in this field.

There is ample scope to increase the efficiency of the scientometric studies in this field in the author and document domains by developing consistent policies and practices in both domains across all the academic databases. In this respect, it seems that authors, journals, and academic databases have a lot to do. Furthermore, the significant gender deficit as in most scientific fields emerges as a public policy issue. The potential deficits on the basis of age, race, disability, and sexuality also need to be explored in this field as in other scientific fields.

The research in this field has boomed since 2009 and 2005 for the population and sample papers, respectively, possibly promoted by the public concerns on global warming, greenhouse gas emissions, and climate change. Furthermore, the recent COVID-19 pandemics and Russian invasion of Ukraine have resulted in a global supply shocks shifting the recent focus of the stakeholders from the crude oil-based fuels to biomass-based fuels such as bioethanol fuels. It is expected that there would be further incentives for the key stakeholders to carry out the research for the non-waste biomass-based bioethanol fuels to increase the ethanol yield and to make it more competitive with the crude oil-based gasoline and petrodiesel fuels. This might be truer for the crude oil- and foreign exchange-deficient countries to maintain the energy and food security at the face of the global supply shocks.

The relatively modest funding rate of 37% and 42% for the sample and population papers, respectively, suggests that funding in this field significantly enhanced the research in this field primarily since 2009, possibly more than doubling in the current decade. However, it is evident that there is ample room for more funding and other incentives to enhance the research in this field further as the research output for the population papers lost its momentum after 2011.

The institutions from the USA and to a lesser extent China, Finland, and Spain have mostly shaped the research in this field. Further, USA, Europe, and to a lesser extent Canada, Sweden, China, Brazil, Japan, Spain, and Finland have been the major producers of the research in this field as the major producers and users of bioethanol fuels. It is evident that these countries have well-developed research infrastructure in bioethanol fuels and their derivatives.

It emerges that ethanol is more popular than bioethanol as a keyword with strong implications for the search strategy. In other words, the search strategy using only bioethanol keyword would not be much helpful. On the other hand, the Scopus keywords are grouped under the five headings: biomass, pretreatments, fermentation, hydrolysis and hydrolysates, and products.

Table 42.9 shows that the most-prolific research fronts for this field are the wood-based bioethanol fuels and to a lesser extent grass-based bioethanol fuels, algal biomass-based bioethanol fuels, sugar feedstock-based bioethanol fuels, and starch feedstock-based bioethanol fuels. Thus, these research fields have substantial importance as the first or third generation bioethanol fuels, complementing the second generation bioethanol fuel research.

Further, Table 42.10 shows that the most-prolific research fronts are the pretreatments of the non-waste feedstocks, followed by the hydrolysis of the non-waste biomass, bioethanol production, hydrolysate fermentation, bioethanol fuel evaluation, and bioethanol fuel utilization.

The first four research fronts dominate the research in this field, while the field of the utilization and evaluation of the non-waste biomass-based bioethanol fuels is relatively a neglected research field. In this context, it is notable that there is ample room for the improvement of the research on social and humanitarian aspects of the research on the bioethanol fuels from the non-waste biomass such as scientometric and user studies.

These findings are thought-provoking in seeking ways to increase non-waste biomass feedstock-based bioethanol yield at the global scale. It is clear that all of these research fronts have public importance and merit substantial funding and other incentives. Further, it is notable that non-waste biomass-based bioethanol fuels have become a core unit of the bioethanol research to make it more competitive with the crude oil-based gasoline and petrodiesel fuels, especially for the USA, Europe, Brazil, and China. It is further notable that the research on the non-waste biomass-based bioethanol emerges as a distinctive research field, complementing the research on the second generation waste biomass-based bioethanol fuels from the agricultural residues and other wastes.

Thus, the scientometric analysis has a great potential to gain valuable insights into the evolution of the research in this field as in other scientific fields especially in the aftermath of the significant global supply shocks such as COVID-19 pandemics and the Russian invasion of Ukraine.

It is recommended that further scientometric studies are carried out for the primary research fronts. It is further recommended that reviews of the most-cited papers are carried out for each primary research front to complement these scientometric studies. Next, the scientometric studies of the hot papers in these primary fields are carried out.

ACKNOWLEDGMENTS

The contribution of the highly cited researchers in the field of the non-waste biomass-based bioethanol fuels has been gratefully acknowledged.

REFERENCES

Aman, V. 2018. Does the Scopus author ID suffice to track scientific international mobility? A case study based on Leibniz laureates. *Scientometrics* 117:705–720.

Angelici, C., B. M. Weckhuysen and P. C. A. Bruijnincx. 2013. Chemocatalytic conversion of ethanol into butadiene and other bulk chemicals. *ChemSusChem* 6:1595–1614.

Antolini, E. 2007. Catalysts for direct ethanol fuel cells. *Journal of Power Sources* 170:1–12.

Antolini, E. 2009. Palladium in fuel cell catalysis. *Energy and Environmental Science* 2:915–931.

Bai, F. W., W. A. Anderson and M. Moo-Young. 2008. Ethanol fermentation technologies from sugar and starch feedstocks. *Biotechnology Advances* 26:89–105.

Beaudry, C. and V. Lariviere. 2016. Which gender gap? Factors affecting researchers' scientific impact in science and medicine. *Research Policy* 45:1790–1817.

Blankenship, K. M. 1993. Bringing gender and race in: US employment discrimination policy. *Gender & Society* 7:204–226.

Bothast, R. J. and M. A. Schlicher. 2005. Biotechnological processes for conversion of corn into ethanol. *Applied Microbiology and Biotechnology* 67:19–25.

Burnham, J. F. 2006. Scopus database: A review. *Biomedical Digital Libraries* 3:1–8.

Carlson, K. M., J. S. Gerber and D. Mueller, et al. 2017. Greenhouse gas emissions intensity of global croplands. *Nature Climate Change* 7:63–68.

Change, C. 2007. Climate change impacts, adaptation and vulnerability. *Science of the Total Environment* 326:95–112.

Dien, B. S., H. J. G. Jung and K. P. Vogel, et al. 2006. Chemical composition and response to dilute-acid pretreatment and enzymatic saccharification of alfalfa, reed canarygrass, and switchgrass. *Biomass and Bioenergy* 30:880–891.

Dirth, T. P. and N. R. Branscombe. 2017. Disability models affect disability policy support through awareness of structural discrimination. *Journal of Social Issues* 73:413–442.

Ebadi, A. and A. Schiffauerova. 2016. How to boost scientific production? A statistical analysis of research funding and other influencing factors. *Scientometrics* 106:1093–1116.

Fang, X., Y. Shen, J. Zhao, X. Bao and Y. Qu. 2010. Status and prospect of lignocellulosic bioethanol production in China. *Bioresource Technology* 101:4814–4819.

Fauci, A. S., H. C. Lane and R. R. Redfield. 2020. Covid-19-navigating the uncharted. *New England Journal of Medicine* 382:1268–1269.

Fernando, S., S. Adhikari, C. Chandrapal and M. Murali. 2006. Biorefineries: Current status, challenges, and future direction. *Energy & Fuels* 20:1727–1737.

Formela, K., A. Hejna, L. Piszczyk, M. R. Saeb and X. Colom. 2016. Processing and structure-property relationships of natural rubber/wheat bran biocomposites. *Cellulose* 23:3157–3175.

Galbe, M. and G. Zacchi. 2002. A review of the production of ethanol from softwood. *Applied Microbiology and Biotechnology* 59:618–628.

Garfield, E. 1955. Citation indexes for science. *Science* 122:108–111.

Gillon, S. 2010. Fields of dreams: Negotiating an ethanol agenda in the Midwest United States. *Journal of Peasant Studies* 37:723–748.

Glanzel, W., J. Leta and B. Thijs. 2006. Science in Brazil. Part 1: A macro-level comparative study. *Scientometrics* 67:67–86.

Gnansounou, E. 2010. Production and use of lignocellulosic bioethanol in Europe: Current situation and perspectives. *Bioresource Technology* 101:4842–4850.

Goh, C. S. and K. T. Lee. 2010. A visionary and conceptual macroalgae-based third-generation bioethanol (TGB) biorefinery in Sabah, Malaysia as an underlay for renewable and sustainable development. *Renewable and Sustainable Energy Reviews* 14:842–848.

Hahn-Hagerdal, B., M. Galbe, M. F. Gorwa-Grauslund, G. Liden and G. Zacchi. 2006. Bio-ethanol - The fuel of tomorrow from the residues of today. *Trends in Biotechnology* 24:549–556.

Hamilton, J. D. 1983. Oil and the macroeconomy since World War II. *Journal of Political Economy* 91:228–248.

Hamilton, J. D. 2003. What is an oil shock? *Journal of Econometrics* 113:363–398.

Hill, J., E. Nelson, D. Tilman, S. Polasky and D. Tiffany. 2006. Environmental, economic, and energetic costs and benefits of biodiesel and ethanol biofuels. *Proceedings of the National Academy of Sciences of the United States of America* 103:11206–11210.

Hill, J., S. Polasky and E. Nelson, et al. 2009. Climate change and health costs of air emissions from biofuels and gasoline. *Proceedings of the National Academy of Sciences of the United States of America* 106:2077–2082.

Hsieh, W. D., R. H. Chen, T. L. Wu and T. H. Lin. 2002. Engine performance and pollutant emission of an SI engine using ethanol-gasoline blended fuels. *Atmospheric Environment* 36:403–410.

Huang, H. J., S. Ramaswamy, U. W. Tschirner and B. V. Ramarao. 2008. A review of separation technologies in current and future biorefineries. *Separation and Purification Technology* 62:1–21.

John, R. P., G. S. Anisha, K. M. Nampoothiri and A. Pandey. 2011. Micro and macroalgal biomass: A renewable source for bioethanol. *Bioresource Technology* 102:186–193.

Jones, T. C. 2012. America, oil, and war in the Middle East. *Journal of American History* 99:208–218.

Kerr, R. A. 2007. Global warming is changing the world. *Science* 316:188–190.

Keshwani, D. R. and J. J. Cheng. 2009. Switchgrass for bioethanol and other value-added applications: A review. *Bioresource Technology* 100:1515–1523.

Kilian, L. 2008. Exogenous oil supply shocks: How big are they and how much do they matter for the US economy? *Review of Economics and Statistics* 90:216–240.

Kilian, L. 2009. Not all oil price shocks are alike: Disentangling demand and supply shocks in the crude oil market. *American Economic Review* 99:1053–69.

Kim, J. 2018. Evaluating author name disambiguation for digital libraries: A case of DBLP. *Scientometrics* 116:1867–1886.

Konur, O. 2000. Creating enforceable civil rights for disabled students in higher education: An institutional theory perspective. *Disability & Society* 15:1041–1063.

Konur, O. 2002a. Access to nursing education by disabled students: Rights and duties of nursing programs. *Nurse Education Today* 22:364–374.

Konur, O. 2002b. Assessment of disabled students in higher education: Current public policy issues. *Assessment and Evaluation in Higher Education* 27:131–152.

Konur, O. 2002c. Access to employment by disabled people in the UK: Is the Disability Discrimination Act working? *International Journal of Discrimination and the Law* 5:247–279.

Konur, O. 2006a. Participation of children with dyslexia in compulsory education: Current public policy issues. *Dyslexia* 12:51–67.

Konur, O. 2006b. Teaching disabled students in higher education. *Teaching in Higher Education* 11:351–363.

Konur, O. 2007a. A judicial outcome analysis of the *Disability Discrimination Act*: A windfall for the employers? *Disability & Society* 22:187–204.

Konur, O. 2007b. Computer-assisted teaching and assessment of disabled students in higher education: The interface between academic standards and disability rights. *Journal of Computer Assisted Learning* 23:207–219.

Konur, O. 2011. The scientometric evaluation of the research on the algae and bio-energy. *Applied Energy* 88:3532–3540.

Konur, O. 2012a. The evaluation of the biogas research: A scientometric approach. *Energy Education Science and Technology Part A: Energy Science and Research* 29:1277–1292.

Konur, O. 2012b. The evaluation of the educational research: A scientometric approach. *Energy Education Science and Technology Part B: Social and Educational Studies* 4:1935–1948.

Konur, O. 2012c. The evaluation of the global energy and fuels research: A scientometric approach. *Energy Education Science and Technology Part A: Energy Science and Research* 30:613–628.

Konur, O. 2012d. The evaluation of the research on the biodiesel: A scientometric approach. *Energy Education Science and Technology Part A: Energy Science and Research* 28:1003–1014.

Konur, O. 2012e. The evaluation of the research on the bioethanol: A scientometric approach. *Energy Education Science and Technology Part A: Energy Science and Research* 28:1051–1064.

Konur, O. 2012f. The evaluation of the research on the biofuels: A scientometric approach. *Energy Education Science and Technology Part A: Energy Science and Research* 28:903–916.

Konur, O. 2012g. The evaluation of the research on the biohydrogen: A scientometric approach. *Energy Education Science and Technology Part A: Energy Science and Research* 29:323–338.

Konur, O. 2012h. The evaluation of the research on the microbial fuel cells: A scientometric approach. *Energy Education Science and Technology Part A: Energy Science and Research* 29:309–322.

Konur, O. 2012i. The scientometric evaluation of the research on the production of bioenergy from biomass. *Biomass and Bioenergy* 47:504–515.

Konur, O. 2015. Current state of research on algal bioethanol. In *Marine Bioenergy: Trends and Developments*, Ed. S. K. Kim and C. G. Lee, pp. 217–244. Boca Raton, FL: CRC Press.

Konur, O., Ed. 2018a. *Bioenergy and Biofuels*. Boca Raton, FL: CRC Press.

Konur, O. 2018b. Bioenergy and biofuels science and technology: Scientometric overview and citation classics. In *Bioenergy and Biofuels*, Ed. O. Konur, pp. 3–63. Boca Raton, FL: CRC Press.

Konur, O. 2019. Cyanobacterial bioenergy and biofuels science and technology: A scientometric overview. In *Cyanobacteria: From Basic Science to Applications*, Ed. A. K. Mishra, D. N. Tiwari and A. N. Rai, pp. 419–442. Amsterdam: Elsevier.

Konur, O. 2020a. The scientometric analysis of the research on the bioethanol production from green macroalgae. In *Handbook of Algal Science, Technology and Medicine*, Ed. O. Konur, pp. 385–401. London: Academic Press.

Konur, O., Ed. 2020b. *Handbook of Algal Science, Technology and Medicine*. London: Academic Press.

Konur, O., Ed. 2021a. *Handbook of Biodiesel and Petrodiesel Fuels: Science, Technology, Health, and Environment*. Boca Raton, FL: CRC Press.

Konur, O., Ed. 2021b. *Handbook of Biodiesel and Petrodiesel Fuels: Science, Technology, Health, and Environment. Volume 1. Biodiesel Fuels: Science, Technology, Health, and Environment*. Boca Raton, FL: CRC Press.

Konur, O., Ed. 2021c. *Handbook of Biodiesel and Petrodiesel Fuels: Science, Technology, Health, and Environment. Volume 2. Biodiesel Fuels based on the Edible and Nonedible Feedstocks, Wastes, and Algae: Science, Technology, Health, and Environment*. Boca Raton, FL: CRC Press.

Konur, O., Ed. 2021d. *Handbook of Biodiesel and Petrodiesel Fuels: Science, Technology, Health, and Environment. Volume 3. Petrodiesel Fuels: Science, Technology, Health, and Environment.* Boca Raton, FL: CRC Press.

Konur, O. 2023a. First generation starch feedstock-based bioethanol fuel fuels: Scientometric study. In *Feedstock-based Bioethanol Fuels. I. Non-Waste Feedstocks: Starch, Sugar, Grass, Wood, Cellulose, Algae, and Biosyngas-based Bioethanol Fuels. Handbook of Bioethanol Fuels Volume 3*, Ed. O. Konur. Boca Raton, FL: CRC Press.

Konur, O. 2023b. First generation sugar feedstock-based bioethanol fuels: Scientometric study. In *Feedstock-based Bioethanol Fuels. I. Non-Waste Feedstocks: Starch, Sugar, Grass, Wood, Cellulose, Algae, and Biosyngas-based Bioethanol Fuels. Handbook of Bioethanol Fuels Volume 3*, Ed. O. Konur. Boca Raton, FL: CRC Press.

Konur, O. 2023c. Grass-based bioethanol fuel production: Scientometric study. In *Feedstock-based Bioethanol Fuels. I. Non-Waste Feedstocks: Starch, Sugar, Grass, Wood, Cellulose, Algae, and Biosyngas-based Bioethanol Fuels. Handbook of Bioethanol Fuels Volume 3*, Ed. O. Konur. Boca Raton, FL: CRC Press.

Konur, O. 2023d. Wood-based bioethanol fuel production: Scientometric study. In *Feedstock-based Bioethanol Fuels. I. Non-Waste Feedstocks: Starch, Sugar, Grass, Wood, Cellulose, Algae, and Biosyngas-based Bioethanol Fuels. Handbook of Bioethanol Fuels Volume 3*, Ed. O. Konur. Boca Raton, FL: CRC Press.

Konur, O. 2023e. Cellulose-based bioethanol fuels: Scientometric study. In *Feedstock-based Bioethanol Fuels. I. Non-Waste Feedstocks: Starch, Sugar, Grass, Wood, Cellulose, Algae, and Biosyngas-based Bioethanol Fuels. Handbook of Bioethanol Fuels Volume 3*, Ed. O. Konur. Boca Raton, FL: CRC Press.

Konur, O. 2023f. Third generation algal bioethanol fuels: Scientometric study. In *Feedstock-based Bioethanol Fuels. I. Non-Waste Feedstocks: Starch, Sugar, Grass, Wood, Cellulose, Algae, and Biosyngas-based Bioethanol Fuels. Handbook of Bioethanol Fuels Volume 3*, Ed. O. Konur. Boca Raton, FL: CRC Press.

Konur, O. 2023g. Bioyngas-based bioethanol fuels: Scientometric study. In *Feedstock-based Bioethanol Fuels. I. Non-Waste Feedstocks: Starch, Sugar, Grass, Wood, Cellulose, Algae, and Biosyngas-based Bioethanol Fuels. Handbook of Bioethanol Fuels Volume 3*, Ed. O. Konur. Boca Raton, FL: CRC Press.

Konur, O. and F. L. Matthews. 1989. Effect of the properties of the constituents on the fatigue performance of composites: A review. *Composites* 20:317–328.

Kruyt, B., D. P. van Vuuren, H. J. de Vries and H. Groenenberg. 2009. Indicators for energy security. *Energy Policy* 37:2166–2181.

Leydesdorff, L. 2000. Is the European Union becoming a single publication system? *Scientometrics* 47:265–280.

Leydesdorff, L. and C. Wagner. 2009. Is the United States losing ground in science? A global perspective on the world science system. *Scientometrics* 78:23-36.0

Leydesdorff, L. and P. Zhou. 2005. Are the contributions of China and Korea upsetting the world system of science? *Scientometrics* 63:617–630.

Li, C., B. Knierim and C. Manisseri, et al. 2010. Comparison of dilute acid and ionic liquid pretreatment of switchgrass: Biomass recalcitrance, delignification and enzymatic saccharification. *Bioresource Technology* 101:4900–4906.

Li, H., S. M. Liu, X. H. Yu, S. L. Tang and C. K. Tang. 2020. Coronavirus disease 2019 (COVID-19): Current status and future perspectives. *International Journal of Antimicrobial Agents* 55:105951.

Lin, Y. and S. Tanaka. 2006. Ethanol fermentation from biomass resources: Current state and prospects. *Applied Microbiology and Biotechnology* 69:627–642.

Ma, X., L. Sun and C. Song. 2002. A new approach to deep desulfurization of gasoline, diesel fuel and jet fuel by selective adsorption for ultra-clean fuels and for fuel cell applications. *Catalysis Today* 77:107–116.

Macedo, I. C., J. E. A. Seabra and J. E. A. R. Silva. 2008. Green house gases emissions in the production and use of ethanol from sugarcane in Brazil: The 2005/2006 averages and a prediction for 2020. *Biomass and Bioenergy* 32:582–595.

Makenete, A. L., W. J. Lemmer and J. Kupka. 2008. The impact of biofuel production on food security: A briefing paper with a particular emphasis on maize-to-ethanol production. *International Food and Agribusiness Management Review* 11:101–110.

Morschbacker, A. 2009. Bio-ethanol based ethylene. *Polymer Reviews* 49:79–84.

Najafi, G., B. Ghobadian and T. Tavakoli, et al. 2009. Performance and exhaust emissions of a gasoline engine with ethanol blended gasoline fuels using artificial neural network. *Applied Energy* 86:630–639.

Negishi, M., Y. Sun and K. Shigi. 2004. Citation database for Japanese papers: A new bibliometric tool for Japanese academic society. *Scientometrics* 60:333–351.

Newman, P. W. G. and J. R. Kenworthy. 1989. Gasoline consumption and cities: A comparison of U.S. cities with a global survey. *Journal of the American Planning Association* 55:24–37.

North, D. C. 1991. Institutions. *Journal of Economic Perspectives* 5:97–112.

Olsson, L. and B. Hahn-Hagerdal. 1996. Fermentation of lignocellulosic hydrolysates for ethanol production. *Enzyme and Microbial Technology* 18:312–331.

Reeves, S. 2014. To Russia with love: How moral arguments for a humanitarian intervention in Syria opened the door for an invasion of the Ukraine. *Michigan State University International Law Review* 23:199.

Sanchez, O. J. and C. A. Cardona. 2008. Trends in biotechnological production of fuel ethanol from different feedstocks. *Bioresource Technology* 99:5270–5295.

Schmer, M. R., K. P. Vogel, R. B. Mitchell and R. K. Perrin. 2008. Net energy of cellulosic ethanol from switchgrass. *Proceedings of the National Academy of Sciences of the United States of America* 105:464–469.

Sun, Y. and J. Cheng. 2002. Hydrolysis of lignocellulosic materials for ethanol production: A review. *Bioresource Technology* 83:1–11.

Taherzadeh, M. J. and K. Karimi. 2007. Enzyme-based hydrolysis processes for ethanol from lignocellulosic materials: A review. *Bioresources* 2:707–738.

Taherzadeh, M. J. and K. Karimi. 2008. Pretreatment of lignocellulosic wastes to improve ethanol and biogas production: A review. *International Journal of Molecular Sciences* 9:1621–1651.

Tahmooresnejad, L., C. Beaudry and A. Schiffauerova. 2015. The role of public funding in nanotechnology scientific production: Where Canada stands in comparison to the United States. *Scientometrics* 102:753–787.

Wingren, A., M. Galbe and G. Zacchi. 2003. Techno-economic evaluation of producing ethanol from softwood: Comparison of SSF and SHF and identification of bottlenecks. *Biotechnology Progress* 19:1109–1117.

Winzer, C. 2012. Conceptualizing energy security. *Energy Policy* 46:36–48.

Wu, F., D. Zhang and J. Zhang. 2012. Will the development of bioenergy in China create a food security problem? Modeling with fuel ethanol as an example. *Renewable Energy* 47:127–134.

Yang, B. and C. E. Wyman. 2008. Pretreatment: The key to unlocking low-cost cellulosic ethanol. *Biofuels, Bioproducts and Biorefining* 2:26–40.

Zabed, H., G. Faruq and J. N. Sahu, et al. 2014. Bioethanol production from fermentable sugar juice. *Scientific World Journal* 2014:957102.

Zhu, J. Y. and X. J. Pan. 2010. Woody biomass pretreatment for cellulosic ethanol production: Technology and energy consumption evaluation. *Bioresource Technology* 101:4992–5002.

Zhu, J. Y., X. J. Pan and R. S. Zalesny. 2010. Pretreatment of woody biomass for biofuel production: Energy efficiency, technologies, and recalcitrance. *Applied Microbiology and Biotechnology* 87:847–857.

43 Non-Waste Biomass-based Bioethanol Fuels

Review

Ozcan Konur
(Formerly) Ankara Yildirim Beyazit University

43.1 INTRODUCTION

The crude oil-based gasoline fuels (Ma et al., 2002; Newman and Kenworthy, 1989) have been widely used in the transportation sector since the 1920s. However, there have been great public concerns over the adverse environmental and human impact of these fuels (Hill et al., 2006, 2009). Hence, biomass-based bioethanol fuels (Hill et al., 2006; Konur, 2012, 2015, 2019, 2020) have increasingly been used in blending gasoline fuels (Hsieh et al., 2002; Najafi et al., 2009), in the fuel cells (Antolini, 2007, 2009), and in the biochemical production (Angelici et al., 2013; Morschbacker, 2009) in a biorefinery context (Fernando et al., 2006; Huang et al., 2008).

However, it is necessary to pretreat the biomass (Alvira et al., 2010; Taherzadeh and Karimi, 2008) to enhance the yield of the bioethanol (Hahn-Hagerdal et al., 2006; Sanchez and Cardona, 2008) prior to the bioethanol fuel production from the feedstocks through the hydrolysis (Sun and Cheng, 2002; Taherzadeh and Karimi, 2007) and fermentation (Lin and Tanaka, 2006; Olsson and Hahn-Hagerdal, 1996) of the biomass and hydrolysates, respectively.

One of the most-studied feedstocks for the bioethanol fuels has been the non-waste biomass at large. The research in the field of the non-waste biomass-based bioethanol fuels has intensified in this context in the key research fronts of the pretreatment (Schwanninger et al., 2004; Sun et al., 2009) and hydrolysis (Kilpelainen et al., 2007; Li et al., 2010) of the non-waste biomass, fermentation (Fu et al., 2011; Larsson et al.,1999a,b) of the non-waste biomass-based hydrolysates, and bioethanol fuel production (Fu et al., 2011; Larsson et al., 1999a,b) and bioethanol fuel evaluation (Pimentel and Patzek, 2005; Schmer et al., 2008).

The research in this field has also intensified for the feedstocks of wood biomass (Larsson et al., 1999a,b; Sun et al., 2009), algal biomass (Harun et al., 2010; Ho et al., 2013), grass biomass (Li et al., 2010; Schmer et al., 2008), sugar feedstock-biomass (Macedo et al., 2008; Zabed et al., 2019), and starch feedstock-biomass (Donohoe et al., 2008; Pimentel and Patzek, 2005). Thus, it emerges as a distinctive research field, complementing the research on the second generation waste biomass-based bioethanol fuels from the agricultural residues and other wastes.

However, it is essential to develop efficient incentive structures (North, 1991) for the primary stakeholders to enhance the research in this field (Konur, 2000, 2002a,b,c, 2006a,b, 2007a,b). Although there has been a number of review papers on the non-waste biomass-based bioethanol fuels (Bai et al., 2008; Galbe and Zacchi, 2002; John et al., 2011), there has been no review of the most-cited 25 papers in this field.

Thus, this book chapter presents a review of the most-cited 25 articles in the field of the non-waste biomass-based bioethanol fuels. Then, it discusses the key findings of these highly influential papers and comments on the future research priorities in this field.

DOI: 10.1201/9781003226451-55

43.2   MATERIALS AND METHODS

The search for this study was carried out using Scopus database (Burnham, 2006) in October 2022.

As a first step for the search of the relevant literature, the keywords were selected using the most-cited first 300 population papers for each non-waste biomass. The selected keyword list was then optimized to obtain a representative sample of papers for the each research field. These seven keyword lists were then integrated to obtain the keyword list for this research field (Konur, 2023a,b,c,d,e,f,g).

As a second step, a sample dataset was used for this study. The first 25 articles with at least 416 citations each were selected for the review study. Key findings from each paper were taken from the abstracts of these papers and were discussed. Additionally, a number of brief conclusions were drawn, and a number of relevant recommendations were made to enhance the future research landscape.

43.3   RESULTS

The brief information about 25 most-cited papers with at least 416 citations each on the non-waste biomass-based bioethanol fuels is given below. The primary research fronts are the pretreatments and hydrolysis of the non-waste biomass and production and evaluation of the non-waste biomass-based bioethanol fuels with 15 and 10 highly cited papers (HCPs), respectively.

43.3.1   Non-Waste Biomass Pretreatment and Hydrolysis

The brief information about 15 most-cited papers on the pretreatments and hydrolysis of non-waste biomass with at least 417 citations each is given below (Table 43.1). There are seven and eight HCPs for the pretreatments and hydrolysis of non-waste biomass, respectively.

43.3.1.1   Pretreatments of the Non-Waste Biomass

There are seven HCPs for the pretreatments of non-waste biomass with at least 531 citations.

Schwanninger et al. (2004) studied the effects of short-time vibratory ball milling on the shape of FTIR spectra of wood and cellulose in a paper with 940 citations. They observed that this milling process had a strong influence on the shape of FTIR spectra of wood and cellulose, even if the samples were milled for only a short time. Further, the mechanical treatment itself, rather than the temperature or particle size or oxidation processes, was the main influencing factor for the observed changes in the structure of the FTIR spectra. They observed the most visible changes in the spectra of cellulose and wood at wave numbers 1,034, 1,059, 1,110, 1,162, 1,318, 1,335, 2,902 cm^{-1} and in the OH-stretching vibration region from 3,200 to 3,500 cm^{-1}. These changes were mainly associated with a decrease in the degree of crystallinity and/or a decrease in the degree of polymerization of the cellulose.

Sun et al. (2009) dissolved and partially delignified yellow pine and red oak in 1-ethyl-3-methylimidazolium acetate ([C$_2$mim]OAc) after mild grinding in a paper with 833 citations. They showed [C$_2$mim]OAc was a better solvent for wood than 1-butyl-3-methylimidazolium chloride ([C$_4$mim]Cl) and that type of wood, initial wood load, particle size, etc., affected dissolution and dissolution rates. For example, red oak dissolved better and faster than yellow pine. They obtained carbohydrate-free lignin and cellulose-rich materials by using the proper reconstitution solvents (e.g., acetone/water 1:1 v/v), and they achieved approximately 26.1% and 34.9% reductions of lignin content in the reconstituted cellulose-rich materials from pine and oak, respectively, in one dissolution/reconstitution cycle. For pine, they recovered 59% of the holocellulose in the original wood in the cellulose-rich reconstituted material, while they recovered 31% and 38% of the original lignin, respectively, as carbohydrate-free lignin and as carbohydrate-bonded lignin in the cellulose-rich materials.

TABLE 43.1
The Pretreatment and Hydrolysis of Non-Waste Biomass

No.	Papers	Biomass	Res. Fronts	Prts.	Yeasts	Parameters	Keywords	Lead Authors	Affil.	Cits
2	Schwanninger et al. (2004)	Wood	Pretreatment	Milling	Na	Pretreatment, temperature, particle size, oxidation processes, FT-IR spectra, wave numbers, crystallinity	Wood, milling	Schwanninger, Manfred 6602877236	Univ. Natr. Res. Life Sci. Austria	940
4	Li et al. (2010)	Switchgrass	Hydrolysis	Acids, IL, enzymes	Na	Pretreatment types, acid hydrolysis, delignification, saccharification, sugar yields, crystallinity, surface area, lignin content, enzymatic hydrolysis	Switchgrass, pretreatment, saccharification, recalcitrance, IL	Singh, Seema* 35264950300	Sandia Natl. Lab. USA	860
5	Sun et al. (2009)	Wood Pine, oak	Pretreatment	IL	Na	IL types, dissolution rates, wood types, delignification, cellulose recovery	Wood, delignification, IL	Rogers, Robin D. 35474829200	Univ. Alabama USA	833
6	Kilpelainen et al. (2007)	Wood Spruce, pine	Hydrolysis	IL, enzymes	Na	IL pretreatment, enzymatic hydrolysis, IL types, wood types	Wood, IL, dissolution	Kilpelainen, Ilkka 7006830888	Univ. Helsinki Finland	832
8	Lee et al. (2009)	Wood	Hydrolysis	IL, enzymes	Na	Enzymatic hydrolysis, IL pretreatment, delignification, cellulose crystallinity index, enzyme recycling	Wood, IL, hydrolysis	Dordick, Jonathan S. 7102545507	Rensselaer Polytech Inst. USA	816
9	Kumar et al. (2009)	Wood Poplar	Hydrolysis	Ammonia, acids, alkali, SO₂, enzymes	Na	Pretreatment types, enzymatic hydrolysis, cellulase adsorption capacity, biomass crystallinity, cellulose degree of polymerization	Poplar, pretreatment	Wyman, Charles E. 7004396809	Univ. Calif. Riverside USA	726
10	Fort et al. (2007)	Wood	Pretreatment	IL	Na	Wood dissolution, IL, cellulose reconstitution, delignification	Wood, IL	Rogers, Robin D. 35474829200	Univ. Alabama USA	725

(Continued)

TABLE 43.1 (*Continued*)
The Pretreatment and Hydrolysis of Non-Waste Biomass

No.	Papers	Biomass	Res. Fronts	Prts.	Yeasts	Parameters	Keywords	Lead Authors	Affil.	Cits
11	Li et al. (2007)	Wood Aspen	Pretreatment	Steam	Na	Lignin depolymerization and repolymerization, delignification	Wood, delignification	Li, Jiebing 7410069979	Rise Res. Inst. Sweden	665
13	Donohoe et al. (2008)	Corn	Pretreatment	Acids	Na	Pretreatment, lignin coalescence and migration, lignin droplets, lignin decompartmentalization and relocalization	Maize, pretreatment	Donohoe, Bryon S. 36688509400	Natl. Renew. Ener. Lab. USA	553
15	Berlin et al. (2006)	Softwood	Hydrolysis	Solvent, enzymes	Na	Enzymatic hydrolysis, lignin–enzyme interactions, enzyme selection and inhibition	Softwood, cellulase	Berlin, Alex 8639650700	Novozymes Biotech Inc. USA	544
16	Mani et al. (2004)	Switchgrass	Pretreatment	Milling	Na	Grinding performance and physical properties, biomass type, energy consumption, calorific value, ash content	Switchgrass, grinding	Tabil, Lope G. 6701349307	Univ. v Canada	531
19	Studer et al. (2011)	Wood Populus	Hydrolysis	Enzymes, LHW	Na	Enzymatic hydrolysis, sugar release and yield, S/G ratio, lignin content, glucose and xylose release	Populus, sugar	Wyman, Charles E. 7004396809	Univ. Calif. Riverside USA	469
20	Esteghlalian et al. (1997)	Poplar, switchgrass	Hydrolysis	Acids	Na	Acid hydrolysis, modeling, optimization xylose yield, xylan content	Poplar, switchgrass, pretreatment	Hashimoto, Andrew G. 7202605473	Univ. Hawaii USA	463
21	Zhu et al. (2009)	Wood Spruce, red pine	Hydrolysis	Acids, sulfite, milling, enzymes	Na	Enzymatic hydrolysis, pretreatments, glucose yield, energy consumption, fermentation inhibitors	Spruce, pine, saccharification	Zhu, Junyong 7405692678	USDA Forest serv. USA	450
24	Sun and Cheng (2005)	Bermudagrass	Hydrolysis	Acids, enzymes	Na	Enzymatic hydrolysis, pretreatment, acid content, residence time, pretreatment severity, glucose and xylose content	Bermudagrass, ethanol, pretreatment	Cheng, Jay J. 15046539600	NC State Univ. USA	417

*, Female; cits., number of citations received for each paper; na, nonavailable; prt, biomass pretreatments.

Fort et al. (2007) dissolved wood in the IL in a paper with 725 citations. They used 1-n-butyl-3-methylimidazolium chloride ([C₄mim]Cl). They presented the dissolution profiles for woods of different hardness. They also showed that cellulose could be readily reconstituted from the IL-based wood liquors in fair yields by the addition of a variety of precipitating solvents. Further, the polysaccharide obtained in this manner was virtually free of lignin and hemicellulose and had characteristics that were comparable to those of pure cellulose samples subjected to similar processing conditions.

Li et al. (2007) studied the lignin depolymerization and repolymerization and its critical role for delignification of aspen wood by steam explosion pretreatment in a paper with 665 citations. They analyzed the lignin portion and observed the competition between lignin depolymerization and repolymerization, and identified the conditions required for these two types of reaction. Further, the addition of 2-naphthol inhibited the repolymerization reaction strongly, resulting in a highly improved delignification by subsequent solvent extraction and an extracted lignin of uniform structure.

Donohoe et al. (2008) visualized lignin coalescence and migration through corn cell walls following thermochemical pretreatment in a paper with 553 citations. In order to better understand the fate of biomass lignins that remained with the solids following dilute acid pretreatment, they tracked lignins on and in biomass cell walls. They observed a range of droplet morphologies that appeared on and within cell walls of pretreated biomass as well as the specific ultrastructural regions that accumulated the droplets. These droplets contained lignin. Further, the thermochemical pretreatments reaching temperatures above the range for lignin phase transition caused lignins to coalesce into larger molten bodies that migrated within and out of the cell wall, and could redeposit on the surface of plant cell walls. In conclusion, this decompartmentalization and relocalization of lignins was likely to be at least as important as lignin removal in the quest to improve the digestibility of biomass for sugar and fuel production.

Mani et al. (2004) evaluated the grinding performance and physical properties of wheat and barley straws, corn stover, and switchgrass in a paper with 531 citations. They ground these feedstocks at two moisture contents were ground using a hammer mill with three different screen sizes (3.2, 1.6, and 0.8 mm) and measured the energy required for grinding these feedstocks. Among these four feedstocks, they found that switchgrass had the highest specific energy consumption (27.6 kW h/t), and corn stover had the least specific energy consumption (11.0 kW h/t) at 3.2 mm screen size. They then determined the physical properties of grinds such as moisture content, geometric mean diameter of grind particles, particle size distribution, and bulk and particle densities. They also developed second- or third-order polynomial models relating bulk and particle densities of grinds to geometric mean diameter within the range of 0.18–1.43 mm. Further, switchgrass had the highest calorific value and the lowest ash content among the biomass species tested.

Esteghlalian et al. (1997) modeled and optimized the dilute-sulfuric-acid pretreatment of corn stover, poplar, and switchgrass in a paper with 463 citations. They pretreated these feedstocks with dilute sulfuric acid (0.6%, 0.9% and 1.2% w/w) at relatively high temperatures (140°C, 160°C and 180°C) in a Parr batch reactor. They then modeled the hydrolysis of hemicellulose to its monomeric constituents and possible degradation of these monomers by a series of first-order reactions. They then determined the kinetic parameters of two mathematical models for predicting the percentage of xylan remaining in the substrate after pretreatment and the net xylose yield in the liquid stream using the actual acid concentration in the reactor after accounting for the neutralization effect of the substrates.

43.3.1.2 Hydrolysis of the Non-Waste Biomass

There are eight HCPs for the pretreatments of non-waste biomass with at least 417 citations. Li et al. (2010) compared the efficiency of dilute acid hydrolysis and dissolution in an ionic liquid (IL) in terms of delignification, saccharification efficiency, and saccharide yields with switchgrass in a paper with 860 citations. When subject to IL pretreatment, they observed that switchgrass exhibited reduced cellulose crystallinity, increased surface area, and decreased lignin content compared to dilute acid pretreatment. Further, the IL pretreatment enabled a significant enhancement in the rate of enzymatic hydrolysis of the cellulose component of switchgrass, with a rate increase of 16.7-fold, and a glucan yield of 96.0% obtained in 24 h. In conclusion, the IL pretreatment offered unique advantages compared to the dilute acid pretreatment process for switchgrass.

Kilpelainen et al. (2007) dissolved wood in ILs in a paper with 832 citations. They found that 1-butyl-3-methylimidazolium chloride ([Bmim]Cl) and 1-allyl-3-methylimidazolium chloride ([Amim]Cl) had good solvating power for Norway spruce sawdust and Norway spruce and Southern pine thermomechanical pulp fibers as these ILs provided solutions which permitted the complete acetylation of the wood. Alternatively, they obtained transparent amber solutions of wood when the dissolution of the same lignocellulosic samples was attempted in 1-benzyl-3-methylimidazolium chloride ([BzMim]Cl). They then digested the cellulose of the regenerated wood to glucose by a cellulase enzymatic hydrolysis treatment. Furthermore, completely acetylated wood was readily soluble in chloroform.

Lee et al. (2009) used 1-ethyl-3-methylimidazolium acetate ([Emim]CH_3COO) to extract lignin from wood flour in a paper with 816 citations. They observed that the cellulose in the pretreated wood flour became far less crystalline without undergoing solubilization. When 40% of the lignin was removed, the cellulose crystallinity index dropped below 45, resulting in >90% of the cellulose in wood flour to be hydrolyzed by *Trichoderma viride* cellulase. They then reused this IL, thereby resulting in a highly concentrated solution of chemically unmodified lignin.

Kumar et al. (2009) characterized corn stover and poplar solids resulting from leading pretreatment technologies in a paper with 726 citations. They performed ammonia fiber expansion (AFEX), ammonia recycled percolation (ARP), controlled pH, dilute acid, flowthrough, lime, and SO_2 pretreatments. They found that lime pretreatment removed the most acetyl groups from both corn stover and poplar, while AFEX removed the least, while the low pH pretreatments depolymerized cellulose and enhanced biomass crystallinity much more than higher pH approaches. Further, the lime pretreated corn stover solids and flow-through pretreated poplar solids had the highest cellulase adsorption capacity, while dilute acid pretreated corn stover solids and controlled pH pretreated poplar solids had the least. Furthermore, enzymatically extracted AFEX lignin preparations for both corn stover and poplar had the lowest cellulase adsorption capacity. SO_2 pretreated solids had the highest surface O/C ratio for poplar, while for corn stover, they observed the highest value for dilute acid pretreatment with a Parr reactor. Although dependent on pretreatment and substrate, along with changes in cross-linking and chemical changes, they reasoned that these pretreatments might also decrystallize cellulose and change the ratio of crystalline cellulose polymorphs.

Berlin et al. (2006) examined the inhibition of seven cellulase preparations, three xylanase preparations, and a β-glucosidase preparation by two purified, particulate lignin preparations derived from softwood using an organosolv pretreatment process followed by enzymatic hydrolysis in a paper with 544 citations. They observed that the two lignin preparations had similar particle sizes and surface areas but differed significantly in other physical properties and in their chemical compositions. Further, the various cellulases differed by up to 3.5-fold in their inhibition by lignin, while the xylanases showed less variability (≤1.7-fold). Of all the enzymes tested, β-glucosidase was least affected by lignin. In conclusion, the selection or engineering of enzymes with reduced lignin interaction offered an alternative means of enzyme improvement as the enzyme performance was reduced during lignocellulose hydrolysis by interaction with lignin or lignin–carbohydrate complex.

Studer et al. (2011) studied the effect of the lignin content on the sugar release in natural poplar samples in a paper with 469 citations. They selected 47 extreme phenotypes across measured lignin content and ratio of syringyl and guaiacyl units (S/G ratio). They tested these samples for total sugar release through enzymatic hydrolysis alone as well as through combined liquid hot water (LHW) pretreatment and enzymatic hydrolysis. They found that the total amount of glucan and xylan released varied widely among samples, with total sugar yields of up to 92% of the theoretical maximum. Further, there was a strong negative correlation between sugar release and lignin content for pretreated samples with an S/G ratio <2.0. For higher S/G ratios, sugar release was generally higher, and the negative influence of lignin was less pronounced. When examined separately, only glucose release was correlated with lignin content and S/G ratio in this manner, whereas xylose release depended on the S/G ratio alone. For enzymatic hydrolysis without pretreatment, sugar release increased significantly with decreasing lignin content below 20%, irrespective of the S/G ratio. Furthermore, certain samples with average lignin content and S/G ratios exhibited exceptional sugar release. In conclusion, factors beyond lignin content and S/G ratio influenced recalcitrance to sugar release.

Zhu et al. (2009) performed the sulfite pretreatment (SPORL) for robust enzymatic saccharification of spruce and red pine in a paper with 450 citations. The process consisted of sulfite pretreatment of wood chips under acidic conditions followed by mechanical size reduction using disk refining. After the SPORL pretreatment of spruce chips with 8%–10% bisulfite and 1.8%–3.7% sulfuric acid on oven dry (od) wood at 180°C for 30 min, they obtained more than 90% cellulose conversion of substrate with enzyme loading of about 14.6 FPU cellulase plus 22.5 CBU β-glucosidase per gram of od substrate after 48 h hydrolysis. Further, the glucose yield from enzymatic hydrolysis of the substrate per 100 g of untreated o.d. spruce wood (glucan content 43%) was about 37 g (excluding the dissolved glucose during pretreatment). On the other hand, the hemicellulose removal was critical as lignin sulfonation for cellulose conversion in the SPORL process. Pretreatment altered the wood chips, which reduced electric energy consumption for size reduction to about 19 Wh/kg o.d. untreated wood, or about 19 g glucose/Wh electricity. Furthermore, the SPORL produced low amounts of fermentation inhibitors, hydroxymethylfurfural (HMF), and furfural, of about 5 and 1 mg/g of untreated o.d. wood, respectively. In addition, they obtained similar results when the SPORL was applied to red pine.

Sun and Cheng (2005) performed the dilute acid pretreatment of rye straw and bermudagrass for ethanol production in a paper with 417 citations. They pretreated the biomass at a solid loading rate of 10% at 121°C with different sulfuric acid concentrations (0.6%, 0.9%, 1.2%, and 1.5%, w/w) and residence times (30, 60, and 90 min). They analyzed the total reducing sugars, arabinose, galactose, glucose, and xylose in the prehydrolysates. In addition, they hydrolyzed the solid residues by cellulases to investigate the enzymatic digestibility. With the increasing acid concentration and residence time, they observed that the amount of arabinose and galactose in the filtrates increased. However, the glucose concentration in the prehydrolysate of rye straw was not significantly influenced by the sulfuric acid concentration and residence time, but it increased in the prehydrolysate of bermudagrass with the increase of pretreatment severity. Further, the xylose concentration in the filtrates increased with the increase of sulfuric acid concentration and residence time. Most of the arabinan, galactan, and xylan in the biomass were hydrolyzed during the acid pretreatment, while the cellulose remaining in the pretreated feedstock was highly digestible by cellulases from *Trichoderma reesei*.

43.3.2 Non-Waste Biomass-based Bioethanol Production and Evaluation

There are 10 HCPs for the production and evaluation of the non-waste biomass-based bioethanol fuels with at least 416 citations each (Table 43.2). Further, there are six and four HCPs for the production and evaluation of the bioethanol fuels, respectively. As the pretreatment and hydrolysis of the non-waste biomass are the fundamental parts of the bioethanol production, these narrated papers often cover these processes too.

43.3.2.1 Non-Waste Biomass-based Bioethanol Production

There are six HCPs for the production of the non-waste biomass-based bioethanol fuels with at least 416 citations each (Table 43.2). These papers also cover the fermentation of the hydrolysates of the non-waste biomass. As the pretreatment and hydrolysis are the fundamental parts of the bioethanol production, these narrated papers often cover these processes too.

TABLE 43.2

The Production and Evaluation of Non-Waste Biomass-based Bioethanol Fuels

No.	Papers	Biomass	Res. Fronts	Prts.	Yeasts	Parameters	Keywords	Lead Authors	Affil.	Cits
1	Pimentel and Patzek (2005)	Corn, switchgrass, wood	Evaluation	Na	Na	Bioethanol energy balance, biodiesel energy balance	Ethanol, corn, switchgrass, wood	Pimentel, David P. 7005471319	Cornell Univ. USA	1020
3	Larsson et al. (1999a)	Softwood Spruce	Production	Acids	S. cerevisiae	Acid hydrolysis severity, fermentation, sugar and ethanol yield, fermentation inhibitors, volumetric productivity	Softwoo.d., fermentation, hydrolysis	Hahn-Hagerdal, Barbel* 7005389381	Lund Univ. Sweden	889
7	Schmer et al. (2008)	Switchgrass	Evaluation	Na	Na	Energy balance, GHG emissions, biomass and ethanol yield, energy yield	Switchgrass, ethanol	Vogel, Kenneth P. 7102498441	Univ. Nebraska-Lincoln USA	831
12	Macedo et al. (2008)	Sugarcane	Evaluation	Na	Na	Energy balance and GHG emissions, sugarcane productivity, ethanol yield, GHG emissions avoidance	Sugarcane, ethanol	Macedo, Isaias C. 6603142465	State Univ. Campinas Brazil	638
14	Wingren et al. (2003)	Softwood	Evaluation	enzymes	Yeasts	Fermentation, techno-economics, SSF, SHF, ethanol production costs, capital costs, ethanol yield	Softwood, ethanol	Zacchi, Guido 7006727748	Lund Univ. Sweden	549

(Continued)

TABLE 43.2 (Continued)
The Production and Evaluation of Non-Waste Biomass-based Bioethanol Fuels

No.	Papers	Biomass	Res. Fronts	Prts.	Yeasts	Parameters	Keywords	Lead Authors	Affil.	Cits
17	Fu et al. (2011)	Switchgrass	Production	Acids, enzymes	C. thermocellum	Ethanol production, biomass engineering, hydrolysis, fermentation, ethanol yield	Switchgrass, ethanol, recalcitrance	Dixon, Richard A. 7402020530	Univ. N. Texas USA	520
18	Pan et al. (2005)	Softwood	Production	Enzymes, solvent	Yeasts	Ethanol production, lignol process, SSF, hydrolysis, fermentation, cellulose hydrolysis, adhesives	Softwood, ethanol	Pan, Xuejun 57203296000	Univ. Wisconsin Madison USA	498
22	Ho et al. (2013)	Microalgae	Production	Enzymes, acids	Yeasts	Ethanol production, enzymatic and acid hydrolysis, fermentation, SSF, SHF, sugar and ethanol yield	Microalgae, bioethanol	Chang, Jo-Shu 8567368700	Tunghai Univ. Taiwan	442
23	Larsson et al. (1999b)	Wood Spruce	Production	Acids	S. cerevisiae	Fermentation, detoxification, fermentation inhibitors	Spruce, detoxification, hydrolyzates	Jonsson, Leif J. 7102349315	Umea Univ. Sweden	430
25	Harun et al. (2010)	Microalgae	Production	Na	S. bayanus	Ethanol production	Microalgal, fermentation, bioethanol	Harun, Razif 35315707300	Univ. Putra Malaysia Malaysia	416

*, Female; cits., number of citations received for each paper; na, nonavailable; prt: biomass pretreatments.

Larsson et al. (1999a) studied the effect of the combined severity (SC) of dilute sulfuric acid hydrolysis of spruce on sugar yield and the fermentability of the hydrolysate by *Saccharomyces cerevisiae* in a paper with 889 citations. When the CS of the hydrolysis conditions increased, they observed that the yield of fermentable sugars increased to a maximum between CS 2.0–2.7 for mannose and 3.0–3.4 for glucose above which it decreased. Further, the decrease in the yield of monosaccharides coincided with the maximum concentrations of furfural and 5-HMF. With the further increase in CS, the concentrations of furfural and 5-HMF decreased, while the formation of formic acid and levulinic acid increased. The yield of ethanol decreased at approximately CS 3, while the volumetric productivity decreased at lower CS. They then assayed the effect of acetic acid, formic acid, levulinic acid, furfural, and 5-HMF on fermentability in model fermentations. Ethanol yield and volumetric productivity decreased with increasing concentrations of acetic acid, formic acid, and levulinic acid. However, furfural and 5-HMF decreased the volumetric productivity but did not influence the final yield of ethanol. The decrease in volumetric productivity was more pronounced when 5-HMF was added to the fermentation, and this compound was depleted at a lower rate than furfural. Further, the inhibition observed in hydrolysates produced in higher CS could not be fully explained by the effect of the furfural, 5-HMF, acetic acid, formic acid, and levulinic acid.

Fu et al. (2011) showed that genetic modification of switchgrass could produce phenotypically normal plants that had reduced thermochemical (≤180°C), enzymatic, and microbial recalcitrance in a paper with 520 citations. They found that the downregulation of the switchgrass caffeic acid O-methyltransferase gene decreased lignin content modestly, reduced the syringyl:guaiacyl lignin monomer ratio (S/G ratio), improved forage quality, and, most importantly, increased the ethanol yield by up to 38% using conventional biomass fermentation processes. Further, the downregulated lines required less severe pretreatment and 300%–400% lower cellulase dosages for equivalent product yields using the simultaneous saccharification and fermentation (SSF) process with yeast. Furthermore, fermentation of diluted acid-pretreated transgenic switchgrass using *Clostridium thermocellum* with no added enzymes showed better product yields than obtained with unmodified switchgrass. Therefore, this apparent reduction in the recalcitrance of transgenic switchgrass had the potential to lower processing costs for ethanol fuels significantly. Alternatively, such modified transgenic switchgrass lines should yield significantly more ethanol fuels per hectare under identical process conditions.

Pan et al. (2005) produced bioethanol fuels and coproducts from softwood in a paper with 498 citations. They prepared pulps with residual lignin ranging from 6.4% to 27.4% (w/w) from mixed softwoods using the Lignol process based on aqueous ethanol organosolv extraction. They evaluated the pulps for bioconversion using enzymatic hydrolysis of the cellulose fraction to glucose and subsequent fermentation to ethanol. They observed that all pulps were readily hydrolyzed without further delignification. Further, more than 90% of the cellulose in low lignin pulps (≤18.4% residual lignin) was hydrolyzed to glucose in 48h using an enzyme loading of 20 filter paper units/g cellulose. Similarly, cellulose in a high lignin pulp (27.4% residual lignin) was hydrolyzed to >90% conversion within 48h using 40 filter paper units/g. The pulps performed well in both sequential and SSF trials indicating an absence of metabolic inhibitors. Lignin extracted during organosolv pulping of softwood was a suitable feedstock for production of lignin-based adhesives and other products due to its high purity, low molecular weight, and abundance of reactive groups. Additional coproducts might also be derived from the hemicellulose sugars and furfural recovered from the water-soluble stream.

Ho et al. (2013) evaluated the potential of using *Chlorella vulgaris* FSP-E as feedstock for bioethanol production via various hydrolysis strategies and fermentation processes in a paper with 442 citations. They found that the enzymatic hydrolysis of this biomass (containing 51% carbohydrate per dry weight) gave a glucose yield of 90.4% (or 0.461 g (g biomass)$^{-1}$). The separate hydrolysis and fermentation (SHF) and SSF processes converted the enzymatic microalgal hydrolysate into ethanol with a 79.9% and 92.3% theoretical yield, respectively. Dilute acidic hydrolysis with 1% sulfuric acid was also very effective in saccharifying this biomass, achieving a glucose yield of nearly 93.6%

from the microalgal carbohydrates at a starting biomass concentration of 50 g/L. Finally, using the acidic hydrolysate of this biomass as feedstock, the SHF process produced ethanol at a concentration of 11.7 g/L and an 87.6% theoretical yield.

Larsson et al. (1999b) compared different methods for the detoxification of lignocellulose hydrolysates of spruce to improve both cell growth and bioethanol production in a paper with 430 citations. They used a dilute-acid hydrolysate of spruce for the all detoxification methods tested. They determined the changes in the concentrations of fermentable sugars and aliphatic acids, furan derivatives, and phenolic compounds and assayed the fermentability of the detoxified hydrolysates. The applied detoxification methods included treatment with alkali (NaOH or calcium hydroxide), treatment with sulfite (0.1% [w/v] or 1% [w/v] at pH 5.5 or 10), evaporation of 10% or 90% of the initial volume; anion exchange (at pH 5.5 or 10), enzymatic detoxification with the phenoloxidase laccase, and detoxification with *Trichoderma reesei*. They found that an ion exchange at pH 5.5 or 10, treatment with laccase, treatment with calcium hydroxide, and treatment with *T. reesei* were the most efficient detoxification methods. On the other hand, the evaporation of 10% of the initial volume and treatment with 0.1% sulfite were the least efficient detoxification methods. Further, the treatment with laccase was the only detoxification method that specifically removed only phenolic compounds, while the anion exchange at pH 10 was the most efficient method for removing all three major groups of inhibitory compounds; however, it also resulted in the loss of fermentable sugars.

Harun et al. (2010) explored the suitability of *Chlorococcum* sp. as a substrate for bioethanol production via *Saccharomyces bayanus* under different fermentation conditions in a paper with 416 citations. They obtained a maximum ethanol concentration of 3.83 g/L from 10 g/L of lipid-extracted microalgal residues. In conclusion, the productivity level of ~38% (w/w) endorsed microalgae as a promising substrate for bioethanol production.

43.3.2.2 Non-Waste Biomass-based Bioethanol Evaluation

There are four HCPs for the evaluation of the non-waste biomass-based bioethanol fuels with at least 549 citations each (Table 43.2).

Pimentel and Patzek (2005) evaluated the energy balance of ethanol fuels from corn, switchgrass, and wood in comparison with biodiesel fuels from soybean and sunflower in a paper with 1,020 citations. They found that energy outputs from ethanol produced using corn, switchgrass, and wood biomass were each less than the respective fossil energy inputs. The same was true for producing biodiesel using soybeans and sunflower. However, the energy cost for producing soybean biodiesel was only slightly negative compared with ethanol production. Ethanol production using corn grains, switchgrass, and wood required 29%, 50%, and 57% more fossil energy than the ethanol fuel produced, respectively. On the other hand, biodiesel production using soybean and sunflower required 27% and 118% more fossil energy than the biodiesel fuel produced, respectively. Further, the energy yield from soy oil per hectare was far lower than the ethanol yield from corn.

Schmer et al. (2008) evaluated the net energy of ethanol fuels from switchgrass in a paper with 831 citations. They managed switchgrass in field trials of 3–9 ha on marginal cropland on ten farms across a wide precipitation and temperature gradient in the mid-continental US to determine net energy and economic costs based on known farm inputs and harvested yields. They summarized the agricultural energy input costs, biomass yield, estimated ethanol output, greenhouse gas (GHG) emissions, and net energy results. Annual biomass yields of established fields averaged 5.2–11.1 Mg/ha with a resulting average estimated net energy yield (NEY) of 60 GJ/ha/y. Switchgrass produced 540% more renewable than nonrenewable energy consumed. Switchgrass monocultures managed for high yield produced 93% more biomass yield and an equivalent estimated NEY than previous estimates from human-made prairies that received low agricultural inputs. Further, the estimated average GHG emissions from cellulosic ethanol derived from switchgrass were 94% lower than estimated GHG from gasoline. This was a baseline study that represented the genetic material and agronomic technology available for switchgrass production in 2000 and 2001, when the fields were planted.

Macedo et al. (2008) evaluated the energy balance and GHG emissions in the production and use of ethanol fuels from sugarcane in Brazil for 2005/2006 for a sample of mills processing up to 100 million tons of sugarcane per year, and proposed a conservative scenario for 2020 in a paper with 638 citations. They found that fossil energy ratio was 9.3 for 2005/2006 and might reach 11.6 in 2020 with technologies already commercial. For anhydrous ethanol production, the total GHG emissions was $436 kg\ CO_2$ eq/m^3 ethanol for 2005/2006, decreasing to $345 kg\ CO_2$ eq/m^3 in the 2020 scenario. Avoided emissions depended on the final use: for E100 use in Brazil, they were (in 2005/2006) $2,181 kg\ CO_2$ eq/m^3 ethanol, and for E25, they were $2,323 kg\ CO_2$ eq/m^3 ethanol (anhydrous). Both values would increase about 26% for the conditions assumed for 2020 mostly due to the large increase in sales of electricity surpluses. They showed the high impact of sugarcane productivity and ethanol yield variation on these balances and of sugarcane bagasse and electricity surpluses on GHG emissions avoidance.

Wingren et al. (2003) performed the technoeconomic evaluation of producing ethanol from softwood in a paper with 549 citations. They evaluated the enzymatic processes involved in the production of ethanol. They compared the SSF and SHF processes. They found that the ethanol production costs for the SSF and SHF processes were 0.57 and 0.63 USD/L (68 and 75 USD/barrel), respectively. The main reason for SSF being lower was that the capital cost was lower and the overall ethanol yield was higher. A major drawback of the SSF process is the problem with recirculation of yeast following the SSF step. Major economic improvements in both SSF and SHF could be achieved by increasing the income from the solid fuel coproduct. This was done by lowering the energy consumption in the process through running the enzymatic hydrolysis or the SSF step at a higher substrate concentration and by recycling the process streams. Running SSF with the use of 8% rather than 5% nonsoluble solid material would result in a 19% decrease in production cost. If after distillation 60% of the stillage stream was recycled back to the SSF step, the production cost would be reduced by 14%. The cumulative effect of these various improvements would result in a production cost of 0.42 USD/L (50USD/barrel) for the SSF process.

43.4   DISCUSSION

43.4.1   INTRODUCTION

The crude oil-based gasoline fuels have been widely used in the transportation sector since the 1920s. However, there have been great public concerns over the adverse environmental and human impact of these fuels. Hence, biomass-based bioethanol fuels have increasingly been used in blending gasoline and petrodiesel fuels, in the fuel cells, and in the biochemical production in a biorefinery context.

However, it is necessary to pretreat the biomass to enhance the yield of the bioethanol prior to the bioethanol fuel production from the feedstocks through the hydrolysis and fermentation of the biomass and hydrolysates, respectively.

One of the most-studied feedstocks for the bioethanol fuels has been the non-waste biomass at large. The research in the field of the non-waste biomass-based bioethanol fuels has intensified in this context in the key research fronts of the pretreatment and hydrolysis of the non-waste biomass, fermentation of the non-waste biomass-based hydrolysates, and production and evaluation of the non-waste biomass-based bioethanol fuels.

The research in this field has also intensified for the feedstocks of wood biomass, algal biomass, grass biomass, sugar feedstock-biomass such as sugarcane, and starch feedstock-biomass such as corn grains. Thus, it emerges as a distinctive research field, complementing the research on the second generation waste biomass-based bioethanol fuels from the agricultural residues and other wastes.

However, it is essential to develop efficient incentive structures for the primary stakeholders to enhance the research in this field. Although there have been a limited number of review papers for this field, there has been no review of the most-cited 25 articles in this field.

Thus, this book chapter presents a review of the most-cited 25 articles on the bioethanol fuel production and evaluation from the non-waste biomass. Then, it discusses the key findings of these highly influential papers and comments on the future research priorities in this field.

As a first step for the search of the relevant literature, the keywords were selected using the most-cited first 300 population papers for each non-waste biomass. The selected keyword list was then optimized to obtain a representative sample of papers for the each research field. These keyword lists were then integrated to obtain the keyword list for this research field (Konur, 2023a,b,c,d,e,f,g).

As a second step, a sample dataset was used for this study. The first 25 articles with at least 416 citations each were selected for the review study. Key findings from each paper were taken from the abstracts of these papers and were discussed. Additionally, a number of brief conclusions were drawn, and a number of relevant recommendations were made to enhance the future research landscape.

Information about the thematic research fronts for the sample papers in the non-waste biomass-based bioethanol fuels is given in Table 43.3. As this table shows, the most prolific research front for this field is the wood biomass feedstocks such as poplar and pine with 68% of the HCPs, followed by the grass biomass such as switchgrass and miscanthus with 28% of these HCPs. The other prolific feedstocks are the sugar feedstock such as sugarcane, starch feedstocks such as corn grains, and algal biomass such as chlamydomonas with 8%, 8%, and 4% of the HCPs, respectively.

Further, the most influential research fronts are the wood and grass biomass with 13 and 12% surplus, respectively, while the sugar feedstocks and algal biomass are the least influential feedstocks with 9% and 5% deficits, respectively.

Information about the thematic research fronts for the sample papers in the non-waste biomass-based bioethanol fuels is given in Table 43.4. As this table shows, the most prolific research fronts for this field are the pretreatment and hydrolysis of the non-waste biomass with 84 and 52% of the HCPs, respectively. The other prolific research fronts are the hydrolysate fermentation and bioethanol production and evaluation with 32%, 28%, and 16% of the sample papers, respectively. However, there is no HCP for the utilization of the bioethanol fuels. Further, the first research front is the most influential research front with 11% surplus, while biomass hydrolysis and bioethanol production are the least influential research fronts with 6% and 5% deficits, respectively.

TABLE 43.3

The Most-Prolific Thematic Research Fronts for the Non-Waste Biomass-based Bioethanol Fuels

No.	Research Fronts	N Paper (%) Review	N Paper (%) Sample	Surplus (%)	N Paper Population (%)
1	Wood-based bioethanol fuels	68	54.6	13.4	57.0
2	Grass-based bioethanol fuels	28	16.3	11.7	10.5
3	Algal biomass-based bioethanol fuels	8	13.3	−5.3	8.0
4	Starch feedstock-based bioethanol fuels	8	8.2	−0.2	14.1
5	Sugar feedstock-based bioethanol fuels	4	12.8	−8.8	11.4
	Sample size	25	196		9,820

N paper (%) sample, the number of papers in the population sample of 196 papers. N paper population: the number of papers in the population sample of 9,820 papers; N paper (%) review, the number of papers in the sample of 25 reviewed papers.

TABLE 43.4

The Most-Prolific Thematic Research Fronts for the Non-Waste Biomass-based Bioethanol Fuels

No.	Research Fronts	N Paper % review	N Paper (%) Sample	Surplus %
1	Biomass pretreatments	84	73.5	10.5
2	Biomass hydrolysis	52	58.2	−6.2
3	Bioethanol production	32	37.2	−5.2
4	Hydrolysate fermentation	28	27.6	0.4
5	Bioethanol fuel evaluation	16	13.8	2.2
6	Bioethanol fuel utilization	0	1.0	−1.0

N Paper (%) review, the number of papers in the sample of 25 reviewed papers; N paper (%) sample, the number of papers in the population sample of 196 papers.

43.4.2 NON-WASTE BIOMASS PRETREATMENT AND HYDROLYSIS

The brief information about 15 most-cited papers on the pretreatments and hydrolysis of non-waste biomass with at least 417 citations each is given below (Table 43.1). Further, there are seven and eight HCPs for the pretreatments and hydrolysis of non-waste biomass, respectively. On the other hand, it is notable that as the Table 43.4 shows, 84% and 52% of these HCPs are related to the pretreatments and hydrolysis of the non-waste biomass, respectively.

These findings show that both pretreatments and hydrolysis of the non-waste biomass are the fundamental processes for the bioethanol production from the non-waste biomass. These narrated studies highlight the importance of the pretreatment and hydrolysis processes for the production of the bioethanol fuels from the non-waste biomass with a high ethanol yield. These pretreatments, primarily enzymatic and chemical pretreatments, fractionate the non-waste biomass and enhance the enzymatic digestibility of the biomass.

43.4.2.1 Pretreatments of the Non-Waste Biomass

There are seven HCPs for the pretreatments of non-waste biomass with at least 531 citations.

Schwanninger et al. (2004) studied the effects of short-time vibratory ball milling on the shape of FTIR spectra of wood and cellulose and observed that this milling process had a strong influence on the shape of FTIR spectra of wood and cellulose. Further, Sun et al. (2009) dissolved and partially delignified yellow pine and red oak in [C$_2$mim]OAc after mild grinding and showed that this IL was a better solvent for wood than [C$_4$mim]Cl and that type of wood, initial wood load, particle size, etc., affected dissolution and dissolution rates.

Fort et al. (2007) dissolved wood in the IL and showed that cellulose could be readily reconstituted from the IL-based wood liquors in fair yields by the addition of a variety of precipitating solvents. Further, Li et al. (2007) studied the lignin depolymerization and repolymerization and its critical role for delignification of aspen wood by steam explosion pretreatment, observed the competition between lignin depolymerization and repolymerization, and identified the conditions required for these two types of reaction.

Donohoe et al. (2008) visualized lignin coalescence and migration through corn cell walls following thermochemical pretreatment and observed a range of droplet morphologies that appeared on and within cell walls of pretreated biomass as well as the specific ultrastructural regions that accumulated the droplets. Further, Mani et al. (2004) evaluated the grinding performance and physical properties of wheat and barley straws, corn stover, and switchgrass and found that switchgrass had the highest specific energy consumption at 3.2 mm screen size. Finally, Esteghlalian et al. (1997) modeled and optimized the dilute-sulfuric-acid pretreatment of corn stover, poplar, and switchgrass and determined the kinetic parameters of two mathematical models for predicting the percentage of xylan remaining in the substrate after pretreatment and the net xylose yield in the liquid stream.

43.4.2.2 Hydrolysis of the Non-Waste Biomass

There are eight HCPs for the pretreatments of non-waste biomass with at least 417 citations.

Li et al. (2010) compared the efficiency of dilute acid hydrolysis and dissolution in an IL with switchgrass and observed that switchgrass exhibited reduced cellulose crystallinity, increased surface area, and decreased lignin content compared to dilute acid pretreatment. Further, Kilpelainen et al. (2007) dissolved wood in ILs and found that [Bmim]Cl and [Amim]Cl had good solvating power for Norway spruce sawdust and Norway spruce and Southern pine thermomechanical pulp fibers.

Lee et al. (2009) used [Emim]CH$_3$COO to extract lignin from wood flour and found that cellulose in the pretreated wood flour became far less crystalline without undergoing solubilization. Further, Kumar et al. (2009) characterized corn stover and poplar solids resulting from leading pretreatment technologies and found that lime pretreatment removed the most acetyl groups from both corn stover and poplar, while AFEX removed the least while the low pH pretreatments depolymerized cellulose and enhanced biomass crystallinity much more than higher pH approaches.

Berlin et al. (2006) examined the inhibition of seven cellulase preparations, three xylanase preparations, and a β-glucosidase preparation by two purified, particulate lignin preparations derived from softwood using an organosolv pretreatment process followed by enzymatic hydrolysis. Further, Studer et al. (2011) studied the effect of the lignin content on the sugar release in natural poplar samples and found that the total amount of glucan and xylan released varied widely among samples, with total sugar yields of up to 92% of the theoretical maximum.

Zhu et al. (2009) performed the SPORL pretreatment for robust enzymatic saccharification of spruce and red pine and obtained more than 90% cellulose conversion of substrate with enzyme loading of about 14.6 FPU cellulase plus 22.5 CBU β-glucosidase per gram of o.d. substrate after 48 h hydrolysis. Further, Sun and Cheng (2005) performed the dilute acid pretreatment of rye straw and bermudagrass for ethanol production, and with the increasing acid concentration and residence time, they observed that the amount of arabinose and galactose in the filtrates increased.

43.4.3 Non-Waste Biomass-based Bioethanol Production and Evaluation

There are 10 HCPs for the production and evaluation of the non-waste biomass-based bioethanol fuels with at least 416 citations each (Table 43.2). Further, there are six and four HCPs for the production and evaluation of the bioethanol fuels, respectively. As the pretreatment and hydrolysis of the non-waste biomass are the fundamental parts of the bioethanol production, these narrated papers often cover these processes too.

It is notable that as the Table 43.4 shows, 32% and 16% of these HCPs are related to the production and evaluation of the bioethanol fuels from the non-waste biomass, respectively. However, there is no HCP on the utilization of these biofuels in diesel or gasoline engines, partially displacing diesel or gasoline fuels.

43.4.3.1 Non-Waste biomass-based Bioethanol Production

There are six HCPs for the production of the non-waste biomass-based bioethanol fuels with at least 416 citations each (Table 43.2). These papers also cover the fermentation of the hydrolysates of the non-waste biomass. As the pretreatment and hydrolysis are the fundamental parts of the bioethanol production, these narrated papers often cover these processes too. It is notable that as the Table 43.4 shows, 32% of these HCPs are related to the production of the bioethanol fuels from the non-waste biomass.

These narrated studies highlight the importance of the pretreatment (primarily chemical, enzymatic, or hydrothermal) and hydrolysis (primarily enzymatic or acid) processes as well as of the fermentation processes (SSF or SHF) on the production of the bioethanol fuels from the non-waste biomass with a high ethanol yield. Further, some fermentation studies focus on the detoxification of the lignocellulosic hydrolysates to improve the ethanol yield.

Larsson et al. (1999a) studied the effect of the combined severity (SC) of dilute sulfuric acid hydrolysis of spruce on sugar yield and the fermentability of the hydrolysate, and when the CS of the hydrolysis conditions increased, they observed that the yield of fermentable sugars increased to a maximum between CS 2.0–2.7 for mannose and 3.0–3.4 for glucose above which it decreased. Further, Fu et al. (2011) showed that genetic modification of switchgrass could produce phenotypically normal plants that had reduced thermal-chemical, enzymatic, and microbial recalcitrance.

Pan et al. (2005) produced ethanol fuels and coproducts from softwood and observed that the pulps performed well in both sequential and SSF trials indicating an absence of metabolic inhibitors. Further, Ho et al. (2013) evaluated the potential of using *C. vulgaris* FSP-E as feedstock for bioethanol production via various hydrolysis strategies and fermentation processes and found that the SHF and SSF processes converted the enzymatic microalgal hydrolysate into ethanol with a 79.9% and 92.3% theoretical yield, respectively.

Larsson et al. (1999b) compared different methods for the detoxification of lignocellulose hydrolysates of spruce to improve both cell growth and ethanol production and found that an ion exchange at pH 5.5 or 10, treatment with laccase, treatment with calcium hydroxide, and treatment with *T. reesei* were the most efficient detoxification methods. Further, Harun et al. (2010) explored the suitability of *Chlorococum* sp. as a substrate for bioethanol production via *S. bayanus* under different fermentation conditions and obtained a maximum ethanol concentration of 3.83 g/L from 10 g/L of lipid-extracted microalgal residues.

43.4.3.2 Non-Waste Biomass-based Bioethanol Evaluation

There are four HCPs for the evaluation of the non-waste biomass-based bioethanol fuels with at least 549 citations each (Table 43.2). It is notable that as the Table 43.4 shows, 16% of these HCPs are related to the evaluation of the bioethanol fuels from the non-waste biomass.

These narrated studies often focus the technoeconomics and environmental impact of the bioethanol fuels from the non-waste biomass. These technoeconomic studies show that the non-waste biomass-based ethanol fuels are cost competitive in relation to the crude oil-based gasoline and diesel fuels, which ethanol fuels partially replace as the ethanol price reached $150 per barrel in 2022 following the invasion of Ukraine by Russia.

Pimentel and Patzek (2005) evaluated the energy balance of ethanol fuels from corn, switchgrass, and wood in comparison with biodiesel fuels from soybean and sunflower in a paper with 1,020 citations and found that energy outputs from ethanol produced using corn, switchgrass, and wood biomass were each less than the respective fossil energy inputs. Further, Schmer et al. (2008) evaluated the net energy of ethanol fuels from switchgrass and found that switchgrass produced 540% more renewable than nonrenewable energy consumed.

Macedo et al. (2008) evaluated the energy balance and GHG emissions in the production and use of ethanol fuels from sugarcane in Brazil for 2005/2006 for a sample of mills processing up to 100 million tons of sugarcane per year, and proposed a conservative scenario for 2020. Further, Wingren et al. (2003) performed the technoeconomic evaluation of producing ethanol from softwood and found that the ethanol production costs for the SSF and SHF processes were 0.57 and 0.63 USD/L (68 and 75 USD/barrel), respectively.

43.5   CONCLUSION AND FUTURE RESEARCH

The brief information about the key research fronts covered by the 25 most-cited papers with at least 416 citations each is given under two primary headings: The pretreatments and hydrolysis of the non-waste biomass and production and evaluation of the bioethanol fuels.

The usual characteristics of these HCPs are that the pretreatments and hydrolysis of the non-waste biomass and fermentation of the resulting hydrolysates are the primary processes for the bioethanol fuel production from non-waste biomass to improve the ethanol yield as the non-waste biomass is one of the most studied feedstocks at large for the bioethanol production especially for the countries with the large farmlands, forests, and crude oil deficiency.

The key findings on these research fronts should be read in the light of the increasing public concerns about climate change, GHG emissions, and global warming as these concerns have been certainly behind the boom in the research on the non-waste biomass-based bioethanol fuels as an alternative to crude oil-based gasoline and petrodiesel fuels in the last decades. It is also a sustainable alternative to food crop-based bioethanol fuels such as corn grain-based bioethanol fuels. The recent supply shocks caused by the COVID-19 pandemics and the Russian invasion of Ukraine also highlight the importance of the production and utilization of the bioethanol fuels as an alternative to the crude oil-based gasoline and diesel fuels.

As Table 43.3 shows, the most-prolific research front for this field is the wood biomass feedstocks such as poplar and pine and to a lesser extent the grass biomass such as switchgrass and miscanthus, sugar feedstocks such as sugarcane, starch feedstocks such as corn grains, and algal biomass such as chlamydomonas. Further, the most influential research fronts are the wood and grass biomass, and to a lesser extent the sugar feedstocks and algal biomass.

As Table 43.4 shows, the most-prolific research fronts for this field are the pretreatment and hydrolysis of the non-waste biomass, and to a lesser extent the hydrolysate fermentation and bioethanol production and evaluation. However, there is no HCP for the utilization of the bioethanol fuels. Further, the first research front is the most influential research front, while biomass hydrolysis and bioethanol production are the least influential research fronts.

These studies emphasize the importance of proper incentive structures for the efficient production of non-waste biomass-based bioethanol fuels in the light of North's institutional framework (North, 1991). In this context, the major producers and users of bioethanol fuels such as the USA and Brazil with vast forests and farmlands have developed strong incentive structures for the efficient non-waste biomass-based bioethanol fuels. In the light of the recent supply shocks caused primarily by the COVID-19 pandemics and Russian invasion of Ukraine, it is expected that the incentive structures such as public funding would be enhanced to increase the share of bioethanol fuels in the global fuel portfolio as a strong alternative to crude oil-based gasoline and petrodiesel fuels.

In this context, it is expected that the most prolific researchers, institutions, countries, funding bodies, and journals in this field would have a first-mover advantage to benefit from such potential incentives. This is especially true for the US stakeholders as the USA has become the global leader in both the production and utilization of second generation bioethanol fuels from the non-waste biomass. It is expected the research would focus more on the algal, wood, and grass biomass-based bioethanol fuels at the expense of the starch and sugar feedstock-based bioethanol fuels due to the large societal concerns about the food security in the future.

It is recommended that further review studies are performed for the primary research fronts of the non-waste biomass-based bioethanol fuels.

ACKNOWLEDGMENTS

The contribution of the highly cited researchers in the field of the non-waste biomass-based bioethanol fuels has been gratefully acknowledged.

REFERENCES

Alvira, P., E. Tomas-Pejo, M. Ballesteros and M. J. Negro. 2010. Pretreatment technologies for an efficient bioethanol production process based on enzymatic hydrolysis: A review. *Bioresource Technology* 101:4851–4861.

Angelici, C., B. M. Weckhuysen and P. C. A. Bruijnincx. 2013. Chemocatalytic conversion of ethanol into butadiene and other bulk chemicals. *ChemSusChem* 6:1595–1614.

Antolini, E. 2007. Catalysts for direct ethanol fuel cells. *Journal of Power Sources* 170:1–12.

Antolini, E. 2009. Palladium in fuel cell catalysis. *Energy and Environmental Science* 2:915–931.

Bai, F. W., W. A. Anderson and M. Moo-Young. 2008. Ethanol fermentation technologies from sugar and starch feedstocks. *Biotechnology Advances* 26:89–105.

Berlin, A., M. Balakshin and N. Gilkes, et al. 2006. Inhibition of cellulase, xylanase and β-glucosidase activities by softwood lignin preparations. *Journal of Biotechnology* 125:198–209.

Burnham, J. F. 2006. Scopus database: A review. *Biomedical Digital Libraries* 3:1–8.

Donohoe, B. S., S. R. Decker, M. P. Tucker, M. E. Himmel and T. B. Vinzant. 2008. Visualizing lignin coalescence and migration through maize cell walls following thermochemical pretreatment. *Biotechnology and Bioengineering* 101:913–925.

Esteghlalian, A., A. G. Hashimoto, J. J. Fenske and M. H. Penner. 1997. Modeling and optimization of the dilute-sulfuric-acid pretreatment of corn stover, poplar and switchgrass. *Bioresource Technology* 59:129–136.

Fernando, S., S. Adhikari, C. Chandrapal and M. Murali. 2006. Biorefineries: Current status, challenges, and future direction. *Energy & Fuels* 20:1727–1737.

Fort, D. A., R. C. Remsing and R. P. Swatloski, et al. 2007. Can ionic liquids dissolve wood? Processing and analysis of lignocellulosic materials with 1-n-butyl-3-methylimidazolium chloride. *Green Chemistry* 9:63–69.

Fu, C., J. R. Mielenz and X. Xiao, et al. 2011. Genetic manipulation of lignin reduces recalcitrance and improves ethanol production from switchgrass. *Proceedings of the National Academy of Sciences of the United States of America* 108:3803–3808.

Galbe, M. and G. Zacchi. 2002. A review of the production of ethanol from softwood. *Applied Microbiology and Biotechnology* 59:618–628.

Hahn-Hagerdal, B., M. Galbe, M. F. Gorwa-Grauslund, G. Liden and G. Zacchi. 2006. Bio-ethanol - The fuel of tomorrow from the residues of today. *Trends in Biotechnology* 24:549–556.

Harun, R., M. K. Danquah and G. M. Forde. 2010. Microalgal biomass as a fermentation feedstock for bioethanol production. *Journal of Chemical Technology and Biotechnology* 85:199–203.

Hill, J., E. Nelson, D. Tilman, S. Polasky and D. Tiffany. 2006. Environmental, economic, and energetic costs and benefits of biodiesel and ethanol biofuels. *Proceedings of the National Academy of Sciences of the United States of America* 103:11206–11210.

Hill, J., S. Polasky and E. Nelson, et al. 2009. Climate change and health costs of air emissions from biofuels and gasoline. *Proceedings of the National Academy of Sciences of the United States of America* 106:2077–2082.

Ho, S. H., S. W. Huang and C. Y. Chen, et al. 2013. Bioethanol production using carbohydrate-rich microalgae biomass as feedstock. *Bioresource Technology* 135:191–198.

Hsieh, W. D., R. H. Chen, T. L. Wu and T. H. Lin. 2002. Engine performance and pollutant emission of an SI engine using ethanol-gasoline blended fuels. *Atmospheric Environment* 36:403–410.

Huang, H. J., S. Ramaswamy, U. W. Tschirner and B. V. Ramarao. 2008. A review of separation technologies in current and future biorefineries. *Separation and Purification Technology* 62:1–21.

John, R. P., G.S. Anisha, K. M. Nampoothiri and A. Pandey. 2011. Micro and macroalgal biomass: A renewable source for bioethanol. *Bioresource Technology* 102:186–193.

Kilpelainen, I., H. Xie and A. King, et al. 2007. Dissolution of wood in ionic liquids. *Journal of Agricultural and Food Chemistry* 55:9142–9148.

Konur, O. 2000. Creating enforceable civil rights for disabled students in higher education: An institutional theory perspective. *Disability & Society* 15:1041–1063.

Konur, O. 2002a. Access to nursing education by disabled students: Rights and duties of nursing programs. *Nurse Education Today* 22:364–374.

Konur, O. 2002b. Assessment of disabled students in higher education: Current public policy issues. *Assessment and Evaluation in Higher Education* 27:131–152.

Konur, O. 2002c. Access to employment by disabled people in the UK: Is the Disability Discrimination Act working? *International Journal of Discrimination and the Law* 5:247–279.

Konur, O. 2006a. Participation of children with dyslexia in compulsory education: Current public policy issues. *Dyslexia* 12:51–67.

Konur, O. 2006b. Teaching disabled students in Higher Education. *Teaching in Higher Education* 11:351–363.

Konur, O. 2007a. A judicial outcome analysis of the *Disability Discrimination Act*: A windfall for the employers? *Disability & Society* 22:187–204.

Konur, O. 2007b. Computer-assisted teaching and assessment of disabled students in higher education: The interface between academic standards and disability rights. *Journal of Computer Assisted Learning* 23:207–219.

Konur, O. 2012. The evaluation of the research on the bioethanol: A scientometric approach. *Energy Education Science and Technology Part A: Energy Science and Research* 28:1051–1064.

Konur, O. 2015. Current state of research on algal bioethanol. In *Marine Bioenergy: Trends and Developments*, Ed. S. K. Kim and C. G. Lee, pp. 217–244. Boca Raton, FL: CRC Press.

Konur, O. 2019. Cyanobacterial bioenergy and biofuels science and technology: A scientometric overview. In *Cyanobacteria: From Basic Science to Applications*, Ed. A. K. Mishra, D. N. Tiwari and A. N. Rai, pp. 419–442. Amsterdam: Elsevier.

Konur, O. 2020. The scientometric analysis of the research on the bioethanol production from green macroalgae. In *Handbook of Algal Science, Technology and Medicine*, Ed. O. Konur, pp. 385–401. London: Academic Press.

Konur, O. 2023a. First generation starch feedstock-based bioethanol fuel fuels: Review. In *Feedstock-based Bioethanol Fuels. I. Non-Waste Feedstocks: Starch, Sugar, Grass, Wood, Cellulose, Algae, and Biosyngas-based Bioethanol Fuels. Handbook of Bioethanol Fuels Volume 3*, Ed. O. Konur. Boca Raton, FL: CRC Press.

Konur, O. 2023b. First generation sugar feedstock-based bioethanol fuels: Review. In *Feedstock-based Bioethanol Fuels. I. Non-Waste Feedstocks: Starch, Sugar, Grass, Wood, Cellulose, Algae, and Biosyngas-based Bioethanol Fuels. Handbook of Bioethanol Fuels Volume 3*, Ed. O. Konur. Boca Raton, FL: CRC Press.

Konur, O. 2023c. Grass-based bioethanol fuel production: Scientometric study. In *Feedstock-based Bioethanol Fuels. I. Non-Waste Feedstocks: Starch, Sugar, Grass, Wood, Cellulose, Algae, and Biosyngas-based Bioethanol Fuels. Handbook of Bioethanol Fuels Volume 3*, Ed. O. Konur. Boca Raton, FL: CRC Press.

Konur, O. 2023d. Wood-based bioethanol fuel production: Scientometric study. In *Feedstock-based Bioethanol Fuels. I. Non-Waste Feedstocks: Starch, Sugar, Grass, Wood, Cellulose, Algae, and Biosyngas-based Bioethanol Fuels. Handbook of Bioethanol Fuels Volume 3*, Ed. O. Konur. Boca Raton, FL: CRC Press.

Konur, O. 2023e. Cellulose-based bioethanol fuels: Scientometric study. In *Feedstock-based Bioethanol Fuels. I. Non-Waste Feedstocks: Starch, Sugar, Grass, Wood, Cellulose, Algae, and Biosyngas-based Bioethanol Fuels. Handbook of Bioethanol Fuels Volume 3*, Ed. O. Konur. Boca Raton, FL: CRC Press.

Konur, O. 2023f. Biosyngas-based bioethanol fuels: Scientometric study. In *Feedstock-based Bioethanol Fuels. I. Non-Waste Feedstocks: Starch, Sugar, Grass, Wood, Cellulose, Algae, and Biosyngas-based Bioethanol Fuels. Handbook of Bioethanol Fuels Volume 3*, Ed. O. Konur. Boca Raton, FL: CRC Press.

Konur, O. 2023g. Third generation algal bioethanol fuels: Scientometric study. In *Feedstock-based Bioethanol Fuels. I. Non-Waste Feedstocks: Starch, Sugar, Grass, Wood, Cellulose, Algae, and Biosyngas-based Bioethanol Fuels. Handbook of Bioethanol Fuels Volume 3*, Ed. O. Konur. Boca Raton, FL: CRC Press.

Kumar, R., G. Mago, V. Balan and C. E. Wyman. 2009. Physical and chemical characterizations of corn stover and poplar solids resulting from leading pretreatment technologies. *Bioresource Technology* 100:3948–3962.

Larsson, S., E. Palmqvist and B. Hahn-Hagerdal, et al. 1999a. The generation of fermentation inhibitors during dilute acid hydrolysis of softwood. *Enzyme and Microbial Technology* 24:151–159.

Larsson, S., A. Reimann, N. O. Nilvebrant and L. J. Jonsson. 1999b. Comparison of different methods for the detoxification of lignocellulose hydrolyzates of spruce. *Applied Biochemistry and Biotechnology - Part A Enzyme Engineering and Biotechnology* 77–79:91–103.

Lee, S. H., T. V. Doherty, R. J. Linhardt and J. S. Dordick. 2009. Ionic liquid-mediated selective extraction of lignin from wood leading to enhanced enzymatic cellulose hydrolysis. *Biotechnology and Bioengineering* 102:1368–1376.

Li, C., B. Knierim and C. Manisseri, et al. 2010. Comparison of dilute acid and ionic liquid pretreatment of switchgrass: Biomass recalcitrance, delignification and enzymatic saccharification. *Bioresource Technology* 101:4900–4906.

Li, J., G. Henriksson and G. Gellerstedt. 2007. Lignin depolymerization/repolymerization and its critical role for delignification of aspen wood by steam explosion. *Bioresource Technology* 98:3061–3068.

Lin, Y. and S. Tanaka. 2006. Ethanol fermentation from biomass resources: Current state and prospects. *Applied Microbiology and Biotechnology* 69:627–642.

Ma, X., L. Sun and C. Song. 2002. A new approach to deep desulfurization of gasoline, diesel fuel and jet fuel by selective adsorption for ultra-clean fuels and for fuel cell applications. *Catalysis Today* 77:107–116.

Macedo, I. C., J. E. A. Seabra and J. E. A. R. Silva. 2008. Green house gases emissions in the production and use of ethanol from sugarcane in Brazil: The 2005/2006 averages and a prediction for 2020. *Biomass and Bioenergy* 32:582–595.

Mani, S., L. G. Tabil and S. Sokhansanj. 2004. Grinding performance and physical properties of wheat and barley straws, corn stover and switchgrass. *Biomass and Bioenergy* 27:339–352.

Morschbacker, A. 2009. Bio-ethanol based ethylene. *Polymer Reviews* 49:79–84.

Najafi, G., B. Ghobadian and T. Tavakoli, et al. 2009. Performance and exhaust emissions of a gasoline engine with ethanol blended gasoline fuels using artificial neural network. *Applied Energy* 86:630–639.

Newman, P. W. G. and J. R. Kenworthy. 1989. Gasoline consumption and cities: A comparison of U.S. cities with a global survey. *Journal of the American Planning Association* 55:24–37.

North, D. C. 1991. Institutions. *Journal of Economic Perspectives* 5:97–112.

Olsson, L. and B. Hahn-Hagerdal. 1996. Fermentation of lignocellulosic hydrolysates for ethanol production. *Enzyme and Microbial Technology* 18:312–331.

Pan, X. J., C. Arato and N. Gilkes, et al. 2005. Biorefining of softwoods using ethanol organosolv pulping: Preliminary evaluation of process streams for manufacture of fuel-grade ethanol and co-products. *Biotechnology and Bioengineering* 90:473–481.

Pimentel, D. and T. W. Patzek. 2005. Ethanol production using corn, switchgrass, and wood; biodiesel production using soybean and sunflower. *Natural Resources Research* 14:65–76.

Sanchez, O. J. and C. A. Cardona. 2008. Trends in biotechnological production of fuel ethanol from different feedstocks. *Bioresource Technology* 99:5270–5295.

Schmer, M. R., K. P. Vogel, R. B. Mitchell and R. K. Perrin. 2008. Net energy of cellulosic ethanol from switchgrass. *Proceedings of the National Academy of Sciences of the United States of America* 105:464–469.

Schwanninger, M., J. C. Rodrigues, H. Pereira and B. Hinterstoisser. 2004. Effects of short-time vibratory ball milling on the shape of FT-IR spectra of wood and cellulose. *Vibrational Spectroscopy* 36:23–40.

Studer, M. H., J. D. DeMartini and M. F. Davis, et al. 2011. Lignin content in natural populus variants affects sugar release. *Proceedings of the National Academy of Sciences of the United States of America* 108:6300–6305.

Sun, N., M. Rahman and Y. Qin, et al. 2009. Complete dissolution and partial delignification of wood in the ionic liquid 1-ethyl-3-methylimidazolium acetate. *Green Chemistry* 11:646–655.

Sun, Y. and J. Cheng. 2002. Hydrolysis of lignocellulosic materials for ethanol production: A review. *Bioresource Technology* 83:1–11.

Sun, Y. and J. J. Cheng. 2005. Dilute acid pretreatment of rye straw and bermudagrass for ethanol production. *Bioresource Technology* 96:1599–1606.

Taherzadeh, M. J. and K. Karimi. 2007. Enzyme-based hydrolysis processes for ethanol from lignocellulosic materials: A review. *Bioresources* 2:707–738.

Taherzadeh, M. J. and K. Karimi. 2008. Pretreatment of lignocellulosic wastes to improve ethanol and biogas production: A review. *International Journal of Molecular Sciences* 9:1621–1651.

Wingren, A., M. Galbe and G. Zacchi. 2003. Techno-economic evaluation of producing ethanol from softwood: Comparison of SSF and SHF and identification of bottlenecks. *Biotechnology Progress* 19:1109–1117.

Zabed, H. M., S. Akter and J. Yun, et al. 2019. Recent advances in biological pretreatment of microalgae and lignocellulosic biomass for biofuel production. *Renewable and Sustainable Energy Reviews* 105:105–128.

Zhu, J. Y., X. J. Pan, G. S. Wang and R. Gleisner. 2009. Sulfite pretreatment (SPORL) for robust enzymatic saccharification of spruce and red pine. *Bioresource Technology* 100:2411–2418.

Part 9

First Generation Starch-based
Bioethanol Fuels

44 First Generation Starch Feedstock-based Bioethanol Fuels
Scientometric Study

Ozcan Konur
(Formerly) Ankara Yildirim Beyazit University

44.1 INTRODUCTION

The crude oil-based gasoline fuels (Ma et al., 2002; Newman and Kenworthy, 1989) have been widely used in the transportation sector since the 1920s. However, there have been great public concerns over the adverse environmental and human impact of these fuels (Hill et al., 2006, 2009). Hence, biomass-based bioethanol fuels (Hill et al., 2006; Konur, 2012e, 2015, 2019, 2020a) have increasingly been used in blending gasoline fuels (Hsieh et al., 2002; Najafi et al., 2009), in the fuel cells (Antolini, 2007, 2009), and in the biochemical production (Angelici et al., 2013; Morschbacker, 2009) in a biorefinery context (Fernando et al., 2006; Huang et al., 2008).

Bioethanol fuels also play a critical role in maintaining the energy security (Kruyt et al., 2009; Winzer, 2012) in the supply shocks (Kilian, 2008, 2009) related to oil price shocks (Hamilton, 2003, 2009), COVID-19 pandemics (Fauci et al., 2020; Li et al., 2020), or wars (Hamilton, 1983; Jones, 2012) in the aftermath of the Russian invasion of Ukraine (Reeves, 2014).

However, it is necessary to pretreat the biomass (Taherzadeh and Karimi, 2008; Yang and Wyman, 2008) to enhance the yield of the bioethanol (Hahn-Hagerdal et al., 2006; Sanchez and Cardona, 2008) prior to the bioethanol production through the hydrolysis (Sun and Cheng, 2002; Taherzadeh and Karimi, 2007) and fermentation (Lin and Tanaka, 2006; Olsson and Hahn-Hagerdal, 1996) of the biomass and hydrolysates, respectively.

One of the most-studied feedstocks for the bioethanol fuels has been the starch feedstocks. The research in the field of the starch feedstock-based bioethanol fuels has intensified in this context in the key research fronts of the pretreatment of the starch feedstocks (Dien et al., 2009; Mojovic et al., 2006; Shigechi et al., 2004), hydrolysis of the starch feedstocks (Dien et al., 2009; Mojovic et al., 2006; Srichuwong et al., 2009), fermentation of the starch feedstock-based hydrolysates (Dien et al. 2009; Graves et al., 2006; Shigechi et al., 2004), bioethanol fuel production (Dien et al., 2009; Graves et al., 2006; Shigechi et al., 2004), bioethanol fuel evaluation (Hertel et al. 2010; Kwiatkowski et al., 2006; Pimentel and Patzek, 2005), and economics (Gardebroek and Hernandez, 2013; Serra et al. 2011; Trujillo-Barrera et al., 2012) of the starch feedstock-based bioethanol fuels. Further, the corn (Hertel et al. 2010; Kwiatkowski et al., 2006; Pimentel and Patzek, 2005), cassava (Dai et al., 2006; Nguyen et al., 2007), and, to a lesser extent, wheat (Dong et al., 2008), sorghum (Dien et al., 2009), potato (Srichuwong et al. 2009; Zhang et al., 2011), and artichoke (Szambelan et al. 2004) have been studied intensively in this context.

However, it is essential to develop efficient incentive structures (North, 1991) for the primary stakeholders to enhance the research in this field (Konur, 2000, 2002a,b,c, 2006a,b, 2007a,b). The scientometric analysis has been used in this context to inform the primary stakeholders

DOI: 10.1201/9781003226451-57

about the current state of the research in a selected research field (Garfield, 1955; Konur, 2011, 2012a,b,c,d,e,f,g,h,i, 2015, 2018b, 2019, 2020a).

As there have been no published scientometric studies in this field, this book chapter presents a scientometric study of the research in the starch feedstock-based bioethanol fuels. It examines the scientometric characteristics of both the sample and population data presenting scientometric characteristics of these both datasets in the order of documents, authors, publication years, institutions, funding bodies, source titles, countries, Scopus subject categories, Scopus keywords, and research fronts.

44.2 MATERIALS AND METHODS

The search for this study was carried out using Scopus database (Burnham, 2006) in July 2022.

As a first step for the search of the relevant literature, the keywords were selected using the first most-cited 200 population papers. The selected keyword list was then optimized to obtain a representative sample of papers for the searched research field. This keyword list is provided in Appendix for future replicative studies.

As a second step, two sets of data were used for this study. First, a population sample of 1,345 papers was used to examine the scientometric characteristics of the population data. Secondly, a sample of 135 most-cited papers, corresponding to 10% of the population papers, was used to examine the scientometric characteristics of these citation classics.

The scientometric characteristics of these both sample and population datasets were presented in the order of documents, authors, publication years, institutions, funding bodies, source titles, countries, Scopus subject categories, Scopus keywords, and research fronts.

Lastly, the key scientometric findings for both datasets were discussed to highlight the research landscape for the starch feedstock-based bioethanol fuels. Additionally, a number of brief conclusions were drawn and a number of relevant recommendations were made to enhance the future research landscape.

44.3 RESULTS

44.3.1 THE MOST-PROLIFIC DOCUMENTS IN THE STARCH FEEDSTOCK-BASED BIOETHANOL FUELS

The information on the types of documents for both datasets is given in Table 44.1. The articles and conference papers, published in journals, dominate both the sample (95%) and population (93%) papers as they are over-represented in the sample papers by 2%. Further, review papers and short surveys have a 2% surplus as they are over-represented in the sample papers as they constitute 4% and 2% of the sample and population papers, respectively.

It is further notable that 96% of the population papers were published in journals while 2% of them each were published in books and book series. Similarly, 99% and 1% of the sample papers were published in the journals and books, respectively.

44.3.2 THE MOST-PROLIFIC AUTHORS IN THE STARCH FEEDSTOCK-BASED BIOETHANOL FUELS

The information about the most-prolific nine authors with at least 3% of sample papers each is given in Table 44.2. The most-prolific authors are W. Michael Ingledew, Akihiko Kondo, Kolothumannil C. Thomas, and Bruce E. Dale with 3.7% of the sample papers each.

On the other hand, the most influential authors are Kolothumannil C. Thomas and Bruce E. Dale with 3.1% surplus each followed by Akihikio Kondo with 2.9% surplus.

The most-prolific institutions for the sample dataset are the Kobe University, Michigan State University, and University of Saskatchewan with two authors each. In total, six institutions house these top authors. On the other hand, the most-prolific country for the sample dataset is Japan with three authors, followed by Canada and the USA with two authors each. In total, only five countries house these top authors.

TABLE 44.1
Documents in the Starch Feedstock-based Bioethanol Fuels

Documents	Sample Dataset (%)	Population Dataset (%)	Surplus (%)
Article	93.3	89.7	3.6
Review	3.7	1.9	1.8
Conference paper	1.5	2.8	−1.3
Book chapter	0.7	2.4	−1.7
Short survey	0.7	0.5	0.2
Letter	0.0	1.3	−1.3
Note	0.0	1.1	−1.1
Book	0.0	0.2	−0.2
Editorial	0.0	0.1	−0.1
Sample size	135	1,345	

Population dataset: the number of papers (%) in the set of the 1,345 population papers; sample dataset: the number of papers (%) in the set of 135 highly cited papers.

TABLE 44.2
Most-Prolific Authors in the Starch Feedstock-based Bioethanol Fuels

No.	Author Name	Author Code	Sample Papers (%)	Population Papers (%)	Surplus	Institution	Country	HI	N	Res. Front
1	Ingledew, W. Michael	7005864542	3.7	1.0	2.7	Univ. Saskatchewan	Canada	37	97	C, W
2	Kondo, Akihiko	57203868143	3.7	0.8	2.9	Kobe Univ.	Japan	78	797	C, S
3	Thomas, Kolothumannil C.	7402627881	3.7	0.6	3.1	Univ. Saskatchewan	Canada	23	27	W
4	Dale, Bruce E.	7201511969	3.7	0.6	3.1	Michigan State Univ.	USA	90	429	C
5	Rakin, Marica*	55903431700	3.0	0.7	2.3	Univ. Belgrade	Serbia	18	60	C
6	Ueda, Mitsuyoshi	7403944406	3.0	0.6	2.4	Kyoto Univ.	Japan	48	409	S
7	Fukuda, Hideki	55425022800	3.0	0.5	2.5	Kobe Univ.	Japan	56	222	C, S
8	Kim, Seungdo	35771061000	3.0	0.4	2.6	Michigan State Univ.	USA	27	57	C
9	Wang, Chengtao	35231632300	3.0	0.3	2.7	Shanghai Jiao Tong Univ.	China	29	311	B

*: female; author code: the unique code given by Scopus to the authors; B: cassava; C: corn; population papers: the number of papers authored in the population dataset; S: starch; sample papers: the number of papers authored in the sample dataset; W: wheat.

There are three primary research fronts for these top authors: bioethanol fuels based on corn, wheat, and starch in general with six, two, and three authors, respectively. On the other hand, there is significant gender deficit (Beaudry and Lariviere, 2016) for the sample dataset as surprisingly only one of these top researchers is female with a representation rate of 11%.

Additionally, there are other authors with the relatively low citation impact and with 0.5%–1.3% of the population papers each: Vijay Singh, Kevin B. Hicks, Argyrios Margaritis, Ljiljana Mojovic, Svetlana Nikolic, Sung-Keun Rhee, Joseph P. Guiraud, Gordon A. Hill, Nhuan J. Nghiem, Dusanka Pejin, David Pimentel, Fengwu Bai, Jennifer B. Dunn, Pierre Galzy, and Chul Ho Kim.

44.3.3 THE MOST-PROLIFIC RESEARCH OUTPUT BY YEARS IN THE STARCH FEEDSTOCK-BASED BIOETHANOL FUELS

Information about papers published between 1970 and 2022 is given in Figure 44.1. This figure clearly shows that the bulk of the research papers in the population dataset were published primarily in the 2000s, 2010s, and the early 2020s with 21%, 46%, and 12% of the population dataset, respectively. Similarly, the publication rates for the 1990s, 1980s, and 1970s were 10%, 8%, and 1% respectively. Additionally, 2% of the population papers were published in the pre-1970s.

Similarly, the bulk of the research papers in the sample dataset were published in the 2000s and 2010s with 55% and 31% of the sample dataset, respectively. Similarly, the publication rates for the 1990s, 1980s, and 1970s were 9, 4, and 0% of the sample papers, respectively.

The most-prolific publication years for the population dataset were 2013, 2010, and 2011 with 6.5%, 5.7%, and 5.9% of the dataset, respectively. Further, 68% of the population papers were published between 2007 and 2022. Similarly, 81% of the sample papers were published between 2003 and 2016 while the most-prolific publication years were 2009, 2008, and 2010 with 15%, 13%, and 11% of the sample papers, respectively.

It is notable that the number of population papers had a rising trend between 2004 and 2011; thereafter, it lost its momentum. Further, there was no sharp rise in the research output for the population papers in 2020 and 2021 due to the supply shocks.

44.3.4 THE MOST-PROLIFIC INSTITUTIONS IN THE STARCH FEEDSTOCK-BASED BIOETHANOL FUELS

Information about the most-prolific 19 institutions publishing papers on the starch feedstock-based bioethanol fuels with at least 2.2% of the sample papers each is given in Table 44.3.

The most-prolific institutions are the USDA Agricultural Research Service, University of Saskatchewan, Michigan State University, and University of California Berkeley with 4.4% of the sample papers each. The other prolific institutions are the Iowa State University, Chinese Academy of Sciences, Shanghai Jiao Tong University, and Kobe University with 3.7% of the sample papers each.

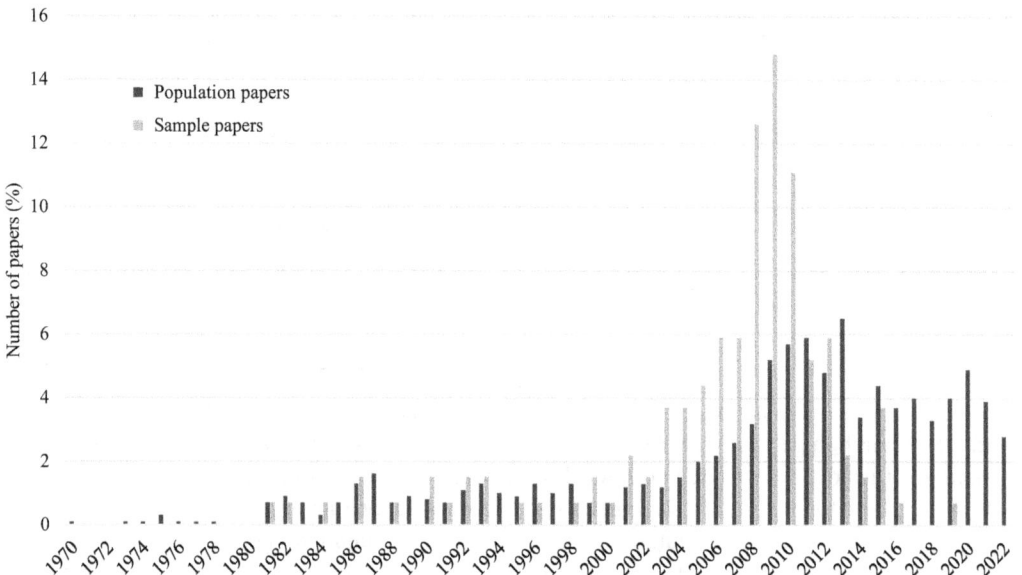

FIGURE 44.1 The research output by years regarding the starch feedstock-based bioethanol fuels.

TABLE 44.3
The Most-Prolific Institutions in the Starch Feedstock-based Bioethanol Fuels

No.	Institutions	Country	Sample Papers (%)	Population Papers (%)	Surplus (%)
1	USDA Agr. Res. Serv.	USA	4.4	2.9	1.5
2	Univ. Saskatchewan	Canada	4.4	2.1	2.3
3	Michigan State Univ.	USA	4.4	1.1	3.3
4	Univ. Calif. Berkeley	USA	4.4	1.0	3.4
5	Iowa State Univ.	USA	3.7	2.4	1.3
6	Chinese Acad. Sci.	China	3.7	1.2	2.5
7	Shanghai Jiao Tong Univ.	China	3.7	1.0	2.7
8	Kobe Univ.	Japan	3.7	0.9	2.8
9	Kyoto Univ.	Japan	3.0	0.7	2.3
10	Iowa State Univ.	USA	3.0	3.0	0.0
11	Univ. Belgrade	Serbia	3.0	1.1	1.9
12	Argonne National Lab.	USA	3.0	0.9	2.1
13	Kansas State Univ.	USA	3.0	0.9	2.1
14	NREL	USA	3.0	0.7	2.3
15	Univ. Tsukuba	Japan	2.2	0.7	1.5
16	King Mongkut's Univ. Technol.	Thailand	2.2	0.4	1.8
17	Carnegie Mellon Univ.	USA	2.2	0.3	1.9
18	Univ. Novi Sad	Serbia	2.2	0.9	1.3
19	State Univ. Campinas	Brazil	2.2	0.7	1.5

Similarly, the top country for these most prolific institutions is the USA with nine institutions. The other prolific countries are Japan, Serbia, and China. In total, only seven countries house these top institutions.

On the other hand, the institution with the most citation impact is the University of California Berkeley with 3.4% surplus, followed by the Michigan State University with 3.3% surplus. The other influential institutions are the Kobe University, Shanghai Jiao Tong University, and Chinese Academy of Sciences with 2.5%–2.8% surplus each.

Additionally, there are other institutions with the relatively low citation impact and with 0.7%–1.3% of the population papers each: Jiangnan University, Western University, Wageningen University & Research, University of Manchester, Dalian University of Technology, Cornell University, University of Sao Paulo, Purdue University, University of Montpellier, University of California Davis, Food Research Institute, University of Putra Malaysia, University of Minnesota Twin Cities, CCS Haryana Agricultural University, University of Nebraska–Lincoln, and University of Life Sciences in Poznan.

44.3.5 THE MOST-PROLIFIC FUNDING BODIES IN THE STARCH FEEDSTOCK-BASED BIOETHANOL FUELS

Information about the most-prolific 11 funding bodies funding at least 1.5% of the sample papers each is given in Table 44.4. Further, only 36% and 30% of the sample and population papers were funded, respectively.

The most-prolific funding bodies are the U.S. Department of Energy and National Natural Science Foundation of China with 5.2% of the sample papers each. The other prolific funding bodies are National Science Foundation, Ministry of Education, Culture, Sports, Science and Technology, and European Commission with 3% of the sample papers each. Further, the National Natural Science Foundation of China is the largest funder of the population papers with a funding rate of 4.5%.

TABLE 44.4

The Most-Prolific Funding Bodies in the Starch Feedstock-based Bioethanol Fuels

No.	Funding bodies	Country	Sample Paper No. (%)	Population Paper No. (%)	Surplus (%)
1	U.S. Department of Energy	USA	5.2	2.2	3.0
2	National Natural Science Foundation of China	China	5.2	4.5	0.7
3	National Science Foundation	USA	3.0	0.9	2.1
4	Ministry of Education, Culture, Sports, Science, and Technology	Japan	3.0	1.0	2.0
5	European Commission	EU	3.0	1.1	1.9
6	Natural Sciences and Engineering Research Council of Canada	Canada	2.2	1.4	0.8
7	Chinese Academy of Sciences	China	2.2	0.6	1.6
8	Ministry of Agriculture, Forestry, and Fisheries	Japan	1.5	0.4	1.1
9	Ministry of Agriculture of China	China	1.5	0.2	1.3
10	Government of West Bengal	India	1.5	0.1	1.4
11	Government of Canada	Canada	1.5	1.0	0.5

On the other hand, the most-prolific country for these top funding bodies is China with three funding bodies, followed by Canada, Japan, and the USA with two funding bodies each. In total, only five countries and the EU house these top funding bodies.

The funding body with the most citation impact is the U.S. Department of Energy with 3% surplus, followed by the National Science Foundation, Ministry of Education, Culture, Sports, Science and Technology, and the European Commission with 1.9%–2.1% surplus, respectively.

The other funding bodies with the relatively low citation impact and with 0.4%–1.7% of the population papers each are the U.S. Department of Agriculture, CAPES Foundation, Ministry of Science and Technology of China, National Council for Scientific and Technological Development, National Key Research and Development Program of China, Office of Science, National Institute of Food and Agriculture, Japan Society for the Promotion of Science, European Regional Development Fund, Fundamental Research Funds for the Central Universities, Sao Paulo Research Foundation, Ministry of Education of China, National High-tech Research and Development Program, Higher Education Discipline Innovation Project, Priority Academic Program Development of Jiangsu Higher Education Institutions, Board for International Development Cooperation, China Postdoctoral Science Foundation, Ministry of Higher Education Malaysia, National Research Foundation of Korea, and Office of Energy Efficiency and Renewable Energy.

44.3.6 THE MOST-PROLIFIC SOURCE TITLES IN THE STARCH FEEDSTOCK-BASED BIOETHANOL FUELS

Information about the most-prolific 16 source titles publishing at least 2.2% of the sample papers each in the starch feedstock-based bioethanol fuels is given in Table 44.5.

The most-prolific source titles are the Bioresource Technology and Biotechnology and Bioengineering with 6.7% of the sample papers each. The other prolific titles are the Biomass and Bioenergy, Applied Microbiology and Biotechnology, and Biochemical Engineering Journal with 5.2% of the sample papers each.

On the other hand, the source title with the most citation impact is the Biochemical Engineering Journal with 4.2% surplus. The other influential titles are the Applied Microbiology and Biotechnology, Applied and Environmental Microbiology, Fuel, Biotechnology and Bioengineering, and Energy Policy with 3.3%–3.6% surplus each.

TABLE 44.5
The Most-Prolific Source Titles in the Starch Feedstock-based Bioethanol Fuels

No.	Source Titles	Sample Papers (%)	Population Papers (%)	Surplus (%)
1	Bioresource Technology	6.7	4.6	2.1
2	Biotechnology and Bioengineering	6.7	3.4	3.3
3	Biomass and Bioenergy	5.2	2.6	2.6
4	Applied Microbiology and Biotechnology	5.2	1.6	3.6
5	Biochemical Engineering Journal	5.2	1.0	4.2
6	Energy Policy	4.4	1.1	3.3
7	Applied and Environmental Microbiology	4.4	0.9	3.5
8	Fuel	4.4	0.9	3.5
9	Applied Biochemistry and Biotechnology	2.2	2.0	0.2
10	Biotechnology for Biofuels	2.2	1.6	0.6
11	Process Biochemistry	2.2	1.6	0.6
12	Enzyme and Microbial Technology	2.2	1.2	1.0
13	Journal of Bioscience and Bioengineering	2.2	0.9	1.3
14	Journal of Industrial Microbiology and Biotechnology	2.2	0.9	1.3
15	Environmental Science and Technology	2.2	0.5	1.7
16	International Journal of Life Cycle Assessment	2.2	0.4	1.8

The other source titles with the relatively low citation impact with 0.6%–2.5% of the population papers each are the Biotechnology Letters, Applied Biochemistry and Biotechnology Part A Enzyme Engineering and Biotechnology, Bioenergy Research, Bioscience Biotechnology and Biochemistry, Journal of Chemical Technology and Biotechnology, Fermentation, World Journal of Microbiology and Biotechnology, Chemical Engineering Transactions, Energy, Journal of Cleaner Production, African Journal of Biotechnology, Biotechnology Progress, Journal of Applied Microbiology, Biofuels Bioproducts and Biorefining, Biotechnology and Bioprocess Engineering, World Journal of Microbiology Biotechnology, Biofuels, Energy Economics, Processes, Renewable and Sustainable Energy Reviews, and Renewable Energy.

44.3.7 THE MOST-PROLIFIC COUNTRIES IN THE STARCH FEEDSTOCK-BASED BIOETHANOL FUELS

Information about the most-prolific 11 countries publishing at least 2.2% of sample papers each in the starch feedstock-based bioethanol fuels is given in Table 44.6.

The most-prolific country is the USA with 45% of the sample papers, followed by China, Japan, and Canada with 13%, 11%, and 8% of the sample papers, respectively. Further, four European countries listed in 44.6 produce 10% and 7% of the sample and population papers, respectively with 3% surplus.

On the other hand, the country with the most citation impact is the USA with 20% surplus. The other influential countries are Japan, Canada, Serbia, and Thailand with 1.9%–4.1% surplus each. Similarly, the countries with the least citation impact are China and India with 0.6% deficit each.

Additionally, there are other countries with relatively low citation impact and with 0.6%–3.6% of the sample papers each: South Korea, the UK, Nigeria, Malaysia, France, Germany, Spain, Indonesia, Italy, Turkey, Iran, Taiwan, Sweden, Australia, South Africa, Mexico, Pakistan, Argentina, Colombia, Israel, Portugal, Belgium, Greece, and Tanzania.

44.3.8 THE MOST-PROLIFIC SCOPUS SUBJECT CATEGORIES IN THE STARCH FEEDSTOCK-BASED BIOETHANOL FUELS

Information about the most-prolific ten Scopus subject categories indexing at least 3.7% of the sample papers each is given in Table 44.7.

TABLE 44.6
The Most-Prolific Countries in the Starch Feedstock-based Bioethanol Fuels

No.	Countries	Sample Papers (%)	Population Papers (%)	Surplus (%)
1	USA	45.2	25.2	20.0
2	China	12.6	13.2	−0.6
3	Japan	11.1	7.0	4.1
4	Canada	8.1	5.4	2.7
5	India	5.9	6.5	−0.6
6	Brazil	5.2	5.1	0.1
7	Thailand	5.2	3.3	1.9
8	Serbia	3.7	1.5	2.2
9	Poland	2.2	2.2	0.0
10	Netherlands	2.2	1.9	0.3
11	Denmark	2.2	1.1	1.1

TABLE 44.7
The Most-Prolific Scopus Subject Categories in the Starch Feedstock-based Bioethanol Fuels

No.	Scopus Subject Categories	Sample Papers (%)	Population Papers (%)	Surplus (%)
1	Environmental Science	45.2	27.0	18.2
2	Biochemistry, Genetics, and Molecular Biology	43.7	40.8	2.9
3	Chemical Engineering	43.0	36.7	6.3
4	Immunology and Microbiology	36.3	31.5	4.8
5	Energy	32.6	23.8	8.8
6	Agricultural and Biological Sciences	17.8	18.2	−0.4
7	Engineering	11.9	11.7	0.2
8	Chemistry	8.9	14.5	−5.6
9	Economics, Econometrics, and Finance	4.4	3.0	1.4
10	Social Sciences	3.7	3.1	0.6

The most-prolific Scopus subject category in the starch feedstock-based bioethanol fuels is Environmental Science with 45% of the sample papers, followed by Biochemistry, Genetics and Molecular Biology, Chemical Engineering, Immunology and Microbiology, and Energy with 44%, 43%, 36%, and 33% of the sample papers, respectively. It is notable that Social Sciences and Economics, Econometrics, and Finance account for 4% of the sample studies each.

On the other hand, the Scopus subject category with the most citation impact is the Environmental Science with 18% surplus, followed by Energy, Chemical Engineering, and Immunology and Microbiology with 9%, 6%, and 5% surplus, respectively. Similarly, the least influential subject category is Chemistry with 6% deficit.

44.3.9 THE MOST-PROLIFIC KEYWORDS IN THE STARCH FEEDSTOCK-BASED BIOETHANOL FUELS

Information about the Scopus keywords used with at least 5.9% or 4.2% of the sample or population papers, respectively, is given in Table 44.8. For this purpose, keywords related to the keyword set given in Appendix are selected from a list of the most-prolific keyword set provided by Scopus database.

TABLE 44.8

The Most-Prolific Keywords in the Starch Feedstock-based Bioethanol Fuels

No.	Keywords	Sample Papers (%)	Population Papers (%)	Surplus (%)
1	Biomass and biomass constituents			
	Zea	40.7	18.3	22.4
	Starch	29.6	22.2	7.4
	Maize	20.0	11.1	8.9
	Manihot	18.5	8.7	9.8
	Corn	17.8	7.4	10.4
	Cassava	14.8	9.0	5.8
	Biomass	12.6	10.3	2.3
	Cellulose	7.4	4.8	2.6
	Wheat	5.9	5.1	0.8
	Solanum	5.9	3.2	2.7
	Triticum	5.2	4.6	0.6
	Sorghum	4.4	4.5	−0.1
	Rice		6.0	−6.0
2	Fermentation			
	Fermentation	46.7	42.8	3.9
	Saccharomyces	37.8	20.7	17.1
	Yeast	28.9	18.5	10.4
	Fungi	10.4	7.2	3.2
	Ethanol fermentation	3.7	5.1	−1.4
3	Hydrolysis and hydrolysates			
	Hydrolysis	17.0	18.9	−1.9
	Saccharification	17.0	13.0	4.0
	Sugar	17.0	11.7	5.3
	Glucose	14.1	11.8	2.3
	Enzymes	13.3	12.0	1.3
	Enzyme activity	11.9	12.0	−0.1
	Enzymatic hydrolysis		5.6	−5.6
4	Products			
	Ethanol	82.2	54.3	27.9
	Alcohol	44.4	26.6	17.8
	Biofuels	40.0	27.4	12.6
	Alcohol production	20.7	10.6	10.1
	Bioethanol	17.8	18.1	−0.3
	Ethanol production	11.9	13.4	−1.5
	Corn–ethanol	6.7	4.8	1.9
	Ethanol yield	6.7	3.9	2.8
	Bio-ethanol production	5.9	6.4	−0.5
5	Evaluation			
	Greenhouse gases	23.0	9.7	13.3
	USA	17.4	5.0	12.4
	Life cycle assessment	16.5	5.5	11.0
	Life cycle	11.9	4.9	7.0

(Continued)

TABLE 44.8 (*Continued*)
The Most-Prolific Keywords in the Starch Feedstock-based Bioethanol Fuels

No.	Keywords	Sample Papers (%)	Population Papers (%)	Surplus (%)
	Gas emissions	11.9	4.0	7.9
	Life cycle analysis	8.3	3.3	5.0
	Land use	8.3	2.9	5.4
	Costs	7.3	3.8	3.5
	Carbon dioxide	7.3	3.8	3.5
	Gasoline	7.3	3.5	3.8
	Environmental impact	7.3	3.2	4.1
	Energy efficiency	7.3	3.0	4.3
	Energy balance	7.3	2.2	5.1
	Global warming	7.3	2.2	5.1

These keywords are grouped under the five headings: biomass, fermentation, hydrolysis and hydrolysates, products, and evaluation.

The most-prolific keywords related to the biomass and biomass constituents are zea, starch, maize, manihot, and corn with 18%–41% of the sample papers each. It is notable there are three prolific keywords related to corn, zea, corn, and maize. Similarly, manihot and cassava are the keywords for cassava while triticum and wheat are the keywords for wheat. Further, solanum is a keyword for potato.

Further, the prolific keywords related to the fermentation are fermentation, saccharomyces, and yeast with 29%–47% of the sample papers each.

The most-prolific keywords related to the hydrolysis and hydrolysates are hydrolysis, saccharification, and sugar with 17% of the sample papers each. Further, the most-prolific keywords related to the products are ethanol, alcohol, biofuels, and alcohol production with 21%–82% of the sample papers each. Finally, the most-prolific keywords related to the evaluation of the bioethanol fuels are greenhouse gases, the USA, and life cycle assessment with 17%–23% of the sample papers each.

On the other hand, the most influential keywords are ethanol, zea, alcohol, saccharomyces, greenhouse gases, biofuels, the USA, life cycle assessment, yeast, corn, and alcohol production with 10%–32% surplus each. Similarly, the most-prolific keywords across all of the research fronts are ethanol, fermentation, alcohol, zea, biofuels, saccharomyces, starch, yeast, greenhouse gases, alcohol production, maize, and manihot with 21%–82% of the sample papers each.

44.3.10 THE MOST-PROLIFIC RESEARCH FRONTS IN THE STARCH FEEDSTOCK-BASED BIOETHANOL FUELS

Information about the research fronts for the sample papers in the starch feedstock-based bioethanol fuels with regard to the starch feedstocks used for the bioethanol production is given in Table 44.9.

As this table shows, the most-prolific starch feedstock is corn with 51% of the sample papers followed by cassava and starch in general with 17% and 12% of the sample papers, respectively. The other prolific feedstocks are wheat, potato, sorghum, and artichoke with 5%–7% of the sample papers, each.

Information about the thematic research fronts for the sample papers in the starch feedstock-based bioethanol fuels is given in Table 44.10. As this table shows, there are five primary research fronts: biomass pretreatments and hydrolysis, hydrolysate fermentation, and production and evaluation of starch feedstock-based bioethanol fuels with 37%, 30%, 42%, 54%, and 39% of the sample papers, respectively.

For the research front of starch feedstock pretreatments, the primary research front is the enzymatic pretreatments of the starch feedstocks with 29% of the sample papers. The most-prolific feedstock used for all these research fronts is corn.

TABLE 44.9

The Most-Prolific Research Fronts For The Starch Feedstock-based Bioethanol Fuels

No.	Research fronts	N Paper (%) Sample
1	Corn	51.1
2	Cassava	17.0
3	Starch	11.9
4	Wheat	7.4
5	Potato	6.7
6	Sorghum	6.7
7	Artichoke	5.2
8	Other starch	3.0
9	Rice	2.2
10	Sago	2.2
11	Triticale	1.5

N paper (%) sample: the number of papers in the population sample of 135 papers.

TABLE 44.10

The Most-Prolific Thematic Research Fronts for the Starch Feedstock-based Bioethanol Fuels

No.	Research Fronts	N Paper (%) Sample
1	Biomass pretreatments	37.0
	Enzymatic pretreatments	28.9
	Chemical pretreatments	4.4
	Feedstock genetic engineering	4.4
	Mechanical pretreatments	4.4
	Yeast microbial engineering	3.0
2	Biomass hydrolysis	29.6
	Enzymatic hydrolysis	28.9
	Acid hydrolysis	0.7
3	Hydrolysate fermentation	42.2
4	Bioethanol production	54.1
5	Bioethanol fuel evaluation	39.3

N paper (%) sample: the number of papers in the population sample of 135 papers.

44.4 DISCUSSION

44.4.1 INTRODUCTION

The crude oil-based gasoline fuels have been widely used in the transportation sector since the 1920s. However, there have been great public concerns over the adverse environmental and human impact of these fuels. Hence, biomass-based bioethanol fuels have increasingly been used in blending gasoline fuels, in the fuel cells, and in the biochemical production in a biorefinery context.

However, it is necessary to pretreat the biomass to enhance the yield of the bioethanol prior to the bioethanol production through the hydrolysis and fermentation. One of the most-studied feedstocks for the bioethanol fuels has been the starch feedstock. The research in the field of the starch feedstock-based bioethanol fuels has intensified in this context in the key research fronts of the pretreatment of the starch feedstocks, hydrolysis of the starch feedstocks, fermentation of the starch feedstock-based hydrolysates, and production, economics, and evaluation of the starch feedstock-based bioethanol fuels. Further, the corn, cassava, and to a lesser extent wheat, sorghum, potato, and artichoke have been studied intensively in this context.

However, it is essential to develop efficient incentive structures for the primary stakeholders to enhance the research in this field. This is especially important to maintain energy security in the cases of supply shocks such as oil shocks, war-related chocks as in the case of Russian invasion of Ukraine or COVID-19 shocks.

The scientometric analysis has been used in this context to inform the primary stakeholders about the current state of the research in a selected research field. As there has been no scientometric study in this field, this book chapter presents a scientometric study of the research in the starch feedstock-based bioethanol fuels. It examines the scientometric characteristics of both the sample and population data presenting scientometric characteristics of these both datasets in the order of documents, authors, publication years, institutions, funding bodies, source titles, countries, Scopus subject categories, Scopus keywords, and research fronts.

As a first step for the search of the relevant literature, the keywords were selected using the first most-cited 200 papers. The selected keyword list was then optimized to obtain a representative sample of papers for the searched research field. A copy of this extended keyword list was provided in Appendix for future replicative studies. Further, a selected list of the keywords is presented in Table 44.8.

As a second step, two sets of data were used for this study. First, a population sample of 1,345 papers was used to examine the scientometric characteristics of the population data. Secondly, a sample of 135 most-cited papers, corresponding to 10% of the population dataset, was used to examine the scientometric characteristics of these citation classics.

The scientometric characteristics of these sample and population datasets were presented in the order of documents, authors, publication years, institutions, funding bodies, source titles, countries, Scopus subject categories, Scopus keywords, and research fronts.

Lastly, the key scientometric findings for both datasets were discussed to highlight the research landscape for starch feedstock-based bioethanol fuels. Additionally, a number of brief conclusions were drawn and a number of relevant recommendations were made to enhance the future research landscape.

44.4.2 The Most-Prolific Documents in the Starch Feedstock-based Bioethanol Fuels

Articles (together with conference papers) dominate both the sample (95%) and population (93%) papers (Table 44.1). Further, review papers and articles have 2% surplus each. The representation of the reviews in the sample papers is relatively modest (4%).

Scopus differs from the Web of Science database in differentiating and showing articles (93%) and conference papers (2%) published in the journals separately. However, it should be noted that these conference papers are also published in journals as articles, compared to those published only in the conference proceedings. Hence, a total number of articles and review papers in the sample dataset are 95% and 4%, respectively.

It is observed during the search process that there has been inconsistency in the classification of the documents in Scopus as well as in other databases such as Web of Science. This is especially relevant for the classification of papers as reviews or articles as the papers not involving a literature review may be erroneously classified as a review paper. There is also a case of review papers being classified as articles. For example, the total number of the reviews in the sample dataset was manually found as nearly 7% compared to 4% as indexed by Scopus, increasing the number of articles and conference papers to 93% for the sample dataset. The close examination of these papers shows that many evaluative studies such as technoeconomic or life cycle studies have often been indexed as the review papers by the Scopus database.

In this context, it would be helpful to provide a classification note for the published papers in the books and journals at the first instance. It would also be helpful to use the document types listed in Table 44.1 for this purpose. Book chapters may also be classified as articles or reviews as an additional classification to differentiate review chapters from the experimental chapters as it is done by the Web of Science. It would be further helpful to additionally classify the conference papers as articles or review papers as well as it is done in the Web of Science database.

44.4.3 The Most-Prolific Authors in the Starch Feedstock-based Bioethanol Fuels

There have been most-prolific nine authors with at least 3% of the sample papers each as given in Table 44.2. These authors have shaped the development of the research in this field.

The most-prolific authors are W. Michael Ingledew, Akihiko Kondo, Kolothumannil C. Thomas, and Bruce E. Dale.

It is important to note the inconsistencies in indexing of the author names in Scopus and other databases. It is especially an issue for the names with more than two components such as 'Blake Sam de Hyun Kondo'. The probable outcomes are 'Kondo, B.S.D.H.', 'de Hyun Kondo, B.S.', or 'Hyun Kondo, B.S.D.'. The first choice is the gold standard of the publishing sector as the last word in the name is taken as the last name. In most of the academic databases such as PUBMED and EBSCO databases, this version is used predominantly. The second choice is a strong alternative while the last choice is an undesired outcome as two last words are taken as the last name. It is good practice to combine the words of the last name by a hyphen: 'Hyun-Kondo, B.S.D.'. It is notable that inconsistent indexing of the author names may cause substantial inefficiencies in the search process for the papers as well as allocating credit to the authors as there are different author entries for each outcome in the databases.

There are also inconsistencies in the shortening Chinese names. For example, 'YangYing Wang' is often shortened as 'Wang, Y.', 'Wang, Y.-Y.', and 'Wang, Y.Y.' as it is done in the Web of Science database as well. However, the gold standard in this case is 'Wang, Y' where the last word is taken as the last name and the first word is taken as a single forename. Nevertheless, it makes sense to use the third option to differentiate Chinese names efficiently: 'Wang, Y.Y.'. In most of the academic databases such as PUBMED and EBSCO, this first version is used predominantly. Therefore, there have been difficulties in locating papers for the Chinese authors. In such cases, the use of the unique author codes provided for each author by the Scopus database has been helpful.

There is also a difficulty in allowing credit for the authors especially for the authors with common names such as 'Wang, X.' in conducting scientometric studies. These difficulties strongly influence the efficiency of the scientometric studies as well as allocating credit to the authors as there are the same author entries for different authors with the same name, for example, 'Wang, X.' in the databases.

In this context, the coding of authors in Scopus database is a welcome innovation compared to the other databases such as Web of Science. In this process, Scopus allocates a unique number to each author in the database (Aman, 2018). However, there might still be substantial inefficiencies in this coding system especially for common names. For example, some of the papers for a certain author may be allocated to another researcher with a different author code. It is possible that Scopus uses a number of software programs to differentiate the author names and the program may not be false-proof (Kim, 2018).

In this context, it does not help that author names are not given in full in some journals and books. This makes difficult to differentiate authors with common names and makes the scientometric studies further difficult in the author domain. Therefore, the author names should be given in all books and journals at the first instance. There is also a cultural issue where some authors do not use their full names in their papers. Instead they use initials for their forenames: 'Kondo, H.J.', 'Kondo', 'Kondo, H.', or 'Kondo, J.' instead of 'Kondo, Hyun Jae'.

There are also inconsistencies in naming of the authors with more than two components by the authors themselves in journal papers and book chapters. For example, 'Kondo, A. P. C.' might be given as 'Kondo, A.' or 'Kondo, A.C.' or 'Kondo, A.P.' or 'Kondo, C' in the journals and books. This also makes the scientometric studies difficult in the author domain. Hence, contributing authors should use their name consistently in their publications.

The other critical issue regarding the author names is the inconsistencies in the spelling of the author names in the national spellings (e.g., Göğüşçağla, Gökçe) rather than in the English spellings (e.g., Goguscagla, Gokce) in Scopus database. Scopus differs from the Web of Science database and many other databases in this respect where the author names are given only in the English spellings. It is observed that national spellings of the author names do not help much in conducting scientometric studies as well in allocating credits to the authors as sometimes there are the different author entries for the English and National spellings in the Scopus database.

The most prolific institutions for the sample dataset are the Kobe University, Michigan State University, and University of Saskatchewan. Further, the most prolific country for the sample dataset is Japan and to a lesser extent Canada and the USA. These findings confirm the dominance of Japan and to a lesser extent Canada and the USA in this field. The primary research fronts are the bioethanol fuels based on corn, wheat, and starch.

It is also notable that there is significant gender deficit for the sample dataset as surprisingly with a representation rate of 11%. This finding is the most thought-provoking with strong public policy implications. Hence, institutions, funding bodies, and policy-makers should take efficient measures to reduce the gender deficit in this field as well as other scientific fields with strong gender deficit. In this context, it is worth to note the level of representation of the researchers from the minority groups in science on the basis of race, sexuality, age, and disability, besides the gender (Blankenship, 1993; Dirth and Branscombe, 2017; Konur, 2000, 2002a,b,c, 2006a,b, 2007a,b).

44.4.4 THE MOST-PROLIFIC RESEARCH OUTPUT BY YEARS IN THE STARCH FEEDSTOCK-BASED BIOETHANOL FUELS

The research output observed between 1970 and 2022 is illustrated in Figure 44.1. This figure clearly shows that the bulk of the research papers in the population dataset were published primarily in the 2000s, 2010s, and early 2020s. Similarly, the bulk of the research papers in the sample dataset were published in the 2000s and 2010s.

These findings suggest that the most-prolific sample and population papers were primarily published in the 2010s. These are the thought-provoking findings as there has been significant research boom in since 2004. In this context, the increasing public concerns about climate change (Change, 2007), greenhouse gas emissions (Carlson et al., 2017), and global warming (Kerr, 2007) have been certainly behind the boom in the research in this field in the last two decades. Furthermore, the supply shocks experiences due to the COVID-19 pandemics might also be behind the research boom in this field since 2019.

It is notable that the number of population papers had a rising trend between 2004 and 2011; thereafter, it lost its momentum. Further, there was no sharp rise in the research output for the population papers in 2020 and 2021 due to the supply shocks.

Based on these findings, the size of the population papers likely to more than double in the current decade provided that the public concerns about climate change, greenhouse gas emissions, and global warming, as well as the supply shocks are translated efficiently to the research funding in this field.

44.4.5   THE MOST-PROLIFIC INSTITUTIONS IN THE STARCH FEEDSTOCK-BASED BIOETHANOL FUELS

The most-prolific 19 institutions publishing papers on the starch feedstock-based bioethanol fuels with at least 2.2% of the sample papers each given in Table 44.3 have shaped the development of the research in this field.

The most-prolific institutions are the USDA Agricultural Research Service, University of Saskatchewan, Michigan State University, University of California Berkeley, and to a lesser extent the Iowa State University, Chinese Academy of Sciences, Shanghai Jiao Tong University, and Kobe University. Similarly, the top countries for these most-prolific institutions are the USA and to a lesser extent Japan, Serbia, and China. In total, only seven countries house these top institutions.

On the other hand, the institutions with the most citation impact are the University of California Berkeley, Michigan State University and to a lesser extent Kobe University, Shanghai Jiao Tong University, and Chinese Academy of Sciences. These findings confirm the dominance of the U.S. and to a lesser extent of Japanese and Chinese institutions for these HCPs.

44.4.6   THE MOST-PROLIFIC FUNDING BODIES IN THE STARCH FEEDSTOCK-BASED BIOETHANOL FUELS

The most-prolific 11 funding bodies funding at least 1.5% of the sample papers each is given in Table 44.4. It is notable that only 36% and 30% of the sample and population papers were funded, respectively.

The most-prolific funding bodies are the U.S. Department of Energy, National Natural Science Foundation of China and to a lesser extent the National Science Foundation, Ministry of Education, Culture, Sports, Science and Technology, and European Commission. The most-prolific countries for these top funding bodies are China and to a lesser extent Canada, Japan, and the USA.

These findings on the funding of the research in this field suggest that the level of the funding, mostly since 2004, is not highly intensive, and nevertheless, it has been largely instrumental in enhancing the research in this field (Ebadi and Schiffauerova, 2016) in light of North's institutional framework (North, 1991). It is also notable that the funding rate in this field is inferior to those in the other research fronts of the bioethanol fuels such as algal bioethanol fuels. Further, it is expected that this modest funding rate would increase in light of the recent supply shocks. Further, it emerges that the USA and China have heavily funded the research on the corn- and cassava-based bioethanol fuels, respectively. However, the falling trend in the research output for the population papers raises questions about the funding after 2011.

44.4.7   THE MOST-PROLIFIC SOURCE TITLES IN THE STARCH FEEDSTOCK-BASED BIOETHANOL FUELS

The most-prolific 16 source titles publishing at least 2.2% of the sample papers each in the starch feed-stock-based bioethanol fuels have shaped the development of the research in this field (Table 44.5).

The most-prolific source titles are the Bioresource Technology, Biotechnology and Bioengineering and to a lesser extent the Biomass and Bioenergy, Applied Microbiology and Biotechnology, and Biochemical Engineering Journal. On the other hand, the source titles with the most citation impact are the Biochemical Engineering Journal and to a lesser extent the Applied Microbiology and Biotechnology, Applied and Environmental Microbiology, Fuel, Biotechnology and Bioengineering, and Energy Policy.

It is notable that these top source titles are primarily related to the bioresources, energy, biotechnology, and microbiology. This finding suggests that Bioresource Technology, Biotechnology and Bioengineering, and the other prolific journals in these fields have significantly shaped the development of the research in this field as they focus primarily on the starch feedstock-based bioethanol fuels with a high yield. In this context, the influence of the top two journals is quite extraordinary with 14% of the sample papers, in total. It is also notable that the energy-related journals have also published papers in the areas of technoeconomics, environmental impact, land use change, economics, and labor relations as the Social Science-related journals.

44.4.8 THE MOST-PROLIFIC COUNTRIES IN THE STARCH FEEDSTOCK-BASED BIOETHANOL FUELS

The most-prolific 11 countries publishing at least 2.2% of the sample papers each have significantly shaped the development of the research in this field (Table 44.6).

The most-prolific countries are the USA and to a lesser extent China, Japan, and Canada. Further, four European countries listed in 44.6 produce 10% and 7% of the sample and population papers, respectively, with 3% surplus.

On the other hand, the countries with the most citation impact are the USA and to a lesser extent Japan, Canada, Serbia, and Thailand. Similarly, the countries with the least impact are China and India.

The close examination of these findings suggests that the USA and, to a lesser extent, China, Japan, Canada, and Europe are the major producers of the research in this field. It is a fact that the USA has been a major player in science (Leydesdorff and Wagner, 2009). The USA has further developed a strong research infrastructure to support its corn- and grass-based bioethanol industry (Gillon, 2010). The USA has been a major producer and user of the corn grain-based bioethanol fuels.

However, China has been a rising mega star in scientific research in competition with the USA and Europe (Leydesdorff and Zhou, 2005). China is also a major player in this field as a major producer of bioethanol (Fang et al., 2010).

Next, Europe has been a persistent player in the scientific research in competition with both the USA and China (Leydesdorff, 2000). Europe has also been a persistent producer of bioethanol along with the USA and Brazil (Gnansounou, 2010).

44.4.9 THE MOST-PROLIFIC SCOPUS SUBJECT CATEGORIES IN THE STARCH FEEDSTOCK-BASED BIOETHANOL FUELS

The most-prolific ten Scopus subject categories indexing at least 3.7% of the sample papers each, respectively, given in Table 44.7 have shaped the development of the research in this field.

The most-prolific Scopus subject categories in the starch feedstock-based bioethanol fuels are Environmental Science and to a lesser extent Biochemistry, Genetics and Molecular Biology, Chemical Engineering, Immunology and Microbiology, and Energy. It is also notable that Social Sciences including Economics and Business account have a relatively strong presence in both sample and population studies.

On the other hand, the Scopus subject categories with the most citation impact are Environmental Science and to a lesser extent Energy, Chemical Engineering, and Immunology and Microbiology. Similarly, the least influential subject category is Chemistry.

These findings are thought provoking suggesting that the primary subject categories are related to energy, environmental science, chemical engineering, genetics, and microbiology as the core of the research in this field concerns with the starch feedstock-based bioethanol fuels. The other finding is that Social Sciences are unusually well represented in both the sample and population papers contrary to the most fields in bioethanol fuels. The social and economic studies account for the field of Social Sciences.

44.4.10 THE MOST PROLIFIC KEYWORDS IN THE STARCH FEEDSTOCK-BASED BIOETHANOL FUELS

A limited number of keywords have shaped the development of the research in this field as shown in Table 44.8 and Appendix. These keywords are grouped under the five headings: biomass, fermentation, hydrolysis and hydrolysates, products, and evaluation.

The most prolific keywords across all of the research fronts are ethanol, fermentation, alcohol, zea, biofuels, saccharomyces, starch, yeast, greenhouse gases, alcohol production, maize, and mani- hot. Similarly, the most influential keywords are ethanol, zea, alcohol, saccharomyces, greenhouse gases, biofuels, the USA, life cycle assessment, yeast, corn, and alcohol production.

These findings suggest that it is necessary to determine the keyword set carefully to locate the relevant research in each of these research fronts. Additionally, the size of the samples for each keyword highlights the intensity of the research in the relevant research areas. These findings also highlight different spelling of some strategic keywords: corn v. maize v. zea, wheat v. triticum, rice v. oryza, barley v. hordeum, rye v. secale, cassava v. manihot, potato v. ipomea v. solanum, and artichoke v. helianthus. It seems both Latin and English forms of these keywords are used by the authors.

44.4.11 The Most-Prolific Research Fronts in the Starch Feedstock-based Bioethanol Fuels

Information about the research fronts for the sample papers in the starch feedstock-based bioethanol fuels with regard to the starch feedstocks used for the bioethanol production is given in Table 44.9. As this table shows, the most-prolific starch feedstock is corn, followed by cassava and starch. The other prolific feedstocks are wheat, potato, sorghum, and artichoke.

Information about the thematic research fronts for the sample papers in the starch feedstock-based bioethanol fuels is given in Table 44.10. As this table shows, there are five primary research fronts: biomass pretreatments and hydrolysis, hydrolysate fermentation, and production and evalu- ation of starch feedstock-based bioethanol fuels. For the research front of starch feedstock pretreat- ments, the primary research front is the enzymatic pretreatments, and the most-prolific feedstock used for all these research fronts is corn.

These findings are thought-provoking in seeking ways to increase starch feedstock-based bio- ethanol yield at the global scale. It is clear that all of these research fronts have public importance and merit substantial funding and other incentives. Further, it is notable that starch feedstock-based bioethanol fuels have become a core unit of the bioethanol research to make it more competitive with the crude oil-based gasoline and diesel fuels, especially for the USA and the other countries with vast farmlands.

In comparison to the other feedstock-based research fronts, it is notable that the production and evaluation of the bioethanol fuels emerge as a primary research front for this field. This suggests that the primary stakeholders have been primarily interested in the production and evaluation of the starch feedstock-based bioethanol fuels such as technoeconomics, life cycle, economics, Social Science, land use, labor, and environment-related studies as a case study for the bioethanol fuels together with algal and corn feedstocks in this field. In this context, the USA and China have been the global leaders in the production and use of the corn- and cassava-based bioethanol fuels.

In the end, these most-cited papers in this field hint that the efficiency of starch feedstock-based bioethanol fuels could be optimized using the structure, processing, and property relationships of these starch feedstocks such as corn, cassava, wheat, potato, sorghum, and artichoke in the fronts of the feedstock pretreatment and hydrolysis, and hydrolysate fermentation (Formela et al., 2016; Konur, 2018a, 2020b, 2021a,b,c,d; Konur and Matthews, 1989).

44.5 CONCLUSION AND FUTURE RESEARCH

The research on the starch feedstock-based bioethanol fuels has been mapped through a scientomet- ric study of both sample (135 papers) and population (1,345 papers) datasets.

The critical issue in this study has been to obtain a representative sample of the research as in any other scientometric study. Therefore, the keyword set has been carefully devised and optimized after a number of runs in the Scopus database. It is a representative sample of the wider population studies. This keyword set was provided in Appendix, and the relevant keywords are presented in Table 44.8. However, it should be noted that it has been very difficult to compile a representative keyword set since this research field has been connected closely with many other fields. Therefore, it has been necessary to compile a keyword list to exclude papers concerned with the other research fields.

It is notable in this context that the research on the production of bioethanol fuels from wastes of the starch feedstocks such as corn stover and wheat straw is closely related to the research on the bioethanol production from these feedstocks themselves, such as corn, cassava, and wheat. Therefore, it is crucial to collect data on these two interconnected research fronts separately. Hence, the studies on the production and evaluation of the wastes of the starch feedstocks for the bioethanol production were presented separately in another section in this volume.

The other issue has been the selection of a multidisciplinary database to carry out the sciento-metric study of the research in this field. For this purpose, Scopus database has been selected. The journal coverage of this database has been notably wider than that of Web of Science and other multi-subject databases.

The key scientometric properties of the research in this field have been determined and discussed in this book chapter. It is evident that a limited number of documents, authors, institutions, publica-tion years, institutions, funding bodies, source titles, countries, Scopus subject categories, Scopus keywords, and research fronts have shaped the development of the research in this field.

There is ample scope to increase the efficiency of the scientometric studies in this field in the author and document domains by developing consistent policies and practices in both domains across all the academic databases. In this respect, it seems that authors, journals, and academic databases have a lot to do. Furthermore, the significant gender deficit as in most scientific fields emerges as a public policy issue. The potential deficits on the basis of age, race, disability, and sexu-ality need also to be explored in this field as in other scientific fields.

The research in this field has boomed since 2004, possibly promoted by the public concerns on global warming, greenhouse gas emissions, and climate change. Furthermore, the recent COVID-19 pandemics and Russian invasion of Ukraine have resulted in a global supply shock shifting the focus of the stakeholders from the crude oil-based fuels to biomass-based fuels such as bioethanol fuels. It is expected that there would be further incentives for the key stakeholders to carry out the research for the starch feedstock-based bioethanol fuels to increase the ethanol yield and to make it more competitive with the crude oil-based gasoline and diesel fuels. This might be truer for the crude oil- and foreign exchange-deficient countries to maintain the energy security at the face of the global supply shocks. However, the stagnation of research output after 2011 raises questions about its cause.

The relatively modest funding rate of 36% and 30% for the sample and population papers, respec-tively, suggests that funding in this field significantly enhanced the research in this field primarily since 2004, possibly more than doubling in the current decade. However, it is evident that there is ample room for more funding and other incentives to enhance the research in this field further as these funding rates are not relatively intensive especially in light of the stagnation of research output after 2011.

The institutions from the USA and to a lesser extent Japan, Serbia, and China have mostly shaped the research in this field. Further, the USA and to a lesser extent China, Japan, Canada, and Europe have been the major producers of the research in this field as the major producers and users of bio-ethanol fuels from different types of biomass such as corn, sugarcane, and grass as well as other types of biomass. It is evident that these countries have well-developed research infrastructure in bioethanol fuels and their derivatives. It is also notable all of these major countries have access to the large farmlands.

It emerges that ethanol is more popular than bioethanol as a keyword with strong implications for the search strategy. In other words, the search strategy using only bioethanol keyword would not be much helpful. The Scopus keywords are grouped under the five headings: biomass, fermentation, hydrolysis and hydrolysates, products, and evaluation. Further, it seems that both Latin and English words for the starch feedstocks are used by the authors again with implications for the search strategy.

As Table 44.9 shows, the most-prolific starch feedstock is corn with 51% of the sample papers followed by cassava and starch feedstocks in general. The other prolific feedstocks are wheat, potato, sorghum, and artichoke. On the other hand, Table 44.10 shows that there are five primary thematic research fronts: biomass pretreatments and hydrolysis, hydrolysate fermentation, and production and evaluation of starch feedstock-based bioethanol fuels.

These findings are thought-provoking in seeking ways to increase bioethanol yield through the starch feedstock-based bioethanol fuels at the global scale. It is clear that all of these research fronts have public importance and merit substantial funding and other incentives. Further, it is notable that starch feedstock-based bioethanol fuels, as a first generation biofuels, have become a core unit of the bioethanol research to make it more competitive with the crude oil-based gasoline and diesel fuels, especially for the countries with large access to the farmlands.

In comparison to the other feedstock-based research fronts, it is notable that production and evaluation of the bioethanol fuels from starch feedstocks emerge as primary research fronts for this field. This suggests that the primary stakeholders have been primarily interested in the evaluation of the starch feedstock-based bioethanol fuels such as technoeconomics, life cycle, economics, Social Science, land use, labor, and environmental impact-related Social Science-based interdisciplinary studies in this field. In this context, the USA and China have been the global leaders in the production and use of the corn- and cassava-based bioethanol fuels.

It should also be noted that as there have been great concerns about the starch feedstock-based bioethanol fuels with regard to the food security (Wu et al., 2012), the research has focused more on the bioethanol fuels from the residual starch feedstocks such as corn stover and wheat straw in recent years with three time stronger sample, compared to the bioethanol fuels from the primary starch feedstocks.

Thus, the scientometric analysis has a great potential to gain valuable insights into the evolution of the research in this field as in other scientific fields especially in the aftermath of the significant global supply shocks such as COVID-19 pandemics and the Russian invasion of Ukraine.

It is recommended that further scientometric studies are carried out for the primary research fronts. It is further recommended that reviews of the most-cited papers are carried out for each primary research front to complement these scientometric studies. Next, the scientometric studies of the hot papers in these primary fields are carried out.

ACKNOWLEDGMENTS

The contribution of the highly cited researchers in the field of the starch feedstock-based bioethanol fuels has been gratefully acknowledged.

APPENDIX: THE KEYWORD SET FOR STARCH FEEDSTOCK-BASED BIOETHANOL FUELS

(((((TITLE (ethanol OR bioethanol) OR TITLE (saccharification OR *hydrolysis OR recalcitrance OR hydrolysate* OR hydrolyzate* OR ferment* OR coferment* OR delignification OR depolymerization OR microwave* OR ultrasound OR grinding OR pretreat* OR {pre-treat*} OR bioorganosolve OR steam* OR {hot water} OR 'hot compressed' OR organosolv* OR afex* OR 'dry grind' OR 'ball milling')) AND TITLE (wheat OR triticum OR corn OR zea OR barley OR hordeum OR rye OR secale OR millet OR sorghum OR penisetum OR potato OR cassava OR manihot OR ipomea OR maize OR sago OR artichoke OR helianthus OR rice OR triticale)) OR (TITLE (ethanol OR bioethanol) AND TITLE (starch OR inulin))) AND NOT (TITLE (cultivar* OR straw OR stover* OR stems OR food* OR toxic* OR germ OR drying OR bran OR protein* OR oil OR shps

OR *hydrogen OR wastewater OR dough OR silage OR 'sweet sorghum' OR diversity OR soil* OR hull* OR nano* OR films OR gluten OR *waste OR peel* OR residu* OR bagasse OR lactic OR xylitol OR gruel OR *cobs OR *cob OR solubles OR ddgs OR *products OR stillage OR properties OR cake OR glumate* OR *cellulosic OR *cellulose OR corngrass* OR inhibitory OR duckweed OR lyase OR liquor OR carboxylic OR deoxynivalenol OR anti* OR succinic OR koko OR butyric OR pulp OR bread OR *toxin OR baking OR butaned* OR liquefact* OR stalks OR flower* OR biogas OR gasification OR garri OR anaerobic OR coproduct OR styrene OR nutrient* OR fiber* OR fibre* OR density OR beer OR diet* OR kisra OR zizania OR durum OR 'distillers' grains' OR rumen OR dehydration OR root OR *butanol OR lipid* OR safety OR seed* OR vitro OR cereal* OR phenolic OR *turonase OR sheep OR sourdough OR flint OR *methane OR beverage OR amarantin OR *char OR acetate OR crantz OR vinegar OR husk* OR *sorption OR nutraceut* OR 21332 OR mortar OR orbital OR two OR fructan* OR solid OR *diesel OR spirit OR furfural OR quinoa OR wine OR *sorber OR lovastatin OR steep OR rheol* OR propan* OR propionic OR itaconic OR hampas OR pigment* OR color* OR *sorbents OR leaves OR sensor OR levulinate OR cel5a) OR SRCTITLE (food* OR animal* OR poultry OR tree* OR postharvest OR carbohydrate* OR mycor* OR materials OR soil* OR biomacro* OR polymer* OR cereal* OR starch OR dairy OR botan* OR nutrit* OR macromol* OR aqua* OR hydrogen OR photo* OR drying OR {crop science} OR phyto* OR rheol* OR lwt* OR brew* OR livestock* OR agr* OR hort* OR anti* OR euphy* OR oil* OR chromat* OR pest* OR zoo* OR analy* OR *sphere OR data OR fibers OR seed* OR store* OR *plasma OR cement* OR pyrolysis OR molecular OR genom* OR metabol* OR asabe OR water OR geo* OR instr* OR sensing OR medi* OR endo* OR alcohol OR biosensors OR insect* OR soft OR elect* OR rice OR potato OR grass* OR plant* OR compounds OR product OR ento-mol*) OR SUBJAREA (medi OR vete OR nurs OR phar OR heal OR neur OR dent OR psyc))) OR ((TITLE (ethanol OR bioethanol) AND TITLE (*diesel OR *hydrogen OR h2 OR *butanol OR bio-gas OR *methane OR biorefinery OR anaerobic OR oil OR lactic OR food OR value OR cellulosic OR prices) AND TITLE (wheat OR triticum OR corn OR zea OR barley OR hordeum OR rye OR secale OR millet OR sorghum OR penisetum OR potato OR cassava OR manihot OR ipomea OR maize OR sago OR artichoke OR helianthus OR rice OR triticale)) AND NOT (TITLE (straw OR lignocellulosic OR stover OR husk OR cobs OR waste* OR bran OR bagasse OR stalk OR sludge OR stillage OR pulp OR sweet OR residue*) OR SRCTITLE (anal* OR data OR plant OR fluid OR oil* OR alcohol OR materials OR animal OR {crop science}) OR SUBJAREA (medi OR phar OR nurs OR heal)))) AND (LIMIT-TO (SRCTYPE, 'j') OR LIMIT-TO (SRCTYPE, 'b') OR LIMIT-TO (SRCTYPE, 'k')) AND (LIMIT-TO (DOCTYPE, 'ar') OR LIMIT-TO (DOCTYPE, 'cp') OR LIMIT-TO (DOCTYPE, 're') OR LIMIT-TO (DOCTYPE, 'ch') OR LIMIT-TO (DOCTYPE, 'no') OR LIMIT-TO (DOCTYPE, 'le') OR LIMIT-TO (DOCTYPE, 'sh') OR LIMIT-TO (DOCTYPE, 'bk') OR LIMIT-TO (DOCTYPE, 'cr') OR LIMIT-TO (DOCTYPE, 'ed')) AND (LIMIT-TO (LANGUAGE, 'English'))

REFERENCES

Aman, V. 2018. Does the Scopus author ID suffice to track scientific international mobility? A case study based on Leibniz laureates. *Scientometrics* 117:705–720.

Angelici, C., B. M. Weckhuysen and P. C. A. Bruijnincx. 2013. Chemocatalytic conversion of ethanol into butadiene and other bulk chemicals. *ChemSusChem* 6:1595–1614.

Antolini, E. 2007. Catalysts for direct ethanol fuel cells. *Journal of Power Sources* 170:1–12.

Antolini, E. 2009. Palladium in fuel cell catalysis. *Energy and Environmental Science* 2:915–931.

Beaudry, C. and V. Lariviere. 2016. Which gender gap? Factors affecting researchers' scientific impact in science and medicine. *Research Policy* 45:1790–1817.

Blankenship, K. M. 1993. Bringing gender and race in: US employment discrimination policy. *Gender & Society* 7:204–226.

Burnham, J. F. 2006. Scopus database: A review. *Biomedical Digital Libraries* 3:1–8.

Carlson, K. M., J. S. Gerber and D. Mueller, et al. 2017. Greenhouse gas emissions intensity of global croplands. *Nature Climate Change* 7:63–68.

Change, C. 2007. Climate change impacts, adaptation and vulnerability. *Science of the Total Environment* 326:95–112.

Dai, D., Z. Hu, G. Pu, H. Li and C. Wang. 2006. Energy efficiency and potentials of cassava fuel ethanol in Guangxi region of China. *Energy Conversion and Management* 47:1686–1699.

Dien, B. S., G. Sarath and J. F. Pedersen, et al. 2009. Improved sugar conversion and ethanol yield for forage sorghum (*Sorghum bicolor* L. Moench) lines with reduced lignin contents. *Bioenergy Research* 2:153–164.

Dirth, T. P. and N. R. Branscombe. 2017. Disability models affect disability policy support through awareness of structural discrimination. *Journal of Social Issues* 73:413–442.

Dong, X., S. Ulgiati, M. Yan, X. Zhang and W. Gao. 2008. Energy and eMergy evaluation of bioethanol production from wheat in Henan Province, China. Energy Policy 36:3882–3892.

Ebadi, A. and A. Schiffauerova. 2016. How to boost scientific production? A statistical analysis of research funding and other influencing factors. *Scientometrics* 106:1093–1116.

Fang, X., Y. Shen, J. Zhao, X. Bao and Y. Qu. 2010. Status and prospect of lignocellulosic bioethanol production in China. *Bioresource Technology* 101:4814–4819.

Fauci, A. S., H. C. Lane and R. R. Redfield. 2020. Covid-19-navigating the uncharted. *New England Journal of Medicine* 382:1268–1269.

Fernando, S., S. Adhikari, C. Chandrapal and M. Murali. 2006. Biorefineries: Current status, challenges, and future direction. *Energy & Fuels* 20:1727–1737.

Formela, K., A. Hejna, L. Piszczyk, M. R. Saeb and X. Colom. 2016. Processing and structure-property relationships of natural rubber/wheat bran biocomposites. *Cellulose* 23:3157–3175.

Gardebroek, C. and M. A. Hernandez. 2013. Do energy prices stimulate food price volatility? Examining volatility transmission between US oil, ethanol and corn markets. *Energy Economics* 40:119–129.

Garfield, E. 1955. Citation indexes for science. *Science* 122:108–111.

Gillon, S. 2010. Fields of dreams: Negotiating an ethanol agenda in the Midwest United States. *Journal of Peasant Studies* 37:723–748.

Gnansounou, E. 2010. Production and use of lignocellulosic bioethanol in Europe: Current situation and perspectives. *Bioresource Technology* 101:4842–4850.

Graves, T., N. V. Narendranath, K. Dawson, K. and R. Power. 2006. Effect of pH and lactic or acetic acid on ethanol productivity by Saccharomyces cerevisiae in corn mash. *Journal of Industrial Microbiology and Biotechnology* 33:469–474.

Hahn-Hagerdal, B., M. Galbe, M. F. Gorwa-Grauslund, G. Liden and G. Zacchi. 2006. Bio-ethanol - The fuel of tomorrow from the residues of today. *Trends in Biotechnology* 24:549–556.

Hamilton, J. D. 1983. Oil and the macroeconomy since World War II. *Journal of Political Economy* 91:228–248.

Hamilton, J. D. 2003. What is an oil shock? *Journal of Econometrics* 113:363–398.

Hamilton, J. D. 2009. Causes and consequences of the oil shock of 2007-08. *Brookings Papers on Economic Activity* 2009:215–261.

Hertel, T. W., A. A. Golub and A. D. Jones, et al. 2010. Effects of US maize ethanol on global land use and greenhouse gas emissions: Estimating market-mediated responses. *BioScience* 60:223–231.

Hill, J., E. Nelson, D. Tilman, S. Polasky and D. Tiffany. 2006. Environmental, economic, and energetic costs and benefits of biodiesel and ethanol biofuels. *Proceedings of the National Academy of Sciences of the United States of America* 103:11206–11210.

Hill, J., S. Polasky and E. Nelson, et al. 2009. Climate change and health costs of air emissions from biofuels and gasoline. *Proceedings of the National Academy of Sciences of the United States of America* 106:2077–2082.

Hsieh, W. D., R. H. Chen, T. L. Wu and T. H. Lin. 2002. Engine performance and pollutant emission of an SI engine using ethanol-gasoline blended fuels. *Atmospheric Environment* 36:403–410.

Huang, H. J., S. Ramaswamy, U. W. Tschirner and B. V. Ramarao. 2008. A review of separation technologies in current and future biorefineries. *Separation and Purification Technology* 62:1–21.

Jones, T. C. 2012. America, oil, and war in the Middle East. *Journal of American History* 99:208–218.

Kerr, R. A. 2007. Global warming is changing the world. *Science* 316:188–190.

Kilian, L. 2008. Exogenous oil supply shocks: How big are they and how much do they matter for the US economy? *Review of Economics and Statistics* 90:216–240.

Kilian, L. 2009. Not all oil price shocks are alike: Disentangling demand and supply shocks in the crude oil market. *American Economic Review*, 99:1053–69.

Kim, J. 2018. Evaluating author name disambiguation for digital libraries: A case of DBLP. *Scientometrics* 116:1867–1886.

Konur, O. 2000. Creating enforceable civil rights for disabled students in higher education: An institutional theory perspective. *Disability & Society* 15:1041–1063.

Konur, O. 2002a. Access to nursing education by disabled students: Rights and duties of nursing programs. *Nurse Education Today* 22:364–374.

Konur, O. 2002b. Assessment of disabled students in higher education: Current public policy issues. *Assessment and Evaluation in Higher Education* 27:131–152.

Konur, O. 2002c. Access to employment by disabled people in the UK: Is the Disability Discrimination Act working? *International Journal of Discrimination and the Law* 5:247–279.

Konur, O. 2006a. Participation of children with dyslexia in compulsory education: Current public policy issues. *Dyslexia* 12:51–67.

Konur, O. 2006b. Teaching disabled students in Higher Education. *Teaching in Higher Education* 11:351–363.

Konur, O. 2007a. A judicial outcome analysis of the *Disability Discrimination Act*: A windfall for the employers? *Disability & Society* 22:187–204.

Konur, O. 2007b. Computer-assisted teaching and assessment of disabled students in higher education: The interface between academic standards and disability rights. *Journal of Computer Assisted Learning* 23:207–219.

Konur, O. 2011. The scientometric evaluation of the research on the algae and bio-energy. *Applied Energy* 88:3532–3540.

Konur, O. 2012a. Prof. Dr. Ayhan Demirbas' scientometric biography. *Energy Education Science and Technology Part A: Energy Science and Research* 28:727–738.

Konur, O. 2012b. The evaluation of the biogas research: A scientometric approach. *Energy Education Science and Technology Part A: Energy Science and Research* 29:1277–1292.

Konur, O. 2012c. The evaluation of the global energy and fuels research: A scientometric approach. *Energy Education Science and Technology Part A: Energy Science and Research* 30:613–628.

Konur, O. 2012d. The evaluation of the research on the biodiesel: A scientometric approach. *Energy Education Science and Technology Part A: Energy Science and Research* 28:1003–1014.

Konur, O. 2012e. The evaluation of the research on the bioethanol: A scientometric approach. *Energy Education Science and Technology Part A: Energy Science and Research* 28:1051–1064.

Konur, O. 2012f. The evaluation of the research on the biofuels: A scientometric approach. *Energy Education Science and Technology Part A: Energy Science and Research* 28:903–916.

Konur, O. 2012g. The evaluation of the research on the biohydrogen: A scientometric approach. *Energy Education Science and Technology Part A: Energy Science and Research* 29:323–338.

Konur, O. 2012h. The evaluation of the research on the microbial fuel cells: A scientometric approach. *Energy Education Science and Technology Part A: Energy Science and Research* 29:309–322.

Konur, O. 2012i. The scientometric evaluation of the research on the production of bioenergy from biomass. *Biomass and Bioenergy* 47:504–515.

Konur, O. 2015. Current state of research on algal bioethanol. In *Marine Bioenergy: Trends and Developments*, Ed. S. K. Kim and C. G. Lee, pp. 217–244. Boca Raton, FL: CRC Press.

Konur, O., Ed. 2018a. *Bioenergy and Biofuels*. Boca Raton, FL: CRC Press.

Konur, O. 2018b. Bioenergy and biofuels science and technology: Scientometric overview and citation classics. In *Bioenergy and Biofuels*, Ed. O. Konur, pp. 3–63. Boca Raton: CRC Press.

Konur, O. 2019. Cyanobacterial bioenergy and biofuels science and technology: A scientometric overview. In *Cyanobacteria: From Basic Science to Applications*, Ed. A. K. Mishra, D. N. Tiwari and A. N. Rai, pp. 419–442. Amsterdam: Elsevier.

Konur, O. 2020a. The scientometric analysis of the research on the bioethanol production from green macroalgae. In *Handbook of Algal Science, Technology and Medicine*, Ed. O. Konur, pp. 385–401. London: Academic Press.

Konur, O., Ed. 2020b. *Handbook of Algal Science, Technology and Medicine*. London: Academic Press.

Konur, O., Ed. 2021a. *Handbook of Biodiesel and Petrodiesel Fuels: Science, Technology, Health, and Environment*. Boca Raton, FL: CRC Press.

Konur, O., Ed. 2021b. *Handbook of Biodiesel and Petrodiesel Fuels: Science, Technology, Health, and Environment. Volume 1. Biodiesel Fuels: Science, Technology, Health, and Environment*. Boca Raton, FL: CRC Press.

Konur, O., Ed. 2021c. *Handbook of Biodiesel and Petrodiesel Fuels: Science, Technology, Health, and Environment. Volume 2. Biodiesel Fuels based on the Edible and Nonedible Feedstocks, Wastes, and Algae: Science, Technology, Health, and Environment*. Boca Raton, FL: CRC Press.

Konur, O., Ed. 2021d. *Handbook of Biodiesel and Petrodiesel Fuels: Science, Technology, Health, and Environment. Volume 3. Petrodiesel Fuels: Science, Technology, Health, and Environment*. Boca Raton, FL: CRC Press.

Konur, O. and F. L. Matthews. 1989. Effect of the properties of the constituents on the fatigue performance of composites: A review. *Composites* 20:317–328.

Kruyt, B., D. P. van Vuuren, H. J. de Vries and H. Groenenberg. 2009. Indicators for energy security. *Energy Policy* 37:2166–2181.

Kwiatkowski, J. R., A. J. McAloon, F. Taylor and D. B. Johnston. 2006. Modeling the process and costs of fuel ethanol production by the corn dry-grind process. *Industrial Crops and Products* 23:288–296.

Leydesdorff, L. 2000. Is the European Union becoming a single publication system? *Scientometrics* 47:265–280.

Leydesdorff, L. and C. Wagner. 2009. Is the United States losing ground in science? A global perspective on the world science system. *Scientometrics* 78:23–36.

Leydesdorff, L. and P. Zhou. 2005. Are the contributions of China and Korea upsetting the world system of science? *Scientometrics* 63:617–630.

Li, H., S. M. Liu, X. H. Yu, S. L. Tang and C. K. Tang. 2020. Coronavirus disease 2019 (COVID-19): Current status and future perspectives. *International Journal of Antimicrobial Agents* 55:105951.

Lin, Y. and S. Tanaka. 2006. Ethanol fermentation from biomass resources: Current state and prospects. *Applied Microbiology and Biotechnology* 69:627–642.

Ma, X., L. Sun and C. Song. 2002. A new approach to deep desulfurization of gasoline, diesel fuel and jet fuel by selective adsorption for ultra-clean fuels and for fuel cell applications. *Catalysis Today* 77:107–116.

Mojovic, L., S. Nikolic, M. Rakin and M. Vukasinovic. 2006. Production of bioethanol from corn meal hydrolyzates. *Fuel* 85:1750–1755.

Morschbacker, A. 2009. Bio-ethanol based ethylene. *Polymer Reviews* 49:79–84.

Najafi, G., B. Ghobadian and T. Tavakoli, et al. 2009. Performance and exhaust emissions of a gasoline engine with ethanol blended gasoline fuels using artificial neural network. *Applied Energy* 86:630–639.

Newman, P. W. G. and J. R. Kenworthy. 1989. Gasoline consumption and cities: A comparison of U.S. cities with a global survey. *Journal of the American Planning Association* 55:24–37.

Nguyen, T. L. T., S. H. Gheewala and S. Garivait. 2007. Energy balance and GHG-abatement cost of cassava utilization for fuel ethanol in Thailand. *Energy Policy* 35:4585–4596.

North, D. C. 1991. Institutions. *Journal of Economic Perspectives* 5:97–112.

Olsson, L. and B. Hahn-Hagerdal. 1996. Fermentation of lignocellulosic hydrolysates for ethanol production. *Enzyme and Microbial Technology* 18:312–331.

Pimentel, D. and T. W. Patzek. 2005. Ethanol production using corn, switchgrass, and wood; biodiesel production using soybean and sunflower. *Natural Resources Research* 14:65–76.

Reeves, S. 2014. To Russia with love: How moral arguments for a humanitarian intervention in Syria opened the door for an invasion of the Ukraine. *Michigan State University International Law Review* 23:199.

Sanchez, O. J. and C. A. Cardona. 2008. Trends in biotechnological production of fuel ethanol from different feedstocks. *Bioresource Technology* 99:5270–5295.

Serra, T., D. Zilberman, J. M. Gil and B. K. Goodwin. 2011. Nonlinearities in the U.S. corn-ethanol-oil-gasoline price system. *Agricultural Economics* 42:35–45.

Shigechi, H., J. Koh and Y. Fujita, et al. 2004. Direct production of ethanol from raw corn starch via fermentation by use of a novel surface-engineered yeast strain codisplaying glucoamylase and α-amylase. *Applied and Environmental Microbiology* 70:5037–5040.

Srichuwong, S., M. Fujiwara and X. Wang, et al. 2009. Simultaneous saccharification and fermentation (SSF) of very high gravity (VHG) potato mash for the production of ethanol. *Biomass and Bioenergy* 33:890–898.

Sun, Y. and J. Cheng. 2002. Hydrolysis of lignocellulosic materials for ethanol production: A review. *Bioresource Technology* 83:1–11.

Szambelan, K., J. Nowak and Z. Czarnecki. 2004. Use of *Zymomonas mobilis* and *Saccharomyces cerevisiae* mixed with *Kluyveromyces fragilis* for improved ethanol production from Jerusalem artichoke tubers. *Biotechnology Letters* 26:845–848.

Taherzadeh, M. J. and K. Karimi. 2007. Enzyme-based hydrolysis processes for ethanol from lignocellulosic materials: A review. *Bioresources* 2:707–738.

Taherzadeh, M. J. and K. Karimi. 2008. Pretreatment of lignocellulosic wastes to improve ethanol and biogas production: A review. *International Journal of Molecular Sciences* 9:1621–1651.

Trujillo-Barrera, A., M. Mallory and P. Garcia. 2012. Volatility spillovers in U.S. crude oil, ethanol, and corn futures markets. *Journal of Agricultural and Resource Economics* 37:247–262.

Winzer, C. 2012. Conceptualizing energy security. *Energy Policy* 46:36–48.

Wu, F., Zhang, D., and Zhang, J. (2012). Will the development of bioenergy in China create a food security problem? Modeling with fuel ethanol as an example. Renewable energy, 47, 127–134.

Yang, B. and C. E. Wyman. 2008. Pretreatment: The key to unlocking low-cost cellulosic ethanol. *Biofuels, Bioproducts and Biorefining* 2:26–40.

Zhang, L., H. Zhao and M. Gan, et al. 2011. Application of simultaneous saccharification and fermentation (SSF) from viscosity reducing of raw sweet potato for bioethanol production at laboratory, pilot and industrial scales. *Bioresource Technology* 102:4573–4579.

45 First Generation Starch Feedstock-based Bioethanol Fuels

Review

Ozcan Konur
(Formerly) Ankara Yildirim Beyazit University

45.1 INTRODUCTION

The crude oil-based gasoline fuels (Ma et al., 2002; Newman and Kenworthy, 1989) have been widely used in the transportation sector since the 1920s. However, there have been great public concerns over the adverse environmental and human impact of these fuels (Hill et al., 2006, 2009). Hence, biomass-based bioethanol fuels (Hill et al., 2006; Konur, 2012, 2015, 2019, 2020) have increasingly been used in blending gasoline fuels (Hsieh et al., 2002; Najafi et al., 2009), in the fuel cells (Antolini, 2007, 2009), and in the biochemical production (Angelici et al., 2013; Morschbacker, 2009) in a biorefinery context (Fernando et al., 2006; Huang et al., 2008).

However, it is necessary to pretreat the biomass (Alvira et al., 2010; Taherzadeh and Karimi, 2008) to enhance the yield of the bioethanol (Hahn-Hagerdal et al., 2006; Sanchez and Cardona, 2008) prior to the bioethanol fuel production from starch feedstocks through the hydrolysis (Sun and Cheng, 2002; Taherzadeh and Karimi, 2007) and fermentation (Lin and Tanaka, 2006; Olsson and Hahn-Hagerdal, 1996) of the biomass and hydrolysates, respectively.

One of the most-studied feedstocks for the bioethanol fuels have been the starch feedstocks. The research in the field of the starch feedstock-based bioethanol fuels has intensified in this context in the key research fronts of the pretreatment of the starch feedstocks (Dien et al., 2009; Mojovic et al., 2006; Shigechi et al., 2004), hydrolysis of the starch feedstocks (Dien et al., 2009; Mojovic et al., 2006; Srichuwong et al., 2009), fermentation of the starch feedstock-based hydrolysates (Dien et al. 2009; Graves et al., 2006; Shigechi et al., 2004), and bioethanol fuel production (Dien et al., 2009; Graves et al., 2006; Shigechi et al., 2004), bioethanol fuel evaluation (Hertel et al. 2010; Kwiatkowski et al., 2006; Pimentel and Patzek, 2005), and economics (Gardebroek and Hernandez, 2013; Serra et al. 2011; Trujillo-Barrera et al., 2012) of the starch feedstock-based bioethanol fuels. Further, the corn (Hertel et al. 2010; Kwiatkowski et al., 2006; Pimentel and Patzek, 2005), cassava (Dai et al., 2006; Nguyen et al., 2007) and to a lesser extent wheat (Dong et al., 2008), sorghum (Dien et al., 2009), potato (Srichuwong et al. 2009; Zhang et al., 2011), and artichoke (Szambelan et al. 2004) have been studied intensively in this context.

However, it is essential to develop efficient incentive structures (North, 1991) for the primary stakeholders to enhance the research in this field (Konur, 2000, 2002a,b,c, 2006a,b, 2007a,b). Although there has been a limited number of review papers on the bioethanol fuel production from starch feedstocks (Bai et al., 2008; Bothast and Schlicher, 2005; Patzek, 2004), there has been no review of the most-cited 25 papers in this field.

Thus, this book chapter presents a review of the most-cited 25 articles in the field of the bioethanol fuel production from the primary starch feedstocks. Then, it discusses the key findings of these highly influential papers and comments on the future research priorities in this field.

DOI: 10.1201/9781003226451-58

45.2 MATERIALS AND METHODS

The search for this study was carried out using Scopus database (Burnham, 2006) in July 2022.

As a first step for the search of the relevant literature, the keywords were selected using the most-cited first 200 population papers. The selected keyword list was then optimized to obtain a representative sample of papers for the searched research field. This final keyword set was provided in Appendix of Konur (2023) for future replication studies.

As a second step, a sample dataset was used for this study. The first 25 articles with at least 97 citations each were selected for the review study. Key findings from each paper were taken from the abstracts of these papers and were discussed. Additionally, a number of brief conclusions were drawn and a number of relevant recommendations were made to enhance the future research landscape.

45.3 RESULTS

The brief information about 25 most-cited papers with at least 97 citations each on the starch feedstock-based bioethanol fuels is given below. The primary research fronts are the production, evaluation, and economics of the starch feedstock-based bioethanol fuels with 7, 15, and 3 highly cited papers (HCPs), respectively.

45.3.1 THE PRODUCTION OF STARCH FEEDSTOCK-BASED BIOETHANOL FUELS

There are seven HCPs for the production of the starch feedstock-based bioethanol fuels (Table 45.1).

Dien et al. (2009) improved sugar conversion and ethanol yield for forage sorghum with reduced lignin contents in a paper with 184 citations. They evaluated forage sorghum plants carrying *brown midrib* (*bmr*) mutations as bioethanol feedstocks: wild type, *bmr*-6, *bmr*-12, and *bmr*-6 *bmr*-12 double mutant. They found that the *bmr*-6 and *bmr*-12 mutations were equally efficient at reducing lignin contents (by 13% and 15%, respectively), and the effects were additive (27%) for the double mutant. Reducing lignin content was highly beneficial for improving biomass conversion yields. They pretreated sorghum samples with dilute acid and hydrolyzed them with cellulase. Glucose yields for the sorghum biomass were improved by 27%, 23%, and 34% for *bmr*-6, *bmr*-12, and the double mutant, respectively, compared to wild type. They used *Saccharomyces cerevisiae* for simultaneous saccharification and fermentation (SSF) to convert hydrolysates into ethanol. Conversion of cellulose to ethanol for the pretreated sorghum was improved by 22%, 21%, and 43% for *bmr*-6, *bmr*-12, and the double mutant compared to wild type, respectively. They observed an increased number of lignin globules in double-mutant tissues as compared to the wild type, suggesting the lignin had become more pliable.

Shigechi et al. (2004) produced ethanol from raw corn starch via fermentation by use of a novel surface-engineered yeast strain in a paper with 169 citations. They used *S. cerevisiae* codisplaying *Rhizopus oryzae* glucoamylase and *Streptococcus bovis* α-amylase by using the C-terminal-half region of α-agglutinin and the flocculation functional domain of Flo1p as the respective anchor proteins. In 72-h fermentation, they observed that this strain produced 61.8 g of ethanol/liter, with 86.5% of theoretical yield from raw corn starch.

Graves et al. (2006) evaluated the effect of pH and lactic or acetic acid as fermentation inhibitors on ethanol productivity by *S. cerevisiae* in corn mash in a paper with 138 citations. The lactic and acetic acid concentrations utilized were 0%, 0.5%, 1.0%, 2.0%, 3.0% and 4.0% w/v, and 0%, 0.1%, 0.2%, 0.4%, 0.8% and 1.6% w/v, respectively. They observed that lactic acid did not completely inhibit ethanol production by the yeast. However, lactic acid at 4% w/v decreased final ethanol concentration in all mashes at all pH levels. In 30% solids, mash set at pH ≤5, lactic acid at 3% w/v reduced ethanol production. In contrast, inhibition by acetic acid increased as the concentration of solids in the mash increased and the pH of the medium declined. Ethanol production was completely inhibited in all mashes set at pH 4 in the presence of acetic acid at concentrations ≥0.8% w/v.

TABLE 45.1
The Production of Starch Feedstock-based Bioethanol Fuels

No.	Papers	Biomass/ Hydrolysate	Prts.	Yeasts	Parameters	Keywords	Lead Author	Affil.	Cits
1	Dien et al. (2009)	Sorghum	Acid cellulase, genetics	S. cerevisiae	Sugar and ethanol yield, biomass engineering, SSF	Sorghum, ethanol	Dien, Bruce S. 6603685796	USDA Agr. Res. Serv. USA	184
2	Shigechi et al. (2004)	Corn	Yeast engineering	S. cerevisiae	Bioethanol production, yeast engineering, ethanol yield	Corn, ethanol	Kondo, Akihiko 57203868143	Kobe Univ. Japan	169
3	Graves et al. (2006)	Corn	Na	S. cerevisiae	Effect of pH and lactic or acetic acid on ethanol productivity	Corn, ethanol	Narendranath, Neelakantam V. 65071105412	Alltech Inc. USA	138
4	Mojovic et al. (2006)	Corn	α-amylase, glucoamylase	S. cerevisiae	Bioethanol production, sugar and ethanol yield, SSF	Corn, bioethanol	Mojovic, Ljiljana* 6603838180	Univ. Belgrade Serbia	121
5	Srichuwong et al. (2009)	Potato	Pectinase, cellulase and hemicellulase	S. cerevisiae	SSF of potato mash, sugar, and ethanol yield	Potato, ethanol	Tokuyasu, Ken 7006344903	Natl. Food Res. Inst. Japan	114
6	Zhang et al. (2011)	Potato	Xylanase	S. cerevisiae	Bioethanol production from raw sweet potato at laboratory, pilot, and industrial scales, ethanol concentration, productivity, and yield.	Potato, bioethanol	Zho, Hai 55715822600	Chinese Acad. Sci. China	107
7	Nikolic et al. (2010)	Corn	Enzymes	S. cerevisiae	Bioethanol production by SSF of corn meal using ultrasound pretreatment, glucose and ethanol yield	Corn, bioethanol	Nikolic, Svetlana 35339599800	Univ. Belgrade Serbia	101

*: Female; Cits.: number of citations received for each paper; Na: not available; Prt: biomass pretreatments.

In 30% solids, mash set at pH 4, final ethanol levels decreased with only 0.1% w/v acetic acid. In conclusion, the inhibitory effects of lactic acid and acetic acid on ethanol production in corn mash fermentation when set at a pH of 5.0–5.5 were not as great as that reported using laboratory media.

Mojovic et al. (2006) produced bioethanol from corn meal hydrolysates in a paper with 121 citations. They performed a two-step enzymatic hydrolysis of corn meal by α-amylase and glucoamylase and further ethanol fermentation of the resulting hydrolysates by *S. cerevisiae*. They optimized the conditions of starch hydrolysis such as substrate and enzyme concentration and the time required for enzymatic action taking into account both the effects of hydrolysis and ethanol fermentation. They found that the corn meal hydrolysates were good substrates for ethanol fermentation by *S. cerevisiae* and obtained the yield of ethanol of more than 80% (w/w) of the theoretical with a satisfactory volumetric productivity P (g/L h). There was no shortage of fermentable sugars during the SSF process. In this process, the savings in energy by carrying out the hydrolysis step at lower temperature (32°C) could be realized, as well as a reduction of the process time for 4 h.

Srichuwong et al. (2009) performed the SSF of very high gravity (VHG) potato mash for the production of ethanol in a paper with 114 citations. This mash contained 304 g/L of dissolved carbohydrates. They ground potato tubers into a viscous mash, and they reduced mash viscosity by the pretreatment with mixed enzyme preparations of pectinase, cellulose, and hemicellulase. The enzymatic pretreatment established the use of VHG mash with a suitable viscosity. Starch in the pretreated mash was liquefied to maltodextrins by the action of thermostable α-amylase at 85°C. They performed the SSF of liquefied mash at 30°C with the simultaneous addition of glucoamylase, *S. cerevisiae*, and ammonium sulfate as a nitrogen source for the yeast. The optimal glucoamylase loading, ammonium sulfate concentration, and fermentation time were 1.65 AGU/g, 30.2 mM, and 61.5 h, respectively. Ammonium sulfate supplementation was necessary to avoid stuck fermentation under VHG condition. Using the optimized condition, they obtained ethanol yield of 16.61% (v/v) which was equivalent to 89.7% of the theoretical yield.

Zhang et al. (2011) developed a bioprocess for bioethanol production from raw sweet potato by *S. cerevisiae* at laboratory, pilot, and industrial scales in a paper with 107 citations. They determined the fermentation mode, inoculum size, and pressure from different gases in laboratory. They observed that the maximum ethanol concentration, average ethanol productivity rate, and yield of ethanol after fermentation in laboratory scale (128.51 g/L, 4.76 g/L/h and 91.4%) were satisfactory with small decrease at pilot scale (109.06 g/L, 4.89 g/L/h and 91.24%) and industrial scale (97.94 g/L, 4.19 g/L/h and 91.27%). When scaled up, the viscosity caused resistance to fermentation parameters, 1.56 AUG/g (sweet potato mash) of xylanase decreased the viscosity from approximately 30,000 to 500 cp. In conclusion, sweet potato was an attractive feedstock for bioethanol production from both the economic and environmental standpoints.

Nikolic et al. (2010) produced bioethanol by SSF of corn meal using ultrasound pretreatment in a paper with 101 citations. They used *S. cerevisiae* var. ellipsoideus in a batch system. They performed an ultrasound pretreatment (at a frequency of 40 kHz) at different sonication times and temperatures, before addition of liquefying enzyme. They selected an optimal duration of the treatment of 5 min and sonication temperature of 60°C. Under the optimum conditions, they obtained an increase of glucose concentration of 6.82% over untreated control sample. They observed that the ultrasound pretreatment could increase the ethanol concentration by 11.15% (compared to the control sample). In this case, they obtained the maximum ethanol concentration of 9.67% w/w (which corresponded to percentage of the theoretical ethanol yield of 88.96%) after 32 h of the SSF process. The ultrasound pretreatment stimulated degradation of starch granules and release of glucose, and thereby accelerated the starch hydrolysis due to the cavitation and acoustic streaming caused by the ultrasonic action.

45.3.2 The Evaluation of the Starch Feedstock-based Bioethanol Fuels

There are 15 HCPs for the evaluation of the starch feedstock-based bioethanol fuels (Table 45.2).

TABLE 45.2

The Evaluation of Starch Feedstock-based Bioethanol Fuels

No.	Papers	Biomass/ Hydrolysate	Parameters	Keywords	Lead Author	Affil.	Cits
1	Pimentel and Patzek (2005)	Corn	Comparative energy balance, bioethanol v. biodiesel, feedstock-type	Corn, ethanol biodiesel	Pimentel, David 7005471319	Cornell Univ. USA	1012
2	Hertel et al. (2010)	Corn	CO_2 emissions due to the indirect LUC for bioethanol production	Maize, ethanol	Hertel, Thomas W. 7006465820	Purdue Univ. USA	371
3	Kwiatkowski et al. (2006)	Corn	Corn dry-grind process, bioethanol production costs, corn price, coproducts	Corn, ethanol	Kwiatkowski, Jason R. 19035260500	USDA Agr. Res. Serv. USA	358
4	Wang et al. (2007)	Corn	Energy consumption and GHG emissions of bioethanol plants fueled with different fuels	Corn, ethanol	Wang, Michael 7406684798	Argonne Natl. Lab. USA	336
5	Wang et al. (2012)	Corn	Energy consumption and GHG emissions of bioethanol fuels, five feedstocks: corn, sugarcane, corn stover, switchgrass, and miscanthus	Corn, ethanol	Wang, Michael 7406684798	Argonne Natl. Lab. USA	270
6	Liska et al. (2009)	Corn	Energy consumption and GHG emissions of bioethanol fuels, gasoline, USA	Corn, ethanol	Cassman, Kenneth G. 7003725994	Univ. Nebraska Lincoln USA	208
7	Kim and Dale (2005)	Corn	Energy consumption and GHG emissions of bioethanol fuels, no tillage, dry and wet milling, N_2O, CO_2 emissions	Corn, ethanol	Dale Bruce E. 7201511969	Michigan State Univ. USA	175
8	Kim and Dale (2002)	Corn	Net energy analysis, system expansion approach, production systems, wet and dry milling	Corn, ethanol	Dale Bruce E. 7201511969	Michigan State Univ. USA	155
9	Hettinga et al. (2009)	Corn	Reductions in US corn ethanol production costs, experience curve approach, corn production, ethanol processing	Corn, ethanol	Hettinga, W. G. 25922162900	Utrecht Univ. Netherlands	133
10	Dai et al. (2006)	Cassava	Energy efficiency and potentials of cassava-based ethanol fuels, cassava ethanol system, energy and renewable energy efficiency	Cassava, ethanol	Dai, Du 10138950200	Shanghai Jiao Tong Univ. China	129
11	Nguyen et al. (2007)	Cassava	Energy balance and GHG-abatement cost of utilization of cassava ethanol fuels in Thailand	Cassava, ethanol	Gheewala, Shabbir H. 6602264724	King Mongkut's Univ. Technol. Thailand	128
12	Wang et al. (2011)	Corn	Energy and GHG emission effects of corn and cellulosic ethanol with technology improvements and LUCs in the USA	Corn, bioethanol	Wang, Michael 7406684798	Argonne Natl. Lab. USA	119
13	Yang et al. (2012)	Corn	Environmental impact of gasoline and E85 with 12 different environmental impacts and regional differences among 19 corn-growing states	Corn, ethanol	Suh, Sangwon 7201479719	Univ. Calif. S. B. USA	107
14	Dunn et al. (2013)	Corn	LUC and GHG emissions from corn and cellulosic ethanol	Corn, ethanol	Dunn, Jennifer B.* 55644969600	Northwestern Univ. USA	106
15	Dong et al. (2008)	Wheat, corn	Energy and emergy evaluation of bioethanol production from wheat and corn	Wheat, bioethanol	Gao, Wangsheng 12244677000	China Agr. Univ. China	104

*: Female; Cits.: number of citations received for each paper; Na: nonavailable; Prt: biomass pretreatments.

Pimentel and Patzek (2005) compared the energy balance of bioethanol fuels from corn, switchgrass, and wood with the biodiesel production from soybean and sunflower in the USA in a paper with 1,012 citations. They found that the energy outputs from bioethanol produced using corn, switchgrass, and wood biomass were each less than the respective fossil energy inputs. The same was true for producing biodiesel using soybeans and sunflower. However, the energy cost for producing soybean biodiesel was only slightly negative compared with bioethanol production. Ethanol production using corn grains, switchgrass, and wood required 29%, 50%, and 57% more fossil energy than the ethanol fuel produced, respectively. Biodiesel production using soybean and sunflower required 27% and 118% more fossil energy than the biodiesel fuel produced, respectively.

Hertel et al. (2010) evaluated the effects of the corn-based bioethanol fuels produced in the USA on global land-use and greenhouse gas (GHG) emissions in a paper with 371 citations. They analyzed the GHG emissions induced by the indirect land-use changes (LUCs). Factoring market-mediated responses and by-product use into their analysis reduced cropland conversion by 72% from the land used for the bioethanol feedstock. Consequently, the associated GHG release estimated in this framework was 800 g of carbon dioxide (CO_2) per megajoule (MJ) or 27 g/MJ/year over 30 years of bioethanol production, or roughly a quarter of the only other published estimate of CO_2 releases attributable to changes in indirect land use. In conclusion, 800 g CO_2 emissions were enough to cancel out the benefits that corn bioethanol had on global warming, thereby limiting its potential contribution in the context of California's Low Carbon Fuel Standard.

Kwiatkowski et al. (2006) modeled the process and costs of bioethanol production by the corn dry-grind process producing 119 million kg/year (40 million gal/year) of bioethanol in the USA in a paper with 258 citations. They found that the cost of producing bioethanol increased from US\$ 0.235 L^{-1} to US\$ 0.365 L^{-1} (US\$ 0.89 gal^{-1} to US\$ 1.38 gal^{-1}) as the price of corn increased from US\$ 0.071 kg^{-1} to US\$ 0.125 kg^{-1} (US\$ 1.80 bu^{-1} to US\$ 3.20 bu^{-1}). Another example gave a reduction from 151 to 140 million L/year as the amount of starch in the feed was lowered from 59.5% to 55% (w/w). In conclusion, this model showed the impact of changes in the process and coproducts of the bioethanol fuels from starch process on the bioethanol production costs.

Wang et al. (2007) evaluated the life-cycle energy and GHG emission impacts of different corn-based bioethanol plant types in the USA in a paper with 336 citations. The majority of corn-based bioethanol plants were powered by natural gas. However, efforts had been made to further reduce the energy used in bioethanol plants or to switch from natural gas to other fuels, such as coal and wood chips. They examined nine corn-based bioethanol plant types, categorized according to the type of process fuels employed, use of combined heat and power, and production of wet distiller grains and solubles as by-products. They found that these ethanol plant types could have distinctly different energy and GHG emission effects on a full fuel-cycle basis. In particular, GHG emission impacts could vary significantly from a 3% increase if coal was the process fuel to a 52% reduction if wood chips were used. In conclusion, in order to achieve energy and GHG emission benefits, researchers should closely examine and differentiate among the types of plants used to produce corn ethanol so that corn ethanol production would move toward a more sustainable path.

Wang et al. (2012) evaluated the well-to-wheels energy consumption and GHG emissions of bioethanol fuels from corn, sugarcane, corn stover, switchgrass, and miscanthus in the USA in a paper with 270 citations. They estimated life-cycle energy consumption and GHG emissions from using ethanol produced from these five feedstocks. They found that even when the highly debated LUC GHG emissions were included, changing from corn to sugarcane and then to cellulosic biomass helped to significantly increase the reductions in energy consumption and GHG emissions from using bioethanol. Relative to gasoline fuels, ethanol from these feedstocks could reduce life-cycle GHG emissions by 19%–48%, 40%–62%, 90%–103%, 77%–97%, and 101%–115%, respectively. They found similar trends with regard to fossil energy benefits for these five bioethanol pathways.

Liska et al. (2009) analyzed the life cycles of corn–ethanol systems accounting for the majority of US capacity to estimate GHG emissions and energy efficiencies on the basis of updated values for crop management and yields, biorefinery operation, and coproduct utilization in a paper with 208 citations. They estimated the direct-effect GHG emissions as 48%–59% reduction compared to gasoline. Ethanol-to-crude oil output/input ratios ranged from 10:1 to 13:1 but could be increased to 19:1 if farmers adopted high-yield progressive crop and soil management practices. An advanced closed-loop biorefinery with anaerobic digestion of corn residues reduced GHG emissions by 67% and increased the net energy ratio to 2.2, from 1.5 to 1.8 for the most common systems. The larger GHG reductions estimated in this study allowed a greater buffer for inclusion of indirect-effect LUC emissions while still meeting regulatory GHG reduction targets. In conclusion, corn–ethanol systems had substantially greater potential to mitigate GHG emissions and reduce dependence on imported crude oil for transportation fuels than reported previously.

Kim and Dale (2005) evaluated the nonrenewable energy consumption and GHG emissions from ethanol produced from no-tilled corn grains in the USA in a paper with 175 citations. They cultivated corn under no-tillage practice (without plowing). The system boundaries included corn production, ethanol production, and the end use of ethanol as a fuel in a midsize passenger car. They allocated the environmental burdens in multi-output biorefinery processes (e.g., corn dry milling and wet milling) to the ethanol product and its various coproducts by the system expansion allocation approach. They found that the nonrenewable energy requirement for producing 1 kg of ethanol is approximately 13.4–21.5 MJ (based on lower heating value), depending on corn milling technologies employed. Thus, the net energy value (NEV) of ethanol was positive where the energy consumed in ethanol production was less than the energy content of the ethanol (26.8 MJ/kg). They estimated nitrous oxide (N_2O) emissions from soil and soil organic carbon levels under corn cultivation by the DAYCENT model. Carbon sequestration rates ranged from 377 to 681 kg C/ha/year and N_2O emissions from soil were 0.5–2.8 kg N/ha/year under no-till conditions. The GHG emissions assigned to 1 kg of ethanol were 260–922 g CO_2 eq. under no-tillage. Using ethanol (E85) fuel in a midsize passenger vehicle could reduce GHG emissions by 41%–61% km^{-1} driven, compared to gasoline-fueled vehicles. In conclusion, using ethanol as a vehicle fuel, therefore, had the potential to reduce nonrenewable energy consumption and GHG emissions.

Kim and Dale (2002) evaluated the system expansion approach to net energy analysis for ethanol production from corn grain in the USA in a paper with 155 citations. Production systems included ethanol production from corn dry milling and corn wet milling, corn grain production (the agricultural system), soybean products from soybean milling (i.e., soybean oil and soybean meal), and urea production. These five product systems were mutually interdependent where all these systems generated products which competed with or displaced all other comparable products in the market place. The displacement ratios between products compared the equivalence of their marketplace functions. The net energy, including transportation to consumers, was 0.56 MJnet/MJ of ethanol from corn grain regardless of the ethanol production technology employed. Using ethanol as a liquid transportation fuel could reduce domestic use of fossil fuels, particularly crude oils. The choice of allocation procedures had the greatest impact on ethanol net energy. Process energy associated with wet milling, dry milling, and the corn agricultural process also significantly influenced the net energy due to the wide ranges of available process energy values. In conclusion, the system expansion approach could completely eliminate allocation procedures in the foreground system of ethanol production from corn grain.

Hettinga et al. (2009) evaluated the reductions in US corn ethanol production costs using the experience curve approach in a paper with 133 citations. They studied costs of dry-grind ethanol production over the time frame 1980–2005. They differentiated the cost reductions between feedstock (corn) production and industrial (ethanol) processing. They found that corn production costs in the US had declined by 62% over 30 years, down to USD100$_{2005}$/tonne in 2005, while corn production volumes almost doubled since 1975. They calculated a progress ratio (PR) of 0.55

indicating a 45% cost decline over each doubling in cumulative production. Higher corn yields and increasing farm sizes were the most important drivers behind this cost decline. Industrial processing costs of ethanol had declined by 45% since 1983, to below USD130$_{2005}$/m³ in 2005 (excluding costs for corn and capital), equivalent to a PR of 0.87. Total ethanol production costs (including capital and net corn costs) had declined approximately 60% from USD800$_{2005}$/m³ in the early 1980s, to USD300$_{2005}$/m³ in 2005. Higher ethanol yields, lower energy use, and the replacement of beverage alcohol-based production technologies have mostly contributed to this substantial cost decline. In addition, the average size of dry-grind ethanol plants increased by 235% since 1990. For the future, they estimated that solely due to technological learning, production costs of ethanol might decline 28%–44%, though this excluded effects of the current rising corn and fossil fuel costs. In conclusion, experience curves were a valuable tool to describe both past and potential future cost reductions in US corn-based ethanol production.

Dai et al. (2006) evaluated the energy efficiency and potentials of cassava-based ethanol fuels in Guangxi region of China in a paper with 129 citations. They used 100 thousand ton ethanol fuel demonstration plant at Qinzhou of Guangxi. They used the NEV and net renewable energy value (NREV) to assess the energy and renewable energy efficiency of the cassava ethanol system during its life cycle. They divided the cassava ethanol system into five subsystems including the cassava plantation/treatment, ethanol conversion, denaturing, refueling, and transportation. They found that the cassava ethanol system was energy and renewable energy efficient as indicated by positive NEV and NREV values that were 7.475 and 7.881 MJ/L, respectively. Cassava ethanol production helped converting the non-liquid fuel into fuel ethanol that could be used for transportation. Through ethanol fuel production, one Joule of crude oil, plus other forms of energy inputs such as coal, could produce 9.8 J of ethanol fuel. Cassava ethanol fuel could substitute for gasoline reducing crude oil imports. With the cassava output in 2003, it could substitute for 166.107 million liters of gasoline, while with the cassava output potential, it could substitute for 618.162 million liters of gasoline. In conclusion, cassava ethanol fuel was more energy efficient than gasoline, petrodiesel fuel, and corn ethanol fuel but less efficient than biodiesel.

Nguyen et al. (2007) evaluated the energy balance and GHG abatement cost of utilization of cassava-based ethanol fuels (CE) in Thailand in a paper with 128 citations. They found positive energy balance of 22.4 MJ/L and net avoided GHG emissions of 1.6 kg CO_2 eq./L for CE and determined that CE would be a good substitute for gasoline, effective in fossil energy saving and GHG reduction. With a GHG abatement cost of USD99 per tonne of CO_2, CE was rather less cost-effective than many other climate strategies relevant to Thailand in the short term.

Wang et al. (2011) evaluated the energy and GHG emission effects of corn and cellulosic ethanol with technology improvements and LUCs in the USA in a paper with 119 citations. With updated simulation results of direct and indirect LUCs and observed technology improvements in the past several years, they performed a life-cycle analysis of ethanol and showed that at present and in the near future, using corn ethanol reduced GHG emission by more than 20%, relative to those of gasoline. On the other hand, second generation ethanol could achieve much higher reductions in GHG emissions. In a broader sense, sound evaluation of US biofuel policies should account for both unanticipated consequences and technology potentials.

Yang et al. (2012) compared the environmental impact of gasoline and E85 taking into consideration 12 different environmental impacts and regional differences among 19 corn-growing states in the USA in a paper with 107 citations. They found that E85 did not outperform gasoline when a wide spectrum of impacts were considered. If the impacts were aggregated using weights developed by the National Institute of Standards and Technology (NIST), overall, E85 generated approximately 6%–108% (23% on average) greater impact compared with gasoline, depending on where corn was produced, primarily because corn production induced significant eutrophication impacts and required intensive irrigation. If GHG emissions from the indirect LUCs were considered, the differences increased to between 16% and 118% (33% on average). In conclusion, replacing gasoline with corn ethanol might only result in shifting the net environmental impacts primarily toward increased eutrophication and greater water scarcity. They recommended that the environmental criteria used in the Energy Independence and Security Act be re-evaluated to include additional categories of environmental impact beyond GHG emissions.

Dunn et al. (2013) evaluated the LUC and GHG emissions from corn and cellulosic ethanol in a paper with 106 citations. They estimated LUC GHG emissions for ethanol from corn, corn stover, switchgrass, and miscanthus. They found that miscanthus and corn ethanol had the lowest ($-10\,$g CO_2e/MJ) and highest ($7.6\,$g CO_2e/MJ) LUC GHG emissions under base case modeling assumptions, respectively. The results for corn ethanol were lower than corresponding results from previous studies. Switchgrass ethanol base case results ($2.8\,$g CO_2e/MJ) were the most influenced by assumptions regarding converted forestlands and the fate of carbon in harvested wood products. They were greater than miscanthus LUC GHG emissions because switchgrass was a lower-yielding crop. Finally, LUC GHG emissions for corn stover were essentially negligible and insensitive to changes in model assumptions. In conclusion, LUC GHG emissions might have a smaller contribution to the overall biofuel life cycle than previously thought. Additionally, they highlight the need for future advances in LUC GHG emissions estimation including improvements to CGE models and aboveground and belowground carbon content data.

Dong et al. (2008) performed energy and emergy evaluation of bioethanol production from wheat in Henan Province, China, in a paper with 104 citations. Energy and emergy indices of ethanol production from wheat and corn in the two agro-industrial systems were as follows: output/input energy ratio, 1.09 (wheat) and 1.19 (corn); transformity of bioethanol, 2.77×10^5 and 1.89×10^5 seJ/J; renewability, 20% and 11%; emergy yield ratio, 1.24 and 1.14; environmental loading ratio, 4.05 and 7.84; and finally emergy sustainability index, 0.31 and 0.15 for wheat and corn, respectively. In conclusion, bioethanol from food crops was not a sustainable source of fuel.

45.3.3 The Economics of the Starch Feedstock-based Bioethanol Fuels

There are three HCPs for the economics of the starch feedstock-based bioethanol fuels (Table 45.3).

Serra et al. (2011) evaluated the nonlinearities in the US corn–ethanol–oil–gasoline price system in a paper with 129 citations. They used a smooth transition vector error correction model to assess price relationships within the US ethanol industry utilizing monthly ethanol, corn, oil, and gasoline prices from 1990 to 2008. They found the existence of long-run relationships among these prices where strong links between energy and food prices were identified.

Gardebroek and Hernandez (2013) evaluated the volatility transmission between US oil, ethanol, and corn prices in the USA between 1997 and 2011 in a paper with 111 citations. They followed a multivariate GARCH approach to evaluate the level of interdependence and the dynamics of volatility across these markets. They found a higher interaction between ethanol and corn markets in recent years, particularly after 2006 when ethanol became the sole alternative fuel for gasoline. They only observed, however, significant volatility spillovers from corn to ethanol prices but not the converse. They also do not find major cross-volatility effects from oil to corn markets. In conclusion, there was no evidence of volatility in energy markets stimulating price volatility in the US corn market.

Trujillo-Barrera et al. (2012) studied the volatility spillovers in US crude oil, ethanol, and corn futures markets in a paper with 97 citations. They found that crude oil spillovers to both corn and ethanol markets were somewhat similar in timing and magnitude, but moderately stronger to the ethanol market. The shares of corn and ethanol price variability directly attributed to volatility in the crude oil market were generally between 10% and 20%, but reached nearly 45% during the financial crisis, when world demand for oil changed dramatically. They also found volatility transmission from the corn to the ethanol market, but not the opposite. In conclusion, these findings provided insights into the extent of volatility linkages among energy and agricultural markets in a period characterized by strong price variability and significant production of corn-based ethanol fuels.

TABLE 45.3
The Economics of Starch Feedstock-based Bioethanol Fuels

No.	Papers	Biomass/ Hydrolysate	Parameters	Keywords	Lead Author	Affil.	Cits
1	Serra et al. (2011)	Corn	US corn–ethanol–oil–gasoline price system, long-run price relationships	Corn, ethanol	Serra, Teresa* 7004120220	Univ. Ill. U. C. USA	129
2	Gardebroek and Hernandez (2013)	Corn	Volatility transmission between US oil, ethanol, and corn prices in the USA between 1997 and 2011	Corn, ethanol	Hernandez, Manuel A. 36701295600	Int. Food Policy Res. Inst. USA	111
3	Trujillo-Barrera, et al. (2012)	Corn	Volatility spillovers in US crude oil, ethanol, and corn futures markets	Corn, ethanol	Trujillo-Barrera, Andres 55369547700	Univ. Idaho USA	97

Cits.: number of citations received for each paper; Na: nonavailable; Prt: biomass pretreatments.

45.4   DISCUSSION

45.4.1   INTRODUCTION

The crude oil-based gasoline fuels have been widely used in the transportation sector since the 1920s. However, there have been great public concerns over the adverse environmental and human impact of these fuels. Hence, biomass-based bioethanol fuels have increasingly been used in blending gasoline and petrodiesel fuels, in the fuel cells, and in the biochemical production in a biorefinery context.

However, it is necessary to pretreat the biomass to enhance the yield of the bioethanol prior to the bioethanol fuel production from feedstocks through the hydrolysis and fermentation of the biomass. One of the most-studied feedstocks for the bioethanol fuels has been the starch feedstocks as the first generation feedstocks. The research in the field of the starch feedstock-based bioethanol fuels has intensified in this context in the key research fronts of the pretreatment and hydrolysis of the starch feedstocks, fermentation of the starch feedstock-based hydrolysates, and production, economics, and evaluation of the starch feedstock-based bioethanol fuels. Further, the corn, cassava, and to a lesser extent wheat, sorghum, potato, and artichoke have been studied intensively in this context.

However, it is essential to develop efficient incentive structures for the primary stakeholders to enhance the research in this field. Although there have been a number of review papers for this field, there has been no review of the most-cited 25 articles in this field. Thus, this book chapter presents a review of the most-cited 25 articles on the bioethanol fuel production from starch feedstocks. Then, it discusses the key findings of these highly influential papers and comments on the future research priorities in this field.

As a first step for the search of the relevant literature, the keywords were selected using the most-cited first 200 population papers. The selected keyword list was then optimized to obtain a representative sample of papers for the searched research field. This keyword list was provided in Appendix of Konur (2023) for future replicative studies.

As a second step, a sample dataset was used for this study. The first 25 articles with at least 97 citations each were selected for the review study. Key findings from each paper were taken from the abstracts of these papers and were discussed. Additionally, a number of brief conclusions were drawn and a number of relevant recommendations were made to enhance the future research landscape.

Information about the research fronts for the sample papers in the bioethanol fuel production from starch feedstocks with regard to the starch feedstocks used in these processes is given in Table 45.4. As this table shows, the most-prolific starch feedstock used in the bioethanol fuels is corn with 80% of the HCPs. The other feedstocks used are cassava, potato, sorghum, and wheat with 4%–8% of the HCPs each.

On the other hand, the corn is the most influential research front with 29% surplus while starch and cassava are the least influential research fronts with 12% and 9% deficits, respectively.

Information about the thematic research fronts for the sample papers in the bioethanol fuel production from starch feedstocks is given in Table 45.5. As this table shows, there are six research fronts for this field: bioethanol fuel evaluation and production, hydrolysate fermentation, biomass pretreatments and hydrolysis, and bioethanol fuel economics with 60%, 28%, 28%, 24%, 20%, and 12% of the HCPs, respectively.

Further, the bioethanol fuel evaluation is the most influential front with 21% surplus. Bioethanol fuel economics is the other influential front with 5% surplus. On the other hand, bioethanol fuel production, hydrolysate fermentation, biomass pretreatments, and biomass hydrolysis are the least influential research fronts with 26%, 14%, 13%, and 10% deficits, respectively.

45.4.2 THE PRODUCTION OF STARCH FEEDSTOCK-BASED BIOETHANOL FUELS

There are seven HCPs for the production of the starch feedstock-based bioethanol fuels (Table 45.1).

These HCPs show a sample of the research on the production of the starch feedstock-based bioethanol fuels. These studies hint that the required steps in the bioethanol production are the pretreatments and the enzymatic hydrolysis of the starch feedstocks, followed by the fermentation of the hydrolysates (Alvira et al., 2010; Taherzadeh and Karimi, 2008).

Dien et al. (2009) improved sugar conversion and ethanol yield for forage sorghum with reduced lignin contents and found that the conversion of cellulose to ethanol for the pretreated sorghum was improved by 22%, 21%, and 43% for *bmr*-6, *bmr*-12, and the double mutant compared to wild type, respectively. Further, Shigechi et al. (2004) produced ethanol from raw corn starch via fermentation by use of a novel surface-engineered yeast strain and observed that this strain produced 61.8 g of ethanol/L, with 86.5% of theoretical yield from raw corn starch.

TABLE 45.4
The Most-Prolific Research Fronts for the Starch Feedstock-based Bioethanol Fuels

No.	Research Fronts	N Paper (%) Review	N Paper (%) Sample	Surplus (%)
1	Corn	80.0	51.1	28.9
2	Cassava	8.0	17.0	−9.0
3	Potato	8.0	6.7	1.3
4	Sorghum	4.0	6.7	−2.7
5	Wheat	4.0	7.4	−3.4
6	Artichoke	0.0	5.2	−5.2
7	Inulin	0.0	1.5	−1.5
8	Other starch	0.0	3.0	−3.0
9	Rice	0.0	2.2	−2.2
10	Sago	0.0	2.2	−2.2
11	Starch	0.0	11.9	−11.9
12	Triticale	0.0	1.5	−1.5

N paper (%) review: the number of papers in the sample of 25 reviewed papers; N paper (%) sample: the number of papers in the population sample of 135 papers.

TABLE 45.5
The Most-Prolific Thematic Research Fronts for the Starch Feedstock-based Bioethanol Fuels

No.	Research Fronts	N Paper (%) Review	N Paper (%) Sample	Surplus (%)
1	Biomass pretreatments	24.0	37.0	−13
	Enzymatic pretreatments	20.0	28.9	−8.9
	Chemical pretreatments	4.0	4.4	−0.4
	Feedstock genetic engineering	4.0	4.4	−0.4
	Yeast microbial engineering	4.0	3.0	1.0
	Mechanical pretreatments	0.0	4.4	−4.4
2	Biomass hydrolysis	20.0	29.6	−9.6
	Enzymatic hydrolysis	20.0	28.9	−8.9
	Acid hydrolysis	0.0	0.7	−0.7
3	Hydrolysate fermentation	28.0	42.2	−14.2
4	Bioethanol fuel production	28.0	54.1	−26.1
5	Bioethanol fuel evaluation	60.0	39.3	20.7
6	Bioethanol fuel economics	12.0	7.2	4.8

N paper (%) review: the number of papers in the sample of 25 reviewed papers; N paper (%) sample: the number of papers in the population sample of 135 papers.

Graves et al. (2006) evaluated the effect of pH and lactic or acetic acid on ethanol productivity by *S. cerevisiae* in corn mash and observed that the inhibitory effects of lactic acid and acetic acid on ethanol production in corn mash fermentation when set at a pH of 5.0–5.5 were not as high. Further, Mojovic et al. (2006) produced bioethanol from corn meal hydrolysates and obtained the yield of ethanol of more than 80% (w/w) of the theoretical with a satisfactory volumetric productivity.

Srichuwong et al. (2009) performed the SSF of VHG potato mash for the production of ethanol and obtained ethanol yield of 16.61% (v/v) which was equivalent to 89.7% of the theoretical yield. Further, Zhang et al. (2011) developed a bioprocess for bioethanol production from raw sweet potato by *S. cerevisiae* at laboratory, pilot, and industrial scales, and observed that the maximum ethanol concentration, average ethanol productivity rate, and yield of ethanol after fermentation in laboratory scale were satisfactory with small decrease at pilot scale and industrial scale. Finally, Nikolic et al. (2010) produced bioethanol by SSF of corn meal using ultrasound pretreatment and observed that the ultrasound pretreatment could increase the ethanol concentration by 11.15% (compared to the control sample).

45.4.3 The Evaluation of the Starch Feedstock-based Bioethanol Fuels

There are 15 HCPs for the evaluation of the starch feedstock-based bioethanol fuels (Table 45.2). These HCPs show a sample of the research on the evaluation of the starch feedstock-based bioethanol fuels.

These studies hint that a large number of variables are considered for the evaluation of these bioethanol fuels: energy balance, GHG emissions, LUC, production costs, environmental and social impacts, economic and environmental performance. It is notable that starch feedstock-based bioethanol fuels have positive evaluations in the most of these variables compared to the gasoline fuels and some other biofuels. Thus, the USA has become a global leader in both the production and utilization of the corn-based bioethanol fuels since the 1970s in the aftermath of the global oil crisis in the early 1970s.

Pimentel and Patzek (2005) compared the energy balance of bioethanol fuels from corn, switch-grass, and wood with the biodiesel production from soybean and sunflower in the USA and found that the energy outputs from bioethanol produced using corn, switchgrass, and wood biomass were each less than the respective fossil energy inputs. Further, Hertel et al. (2010) evaluated the effects of the corn-based bioethanol fuels produced in the USA on global land-use and GHG emissions and found that factoring market-mediated responses and by-product use into their analysis reduced cropland conversion by 72% from the land used for the bioethanol feedstock.

Kwiatkowski et al. (2006) modeled the process and costs of bioethanol production by the corn dry-grind process producing 119 million kg/year (40 million gal/year) of bioethanol in the USA and found that the cost of producing bioethanol increased as the price of corn increased. Further, Wang et al. (2007) evaluated the life-cycle energy and GHG emission impacts of different corn-based bio-ethanol plant types in the USA and found that the ethanol plant types could have distinctly different energy and GHG emission effects on a full fuel-cycle basis.

Wang et al. (2012) evaluated the well-to-wheels energy consumption and GHG emissions of bio-ethanol fuels from corn, sugarcane, corn stover, switchgrass, and miscanthus, and found that even when the highly debated LUC GHG emissions were included, changing from corn to sugarcane and then to cellulosic biomass helped to significantly increase the reductions in energy consumption and GHG emissions from using bioethanol. Further, Liska et al. (2009) analyzed the life cycles of corn–ethanol systems accounting for the majority of US capacity to estimate GHG emissions and energy efficiencies on the basis of updated values for crop management and yields, biorefinery operation, and coproduct utilization, and estimated the direct-effect GHG emissions as 48%–59% reduction compared to gasoline.

Kim and Dale (2005) evaluated the nonrenewable energy consumption and GHG emissions from ethanol produced from no-tilled corn grain in the USA and found that the NEV of ethanol was positive where the energy consumed in ethanol production was less than the energy content of the ethanol (26.8 MJ/kg). Further, Kim and Dale (2002) evaluated the system expansion approach to net energy analysis for ethanol production from corn grains in the USA and found that the net energy, including transportation to consumers, was 0.56 MJnet/MJ of ethanol from corn grain regardless of the ethanol production technology employed.

Hettinga et al. (2009) evaluated the reductions in US corn ethanol production costs using the expe-rience curve approach and found that corn production costs in the USA had declined by 62% over 30 years, down to USD 100_{2005}/tonne in 2005, while corn production volumes almost doubled since 1975. Further, Dai et al. (2006) evaluated the energy efficiency and potentials of cassava-based ethanol fuels in Guangxi region of China and found that the cassava ethanol system was energy and renewable energy efficient.

Nguyen et al. (2007) evaluated the energy balance and GHG-abatement cost of utilization of cassava-based ethanol fuels (CE) in Thailand, found positive energy balance and net avoided GHG emissions, and determined that CE would be a good substitute for gasoline, effective in fossil energy saving and GHG reduction. Further, Wang et al. (2011) evaluated the energy and GHG emission effects of corn and cellulosic ethanol with technology improvements and LUCs in the USA and showed that at present and in the near future, using corn ethanol reduced GHG emission by more than 20%, relative to those of gasoline.

Yang et al. (2012) compared the environmental impact of gasoline and E85 taking into consideration 12 different environmental impacts and regional differences among 19 corn-growing states in the USA and found that E85 did not outperform gasoline when a wide spectrum of impacts were considered. Further, Dunn et al. (2013) evaluated the LUC and GHG emissions from corn and cellulosic ethanol and found that miscanthus and corn ethanol had the lowest and highest LUC GHG emissions under base case modeling assumptions, respectively. Finally, Dong et al. (2008) performed energy and emergy evaluation of bioethanol production from wheat in Henan Province, China, and determined the energy and emergy indices of ethanol production from wheat and corn in the two agro-industrial systems.

45.4.4 THE ECONOMICS OF STARCH FEEDSTOCK-BASED BIOETHANOL FUELS

There are three HCPs for the economics of the starch feedstock-based bioethanol fuels (Table 45.3).

Serra et al. (2011) evaluated the nonlinearities in the US corn–ethanol–oil–gasoline price system and found the existence of long-run relationships among these prices where strong links between energy and food prices were identified. Further, Gardebroek and Hernandez (2013) evaluated the volatility transmission between US oil, ethanol, and corn prices in the USA between 1997 and 2011 and found a higher interaction between ethanol and corn markets in recent years, particularly after 2006 when ethanol became the sole alternative fuel for gasoline. Finally, Trujillo-Barrera et al. (2012) studied the volatility spillovers in US crude oil, ethanol, and corn futures markets and found that crude oil spillovers to both corn and ethanol markets were somewhat similar in timing and magnitude, but moderately stronger to the ethanol market.

45.5 CONCLUSION AND FUTURE RESEARCH

The brief information about the key research fronts covered by the 25 most-cited papers with at least 97 citations each is given under three primary headings: production, evaluation, and economics of the bioethanol fuels.

The usual characteristics of these HCPs are that the pretreatments and hydrolysis of the starch feedstocks and fermentation of the resulting hydrolysates are the primary processes for the bioethanol fuel production from starch feedstocks to improve the ethanol yield as the starch feedstocks are one of the most-studied feedstocks for the bioethanol production especially for the countries with the large farmlands such as the USA and Canada.

The key findings on these research fronts should be read in light of the increasing public concerns about climate change, GHG emissions, and global warming as these concerns have been certainly behind the boom in the research on the bioethanol fuel production from starch feedstocks as an alternative to crude oil-based gasoline and petrodiesel fuels in the last decades. However, it is a least sustainable alternative as first generation bioethanol fuels compared to the waste (second generation) and algae (third generation)-based bioethanol fuels since it increases concerns about the food security (Mechlem, 2004) as these feedstocks are also used as foods and feeds for the humans (Gwirtz and Garcia-Casal, 2014) and animals (Lakshmi et al., 2017), respectively (Gardebroek and Hernandez, 2013; Serra et al., 2011; Trujillo-Barrera et al., 2012). The recent supply shocks caused by the COVID-19 pandemics and the Russian invasion of Ukraine also highlight the importance of the production and utilization of the bioethanol fuels as an alternative to the crude oil-based gasoline and petrodiesel fuels.

As Table 45.4 shows, the most-prolific starch feedstock used in the bioethanol fuels is corn with 80% of the HCPs. The other feedstocks used are cassava, potato, sorghum, and wheat. Similarly, as Table 45.5 shows, there are six thematic research fronts for this field: bioethanol fuel evaluation and production, hydrolysate fermentation, biomass pretreatments and hydrolysis, and bioethanol fuel economics. Further, bioethanol fuel evaluation is the most influential front, followed by bioethanol fuel economics while bioethanol fuel production, hydrolysate fermentation, biomass pretreatments, and biomass hydrolysis are the least influential research fronts.

These studies emphasize the importance of proper incentive structures for the efficient bioethanol fuel production from starch feedstocks in light of North's institutional framework (North, 1991). In this context, the major producers and users of bioethanol fuels such as the USA and Canada with vast farmlands have developed strong incentive structures for the efficient bioethanol fuel production from starch feedstocks. In light of the supply shocks caused primarily by the COVID-19 pandemics and Russian invasion of Ukraine, it is expected that the incentive structures such as public funding would be enhanced to increase the share of bioethanol fuels in the global fuel portfolio as a strong alternative to crude oil-based gasoline and petrodiesel fuels.

In this context, it is expected that the most-prolific researchers, institutions, countries, funding bodies, and journals in this field would have a first-mover advantage to benefit from such potential incentives. This is especially true for the US stakeholders as the USA has become the global leader in both the production and utilization of starch feedstock-based bioethanol fuels.

It is recommended that such review studies are performed for the primary research fronts of the bioethanol fuel production from starch feedstocks.

ACKNOWLEDGMENTS

The contribution of the highly cited researchers in the field of the bioethanol fuels from starch feedstocks has been gratefully acknowledged.

REFERENCES

Alvira, P., E. Tomas-Pejo, M. Ballesteros and M. J. Negro. 2010. Pretreatment technologies for an efficient bioethanol production process based on enzymatic hydrolysis: A review. *Bioresource Technology* 101:4851–4861.

Angelici, C., B. M. Weckhuysen and P. C. A. Bruijnincx. 2013. Chemocatalytic conversion of ethanol into butadiene and other bulk chemicals. *ChemSusChem* 6:1595–1614.

Antolini, E. 2007. Catalysts for direct ethanol fuel cells. *Journal of Power Sources* 170:1–12.

Antolini, E. 2009. Palladium in fuel cell catalysis. *Energy and Environmental Science* 2:915–931.

Bai, F. W., W. A. Anderson and M. Moo-Young. 2008. Ethanol fermentation technologies from sugar and starch feedstocks. *Biotechnology Advances* 26:89–105.

Bothast, R. J. and M. A. Schlicher. 2005. Biotechnological processes for conversion of corn into ethanol. *Applied Microbiology and Biotechnology* 67:19–25.

Burnham, J. F. 2006. Scopus database: A review. *Biomedical Digital Libraries* 3:1–8.

Dai, D., Z. Hu, G. Pu, H. Li and C. Wang. 2006. Energy efficiency and potentials of cassava fuel ethanol in Guangxi region of China. *Energy Conversion and Management* 47:1686–1699.

Dien, B. S., G. Sarath and J. F. Pedersen, et al. 2009. Improved sugar conversion and ethanol yield for forage sorghum (*Sorghum bicolor* L. Moench) lines with reduced lignin contents. *Bioenergy Research* 2:153–164.

Dong, X., S. Ulgiati, M. Yan, X. Zhang and W. Gao. 2008. Energy and emergy evaluation of bioethanol production from wheat in Henan Province, China. Energy Policy 36:3882–3892.

Dunn, J. B., S. Mueller, H. Y. Kwon and M. Q. Wang. 2013. Land-use change and greenhouse gas emissions from corn and cellulosic ethanol. *Biotechnology for Biofuels* 6:51.

Fernando, S., S. Adhikari, C. Chandrapal and M. Murali. 2006. Biorefineries: Current status, challenges, and future direction. *Energy & Fuels* 20:1727–1737.

Gardebroek, C. and M. A. Hernandez. 2013. Do energy prices stimulate food price volatility? Examining volatility transmission between US oil, ethanol and corn markets. *Energy Economics* 40:119–129.

Graves, T., N. V. Narendranath, K. Dawson, K. and R. Power. 2006. Effect of pH and lactic or acetic acid on ethanol productivity by *Saccharomyces cerevisiae* in corn mash. *Journal of Industrial Microbiology and Biotechnology* 33:469–474.

Gwirtz, J. A. and M. N. Garcia-Casal. 2014. Processing maize flour and corn meal food products. *Annals of the New York Academy of Sciences* 1312:66–75.

Hahn-Hagerdal, B., M. Galbe, M. F. Gorwa-Grauslund, G. Liden and G. Zacchi. 2006. Bio-ethanol - The fuel of tomorrow from the residues of today. *Trends in Biotechnology* 24:549–556.

Hertel, T. W., A. A. Golub and A. D. Jones, et al. 2010. Effects of US maize ethanol on global land use and greenhouse gas emissions: Estimating market-mediated responses. *BioScience* 60:223–231.

Hettinga, W. G., H. M. Junginger and S. C. Dekker, et al. 2009. Understanding the reductions in US corn ethanol production costs: An experience curve approach. *Energy Policy* 37:190–203.

Hill, J., E. Nelson, D. Tilman, S. Polasky and D. Tiffany. 2006. Environmental, economic, and energetic costs and benefits of biodiesel and ethanol biofuels. *Proceedings of the National Academy of Sciences of the United States of America* 103:11206–11210.

Hill, J., S. Polasky and E. Nelson, et al. 2009. Climate change and health costs of air emissions from biofuels and gasoline. *Proceedings of the National Academy of Sciences of the United States of America* 106:2077–2082.

Hsieh, W. D., R. H. Chen, T. L. Wu and T. H. Lin. 2002. Engine performance and pollutant emission of an SI engine using ethanol-gasoline blended fuels. *Atmospheric Environment* 36:403–410.

Huang, H. J., S. Ramaswamy, U. W. Tschirner and B. V. Ramarao. 2008. A review of separation technologies in current and future biorefineries. *Separation and Purification Technology* 62:1–21.

Kim, S. and B. E. Dale. 2002. Allocation procedure in ethanol production system from corn grain: I. System expansion. *International Journal of Life Cycle Assessment* 7:237–243.

Kim, S. and B. E. Dale. 2005. Environmental aspects of ethanol derived from no-tilled corn grain: Nonrenewable energy consumption and greenhouse gas emissions. *Biomass and Bioenergy* 28:475–489.

Konur, O. 2000. Creating enforceable civil rights for disabled students in higher education: An institutional theory perspective. *Disability & Society* 15:1041–1063.

Konur, O. 2002a. Access to nursing education by disabled students: Rights and duties of nursing programs. *Nurse Education Today* 22:364–374.

Konur, O. 2002b. Assessment of disabled students in higher education: Current public policy issues. *Assessment and Evaluation in Higher Education* 27:131–152.

Konur, O. 2002c. Access to employment by disabled people in the UK: Is the Disability Discrimination Act working? *International Journal of Discrimination and the Law* 5:247–279.

Konur, O. 2006a. Participation of children with dyslexia in compulsory education: Current public policy issues. *Dyslexia* 12:51–67.

Konur, O. 2006b. Teaching disabled students in higher education. *Teaching in Higher Education* 11:351–363.

Konur, O. 2007a. A judicial outcome analysis of the *Disability Discrimination Act*: A windfall for the employers? *Disability & Society* 22:187–204.

Konur, O. 2007b. Computer-assisted teaching and assessment of disabled students in higher education: The interface between academic standards and disability rights. *Journal of Computer Assisted Learning* 23:207–219.

Konur, O. 2012. The evaluation of the research on the bioethanol: A scientometric approach. *Energy Education Science and Technology Part A: Energy Science and Research* 28:1051–1064.

Konur, O. 2015. Current state of research on algal bioethanol. In *Marine Bioenergy: Trends and Developments*, Ed. S. K. Kim and C. G. Lee, pp. 217–244. Boca Raton, FL: CRC Press.

Konur, O. 2019. Cyanobacterial bioenergy and biofuels science and technology: A scientometric overview. In *Cyanobacteria: From Basic Science to Applications*, Ed. A. K. Mishra, D. N. Tiwari and A. N. Rai, pp. 419–442. Amsterdam: Elsevier.

Konur, O. 2020. The scientometric analysis of the research on the bioethanol production from green macroalgae. In *Handbook of Algal Science, Technology and Medicine*, Ed. O. Konur, pp. 385–401. London: Academic Press.

Konur, O. 2023. First generation starch feedstock-based bioethanol fuel fuels: Scientometric study. In *Feedstock-based Bioethanol Fuels. I. Non-Waste Feedstocks: Starch, Sugar, Grass, Wood, Cellulose, Algae, and Biosyngas-based Bioethanol Fuels. Handbook of Bioethanol Fuels Volume 3*, Ed. O. Konur. Boca Raton, FL: CRC Press.

Kwiatkowski, J. R., A. J. McAloon, F. Taylor and D. B. Johnston. 2006. Modeling the process and costs of fuel ethanol production by the corn dry-grind process. *Industrial Crops and Products* 23:288–296.

Lakshmi, R. S., K. Kumari and P. Reddy. 2017. Corn germ meal (CGM)-potential feed ingredient for livestock and poultry in India-a review. *International Journal of Livestock Research* 7:39–50.

Lin, Y. and S. Tanaka. 2006. Ethanol fermentation from biomass resources: Current state and prospects. *Applied Microbiology and Biotechnology* 69:627–642.

Liska, A. J., H. S. Yang and V. R. Bremer, et al. 2009. Improvements in life cycle energy efficiency and greenhouse gas emissions of corn-ethanol. *Journal of Industrial Ecology* 13:58–74.

Ma, X., L. Sun and C. Song. 2002. A new approach to deep desulfurization of gasoline, diesel fuel and jet fuel by selective adsorption for ultra-clean fuels and for fuel cell applications. *Catalysis Today* 77:107–116.

Mechlem, K. 2004. Food Security and the right to food in the discourse of the United Nations. *European Law Journal*, 10:631–648.

Mojovic, L., S. Nikolic, M. Rakin and M. Vukasinovic. 2006. Production of bioethanol from corn meal hydrolysates. *Fuel* 85:1750–1755.

Morschbacker, A. 2009. Bio-ethanol based ethylene. *Polymer Reviews* 49:79–84.

Najafi, G., B. Ghobadian and T. Tavakoli, et al. 2009. Performance and exhaust emissions of a gasoline engine with ethanol blended gasoline fuels using artificial neural network. *Applied Energy* 86:630–639.

Newman, P. W. G. and J. R. Kenworthy. 1989. Gasoline consumption and cities: A comparison of U.S. cities with a global survey. *Journal of the American Planning Association* 55:24–37.

Nguyen, T. L. T., S. H. Gheewala and S. Garivait. 2007. Energy balance and GHG-abatement cost of cassava utilization for fuel ethanol in Thailand. *Energy Policy* 35:4585–4596.

Nikolic, S., L. Mojovic, L., M. Rakin, D. Pejin and J. Pejin. 2010. Ultrasound-assisted production of bioethanol by simultaneous saccharification and fermentation of corn meal. *Food Chemistry* 122:216–222.

North, D. C. 1991. Institutions. *Journal of Economic Perspectives* 5:97–112.

Olsson, L. and B. Hahn-Hagerdal. 1996. Fermentation of lignocellulosic hydrolysates for ethanol production. *Enzyme and Microbial Technology* 18:312–331.

Patzek, T. W. 2004. Thermodynamics of the corn-ethanol biofuel cycle. *Critical Reviews in Plant Sciences* 23:519–567.

Pimentel, D. and T. W. Patzek. 2005. Ethanol production using corn, switchgrass, and wood; biodiesel production using soybean and sunflower. *Natural Resources Research* 14:65–76.

Sanchez, O. J. and C. A. Cardona. 2008. Trends in biotechnological production of fuel ethanol from different feedstocks. *Bioresource Technology* 99:5270–5295.

Serra, T., D. Zilberman, J. M. Gil and B. K. Goodwin. 2011. Nonlinearities in the U.S. corn-ethanol-oil-gasoline price system. *Agricultural Economics* 42:35–45.

Shigechi, H., J. Koh and Y. Fujita, et al. 2004. Direct production of ethanol from raw corn starch via fermentation by use of a novel surface-engineered yeast strain codisplaying glucoamylase and α-amylase. *Applied and Environmental Microbiology* 70:5037–5040.

Srichuwong, S., M. Fujiwara and X. Wang, et al. 2009. Simultaneous saccharification and fermentation (SSF) of very high gravity (VHG) potato mash for the production of ethanol. *Biomass and Bioenergy* 33:890–898.

Sun, Y. and J. Cheng. 2002. Hydrolysis of lignocellulosic materials for ethanol production: A review. *Bioresource Technology* 83:1–11.

Szambelan, K., J. Nowak and Z. Czarnecki. 2004. Use of *Zymomonas mobilis* and Saccharomyces cerevisiae mixed with *Kluyveromyces fragilis* for improved ethanol production from Jerusalem artichoke tubers. *Biotechnology Letters* 26:845–848.

Taherzadeh, M. J. and K. Karimi. 2007. Enzyme-based hydrolysis processes for ethanol from lignocellulosic materials: A review. *Bioresources* 2:707–738.

Taherzadeh, M. J. and K. Karimi. 2008. Pretreatment of lignocellulosic wastes to improve ethanol and biogas production: A review. *International Journal of Molecular Sciences* 9:1621–1651.

Trujillo-Barrera, A., M. Mallory and P. Garcia. 2012. Volatility spillovers in U.S. crude oil, ethanol, and corn futures markets. *Journal of Agricultural and Resource Economics* 37:247–262.

Wang, M., J. Han, J. B. Dunn, H. Cai and A. Elgowainy. 2012. Well-to-wheels energy use and greenhouse gas emissions of ethanol from corn, sugarcane and cellulosic biomass for US use. *Environmental Research Letters* 7:045905.

Wang, M., M. Wu and H. Huo. 2007. Life-cycle energy and greenhouse gas emission impacts of different corn ethanol plant types. *Environmental Research Letters* 2:024001.

Wang, M. Q., J. Han and Z. Haq, et al. 2011. Energy and greenhouse gas emission effects of corn and cellulosic ethanol with technology improvements and land use changes. *Biomass and Bioenergy* 35:1885–1896.

Yang, Y., J. Bae, J. Kim and S. Suh. 2012. Replacing gasoline with corn ethanol results in significant environmental problem-shifting. *Environmental Science and Technology* 46:3671–3678.

Zhang, L., H. Zhao and M. Gan, et al. 2011. Application of simultaneous saccharification and fermentation (SSF) from viscosity reducing of raw sweet potato for bioethanol production at laboratory, pilot and industrial scales. *Bioresource Technology* 102:4573–4579.

46 Corn Pericarp Bioethanol through Steam Explosion and Periodic Peristalsis

Mst. Husne Ara Khatun, Hongzhang Chen, and Lan Wang
Chinese Academy of Sciences

46.1 CORN PERICARP: A PROSPECTIVE RENEWABLE SUGAR SUPPLY IN CORN ETHANOL INDUSTRY

There is evolution taking place at the corn ethanol plant worldwide, which is experiencing numerous changes. Corn ethanol manufacturers have more expectations than just primary ethanol production and want more value-added products from their existing facilities. They are seeking a fully optimized integrated system that heightens the value products of corn fiber. Corn fiber is described as captive feedstock for cellulose ethanol with less capital cost.

Bioethanol is prevalent in biofuel industries, mainly produced from corn starch through the dry-mill process and most industrialized in the USA and China (Li et al., 2019). According to Statista, the worldwide corn production was 1.2 billion metric tons in 2019–2020. According to the data, there are approximately 132 million tons of corn pericarp feedstock (1.1 billion metric tons corn production worldwide in the year 2018–2019) (Abdelaziz et al., 2016) available that could produce 15.45 billion gallons of cellulosic ethanol (Biorefineries Blog, 2017b). The USA is the top producer and consumer (about 12.3 billion bushels of corn) while China, the runner-up, produced and consumed about 11 billion and million bushels of corn, respectively (Shahbandeh, 2021). Corn is ethanol-producing essential grain in China, which is sharing 2.7 mn t among 86.8 mn t ethanol production worldwide. The country has not encouraged fossil fuel ethanol production and expects carbon pollution to be reduced by 2030, and its push on bioethanol could assist China to reach this goal and reduce its dependence on crude imports, which will help to increment its corn ethanol production (Teo, 2020).

Emergent demand for renewable energy is opening up the potential for cellulose and hemicellulose ethanol to become a reality, but it needs significant support for development. It could help explore the possible use of lignocellulosic residues already present at biomass ethanol plants to minimize the biomass's total expense to ethanol. Despite having well-developed technology to turn corn starch into ethanol, cellulose- and hemicellulose-enriched corn pericarp, which makes up about 7%–25% of the total sugar in corn kernel fiber, is not being utilized in the current process. The conversion of corn pericarp's cellulose and hemicellulose has great importance in the field, which not only increases the ethanol yield from cellulose and hemicellulose but also enhances starch conversion, reduces the viscosity of stream, and improves the protein content of distiller's dried grains with solubles (DDGS) (Gaspar et al., 2007).

46.2 TECHNOLOGY BEING USED IN CORN BIOETHANOL INDUSTRY

Recently, corn pericarp ethanol refineries are becoming popular to jack up the conversion to cellulosic biofuels. The number of corn fiber ethanol plants established, namely D3MAX, ICM, Quad County Corn Processors, and Edeniq, and their technology for corn fiber ethanol production HPGs put into practice.

DOI: 10.1201/9781003226451-59

The Distillers MAX (D3MAX) (Biorefineries Blog, 2017b; D3MAX, 2017a; Yancey, 2017) (called Bolt-on Technology) process uses advanced technology (patented) to convert residual starch and corn kernel fiber (all five and six-carbon sugars of starch, cellulose, xylose, and arabinose) into monomeric sugars in the form of wet cake by pretreatment and enzymatic hydrolysis, which ferment to cellulosic ethanol. The D3MAX technology cuts the volume of DDGS by about 20% by lowering the fiber content, increase corn oil recovery by 50%, and generating high-protein grain streams (HPGS) by about 40%, which are ideal for feed for mono-gastric animals like swine and poultry, and others as cattle and dairy. Several significant attributes of the D3MAX process, which cooked or pretreated wet cakes, can be retreated under mild pretreatment conditions (much lower temperature, pressure, and pH), and also wet cake allowing much greater total solid loads, which lowering capital, resource, and operating costs and reducing dryer energy by 20% (D3MAX, 2017a,b; Yancey, 2017).

The generation 1.5 grain fiber to cellulosic ethanol technology of ICM (Biorefineries Blog, 2017a; ICM, 2016) combines a method with existing ethanol plants to transform corn fiber to cellulosic ethanol. By integrating mechanical, chemical, and biological methods, the road of corn fiber to cellulosic ethanol achieved with the patented technology of ICM is called selective milling technology (SMT) and fiber separation technology (FST). SMT allows starch and oil more usable by milling selectively while the fiber is obtained from the stream by counter-flow washing by using FST. A dilute acid pretreatment is applied to break down the fiber stream. It releases the cellulose, which converts to sugars with an enzyme mixture by Novozymes and then cellulosic ethanol fermentation with advanced yeast DSM (ferment both the pentose and hexose sugars).

Cellerate, the Quad County Corn Processors (QCCP), is a partnership with the Syngenta 'Enogen Corn Enzyme Technology' (Biorefineries Blog, 2017b), which enables the processing the cellulose for ethanol production from the corn whole stillage fiber to capture the residual starch, sugars, and cellulosic components in a second fermentation, and has a 6% increase in ethanol yield and improve HPGS and corn oil yield. This process does not modify the conventional corn ethanol process. Syngenta has invented Enogen corn enzyme processing directly applied to the grain.

Edeniq's technique is an *in situ* conversion of corn kernel fiber into cellulosic ethanol by using cellulase and intellulose. Accessibility of the enzyme to starch and corn kernel fiber increases by mechanical milling by producing high shear strength using colloid mill technology (Perkins, 2018).

Fluid Quip Technologies (FQT) (FQT, 2017) has commercialized several patented technologies that are focused on improving the base corn-to-ethanol dry grind process, such as maximized stillage co-products (MSC), thin stillage clarity (TSC), selective grind technology (SGT), fiber by-pass separator (FBP), low-cost sugar stream (CST), and BOS corn oil recovery. The MSC technology yields a 50% protein product in high demand from alternativehigh-value protein markets and is a unique combination of spent brewer's yeast and corn gluten meal. TSC mainly extracts insoluble solids from the spent mash, supplying fine mash for starch content to improve. The SGT is used to expose more starch for conversion to ethanol, and it opens the germ to release more corn oil. FBP produces pure fiber that is an excellent source for cellulosic ethanol and other biochemical and generates HPGS.

CST produces a low-cost sugar stream for biochemical and high-value corn oil stream, animal feed, and HPGS, while the BOS system allows more oil extraction from liquefaction before fermentation. Fluid Quip Technologies' patented distillation technology offers versatility to corn ethanol processing with 40% less steam consumption in new and developed plants. Although mechanical milling has functional advantages, it does not completely open up the corn fiber structure, so the conversion rate is not optimal. All current corn fiber structure technologies require pretreatment to promote the effective enzymatic of cellulose and hemicellulose conversion to fermentable sugars, which are expensive and require additional equipment and energy (Li et al., 2019). Thus, the summary of commercial technologies for corn kernel fiber conversion is presented in Table 46.1.

46.3   CURRENT STATUS AND TECHNICAL PROSPECT OF HIGH SOLID CORN PERICARP BIOETHANOL

Cellulose, a major sugar component in the corn kernel, is not extracted in the current process. The potential of steam explosion as a physicochemical fractionation and defibrillation mechanism is recognized long before. Since then, SE has commonly used pretreatment technology to split apart the cell wall structure and extract biochemical and bionutrients from lignocellulosic materials. The steam-exploded wheat bran has shown the best recovery of soluble dietary fiber (2.08-fold higher than untreated one) at the optimum condition of 0.8 MPa for 5 min, and then, it began to decrease with severity (Sui et al., 2018). Bura et al. (2002a) employed several degrees of severity of steam explosion and found the optimum condition at 190°C for 5 min after exposure to 3% SO_2 (db), which recovered more than 50.3% of the original hemicellulose-derived sugars in water soluble with 22% xylose and arabinose, and 16% glucose at 6% SO_2 impregnation, while 98% glucose, 62% xylose, 56% arabinose, and 80% total sugars at 6% SO_2 impregnation recovered from unwashed SE treated corn fiber and lignin 29.4% in water-insoluble fractions (acid-soluble and insoluble). The amount of sugar derived product 5-HMF and furfural at this condition 0.24% and 1.17%, respectively. A steam explosion combined with acid hydrolysis increases the pore volume, specific surface area, reduces particle size, and makes the enzyme more available.

The conversion of corn cellulose is an important topic in the field of biochemistry (Table 46.2). Li et al. (2019) researched the corn fiber's sugar C_6/C_5 to produce bioethanol through an *in situ* acid pretreatment in the dry-mill process. Result has shown recycling of 40% hydrolysate increased 4.6%–7.9% total ethanol yield with 56.0%–69.2% cellulose conversion than a regular-dry mill. However, they have not found any benefit of recycling ratio over 40% led to incomplete fermentation, which indicates inhibitor production from acid pretreatment even in mild severity. The addition of cellulase increased ethanol, cellulose, and starch yields by 1.8%, 19.8%, and 99.0% relative to conventional SSF without cellulase (Li et al., 2018). Cellulase degrades cellulose and synergizes with glucoamylase to hydrolyze resistant starch by disrupting the substrate's cell wall structure and stimulating the release of contents, including starch and protein helps to increase starch ethanol yield. It also helps to minimize fermentation broth's viscosity and improves its rheological property (Knutsen et al., 2010), making it easier to handle during saccharification and fermentation (Li et al., 2018).

However, Kim et al. (2017) designed a dual impeller overhead mixer that provided a better performance 98% cellulose conversion of corn pericarp with multifect pectinase (27 mg protein/solid) at 15% (w/v) for shaking incubator 72%, where corn pericarp (cracked 5.1 mm–200 gm/800 mL buffer) enzymatically fractionated with Spezyme CP (0.84 FPU/g solids) and Promod 144 GL (0.9 gm protein/g solids) and incubated at 50°C for 36h at 100 rpm in an up-flow impeller, finally washed-up and screened with through 20 mesh size. They also performed the enzymatic hydrolysis of hemicellulose fraction with a high amount of hemicellulase enzyme (mainly xylanase and β-xylosidase), but without a pretreatment, it was difficult for the enzyme to reach the substrate, resulting in negligible xylan hydrolysis (Kim et al., 2017), because the complex inter-linkages with arabinan and ferulic acid bridge shield the pericarp xylan from enzyme penetration (Dien et al., 2008). The authors also mentioned that the ground pericarp (0.85 mm) released 22 times more total phenol (0.44 mg/L) than pericarp half-size (0.02 mg/L), which reflects the chemical compositions and enzymatic hydrolysis (Kim et al., 2017).

In contrast, ground fiber recovered from whole stillage by sieving process and pretreated with 0.5% v/v H_2SO_4 at 150°C for 20 min in a fluidized sand bath reactor and hydrolysis conducted at 10% solid loading with 5 FPU cellulase enzyme in an incubator shaker at 62°C and 150 rpm for 72 h showed higher glucose, xylose, and arabinose yields after hydrolysis than whole stillage, 90.78%, 92.93%, and 76.99%, respectively (Kim et al., 2016). Grohmann and Bothast (1997) investigated the dilute H_2SO_4 (100°C–160°C) on corn fiber to fractionate the structure and saccharify with enzyme cellulase and glucoamylase at 45°C after partial neutralization, which achieved a high 85% monomeric sugars conversion. Osborn and Chen (1984) reported that the destarched corn hull's hemicellulose fraction is readily hydrolyzable with dilute H_2SO_4 acid at 135°C after 5 min heat treatment steam.

TABLE 46.1

Summary of Commercial Technologies for Corn Kernel Fiber Conversion

Company	Technology	Category	Ethanol Conversion	Stage of Commercialization	Refs.
Edeniq	Edeniq/Intellulose	In situ process	Starch-not applicable, cellulose up to 2.5%	15 plants with EPA approval	Perkins (2018)
ICM	Selective Milling Technology (SMT) & Fiber Separation Technology (FST)	Fiber separation	Starch-up to 3%, cellulose not reported	A new plant built in Colwich, KS (ELEMENT, LLC, a collaborative project involving ICM and The Andersons, Maumee, OH	ICM (2016)
Fluid Quip Process Technology	Selective Grind Technology (SGT) & Fiber BY- Pass (FBP)	Fiber separation	Starch-up to 3%, cellulose up to 4.5%–8%	No public information	FQT (2017)
Quad County Corn Processors	QCCP Technology	Post-distillation residual solids (whole stillage)	Starch-not applicable, cellulose up to 9%	Only QCCP	QQCP (2021)
D3MAX	D3MAX	Post-distillation residual solids (wet cake)	Starch-not applicable, cellulose- not reported	Only Ace ethanol	D3MAX (2017a)

The fermentability of the destarched and washed residual fibrous material treated with alkaline 1% KOH and 2% NaOH (120°C, 1 h, 2 bar) and incubated at 37°C for 48 h and supplemented with 40 μL enzyme solution/g of cellulose of cellulase and 8 mL enzyme solution/g of cellulose of β-glucosidase and 5% w/w yeast, achieved ethanol yields of about 96.9% and 94.5%, respectively, (Gaspar et al. 2007). Moniruzzaman et al. (1997) optimized the ammonia fiber explosion (AFEX) to fractionate corn fiber (150% moisture content and ammonia and dry corn fiber ratio 1:1) in a reactor at 90°C, 30 min, and 200 PSI, and hydrolyzed with enzyme mixture (α-amylase, glucoamylase, cellulase, hemicellulase, and β-glucosidase), which produced 80% sugar and about 30%–40% of xylan degradation products, xylooligosaccharides. The corn fiber pretreated with liquid hot water pretreatment at 160°C for 5 and 20 min, converted to ethanol, was 76.6% and above 90% higher than untreated fiber (Kurambhatti et al., 2018; Mosier et al., 2005), and disk milling was not successful in increasing conversion (Kurambhatti et al., 2018). From long time, researchers have been working tirelessly to build corn pericarp ethanol plant, but there are many obstacles ahead before get-together of corn pericarp's all-inclusive practice in corn ethanol plant can become a reality, including the high-cost cellulase enzymes, the lack of efficacy of hemicellulases to degrade corn fiber xylans, and the development of inhibitory substances during pretreatment.

46.4 CHALLENGES IN CORN PERICARP BIOETHANOL INDUSTRY

46.4.1 RECALCITRANCE OF CORN PERICARP PHYSIOCHEMICAL PROPERTIES

The corn pericarp's cell wall is a complex, multilayered structure made up of complete and partial cells consisting of parenchyma tissue and vascular strands (Kiesselbach and Walker, 1952) (Figure 46.1). The outer pericarp comes to be composed of elongated cells with thick walls having large simple pits. The outer periphery of corn, named as corn pericarp, comprises a complex mixture of polysaccharides including cellulose, arabinoxylans, and β-glucans, as well as proteins and

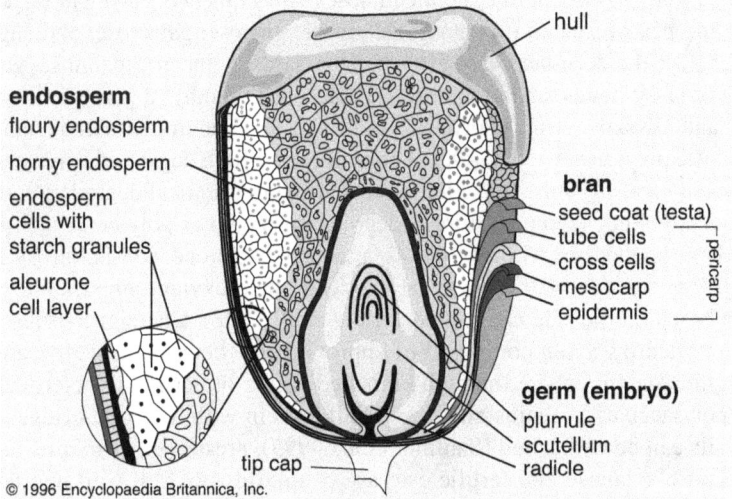

FIGURE 46.1 Structural arrangement of corn grain (Britannica, 2021). Corn grain has primary sections of endosperm, hull, bran, germ (embryo) and tip cap. Further, endosperm has floury and horny endosperms and bran has seed coat, tube cells, cross cells, mesocarp, and epidermis while germ has plumule, scutellum, and radicle.

esterified phenolic acids as a polymer compound; their contribution is essential both in the creation of the structural structure with which grain arrange and in their direct influence on its susceptibility bioprocessing of goods as a physical barrier for access to enzymes generated outside the cell (Evers et al., 1999). A better understanding of the corn pericarp's key components is essential to establish an effective bioconversion method. The plant cell wall comprises hemicellulose (20%–30%), which plays a significant role in the cell wall's structural stability; corn fiber has high hemicellulose content. Cellulose is embedded in the hemicellulose, while lignin binds the entire system and creates an intertwined tissue and cell network. Despite having an outstanding carbon source in corn pericarp, one of the foremost challenges is avoiding the comprehensive utilization of its elements due to the structural complexity, especially heteroxylan and cell wall networks (Perkins, 2018), which provides extreme resistance enzymatic attack (Kale et al., 2018). Conversion of cellulosic feedstocks into biofuels is tricky as they are incredibly recalcitrant, usually addressed with thermochemical pretreatment and a then substantial amount of cellulase and hemicellulase enzymes to unlock soluble carbohydrates. The comprehensive utilization concept demanded high ethanol production from available carbohydrates; structure and linkages generate challenges for even specialized xylanolytic organisms (Beri et al., 2020).

Corn pericarp is a mixture of adherent starch 11%–23% and cellulosic carbohydrate of 12%–25% cellulose, 18%–28% xylan, and 11%–19% arabinan (Bura et al., 2002b; Kim et al., 2017). Cellulose is a linear polymer of β-1, 4-linked D-glucopyranose molecules that serves as a primary structural component in plant cell walls (Vermerris and Abril, 2015). The amount of adherent starch differs according to the processing plant but should be between 15% and 20% or higher every day (Leathers, 2003). Amylose and amylopectin are two starch granule glucan polymers, composing 25%–30% and 70%–75% starch, linked by α-$(1\rightarrow4)$ glycosidic linkages of the linear structure of glucose molecules (amylose), and a branched-chain of the α-d-glucose molecule with α-$(1\rightarrow6)$ glycosidic linkage and α-$(1\rightarrow4)$ glycosidic-OH linear regions (amylopectin), respectively (Liu, 2020). On the other hand, Kiesselbach and Walker (1952) confirmed through microchemical experiments on the walls of xylem vessels that give the lignin reaction only a slight presence in the pericarp cells wall, and they are primarily composed of cellulose. Corn pericarp was commonly thought to be exceptionally small in lignin, but glucose chains formed into crystalline fibrils, nearly impenetrable to water or enzymes.

The cell wall heteropolysaccharide, hemicelluloses, are branched polymers of pentose (xylose and arabinose) and hexose sugars (mannose, rhamnose, glucose, galactose, and sugar acids) that makeup up to 35% of the corn pericarp. They include xylose, glucose, mannose, galactose in the backbone chain of-(1–4) bonds, and arabinose, galactose 4-O-methyl-d-glucuronic acid in the side chain (Scheller and Ulvskov, 2010; Olorunsola et al., 2018). Glucuronoarabinoxylan is the principally hemicellulosic polysaccharide (GAX) of corn fiber; the high degree of polymerization and its side chains' diversity and structure make it too recalcitrant to chemical degradation that demanded a synergistic cocktail or coenzyme mixture to degrade this complex polymer.

Xylans are heteropolymers with a β-(1–4) attached backbone consisting of D-xylose that makes up to 30% of the cell wall content. Homoxylan, heteroxylan, and arabinoxylan are subtypes of xylans based on the side chain (Olorunsola et al., 2018). Chateigner-Boutin et al. (2016) found that corn pericarp's xylan composed of approximately heteroxylans 50% and ferulic acid 5% with low lignin content, where the high branched nature of corn bran heteroxylan creating a network of its polysaccharide chains and a cell wall protein with diferulic bridges in which cellulose microfibrils can be embedded (Saulnier et al., 1995), presumably explains its resistance to degradation by pure xylanase and ferulic esterase or mixtures of cell wall degrading enzymes (Faulds and Williamson, 1994; Faulds et al., 1995, 1997). Amado and Neukom (1985) mentioned that a non-carbohydrate molecule is relatively abundant in cereal cell walls as ferulic acid and a carbonic phenol acid intimately associated with arabinoxylans and esterified to the primary alcoholic group of the arabinose side chain (Evers et al., 1999). Further, protein-polysaccharide interactions may explain CP's heteroxylans' insolubility, but difurilic acid bridges strengthen the general protein-polysaccharide network instead of insolubilizing a particular cell wall population of heteroxylans (Saulnier et al., 1995). Authors have found that these parameters contributed most to agriculture (pathogen resistance, silage digestibility), industries (bread-making quality, biomass conversion), and human health (dietary fiber) (Chateigner-Boutin et al., 2016). The remaining portion of corn pericarp is protein (11%–12%), oils (3%–4%), lipids (2.5%), and ash (less than 1%), which vary due to the content of tip cap, kernel, mainly depends on the distinct milling process (Bura et al., 2002b).

However, some of these side chains' structural arrangement is not yet fully clarified, and this understanding is essential to classify the hydrolytic enzyme responsible for their deconstruction. No investigations are known to have been performed on the study of the physical configuration of several layers of tissues and cells and their alteration after pretreatment the existence of tissues and cells. There is space for further research to open up the complexity of the physical structure of CP.

46.4.2 Corn Pericarp Inopportune Utilization and Its Potentiality in Bioindustry

The concept of biorefinery is the full usage of bioresources to promote energy independence, support domestic agriculture, and reduce the emission of global warming, greenhouse gases, and the increasing availability of bioresources for nonfood applications by improving food chain efficiency in the world. Despite having an outstanding carbon source in corn pericarp (CP), one of the foremost challenges is the comprehensive utilization of CP's polysaccharides, owing to the structural complexity, especially heteroxylan together with cell wall networks (Saulnier et al., 1995) providing extreme resistance to enzymatic attack (Kale et al., 2018).

The earlier practice of corn pericarp usage was minimal and mostly used as a low-value by-product as boiler fuel, animal, poultry, and swine diets with corn protein. Corn pericarp ethanol refineries have recently gained popularity as a way to speed up the conversion to cellulosic biofuels (Dien et al., 2005) along with producing high-value food components or additives, including corn fiber oil, film, gum, and gels, xylooligosaccharides, ferulic acid; food chemicals such as vanillin, and xylitol, (Rose et al., 2010) nutraceutical components, and the paper industry as an emulsifier Yadav et al. (2007, 2009, 2012), and flexible transparent films (Yoshida et al., 2013). Corn fiber oil produces large amounts of ferulate phytosterol esters (Moreau et al., 2005, 2010), which have nutraceutical benefits and currently shows to minimize blood serum cholesterol (Moreau et al., 1998, 2009). Phenolic acids, richly found in corn bran, can be used as preservatives, as crosslinking agents, as supplements, or as precursors to natural vanillin development. Corn heteroxylan is an attractive ingredient of potential prebiotic, and corn arabinoxylan is a functional food (Crittenden and Payne, 1996; Playne and Crittenden, 1997).

There have been various technological challenges facing the marketing of processes for isolating or processing valued products from corn bran or corn fiber, so more research should conduct to increase these goods' production. Owing to the fractionation of the corn before fermentation of the coproduct, the high-protein grain stream (HPSG) is equivalent in the formulation to the soya bean meal rather than DDGS. Besides, zein is a high-value corn protein extracted from HPGS, approximately 12–14 kg zein can be recovered from per metric ton of corn.

46.4.3 TECHNOLOGICAL INCOMPATIBILITY CHALLENGES

Biomass has a relatively well-ordered cell wall framework that has remedied the pretreatment phase to interrupt the entangled physiochemical network. Pretreatment must preferably conduct under conditions that allow the different biomass components to fractionate and extract efficiently, and the cellulosic fraction is opened to increase the rate and amount of enzyme hydrolysis. Pretreatment methods have usually been classified by, or a mixture of, biological, physical, and chemical systems. However, much of the operational costs still go to pretreatment for bioconversion. Emerging effective and productive pretreatment and its optimum operating conditions will overcome the burden of operational costs by increasing the physicochemical fractionation rate and yield by minimizing energy consumption, forming inhibitors, and without deforming the native cellulosic fractions with no added chemical and less or no water.

CP is a prominent substrate for cellulosic ethanol, and its physicochemical impediments for efficient bioconversion, could be overcome by applying particular pretreatment according to the necessities. Each pretreatment has its distinct characteristics of component separation, cell/tissue separation, or particle size reduction. Table 46.2 shows some of the pretreatments that have recently been used in corn pericarp and present with a subsequent enzymatic hydrolysis and fermentation process. However, several pretreatments are applied to integrate corn pericarp cellulosic ethanol into the mainstream of corn ethanol production such as chemical (dilute alkali (Gaspar et al., 2005; Yoshida et al., 2012) or acid pretreatment (Dien et al., 2005; Kim et al., 2016), physical (sieving (Kim et al., 2016), wet/disk milling (Kurambhatti et al., 2018; Li et al., 2018), liquid hot water (Kim et al., 2008; Kurambhatti et al., 2018; Mosier et al., 2005), pressure cooking (Weil et al., 1998), physicochemical (steam/ammonia fiber expansion) (Allen et al., 2001; Bura et al., 2002b; Kim et al., 2008). As we might be aware, the chemical neutralization and down-streaming processes are not environmentally friendly. Physical pretreatment methods such as mechanical grinding, on the other hand, cannot break the entangled network of fiber structure, especially in heteroxylan. In light of these facts, a perfect permutation of substrate, process, and processor will solve the issues mentioned earlier and recover the full-sugar conversion.

The key factor influencing the economic viability of high-solids enzymatic hydrolysis is the energy consumption of mixing, which determined by the mode of mixing, the size of the reactor and impeller, the mixing speed, the hydrolysis mixture property, etc. (Koppram et al., 2014; Modenbach and Nokes, 2013; Zhang et al., 2010), whereas the low solid load is uneconomic (below 15%). As the solid loading rate increases, the amount of power required to mix the hydrolysis solution increases to an unsustainable level for stirred tank reactors (Modenbach and Nokes, 2013; Palmqvist et al., 2011; Palmqvist and Liden, 2012). Therefore, new sustainability approaches must be built to promote green bioindustry in both the academic and industrial fields. Researchers have recently developed various groundbreaking biomass bioconversion reactors with a high solid material, focusing mainly, on enzymatic hydrolysis and the SSF method to produce ethanol.

TABLE 46.2

Comparative Study of Different Technology Recently Applied in Corn Pericarp' Sugar/Bioethanol

Substrate	Pretreatment	System	Solid Loading	Hydrolysis Enzyme	Temp	Time	RPM	Yield	Refs.
				Hydrolysis System and Condition					
Ground whole stillage, pretreatment, NDF/corn fiber	1% Tween 80, 0.5% v/v H_2SO_4, 150°C, 20min	Fluidized sand bath, incubator shaker	10%	Protease, cellulase	48°C, 62°C	2h, 72h	Mix once per 30min, 150	90.78% Glucose, 92.93% xylose & 76.99%, arabinose	Kim et al. (2016)
Ground corn	50% Urea (0.4%)	Incubator Shaker	32%	α-amylase, glucoamylase, cellulase, hemicellulase	85°C, 32°C	73h	150	9.01% glucose increased	Kurambhatti et al., 2018
Corn flour, corn fiber	Urea (1.28 g/kg mash)	Water bath shaker	30%	Alpha-amylase, glucoamylase, cellulase	85°C, 60°C, 50°C	196h	150	Starch 99%, cellulose 19.8% conversion	Li et al. (2018)
Cracked corn, corn pericarp	Enzymatic fractionation, no pretreatment (unground-Cracked 5.1mm)	Up-flow Impeller, a dual impeller overhead mixer in a water bath	25% (w/v) 15%	SpezymeCP cellulase, hemicellulose Promod GL-Protease Multifect pectinase, cellulase, hemicellulose	50°C, 50°C	36h, 7h	100, 290	Cellulose 98% conversion	Kim et al. (2017)
Corn fiber	Steam explosion (190°C) for 5 with 3% SO_2	Shaker	2%	α-amylase, glucoamylase, cellulase post hydrolysis, (3% H_2SO_4, 1hr, 120°C)	80°C, 65°C, 45°C	72h	200	81% all polysaccharide, 90%–96% ethanol	Bura et al. (2002a,b)
Corn pericarp	Steam explosion, 0.8MPa (185°C) for 5 min (unground)	Periodic peristalsis enzymatic hydrolysis system (SE-PPEH)	30%	A-amylase, glucoamylase, cellulase	80°C, 65°C, 45°C	48h	70	97% glucan and 87% xylan conversion	Khatun et al. (2020)

Another critical factor affecting enzymatic hydrolysis, especially, is water constraint at high-solids loadings (Liu and Chen, 2015). Water can be found in and around cellulose in lignocellulosic biomass in several states and locations, creating various water pools that restrict the availability of water, which has a significantly adverse effect on the diffusion processes and the reactions, and hence the efficiency of enzymatic hydrolysis, particularly at high-solids loadings. If the solids content of substrate was more than 15% (w/w) at the start of the enzymatic hydrolysis, the external-particle system is almost no, or almost no free-flowing water and soluble species may hinder the mobility of water, and the substance became increasingly viscous as time passes (Modenbach and Nokes, 2013).

46.5   SEPP CORN PERICARP BIOCONVERSION: A NOVEL APPROACH

In 2012, our research groups developed an eco-friendly steam explosion technology by originating a series of innovative methods to produce bioproducts and applied for several patents (steam explosion and periodic peristalsis, SEPP) (Chen, 2006, 2007, 2008, 2009, 2011a,b, 2015; Chen et al., 2009, 2011a,b; Li and Chen, 2014). Steam explosion is a nonpolluting and low-pressure pretreatment, which can realize the sufficing of energy demands. It can be dropping chemical addition and restraining the formation of inhibitors that induced fermentative microorganisms (Chen, 2015; Li and Chen, 2014). The steam explosion process's allocation is swelling, cooking, fragmentation, and disruption to accelerate lignocellulose's degradation to small compounds. In this entire process, water is the only heat transfer medium related to pretreatment energy consumption and the severity of refining efficiency (Sui and Chen, 2015). The steam explosion's performance depends on several dynamics such as loading coefficient (loading pattern, chip size, moisture content) and mechanical property. SE is a technologically and economically feasible pretreatment solution for corn pericarp due to its lower severity requirements for transformed CP to recover total sugars minimal inhibitor concentrations (Bura et al., 2002a, 2002b).

The theory of periodic peristalsis agitation derives from the bionic principle of stomach and intestine in a constant cycle of expansion and contraction that mix the food and digestive juices (mostly digestive enzyme) (Chen, 2018) in PPAB (Chen and Xi, 2012), the biomass substrate, and the reactor's parts isolated by an elastic casing. The elastic bag's substrate movement directs it along with the elastic bag and periodic expansion and contraction stimulation. When the elastic bag extends externally, the substrate's movement direction moves along with the flow to create a natural acting force perpendicular to the movement direction, and reverse force occurs when the elastic bag contracted. The reactor materials alternately do peristaltic movement under the periodic normal acting force due to the periodic expansion and contraction of the system's materials, enhancing heat and mass transfer (Chen and Xi, 2012). Xia et al. (2015) investigated the mechanism of interaction of periodic peristalsis agitation effects on cell physiology. They reported higher-level hydrodynamic damage on cell growth due to high shear force from traditional agitation.

In contrast, the cell membrane rearranges its metabolism to resist shear forces by modulating membrane fluidity via ATP, glycolysis, and acid synthesis at the expense of the pentose phosphate pathway. On the other hand, Periodic peristalsis agitation causes less hydrodynamic damage to cells by enhancing heat and mass transfer by high turbulence frequency rather than high velocity or operating under a favorable environment. Furthermore, the increase in cell permeability due to periodic peristalsis induces the fluctuation of extracellular pressure influencing cell flux distribution (Xia et al., 2015). Periodic peristalsis reduces water constraint (Liu and Chen, 2015), the initial viscosity of hydrolysis mixture, shorten the time of transition point, the denaturation effects of enzymes (Liu and Chen, 2015). Periodic peristalsis can produce high solid conversion by reducing the solid's effect by strengthening the mass and heat transfer, promoting the accessibility of enzymes, and increasing the substrate rate fractionation (Liu and Chen, 2016a,b,c). This feature makes periodic-peristole agitation unique from conventional agitation methods.

46.5.1 STEAM EXPLOSION ON THE PHYSICAL AND CHEMICAL PROPERTIES ALTERATION OF CORN PERICARP

Steam explosion (SE) is an efficient and productive pretreatment, allowing for creating porous structures and an increase in reactive/specific surface area during the steam explosion, thus improving mass transfer rate and enzyme accessibility (Figure 46.2) (Chen and Wang, 2016; Zhao and Chen, 2013).

The untreated CP's intact, textile-like smooth surface appeared to be cellulosic fiber with starch granules surrounding the surfaces while the steam explosion substantially ruptured the structure and shortened the fiber size of corn pericarp. It became more porous, erratic, and honeycomb-like on the surface and transitional layer, which may be facilities more enzyme and solvent access to polysaccharides as they are degraded into glucose (Khatun et al., 2020) allowing for a more efficient fermentation process. The starch layer seemed to be in the melted state, and also the node like effects appeared probably cellulosic material, mainly hemicellulose. Due to the tremendous explosion and shearing force of SE, the hemicellulose and lignin wrapping around cellulose may soften and most likely all hydrogen bonds between the component structure of CP will be disrupted (Berry and Roderick, 2005; Sui et al., 2018).

Several alternative technologies apply to observe the structural alteration of the DDGS in the literature (Wang et al., 2009), a smooth and texture-like structure found in untreated one with a significant number of starch granules, and the same pattern is often seen in corn pericarp (Ibrahim et al., 2019; Liu et al., 2017). The flocks of tiny pores formed and only a few starches intact on the surfaces after a hot water pretreatment (160°C for 30 min) when AEW-acidic electrolyte water (H_2SO_4-pH 2.7) treated samples distinguished by more structural alterations, large cracks detected on the fiber matrix; these porous surfaces can increase the enzyme accessibility and digestibility (Liu et al., 2017). In comparison with hydrothermal treatment of maize bran using twin-screw extrusion at 180°C (Liu et al., 2017) the SEM image of SECP revealed a significantly greater degree of structural fractionation. Additionally, Berry and Roderick (2005) discussed the plant-water relations and fiber saturation points where they noted that the cell wall molecules are held together by a strong hydrogen bond, forming a rigid structure that is inaccessible to water molecules, but the amorphous regions are capable of forming a bond with water molecules. Also, Browning (1963) notes that when tissues come into contact with liquid or mist, water molecules establish strong hydrogen bonds with the exposed hydroxyl groups on the cell wall molecules.

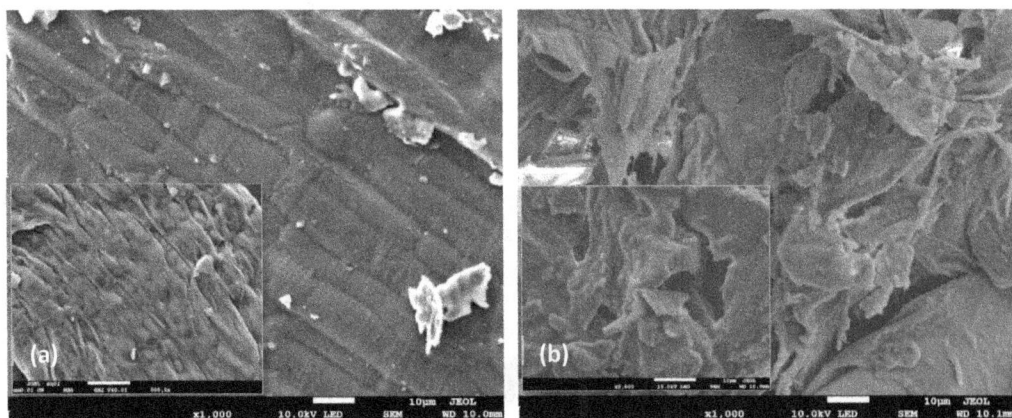

FIGURE 46.2 Structure alteration of (a) untreated corn pericarp into (b) steam-exploded corn pericarp: Scanning electron micrograph (1,000x and 2,000x magnification). These micrographs show the effect of the steam explosion pretreatment on the microstructure of corn pericarp: Creation of porous structures and an increase in reactive/specific surface area during the steam explosion.

Corn pericarp composition varies significantly depending on the source of the material and the analytical procedures employed to assess the content (Table 46.3). Adherent starch levels vary by manufacturing facility and day-to-day, but can be as high as 15%–20% (Leathers, 2003). The steam-exploded corn pericarps (SECPs), on the other hand, have exhibited a greater degree of compositional alteration than untreated corn pericarps (UCPs). With increasing severity, the glucan level ranged between 25.44% and 36.14% (cellulose) and 21.98%–17.70% (starch). SECPs recovered more total glucans by 20.21%–36.47% as compared to UCP. The results of the experiment demonstrated that pretreatment with SE enhanced the release of adherent and bound starch granules from the complicated skeletal structure of cellulose in milder conditions than in severe conditions, while milder conditions demonstrated greater glucan recovery from cellulose.

However, SECPs contained lower hemisugars (xylose and arabinose) than UCPs. SECPS had a xylan content less than or almost equal to UCP (22.62%), ranging from 19.74% to 23.22%, and an arabinan content steadily declined (13.86%–11.08%) with higher severity, compared with UCP (14.34%). The results indicated that the xylan content was higher under low pressure and for a shorter retention time (5 min) than under higher pressure and for a longer retention time (10 min). This conclusion is because xylan reacts more rapidly than glucan and also hydrolyzes for a longer amount of time during SE, whereas glucan is more resistant to harsh circumstances than xylan.

46.5.2 PERIODIC PERISTALSIS ON THE STEAM-EXPLODED CORN PERICARP

Periodic peristalsis (PP) boosted the effectiveness of enzymatic hydrolysis than the static state and in an incubator shaker with high-solids content (Figure 46.3). In comparison with the incubator shaker hydrolysis method, periodic peristalsis enzymatic hydrolysis (PPEH) improved glucan conversion by reducing hardness and completing operations before 24 h. The generation of butanol in periodic-peristole was increased by 2.06-fold in the typical Rushton impeller and by 1.19-fold in the stationary state (Xia et al., 2015). Khatun et al. (2021) found that periodic peristalsis stimulated glucan and xylan hydrolysis almost double as much as a water bath shaker in unground steam-exploded corn pericarp. PP had a better effect on 30% solid load than 25% solid load compared with water bath shaker. Periodic peristalsis shortened the point of transition from a solid to a slurry state, resulting in a decrease in viscosity during the initial stage of hydrolysis. Periodic peristalsis at a low speed protects enzyme proteins against denaturation caused by high-speed shear forces, which are responsible for the loss of binding capacity and enzyme activity in a shaking system (Liu and Chen, 2015, 2016b).

Without pretreatment, the glucan and xylan in ground corn pericarp were hydrolyzed to 76.80% and 22.10% of their raw composition, respectively, in a shaker hydrolysis system. The steam-exploded CP with the same particle size achieved better outcomes, increasing by 18% and 218% of

TABLE 46.3
Chemical Composition of Untreated and Steam-Exploded Corn Pericarp (% Dry Basis)

	% of Sugars Compositions						
SE Severity (Log R0)	Glucan from Starch	Glucan from Cellulose	Total Glucan	Xylan	Arabinan	Total Sugars	% Total Sugars Recovery
2.47	21.98	25.44	47.42	20.93	13.86	82.20	7.16
2.77	21.20	26.91	48.10	22.30	12.57	82.98	8.17
2.91	20.36	30.20	50.56	23.22	12.61	86.39	12.62
3.21	18.20	34.84	53.04	20.79	12.14	85.97	12.07
3.05	18.77	34.	53.52	21.63	12.49	87.63	14.24
3.36	17.70	36.14	53.84	19.74	11.08	84.66	10.36
UCP	19.70	19.75	39.45	22.62	14.34	76.71	

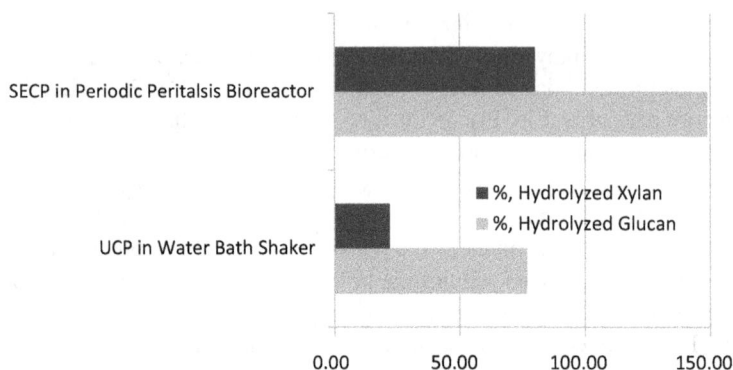

FIGURE 46.3 The hydrolyzed glucan and xylan yield on the basis of the untreated corn pericarp and comparative demonstration of periodic peristalsis and water bath shaker system. It shows steam-exploded corn pericarp in periodic peristalsis and untreated corn pericarp in water bath shaker for both hydrolyzed xylan and glucan. It shows that periodic peristalsis boosted the effectiveness of enzyme hydrolysis than the static state and in an incubator shaker with high-solids content.

UCP raw composition, respectively, and by 148% and 80% of their raw composition, respectively. Indication of the importance of SE pretreatment to break down the complex structural and chemical network, particularly heteroxylan and cell wall networks (Saulnier et al., 1995) exhibiting a high level of resistance to enzymatic attack (Kale et al., 2018). Periodic peristalsis generates vertical pressure directly on porous particles, compressing them. The pressure is sufficient to force the constrained water to migrate, and the steam-exploded substrate transformed from a porous medium to a suspension of solids via an enzymatic hydrolysis progression (Liu and Chen, 2016a).

The maximum amount of hydrolyzed glucan and xylan was found in unground SECP and was raised by 22.62% and 24.28%, respectively, when compared to ground SECP. The results suggested that grinding was not an effective strategy for increasing bioconversion efficiency (Khatun et al., 2020; Kim et al., 2017). On the other hand, an increase in the initial viscosity of the digestibility mixture of ground UCP, which seems unable to resolve in a shaker hydrolysis system with a high solid load, affected mass and heat transfer efficiency, hence reducing glucan and xylan conversion (Liu and Chen, 2016a,b, Liu et al., 2013). However, with a 15 FPU/g solid enzyme dose in SE conditions of 0.8 MPa/5 min, the conversion of glucan and xylan achieved around 97% and 87%, respectively.

The increased sugar recovery rate in SE-PPEH was attributed to the released free water at high solid loading (Selig et al., 2014). Further, steam explosion (SE) improved the interaction of matrix and water, as well as improved the accessibility of polysaccharides to enzymes, due to the disruption of structure, which facilitates the release of constraint water (Chen, 2011a,b) and periodicity. The increased sugar recovery rate in SE-PPEH was attributed to the released free water at high solid loading (Selig et al., 2014). Further, steam explosion improved the interaction of matrix and water, as well as improved the accessibility of polysaccharides to enzymes, due to the disruption of structure, which facilitates the release of constraint water. Finally, periodic peristalsis (PP) assisted in releasing more confined water and boosted mass and heat transfer efficiency in comparison with water bath shakers (Liu and Chen 2016a,b).

46.6 FUTURE PERSPECTIVE AND CONCLUSIONS

Corn, corn stover, and corn pericarp could be utilized to produce bioethanol as part of an integrated biorefinery concept that combines bioresources, bioconversion technologies, and bioproducts. The combination of steam explosion and periodic peristalsis (SEPP) is a novel approach to producing value-added intermediate products alongside biofuels while reducing chemical complexity and utility costs. SEPP has the ability to considerably increase the enzymatic digestibility of its total polysaccharides. The boosting impact of SEPP can replace the energy-intensive conventional sub-process of grinding by reducing viscosity and maintaining structural porosity. This type of integrated approach can be implemented in existing corn-to-ethanol plant's to take the benefits of the infrastructure, resources, technology, and expertise. As a result, a more environment-friendly and economical viable biorefinery with zero emissions can be achieved.

REFERENCES

Abdelaziz, O. Y., D. P. Brink and J. Prothmann, et al. 2016. Biological valorization of low molecular weight lignin. *Biotechnology Advances* 34:1318–1346.

Allen, S. G., D. Schulman and J. Lichwa, et al. 2001. A comparison between hot liquid water and steam fractionation of corn fiber. *Industrial & Engineering Chemistry Research* 40:2934–2941.

Amado, R. and H. Neukom. 1985. Minor constituents of wheat flour: The pentosans, *Progress in Biotechnology* 1:241–251.

Beri, D., W. S. York, L. R. Lynd, M. J. Pena and C. D. Herring. 2020. Development of a thermophilic coculture for corn fiber conversion to ethanol. *Nature Communications* 11:1–11.

Berry, S. L. and M. L. Roderick. 2005. Plant-water relations and the fibre saturation point. *New Phytologist* 168:25–37.

Biorefineries Blog. 2017a. *Corn Fiber Ethanol - Examining 1.5G Technologies.* https://biorrefineria.blogspot. com/2017/04/corn-fiber-ethanol-examining-1.5g-technologies-biorefineries.html

Biorefineries Blog. 2017b. *Corn Fiber Ethanol-Examining 1.5G Technologies.* https://biorrefineria.blogspot. com/2017/04/corn-fiber-ethanol-examining-1.5g-technologies-biorefineries.html

Britannica. 2021. *Cereal Processing Technology.* https://www.britannica.com/technology/cereal-processing/ Nonwheat-cereals#ref50113.

Browning, B. L. 1963. The wood-water relationship. In *The Chemistry of Woo*, Ed. B. L. Browning, New York: Interscience Publishers.

Bura, R., R. J. Bothast, S. D. Mansfield and J. N. Saddler. 2002a. Optimization of SO₂-catalyzed steam pretreatment of corn fiber for ethanol production. In *Biotechnology for Fuels and Chemicals*, Ed. J. D. McMillan, J. R. Mielenz, W. S. Adney and K. C. Klasson, pp. 319–335. Cham: Springer.

Bura, R., S. D. Mansfield, J. N. Saddler and R. J. Bothast. 2002b. SO₂-catalyzed steam explosion of corn fiber for ethanol production. In *Biotechnology for Fuels and Chemicals*, Ed. J. D. McMillan, J. R. Mielenz, W. S. Adney and K. C. Klasson, pp. 59–72. Cham: Springer.

Chateigner-Boutin, A. L., J. J. Ordaz-Ortiz and C. Alvarado, et al. 2016. Developing pericarp of maize: A model to study arabinoxylan synthesis and feruloylation, *Frontiers in Plant Science* 7:1476.

Chen, H. Z. 2006. *A Method to Produce Straw Polyol by Liquefaction.* Chinese patent. 200610114197.7 [P].

Chen, H. Z. 2007. *A Method for the Production of Nano-Silica from Straw.* Chinese patent. 200710062669.3 [P].

Chen, H. Z. 2008. *A Method to Selectively Liquefy Lignocellulose by Phenol.* Chinese patent. 200810102984.9 [P].

Chen, H. Z. 2009. *A Method for the Acetone-Butanol Fermentation using Xylose from Steam-Exploded Straw and the Extraction of Residues.* Chinese patent.

Chen, H. Z. 2011a. *A Method for Direct Preparation of Furfural by Acid-Free Self-Catalysis of Water Washed Solution of Steam Exploded Straw.* Chinese patent. 200810134421.8 [P].

Chen, H. Z. 2011b. *A Method for the Production of Furfural, Polyether Polyols, Phenolic Resins and Nano-Silica from Straw.* Chinese patent. 201110034318.8 [P].

Chen, H. Z., Ed. 2015. *Lignocellulose Biorefinery Engineering: Principles and Applications.* Cambridge: Woodhead Publishing.

Chen, H. Z. 2018. Periodic intensification principles and methods of high-solid and multi-phase bioprocesses. In *High-solid and Multi-phase Bioprocess Engineering*, Ed. H. Z. Chen, pp. 173–241. Cham: Springer.

Chen, H. Z., M. Baohua and S. Xiaodong. 2009. *A Method for the Acid Hydrolysis of Straw and Separation of Furfural from the Hydrolyzate.* Chinese patent. 200910088098.X [P].

Chen, H. Z., G. Wang, J. Qi and Y. Zheng. 2011b. *A Method to Fractionate Lignin Sulfonate from Paper-Making Red Liquid and Convert it into High Value Products.* Chinese patent. 201110393445.7 [P].

Chen, H. Z. and L. Wang. 2016. *Technologies for Biochemical Conversion of Biomass.* London: Academic Press.

Chen, H. Z. and F. Xi. 2012. *Periodic Peristaltic Agitation Method.* Chinese patent.

Chen, H. Z., Z. Yanmin and Q. Jianhua. 2011a. *A Method for the Utilization of Sugar Component in Paper-Making Red Liquid.* Chinese Patent. 201110393509.3 [P].

Crittenden, R. A. and M. J. Playne. 1996. Production, properties and applications of food-grade oligosaccharides. *Trends in Food Science & Technology* 7:353–361.

D3MAX. 2017a. *D3Max Pilot Test Results at ACE Ethanol Exceed Expectations*. BBI International. D3Max pilot test results at ACE Ethanol exceed expectations I EthanolProducer.com

D3MAX. 2017b. *The New Industry Standard for Energy Efficiency, Yield and D3RIN Production*.

Dien, B. S., D. B. Johnston, K. B. Hicks, M. A. Cotta and V. Singh. 2005. Hydrolysis and fermentation of pericarp and endosperm fibers recovered from enzymatic corn dry-grind process. *Cereal Chemistry* 82:616–620.

Dien, B.S., E. A. Ximenes and P. J. O'Bryan, et al. 2008. Characterization for hydrolysis of AFEX and liquid hot-water pretreated distillers' grains and their conversion to ethanol. *Bioresource Technology* 99:5216–5225.

Evers, A., L. O'Brien and A. Blakeney. 1999. Cereal structure and composition. *Australian Journal of Agricultural Research* 50:629–650.

Faulds, C. B., B. Bartolome and G. Williamson. 1997. Novel biotransformations of agro-industrial cereal waste by ferulic acid esterases. *Industrial Crops and Products* 6:367–374.

Faulds, C. B., P. A. Kroon, L. Saulnier, J. F. Thibault and G. Williamson. 1995. Release of ferulic acid from maize bran and derived oligosaccharides by *Aspergillus niger* esterases. *Carbohydrate Polymers* 27:187–190.

Faulds, C. B. and G. Williamson. 1994. Purification and characterization of a ferulic acid esterase (FAE-III) from *Aspergillus niger:* Specificity for the phenolic moiety and binding to microcrystalline cellulose. *Microbiology* 140:779–787.

FQT. 2017. *Proven Technologies - Fluid Quip Technologies*. https://fluidquiptechnologies.com/

Gaspar, M., T. Juhasz, Z. Szengyel and K. Reczey. 2005. Fractionation and utilisation of corn fibre carbohydrates. *Process Biochemistry* 40:1183–1188.

Gaspar, M., G. Kalman and K. Reczey. 2007. Corn fiber as a raw material for hemicellulose and ethanol production. *Process Biochemistry* 42:1135–1139.

Grohmann, K. and R. J. Bothast. 1997. Saccharification of corn fibre by combined treatment with dilute sulphuric acid and enzymes. *Process Biochemistry* 32:405–415.

Ibrahim, M., S. M. Sapuan, E. Zainudin and M. M. Zuhri. 2019. Extraction, chemical composition, and characterization of potential lignocellulosic biomasses and polymers from corn plant part. *BioResources* 14:6485–6500.

ICM. 2016. *ICM Advances Pathway to Cellulosic Ethanol*. https://ethanolproducer.com/articles/13450/icm-advances-pathway-to-cellulosic-ethanol

Kale, M. S., M. P. Yadav, H. K. Chau and A. T. Hotchkiss Jr. 2018. Molecular and functional properties of a xylanase hydrolysate of corn bran arabinoxylan. *Carbohydrate Polymers* 181:119–123.

Khatun, M. H. A., L. Wang and H. Z. Chen. 2020. High solids all-inclusive polysaccharide hydrolysis of steam-exploded corn pericarp by periodic peristalsis. *Carbohydrate Polymers* 246:116483.

Khatun, M. H. A., L. Wang and H. Z. Chen. 2021. *Steam Explosion of Corn Pericarp and Corn Stover: High-Solid Enzymatic Hydrolysis of Steam-Exploded Corn Pericarp by Periodic Peristalsis and Techno-Economic Analysis of Fractionation of Steam-Exploded Corn Stover*. Beijing: University of Chinese Academy of Sciences.

Kiesselbach, T. A. and E. R. Walker. 1952. Structure of certain specialized tissue in the kernel of corn. *American Journal of Botany* 39:561–569.

Kim, D., D. Orrego, E. A. Ximenes and M. R. Ladisch. 2017. Cellulose conversion of corn pericarp without pretreatment. *Bioresource Technology* 245:511–517.

Kim, S. M., S. Li and S. C. Pan, et al. 2016. A whole stillage sieving process to recover fiber for cellulosic ethanol production. *Industrial Crops and Products* 92:271–276.

Kim, Y., R. Hendrickson and N. S. Mosier, et al. 2008. Enzyme hydrolysis and ethanol fermentation of liquid hot water and AFEX pretreated distillers' grains at high-solids loadings. *Bioresource Technology* 99:5206–5215.

Knutsen, J. S. and M. W. Liberatore. 2010. Rheology modification and enzyme kinetics of high solids cellulosic slurries. *Energy & Fuels* 24:3267–3274.

Koppram, R., E. Tomas-Pejo, C. Xiros and L. Olsson. 2014. Lignocellulosic ethanol production at high-gravity: Challenges and perspectives. *Trends in Biotechnology* 32:46–53.

Kurambhatti, C. V., D. Kumar, K. D. Rausch, M. E. Tumbleson and V. Singh. 2018. Ethanol production from corn fiber separated after liquefaction in the dry grind process. *Energies* 11:2921.

Leathers, T. D. 2003. Bioconversions of maize residues to value-added coproducts using yeast-like fungi. *FEMS Yeast Research* 3:133–140.

Li, G. and H. Chen. 2014. Synergistic mechanism of steam explosion combined with fungal treatment by *Phellinus baumii* for the pretreatment of corn stalk. *Biomass and Bioenergy* 67:1–7.

Li, X., S. Chen, H. Huang and M. Jin. 2018. *In-situ* corn fiber conversion improves ethanol yield in corn dry-mill process. *Industrial Crops and Products* 113:217–224.

Li, X., Z. Xu, J. Yu, H. Huang and M. Jin. 2019. *In situ* pretreatment during distillation improves corn fiber conversion and ethanol yield in the dry mill process. *Green Chemistry* 21:1080–1090.

Liu, H. M., H. Y. Li and A. C. Wei. 2017. Enhanced polysaccharides yield obtained from hydrothermal treatment of corn bran via twin-screw extrusion. *Bioresources* 12:3933–3947.

Liu, S. 2020. *Bioprocess Engineering: Kinetics, Sustainability, and Reactor Design*. Amsterdam: Elsevier.

Liu, Z. H. and H. Z. Chen. 2015. Xylose production from corn stover biomass by steam explosion combined with enzymatic digestibility. *Bioresource Technology* 193:345–356.

Liu, Z. H. and H. Z. Chen. 2016a. Periodic peristalsis releasing constrained water in high solids enzymatic hydrolysis of steam exploded corn stover. *Bioresource Technology* 205:142–152.

Liu, Z. H. and H. Z. Chen. 2016b. Periodic peristalsis enhancing the high solids enzymatic hydrolysis performance of steam exploded corn stover biomass. *Biomass and Bioenergy* 93:13–24.

Liu, Z. H. and H. Z. Chen. 2016c. Mechanical property of different corn stover morphological fractions and its correlations with high solids enzymatic hydrolysis by periodic peristalsis. *Bioresource Technology* 214:292–302.

Liu, Z. H., L. Qin and F. Pang, et al. 2013. Effects of biomass particle size on steam explosion pretreatment performance for improving the enzyme digestibility of corn stover. *Industrial Crops and Products* 44:176–184.

Modenbach, A. A. and S. E. Nokes. 2013. Enzymatic hydrolysis of biomass at high-solids loadings-a review. *Biomass and Bioenergy* 56:526–544.

Moniruzzaman, M., B. E. Dale, R. B. Hespell and R. J. Bothast. 1997. Enzymatic hydrolysis of high-moisture corn fiber pretreated by AFEX and recovery and recycling of the enzyme complex. *Applied Biochemistry and Biotechnology* 67:113–126.

Moreau, R. A., K. B. Hicks, D. B. Johnston and N. P. Laun. 2010. The composition of crude corn oil recovered after fermentation via centrifugation from a commercial dry grind ethanol process. *Journal of the American Oil Chemists' Society* 87:895–902.

Moreau, R. A., K. B. Hicks, R. J. Nicolosi and R. A. Norton. 1998. *Corn Fiber Oil its Preparation and Use*. U.S. Patent No. 5,843,499. Washington, DC: U.S. Patent and Trademark Office.

Moreau, R.A. and M. E. Hums. 2005. Corn oil. In *Bailey's Industrial Oil and Fat Products*, Ed. F. Shahidi, pp. 169–186. Hoboken, NJ: Wiley.

Moreau, R. A., V. Singh, M. J. Powell and K. B. Hicks. 2009. Corn kernel oil and corn fiber oil. In *Gourmet and Health-Promoting Specialty Oils*, Ed. R. A. Moreau and A. Kamal-Eldin, pp. 409–431. London: Academic Press.

Mosier, N. S., R. Hendrickson and M. Brewer, et al. 2005. Industrial scale-up of pH-controlled liquid hot water pretreatment of corn fiber for fuel ethanol production. *Applied Biochemistry and Biotechnology* 125:77–97.

Olorunsola, E. O., E. I. Akpabio, M. O. Adedokun and D. O. Ajibola. 2018. Emulsifying properties of hemicelluloses. In *Science and Technology behind Nanoemulsions*, Ed. S. Karakus. London: IntechOpen.

Osborn, D. and L. Chen. 1984. Corn hull hydrolysis using glucoamylase and sulfuric acid. *Starch-Starke* 36:393–395.

Palmqvist, B. and G. Liden. 2012. Torque measurements reveal large process differences between materials during high solid enzymatic hydrolysis of pretreated lignocellulose. *Biotechnology for Biofuels* 5:1–9.

Palmqvist, B., M. Wiman and G. Liden. 2011. Effect of mixing on enzymatic hydrolysis of steam-pretreated spruce: A quantitative analysis of conversion and power consumption. *Biotechnology for Biofuels* 4:1–8.

Perkins, J. 2018. Processing pathway: EPA approves Edeniq's cellulosic technology at 6 plants. *Biofuels Journal* 2018:44–45.

Playne, M. and R. Crittenden. 1997. Commercially available oligosaccharides. *International Dairy Federation* 1997:10–22.

QQCP. 2021. *Quad County Corn Processors, IOWA'S First Commercial Cellulosic Ethanol Production*. https://iowarfa.org/2014/07/quad-county-corn-processors-produces-first-gallons-of-cellulosic-ethanol-in-iowa/

Rose, D. J., G. E. Inglett and S. X. Liu. 2010. Utilisation of corn (*Zea mays*) bran and corn fiber in the production of food components. *Journal of the Science of Food and Agriculture* 90:915–924.

Saulnier, L., C. Marot, E. Chanliaud and J. F. Thibault. 1995. Cell wall polysaccharide interactions in maize bran. *Carbohydrate Polymers* 26:279–287.

Scheller, H. V. and P. Ulvskov. 2010. Hemicelluloses. *Annual Review of Plant Biology* 61:263–289.

Selig, M. J., L. G. Thygesen and C. Felby. 2014. Correlating the ability of lignocellulosic polymers to constrain water with the potential to inhibit cellulose saccharification. *Biotechnology for Biofuels* 7:1–10.

Shahbandeh, M. 2021. *Corn Production Worldwide 2019/2020, By Country*, Statista. https://www.statista.com/statistics/254294/distribution-of-global-corn-production-by-country-2012/

Sui, W. and H. Z. Chen. 2015. Water transfer in steam explosion process of corn stalk. *Industrial Crops and Products* 76:977–986.

Sui, W., X. Xie, R. Liu, T. Wu and M. Zhang. 2018. Effect of wheat bran modification by steam explosion on structural characteristics and rheological properties of wheat flour dough. *Food Hydrocolloids* 84:571–580.

Teo, K. 2020. *China Slows Home-Grown Ethanol Push.* Argus Blog. https://www.argusmedia.com/en/ news/2136320-china-slows-homegrown-ethanol-push

Vermerris, W. and A. Abril. 2015. Enhancing cellulose utilization for fuels and chemicals by genetic modification of plant cell wall architecture. *Current Opinion in Biotechnology* 32:104–112.

Wang, B., T. Ezeji, Z. Shi, H. Feng and H. Blaschek. 2009. Pretreatment and conversion of distiller's dried grains with solubles for acetone-butanol-ethanol (ABE) production. *Transactions of the ASABE* 52:885–892.

Weil, J. R., A. Sarikaya and S. L. Rau, et al. 1998. Pretreatment of corn fiber by pressure cooking in water. *Applied Biochemistry and Biotechnology* 73:1–17.

Xia, M. L., L. Wang, Z. X. Yang and H. Z. Chen. 2015. Periodic-peristole agitation for process enhancement of butanol fermentation. *Biotechnology for Biofuels* 8:1–15.

Yadav, M. P., D. B. Johnston and K. B. Hicks. 2007. Structural characterization of corn fiber gums from coarse and fine fiber and a study of their emulsifying properties. *Journal of Agricultural and Food Chemistry* 55:6366–6371.

Yadav, M. P., D. B. Johnston and K. B. Hicks. 2009. Corn fiber gum: New structure/function relationships for this potential beverage flavor stabilizer. *Food Hydrocolloids* 23:1488–1493.

Yadav, M. P., G. D. Strahan, S. Mukhopadhyay, A. T. Hotchkiss and K. B. Hicks. 2012. Formation of corn fiber gum-milk protein conjugates and their molecular characterization. *Food Hydrocolloids* 26:326–333.

Yancey, M. 2017. *D3MAX Technology Deployment Update.* https://www.bbiinternational.com/files/docs/ Yancey_M_0417.pdf

Yoshida, T., W. Dwianto, Y. Honda, H. Uyama and J. I. Azuma 2013. Removal of arabinose substituents from corn pericarp arabinoxylan. *Wood Research Journal* 4:41–45.

Yoshida, T., M. Sakamoto and J. Azuma. 2012. Extraction of hemicelluloses from corn pericarp by the NaOH-urea solvent system. *Procedia Chemistry* 4:294–300.

Zhang, J., D. Chu and J. Huang, et al. 2010. Simultaneous saccharification and ethanol fermentation at high corn stover solids loading in a helical stirring bioreactor. *Biotechnology and Bioengineering* 105:718–728.

Zhao, J. and H. Z. Chen. 2013. Correlation of porous structure, mass transfer and enzymatic hydrolysis of steam exploded corn stover. *Chemical Engineering Science* 104:1036–1044.

Part 10

First Generation Sugar Feedstock-based Bioethanol Fuels

47 First Generation Sugar Feedstock-based Bioethanol Fuels

Scientometric Study

Ozcan Konur
(Formerly) Ankara Yildirim Beyazit University

47.1 INTRODUCTION

The crude oil-based gasoline fuels (Ma et al., 2002; Newman and Kenworthy, 1989) have been widely used in the transportation sector since the 1920s. However, there have been great public concerns over the adverse environmental and human impact of these fuels (Hill et al., 2006, 2009). Hence, biomass-based bioethanol fuels (Hill et al., 2006; Konur, 2012e, 2015, 2019, 2020a) have increasingly been used in blending gasoline fuels (Hsieh et al., 2002; Najafi et al., 2009), fuel cells (Antolini, 2007, 2009), and biochemical production (Angelici et al., 2013; Morschbacker, 2009) in a biorefinery context (Fernando et al., 2006; Huang et al., 2008).

Bioethanol fuels also play a critical role in maintaining the energy security (Kruyt et al., 2009; Winzer, 2012) in the supply shocks (Kilian, 2008, 2009) related to oil price shocks (Hamilton, 2003, 2009), COVID-19 pandemic (Fauci et al., 2020; Li et al., 2020) or wars (Hamilton, 1983; Jones, 2012) in the aftermath of the Russian invasion of Ukraine (Reeves, 2014).

However, it is necessary to pretreat the biomass (Taherzadeh and Karimi, 2008; Yang and Wyman, 2008) to enhance the yield of the bioethanol (Hahn-Hagerdal et al., 2006; Sanchez and Cardona, 2008) prior to the bioethanol production through the hydrolysis (Sun and Cheng, 2002; Taherzadeh and Karimi, 2007) and fermentation (Lin and Tanaka, 2006; Olsson and Hahn-Hagerdal, 1996) of the biomass and hydrolysates, respectively.

One of the most studied feedstocks for the bioethanol fuels has been the sugar feedstocks. The research in the field of the sugar feedstock-based bioethanol fuels has intensified in this context in the key research fronts of the pretreatment of the sugar feedstocks (Chandel et al., 2009, 2011; Siqueira et al., 2011), hydrolysis of the sugar feedstocks (Chandel et al., 2009, 2011; Siqueira et al., 2011), fermentation of the sugar feedstock-based hydrolysates (Chandel et al., 2009; Laopaiboon et al., 2007; Renouf et al., 2008), and production (Kim and Day, 2011; Laopaiboon et al., 2009; Gnansounou et al., 2005) and evaluation (de Carvalho Macedo et al., 2008; Goldemberg et al., 2008; Martinelli et al., 2011) of the sugar feedstock-based bioethanol fuels. Further, the sugar cane (Goldemberg et al., 2008; de Carvalho Macedo et al., 2008; Martinelli and Filoso, 2008), sweet sorghum (Gnansounou et al., 2005; Kim and Day, 2011; Laopaiboon et al., 2007), and, to a lesser extent, sugar beet (Halleux et al., 2008; Ogbonna et al., 2001; Renouf et al., 2008) and energy cane (Kim and Day, 2011) have been intensively studied in this context.

However, it is essential to develop efficient incentive structures (North, 1991) for the primary stakeholders to enhance the research in this field (Konur, 2000, 2002a,b,c, 2006a,b, 2007a,b). The scientometric analysis has been used in this context to inform the primary stakeholders about the current state of the research in a selected research field (Garfield, 1955; Konur, 2011, 2012a,b,c,d,e,f,g,h,i, 2015, 2018b, 2019, 2020a).

DOI: 10.1201/9781003226451-61

As there have been no published scientometric studies in this field, this book chapter presents a scientometric study of the research in the sugar feedstock-based bioethanol fuels. It examines the scientometric characteristics of both the sample and population data presenting the scientometric characteristics of these two datasets in the order of documents, authors, publication years, institutions, funding bodies, source titles, countries, Scopus subject categories, Scopus keywords, and research fronts.

47.2 MATERIALS AND METHODS

The search for this study was carried out using Scopus database (Burnham, 2006) in July 2022.

As a first step for the search of the relevant literature, the keywords were selected using the first 200 most cited population papers. The selected keyword list was then optimized to obtain a representative sample of papers for the searched research field. This keyword list was provided in the Appendix for future replicative studies.

As a second step, two sets of data were used for this study. First, a population sample of 1,093 papers was used to examine the scientometric characteristics of the population data. Secondly, a sample of 109 most cited papers, corresponding to 10% of the population papers, was used to examine the scientometric characteristics of these citation classics.

The scientometric characteristics of these both sample and population datasets were presented in the order of documents, authors, publication years, institutions, funding bodies, source titles, countries, Scopus subject categories, Scopus keywords, and research fronts.

Lastly, the key scientometric findings for both datasets were discussed to highlight the research landscape for the sugar feedstock-based bioethanol fuels. Additionally, a number of brief conclusions were drawn and a number of relevant recommendations were made to enhance the future research landscape.

47.3 RESULTS

47.3.1 The Most Prolific Documents in the Sugar Feedstock-based Bioethanol Fuels

The information on the types of documents for both datasets is given in Table 47.1. The articles and conference papers, published in journals, dominate both the sample (89%) and population (87%) papers as they are over-represented in the sample papers by 2%. Further, review papers and short surveys have a 5% surplus as they are over-represented in the sample papers as they constitute 10% and 5% of the sample and population papers, respectively.

TABLE 47.1
Documents in the Sugar Feedstock-based Bioethanol Fuels

Documents	Sample Dataset (%)	Population Dataset (%)	Surplus (%)
Article	86.2	84.4	1.8
Review	9.2	4.2	5.0
Conference paper	2.8	2.4	0.4
Book chapter	0.9	5.8	−4.9
Short Survey	0.9	0.3	0.6
Note	0.0	1.5	−1.5
Letter	0.0	0.8	−0.8
Book	0.0	0.4	−0.4
Editorial	0.0	0.1	−0.1
Sample size	109	1,093	

Population dataset, The number of papers (%) in the set of the 1,093 population papers; Sample dataset, The number of papers (%) in the set of 109 highly cited papers.

It is further notable that 93% of the population papers were published in journals while 6% and 1% of them were published in books and book series, respectively. Similarly, 99% and 1% of the sample papers were published in the journals and books, respectively.

47.3.2   THE MOST PROLIFIC AUTHORS IN THE SUGAR FEEDSTOCK-BASED BIOETHANOL FUELS

The information about the most prolific 11 authors with at least 2.8% of sample papers each is given in Table 47.2. The most prolific authors are Lakkana Laopaiboon, Pattana Laopaiboon, Joaquim E. A. Seabra, and Arnaldo Walter with 4.6% of the sample papers each, followed by Antonio Bonomi, Jose Goldemberg, and Otavio Cavalett with 3.7% of the sample papers each.

On the other hand, the most influential author is Andre P. C. Faaij with 3.9% surplus, followed by Joaquim E. A. Seabra and Arnaldo Walter with 3.6% surplus each.

The most prolific institution for the sample dataset is the University of Sao Paulo with three authors, followed by the National Bioethanol Science and Technology Laboratory (CTBE), Khon Kaen University, and State University of Campinas with two authors each. In total, six institutions house these top authors. On the other hand, the most prolific country for the sample dataset is Brazil with seven authors, followed by Thailand with two authors. In total, only four countries house these top authors.

There are two primary research fronts for these top authors: Bioethanol fuels based on sweet sorghum (SS) and sugar cane (SC) with two and nine authors, respectively. On the other hand, there is significant gender deficit (Beaudry and Lariviere, 2016) for the sample dataset as surprisingly only two of these top researchers are female with a representation rate of 18%.

TABLE 47.2
Most Prolific Authors in the Sugar Feedstock-based Bioethanol Fuels

No.	Author Name	Author Code	Sample Papers (%)	Population Papers (%)	Surplus	Institution	Country	HI	N	Res. Front
1	Laopaiboon, Lakkana*	6506483310	4.6	2.8	1.8	Khon Kaen Univ.	Thailand	20	54	SS
2	Laopaiboon, Pattana	14064947400	4.6	2.7	1.9	Khon Kaen Univ.	Thailand	20	51	SS
3	Seabra, Joaquim E. A.	23390441500	4.6	1.0	3.6	State Univ. Campinas	Brazil	22	55	SC
4	Walter, Arnaldo	7102205549	4.6	1.0	3.6	State Univ. Campinas	Brazil	21	84	SC
5	Faaij, Andre P. C.	6701681600	4.6	0.7	3.9	Univ. Groningen	Netherlands	85	360	SC
6	Bonomi, Antonio	7004767629	3.7	1.3	2.4	CTBE	Brazil	33	104	SC
7	Goldemberg, Jose	7003862006	3.7	0.7	3.0	Univ. Sao Paulo	Brazil	36	172	SC
8	Cavalett, Otavio	9846419000	3.7	0.6	3.1	Norwegian Univ. Sci. Technol.	Norway	29	76	SC
9	Coelho, Suani T.*	7003751382	2.8	0.5	2.3	Univ. Sao Paulo	Brazil	18	67	SC
10	Martinelli, Luiz A.	7102366222	2.8	0.5	2.3	Univ. Sao Paulo	Brazil	63	254	SC
11	Galdos, MV	26534213300	2.8	0.4	2.4	CTBE	Brazil	19	38	SC

Author code, the unique code given by Scopus to the authors; Population papers, the number of papers authored in the population dataset; Sample papers, the number of papers authored in the sample dataset; SC, Sugar cane; SS, Sweet sorghum. *: Female.

Additionally, there are other authors with the relatively low citation impact and with 0.5%–1.0% of the population papers each: William R. Gibbons, Carl A. Westby, Alexandre Szklo, Prasit Jaisil, Naulchan Khongsay, Preekamol Khanrit, Manuel R. L. V. Leal, Niphaphat Phukoetphim, Derick D. Quintino, Tianwei Tan, Marcos S. Buckeridge, Marcelo Modesto, Silvia A. Nebra, Carlos Rolz, Pornthap Thanonkeo, Henrique V. de Amorim, Di Cai, Adriano V. Ensinas, Gabriel Granco, Shimei Li, Miguel A. Mutton, Satoshi Ohara, Carlos E. V. Rossell, and Gerd Sparovek.

47.3.3 THE MOST PROLIFIC RESEARCH OUTPUT BY YEARS IN THE SUGAR FEEDSTOCK-BASED BIOETHANOL FUELS

Information about papers published between 1970 and 2022 is given in Figure 47.1. This figure clearly shows that the bulk of the research papers in the population dataset were published primarily in the 2010s and the early 2020s with 60% and 13% of the population dataset, respectively. Similarly, the publication rates for the 2000s, 1990s, 1980s, and 1970s were 16%, 4%, 6%, and 1% respectively. Additionally, 0.6% of the population papers were published in the pre-1970s.

Similarly, the bulk of the research papers in the sample dataset were published in the 2000s and 2010s with 39% and 50% of the sample dataset, respectively. Similarly, the publication rates for the 1990s, 1980s, and 1970s were 6, 4, and 0% of the sample papers, respectively.

The most prolific publication years for the population dataset were 2012 and 2018 with 6.9% and 6.8% of the dataset, respectively. Further, 84% of the population papers were published between 2007 and 2022. Similarly, 80% of the sample papers were published between 2007 and 2017 while the most prolific publication years were 2008, 2009, and 2011 with 14% of the sample papers each. There is a rising trend for the population papers between 2006 and 2012, and thereafter a consolidation after 2012 around 6% level for each year losing its momentum.

47.3.4 THE MOST PROLIFIC INSTITUTIONS IN THE SUGAR FEEDSTOCK-BASED BIOETHANOL FUELS

Information about the most prolific 11 institutions publishing papers on the sugar feedstock-based bioethanol fuels with at least 2.8% of the sample papers each is given in Table 47.3.

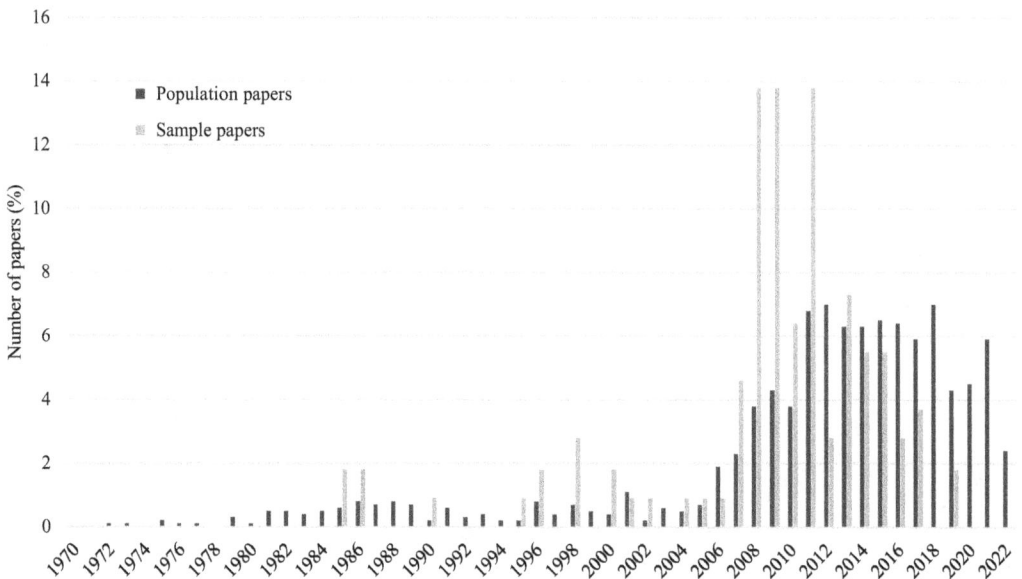

FIGURE 47.1   The research output by years regarding the sugar feedstock-based bioethanol fuels.

TABLE 47.3

The Most Prolific Institutions in the Sugar Feedstock-based Bioethanol Fuels

No.	Institutions	Country	Sample Papers (%)	Population Papers (%)	Surplus (%)
1	State Univ. Campinas	Brazil	15.6	6.1	9.5
2	Univ. Sao Paulo	Brazil	12.8	11.0	1.8
3	Biorenewables Natl. Lab.	Brazil	8.3	3.0	5.3
4	Khon Kaen Univ.	Thailand	4.6	4.1	0.5
5	Univ. Utrecht	Netherlands	3.7	0.8	2.9
6	Fed. Univ. Rio de Janeiro	Brazil	2.8	3.8	−1.0
7	Texas A&M Univ.	USA	2.8	1.4	1.4
8	USDA Agric. Res. Serv.	USA	2.8	1.1	1.7
9	Iowa State Univ.	USA	2.8	0.9	1.9
10	Univ. Ill. U. C.	USA	2.8	0.8	2.0
11	Fermentec Inc.	Brazil	2.8	0.3	2.5

The most prolific institution is the State University of Campinas with 15.6% of the sample papers, followed by the University of Sao Paulo with 12.8% of the sample papers. The other prolific institutions are the Biorenewables National Laboratory, Khon Kaen University, and University of Utrecht with 3.7%–8.3% of the sample papers each.

Similarly, the top countries for these most prolific institutions are Brazil and the USA with five and four institutions. In total, only four countries house these top institutions.

On the other hand, the institution with the most impact is the State University of Campinas with 9.5% surplus, followed by the Biorenewables National Laboratory with 5.3% surplus. The other influential institutions are the University of Utrecht, Fermentec Inc., and University of Illinois Urbana Champaign with 2.0%–2.9% surplus each.

Additionally, there are other institutions with the relatively low citation impact and with 0.6%–2.7% of the population papers each: Sao Paulo State University, Kansas State University, Federal University of Sao Carlos, Federal Center for Technological Education, South Dakota State University, Beijing University of Chemical Technology, University of Tokyo, North Dakota State University, Royal Institute of Technology, Punjab Agricultural University, Agricultural Research Inc., Federal University of Vicosa, Tsinghua University, University of Novi Sad, Wageningen University and Research, Cornell University; Federal University of Parana, Chinese Academy of Sciences, Kasetsart University, University of Brasília, Shanghai Jiao Tong University, Pontifical Catholic University of Rio de Janeiro, Federal University of Southern Rio Grande, University of Florida, Federal University of Itajuba, University of Nebraska–Lincoln, State University of Maringa, University of Queensland, University of Valley of Guatemala, and Chalmers University of Technology.

47.3.5 THE MOST PROLIFIC FUNDING BODIES IN THE SUGAR FEEDSTOCK-BASED BIOETHANOL FUELS

Information about the most prolific eight funding bodies funding at least 2.8% of the sample papers each is given in Table 47.4. Further, only 46% and 36% of the sample and population papers were funded, respectively.

The most prolific funding body is the Sao Paulo Research Foundation with 9.2% of the sample papers, followed by the National Council for Scientific and Technological Development and European Commission with 6.4% and 5.5% of the sample papers, respectively. On the other hand, the most prolific country for these top funding bodies is Brazil with four funding bodies, followed by China with two funding bodies. In total, only three countries and the EU house these top funding bodies.

TABLE 47.4

The Most Prolific Funding Bodies in the Sugar Feedstock-based Bioethanol Fuels

No.	Funding Bodies	Country	Sample Paper No. (%)	Population Paper No. (%)	Surplus (%)
1	Sao Paulo Research Foundation	Brazil	9.2	6.0	3.2
2	National Council for Scientific and Technological Development	Brazil	6.4	8.6	−2.2
3	European Commission	EU	5.5	1.0	4.5
4	CAPES Foundation	Brazil	4.6	5.9	−1.3
5	Ministry of Science, Technology and Innovation	Brazil	4.6	3.7	0.9
6	Ministry of Science and Technology of China	China	3.7	1.3	2.4
7	U.S. Department of Energy	USA	3.7	0.8	2.9
8	National Natural Science Foundation of China	China	2.8	1.7	1.1

The funding body with the most citation impact is the European Commission with 4.5% surplus, followed by the Sao Paulo Research Foundation and the U.S. Department of Energy with 3.3% and 2.9% surplus, respectively.

The other funding bodies with the relatively low citation impact and with 0.6%–2% of the population papers each are the Khon Kaen University, U.S. Department of Agriculture, Japan Society for the Promotion of Science, Ministry of Education, Culture, Sports, Science and Technology, National Council of Science and Technology, National Key Research and Development Program of China, National Research Council of Thailand, National Science Foundation, Funding Authority for Studies and Projects, Thailand Research Fund, and National Agency of Petroleum, Natural Gas and Biofuels.

47.3.6 THE MOST PROLIFIC SOURCE TITLES IN THE SUGAR FEEDSTOCK-BASED BIOETHANOL FUELS

Information about the most prolific 17 source titles publishing at least 1.8% of the sample papers each in the sugar feedstock-based bioethanol fuels is given in Table 47.5.

The most prolific source title is the Biomass and Bioenergy with 12.8% of the sample papers, followed by Bioresource Technology with 11% of the sample papers. The other prolific titles are the Energy Policy, Industrial Crops and Products, Renewable and Sustainable Energy Reviews, and Applied Energy with 4.6%–5.5% of the sample papers each.

On the other hand, the source titles with the most citation impact are the Biomass and Bioenergy and Bioresource Technology with 7.8% surplus each. The other influential titles are the Applied Energy, Energy Policy, and International Journal of Life Cycle Assessment with 3.1%–3.6% surplus each. Similarly, the source titles with the least impact are the Biofuels Bioproducts and Biorefining and Journal of Cleaner Production with 0.3% and 0.5% surplus, respectively.

The other source titles with the relatively low citation impact with 0.5%–4.1% of the population papers each are the Sugar Tech, International Sugar Journal, Biotechnology Letters, Energies, Renewable Energy, Biofuels, Chemical Engineering Transactions, GCB Bioenergy, Energy and Fuels, Sustainability, African Journal of Biotechnology, Bioenergy Research, SAE Technical Papers, Transactions of the American Society of Agricultural Engineers, Zuckerindustrie, Antonie Van Leeuwenhoek International Journal of General and Molecular Microbiology, Applied Biochemistry and Biotechnology, Biocatalysis and Agricultural Biotechnology, Biomass, Brazilian Journal of Chemical Engineering, Chemical and Engineering News, Chiang Mai Journal of Science, Computer Aided Chemical Engineering, Electronic Journal of Biotechnology, Energy Economics, Energy Sources Part A Recovery Utilization and Environmental Effects, Environmental Science and Technology, and Fermentation.

TABLE 47.5
The Most Prolific Source Titles in the Sugar Feedstock-based Bioethanol Fuels

No.	Source Titles	Sample Papers (%)	Population Papers (%)	Surplus (%)
1	Biomass and Bioenergy	12.8	5.0	7.8
2	Bioresource Technology	11.0	3.2	7.8
3	Energy Policy	5.5	2.0	3.5
4	Industrial Crops and Products	4.6	2.2	2.4
5	Renewable and Sustainable Energy Reviews	4.6	1.9	2.7
6	Applied Energy	4.6	1.0	3.6
7	Energy	3.7	1.3	2.4
8	International Journal of Life Cycle Assessment	3.7	0.6	3.1
9	Journal of Cleaner Production	2.8	2.3	0.5
10	Nature Climate Change	2.8	0.5	2.3
11	Biofuels Bioproducts and Biorefining	1.8	1.5	0.3
12	Biotechnology and Bioengineering	1.8	0.9	0.9
13	Biotechnology for Biofuels	1.8	0.8	1.0
14	Applied Microbiology and Biotechnology	1.8	0.5	1.3
15	Applied and Environmental Microbiology	1.8	0.4	1.4
16	American Journal of Agricultural Economics	1.8	0.3	1.5
17	Fuel Processing Technology	1.8	0.2	1.6

TABLE 47.6
The Most Prolific Countries in the Sugar Feedstock-based Bioethanol Fuels

No.	Countries	Sample Papers (%)	Population Papers (%)	Surplus (%)
1	Brazil	38.5	35.8	2.7
2	USA	27.5	17.7	9.8
3	China	7.3	5.7	1.6
4	Netherlands	6.4	2.5	3.9
5	India	5.5	5.2	0.3
6	Thailand	5.5	5.1	0.4
7	Australia	4.6	2.4	2.2
8	France	4.6	2.0	2.6
9	UK	3.7	3.5	0.2
10	Germany	3.7	2.1	1.6
11	Canada	3.7	1.6	2.1
12	Switzerland	3.7	0.9	2.8

47.3.7 THE MOST PROLIFIC COUNTRIES IN THE SUGAR FEEDSTOCK-BASED BIOETHANOL FUELS

Information about the most prolific 12 countries publishing at least 3.7% of sample papers each in the sugar feedstock-based bioethanol fuels is given in Table 47.6.

The most prolific country is Brazil with 39% of the sample papers, followed by the USA with 28% of the sample papers. The other prolific countries are China, Netherlands, India, Thailand, Australia, and France with 4.6%–7.3% of the sample papers each. Further, five European countries listed in Table 47.6 produce 22% and 11% of the sample and population papers, respectively.

On the other hand, the country with the most citation impact is the USA with 10% surplus. The other influential countries are Netherlands, Switzerland, Brazil, and France with 2.6%–3.9%

surplus each. Similarly, the countries with the least citation impact are the UK, India, and Thailand with 0.2%–0.4% surplus each.

Additionally, there are other countries with relatively low citation impact and with 0.5%–4.2% of the sample papers each: Japan, Mexico, Sweden, Spain, Guatemala, Iran, South Africa, Italy, Portugal, Belgium, Greece, Nigeria, Serbia, Austria, Colombia, Poland, Turkey, Philippines, Denmark, Pakistan, Chile, Romania, and S. Korea.

47.3.8 THE MOST PROLIFIC SCOPUS SUBJECT CATEGORIES IN THE SUGAR FEEDSTOCK-BASED BIOETHANOL FUELS

Information about the most prolific eight Scopus subject categories indexing at least 6.4% of the sample papers each is given in Table 47.7.

The most prolific Scopus subject category in the sugar feedstock-based bioethanol fuels is Environmental Science with 62% of the sample papers, followed by Energy with 54% of the sample papers. The other prolific subject categories are Agricultural and Biological Sciences and Chemical Engineering with 29% and 23% of the sample papers, respectively. It is notable that Social Sciences including Economics and Business account for 11% and 18% of the sample and population studies.

On the other hand, the Scopus subject category with the most citation impact is the Environmental Science 28% surplus, followed by Energy with 21% surplus. Similarly, the least influential subject categories are Agricultural and Biological Sciences, Engineering, Biochemistry, Genetics and Molecular Biology, and Social Sciences with 0.7%–4.5% deficit, respectively.

47.3.9 THE MOST PROLIFIC KEYWORDS IN THE SUGAR FEEDSTOCK-BASED BIOETHANOL FUELS

Information about the Scopus keywords used with at least 6.4% or 2.7% of the sample or population papers, respectively, is given in Table 47.8. For this purpose, keywords related to the keyword set given in the appendix are selected from a list of the most prolific keyword set provided by Scopus database.

These keywords are grouped under the five headings: Biomass, fermentation, hydrolysis and hydrolysates, products, and evaluation.

The most prolific keywords related to the biomass and biomass constituents are sugar cane and sugarcane with 35% and 30% of the sample papers, respectively, followed by sweet sorghum, saccharum, biomass in general, and sugar beet with 11%–26% of the sample papers each. Further, the prolific keyword related to the fermentation is fermentation with 40% of the sample papers, followed by yeast and saccharomyces with 25% and 21% of the sample papers, respectively.

TABLE 47.7
The Most Prolific Scopus Subject Categories in the Sugar Feedstock-based Bioethanol Fuels

No.	Scopus Subject Categories	Sample Papers (%)	Population Papers (%)	Surplus (%)
1	Environmental Science	61.5	33.1	28.4
2	Energy	54.1	32.7	21.4
3	Agricultural and Biological Sciences	29.4	33.9	−4.5
4	Chemical Engineering	22.9	21.0	1.9
5	Biochemistry, Genetics, and Molecular Biology	15.6	16.6	−1.0
6	Immunology and Microbiology	15.6	11.7	3.9
7	Engineering	11.9	13.6	−1.7
8	Social Sciences	6.4	7.1	−0.7

TABLE 47.8
The Most Prolific Keywords in the Sugar Feedstock-based Bioethanol Fuels

No.	Keywords	Sample Papers (%)	Population Papers (%)	Surplus (%)
1	Biomass and biomass constituents			
	Sugar cane	34.9	17.7	17.2
	Sugarcane	30.3	16.3	14.0
	Sweet sorghum	25.7	15.7	10.0
	Saccharum	15.6	9.0	6.6
	Biomass	12.8	6.4	6.4
	Sugar beet	11.0	8.2	2.8
	Cellulose	10.1	3.8	6.3
	Lignin	8.3	3.1	5.2
	B-vulgaris	7.3	6.4	0.9
	Sweet sorghum	3.7	6.7	−3.0
2	Fermentation			
	Fermentation	40.4	24.5	15.9
	Yeast	24.8	12.3	12.5
	Saccharomyces	21.1	10.9	10.2
	Ethanol fermentation	6.4	4.7	1.7
3	Hydrolysis and hydrolysates			
	Sugar	51.4	19.5	31.9
	Hydrolysis	9.2	4.1	5.1
	Glucose	7.3	3.8	3.5
	Enzyme activity	6.4	3.1	3.3
4	Products			
	Ethanol	84.4	54.8	29.6
	Biofuels	55.0	28.9	26.1
	Alcohol	26.6	13.4	13.2
	Bioethanol	25.7	17.9	7.8
	Ethanol production	25.7	14.6	11
	Ethanol fuels	13.8	5.3	9
	Alcohol production	11.0	5.0	6.0
	Ethanol yield	10.1	2.9	7
	Bio-ethanol production	9.2	6.5	3
	Biofuel production	6.4	3.5	3
	Bioenergy	3.7	3.2	1
5	Evaluation			
	Brazil	36.7	18.0	19
	Greenhouse gases	21.1	10.0	11
	Life cycle assessment	17.4	6.8	11
	Environmental impact	11.0	3.9	7
	Economics	10.1	3.8	6
	Land use	10.1	3.4	7
	Global warming	9.2	2.3	7
	Gas emissions	8.3	3.4	5

(Continued)

TABLE 47.8 (*Continued*)

The Most Prolific Keywords in the Sugar Feedstock-based Bioethanol Fuels

No.	Keywords	Sample Papers (%)	Population Papers (%)	Surplus (%)
	Cost-benefit analysis	8.3	1.6	7
	Carbon dioxide	7.3	3.8	4
	Sustainable development	6.4	4.3	2
	Sustainability	6.4	4.0	2
	USA	6.4	2.4	4
	Biodiversity	6.4	1.8	5
	Costs	5.5	4.9	1
	Energy efficiency		3.3	−3
	Energy policy		2.9	−3

The most prolific keyword related to the hydrolysis and hydrolysates is sugar with 51% of the sample papers. Further, the most prolific keyword related to the products is ethanol with 84% of the sample papers, followed by biofuels, alcohol, bioethanol, and ethanol production with 26%–55% of the sample papers each. Finally, the most prolific keywords related to the evaluation of the bioethanol fuels is Brazil with 27% of the sample papers, followed by greenhouse gases, life cycle assessment, environmental impact, economics, and land use with 10%–21% of the sample papers each.

On the other hand, the most influential keywords are sugar, ethanol, biofuels, Brazil, sugar cane, fermentation, sugarcane, alcohol, yeast, ethanol production, greenhouse gases, life cycle assessment, saccharomyces, and sweet sorghum with 10%–32% surplus each. Similarly, the most prolific keywords across all of the research fronts are ethanol, biofuels, sugar, fermentation, Brazil, sugar cane, sugarcane, alcohol, bioethanol, sweet sorghum, ethanol production, yeast, saccharomyces, and greenhouse gases with 21%–84% of the sample papers each.

47.3.10 THE MOST PROLIFIC RESEARCH FRONTS IN THE SUGAR FEEDSTOCK-BASED BIOETHANOL FUELS

Information about the research fronts for the sample papers in the sugar feedstock-based bioethanol fuels with regard to the sugar feedstocks used for the bioethanol production is given in Table 47.9.

As this table shows, there are two primary research fronts for this field with regard to the feedstocks: sugar cane and sweet sorghum with 73% and 30% of the sample papers, respectively. The other sugar feedstocks used in these studies are sugar beet and energy cane with 7% and 4% of the sample papers, respectively.

Information about the thematic research fronts for the sample papers in the sugar feedstock-based bioethanol fuels is given in Table 47.10. As this table shows, there are five primary research fronts: biomass pretreatments and hydrolysis, hydrolysate fermentation, sugar feedstock-based bioethanol fuel production and evaluation with 13%, 10%, 29%, 36%, and 75% of the sample papers, respectively.

For the research front of sugar feedstock pretreatments, the primary research fronts are the chemical and enzymatic pretreatments with 8% and 10% of the sample papers, respectively. The most prolific feedstock used for the fermentation and bioethanol production is sweet sorghum with 19% and 23% of the sample papers, respectively. Similarly, the most prolific feedstock for the bioethanol evaluation is sugar cane with 64% of the sample papers.

TABLE 47.9
The Most Prolific Research Fronts for the Sugar Feedstock-based Bioethanol Fuels

No.	Research Fronts	N Paper (%) Sample
1	Sugar cane	73.4
2	Sweet sorghum	30.3
3	Sugar beet	7.3
4	Energy cane	3.7

N paper (%) sample, The number of papers in the population sample of 109 papers.

TABLE 47.10
The Most Prolific Thematic Research Fronts for the Sugar Feedstock-based Bioethanol Fuels

No.	Research Fronts	N Paper (%) Sample
1	Biomass Pretreatments	12.8
	Hydrothermal pretreatments	0.9
	Sweet sorghum	0.9
	Chemical pretreatments	8.3
	Acid pretreatment	4.6
	Sugarcane	3.7
	Sweet sorghum	0.9
	Alkaline pretreatment	3.7
	Sweet sorghum	1.8
	Sugar cane	1.8
	Enzymatic pretreatments	10.1
	Feedstock genetic engineering	1.8
	Sugar cane	1.8
2	Biomass Hydrolysis	10.1
	Enzymatic hydrolysis	10.1
	Sweet sorghum	5.5
	Sugar cane	3.7
	Energy cane	0.9
3	Hydrolysate Fermentation	28.5
	Sweet sorghum	19.3
	Sugar cane	6.4
	Sugar beet	2.8
4	Bioethanol Production	35.7
	Sweet sorghum	22.9
	Sugar cane	7.3
	Sugar beet	3.7
	Energy cane	1.8
	Bioethanol Fuel Evaluation	75.2
	Sugar cane	64.2
	Sweet sorghum	6.4
	Sugar beet	2.8
	Sugar feedstocks in general	1.8

N paper (%) sample, The number of papers in the population sample of 109 papers.

47.4 DISCUSSION

47.4.1 Introduction

The crude oil-based gasoline fuels have been widely used in the transportation sector since the 1920s. However, there have been great public concerns over the adverse environmental and human impact of these fuels. Hence, biomass-based bioethanol fuels have increasingly been used in blending gasoline fuels, fuel cells, and biochemical production in a biorefinery context.

However, it is necessary to pretreat the biomass to enhance the yield of the bioethanol prior to the bioethanol production through the hydrolysis and fermentation. One of the most studied feedstocks for the bioethanol fuels have been the sugar feedstocks, such as sugar cane, sugar beet, sweet sorghum, and energy cane. The research in the field of the sugar feedstock-based bioethanol fuels has intensified in this context in the key research fronts of the pretreatment and hydrolysis of the sugar feedstocks, fermentation of the sugar feedstock-based hydrolysates, and production and evaluation of the sugar feedstock-based bioethanol fuels. Further, sugar cane, sweet sorghum, and, to a lesser extent, sugar beet and energy cane have been intensively studied in this context.

However, it is essential to develop efficient incentive structures for the primary stakeholders to enhance the research in this field. This is especially important to maintain energy security in the cases of supply shocks such as oil price shocks and war-related shocks as in the case of Russian invasion of Ukraine or COVID-19 pandemics.

The scientometric analysis has been used in this context to inform the primary stakeholders about the current state of the research in a selected research field. As there has been no scientometric study in this field, this book chapter presents a scientometric study of the research in the sugar feedstock-based bioethanol fuels. It examines the scientometric characteristics of both the sample and population data presenting scientometric characteristics of these two datasets in the order of documents, authors, publication years, institutions, funding bodies, source titles, countries, Scopus subject categories, Scopus keywords, and research fronts.

As a first step for the search of the relevant literature, the keywords were selected using the first most cited 200 papers. The selected keyword list was then optimized to obtain a representative sample of papers for the searched research field. A copy of this keyword list was provided in the appendix for future replicative studies. Further, a selected list of the keywords was presented in Table 47.8.

As a second step, two sets of data were used for this study. First, a population sample of 1,093 papers was used to examine the scientometric characteristics of the population data. Secondly, a sample of 109 most cited papers, corresponding to 10% of the population dataset was used to examine the scientometric characteristics of these citation classics.

The scientometric characteristics of these sample and population datasets were presented in the order of documents, authors, publication years, institutions, funding bodies, source titles, countries, Scopus subject categories, Scopus keywords, and research fronts.

Lastly, the key scientometric findings for both datasets were discussed to highlight the research landscape for sugar feedstock-based bioethanol fuels. Additionally, a number of brief conclusions were drawn and a number of relevant recommendations were made to enhance the future research landscape.

47.4.2 The Most Prolific Documents in the Sugar Feedstock-based Bioethanol Fuels

Articles (together with conference papers) dominate both the sample (89%) and population (87%) papers (Table 47.1). Further, review papers and articles have a surplus (5%) and deficit (2%). respectively. The representation of the reviews in the sample papers is relatively significant (10%).

Scopus differs from the Web of Science database in differentiating and showing articles (86%) and conference papers (3%) published in the journals separately. However, it should be noted that these conference papers are also published in journals as articles, compared with those published only in the conference proceedings. Hence, the total number of articles and review papers in the sample dataset are 89% and 10%, respectively.

It is observed during the search process that there has been inconsistency in the classification of the documents in Scopus as well as in other databases such as Web of Science. This is especially relevant for the classification of papers as reviews or articles as the papers not involving a literature review may be erroneously classified as a review paper. There is also a case of review papers being classified as articles. For example, the total number of the reviews in the sample dataset was manually found as nearly 6% compared to 10% as indexed by Scopus, increasing the number of articles and conference papers to 94% for the sample dataset. The close examination of these papers shows that many evaluative studies such as technoeconomic or life cycle studies have often been indexed as the review papers by the Scopus database.

In this context, it would be helpful to provide a classification note for the published papers in the books and journals at the first instance. It would also be helpful to use the document types listed in Table 47.1 for this purpose. Book chapters may also be classified as articles or reviews as an additional classification to differentiate review chapters from the experimental chapters as it is done by the Web of Science. It would be further helpful to additionally classify the conference papers as articles or review papers as well as it is done in the Web of Science database.

47.4.3 THE MOST PROLIFIC AUTHORS IN THE SUGAR FEEDSTOCK-BASED BIOETHANOL FUELS

There have been most prolific 11 authors with at least 2.8% of the sample papers each as given in Table 47.2. These authors have shaped the development of the research in this field.

The most prolific authors are Lakkana Laopaiboon, Pattana Laopaiboon, Joaquim E. A. Seabra, Arnaldo Walter and to a lesser extent Antonio Bonomi, Jose Goldemberg, and Otavio Cavalett.

It is important to note the inconsistencies in indexing of the author names in Scopus and other databases. It is especially an issue for the names with more than two components such as 'Blake Sam de Hyun Seabra'. The probable outcomes are 'Seabra, B.S.D.H.', 'de Hyun Seabra, B.S.', or 'Hyun Seabra, B.S.D.'. The first choice is the gold standard of the publishing sector as the last word in the name is taken as the last name. In most of the academic databases such as PUBMED and EBSCO databases, this version is used predominantly. The second choice is a strong alternative while the last choice is an undesired outcome as two last words are taken as the last name. It is good practice to combine the words of the last name by a hyphen: 'Hyun-Seabra, B.S.'. It is notable that inconsistent indexing of the author names may cause substantial inefficiencies in the search process for the papers as well as allocating credit to the authors as there are different author entries for each outcome in the databases.

There are also inconsistencies in the shortening Chinese names. For example. 'YangYing Zhang' is often shortened as 'Zhang, Y.', 'Zhang, Y.-Y.', and 'Zhang, Y.Y.' as it is done in the Web of Science database as well. However, the gold standard in this case is 'Zhang, Y' where the last word is taken as the last name and the first word is taken as a single forename. In most of the academic databases such as PUBMED and EBSCO, this first version is used predominantly. Nevertheless, it is helpful to use the third option to differentiate Chinese names efficiently: 'Zhang, Y.Y.'. Therefore, there have been difficulties in locating papers for the Chinese authors. In such cases, the use of the unique author codes provided for each author by the Scopus database has been helpful.

There is also a difficulty in allowing credit for the authors especially for the authors with common names such as 'Zhang, X.' in conducting scientometric studies. These difficulties strongly influence the efficiency of the scientometric studies as well as allocating credit to the authors as there are the same author entries for different authors with the same name. e.g. 'Zhang, X.' in the databases.

In this context, the coding of authors in Scopus database is a welcome innovation compared with the other databases such as Web of Science. In this process, Scopus allocates a unique number to each author in the database (Aman, 2018). However, there might still be substantial inefficiencies in this coding system especially for common names. For example, some of the papers for a certain author maybe allocated to another researcher with a different author code. It is possible that Scopus uses a number of software programs to differentiate the author names and the program may not be false-proof (Kim, 2018).

In this context, it does not help that author names are not given in full in some journals and books. This makes difficult to differentiate authors with common names and makes the scientometric studies further difficult in the author domain. Therefore, the author names should be given in all books and journals at the first instance. There is also a cultural issue where some authors do not use their full names in their papers. Instead they use initials for their forenames: 'Seabra, H.J.', 'Seabra, 'Seabra, H.', or 'Seabra, J.' instead of 'Seabra, Hyun Jae'.

There are also inconsistencies in naming of the authors with more than two components by the authors themselves in journal papers and book chapters. For example. 'Seabra, A.P.C.' might be given as 'Seabra, A.' or 'Seabra, A.C.' or 'Seabra, A.P.' or 'Seabra, C' in the journals and books. This also makes the scientometric studies difficult in the author domain. Hence, contributing authors should use their name consistently in their publications.

The other critical issue regarding the author names is the inconsistencies in the spelling of the author names in the national spellings (e.g. Göğüşçığıl, Gökçe) rather than in the English spellings (e.g. Goguscigil, Gokce) in Scopus database. Scopus differs from the Web of Science database and many other databases in this respect where the author names are given only in the English spellings. It is observed that national spellings of the author names do not help much in conducting scientometric studies as well in allocating credits to the authors as sometimes there are the different author entries for the English and National spellings in the Scopus database.

The most prolific institutions for the sample dataset are the University of Sao Paulo and to a lesser extent National Bioethanol Science and Technology Laboratory (CTBE), Khon Kaen University, and State University of Campinas. Further, the most prolific country for the sample dataset is Brazil and to a lesser extent Thailand. These findings confirm the dominance of Brazil and to a lesser extent Thailand in this field. The primary research fronts are the bioethanol fuels based on sweet sorghum (SS) and sugar cane (SC).

It is also notable that there is significant gender deficit for the sample dataset as surprisingly with a representation rate of 18%. This finding is the most thought-provoking with strong public policy implications. Hence, institutions, funding bodies, and policy makers should take efficient measures to reduce the gender deficit in this field as well as other scientific fields with strong gender deficit. In this context, it is worth noting the level of representation of the researchers from the minority groups in science on the basis of race, sexuality, age, and disability, besides the gender (Blankenship, 1993; Dirth and Branscombe, 2017; Konur, 2000, 2002a,b,c, 2006a,b, 2007a,b).

47.4.4 THE MOST PROLIFIC RESEARCH OUTPUT BY YEARS IN THE SUGAR FEEDSTOCK-BASED BIOETHANOL FUELS

The research output observed between 1970 and 2022 is illustrated in Figure 47.1. This figure clearly shows that the bulk of the research papers in the population dataset were published primarily in the 2010s and early 2020s. Similarly, the bulk of the research papers in the sample dataset were published in the 2000s and 2010s.

These findings suggest that the most prolific sample and population papers were primarily published in the 2010s. These are the thought-provoking findings as there has been significant research boom in since 2008. In this context, the increasing public concerns about climate change (Change, 2007), greenhouse gas emissions (Carlson et al., 2017), and global warming (Kerr, 2007) have been certainly behind the boom in the research in this field in the last two decades. Furthermore, the supply shocks experiences due to the COVID-19 pandemic might also be behind the research boom in this field since 2019.

Based on these findings, the size of the population papers likely to be more than double in the current decade, provided that the public concerns about climate change, greenhouse gas emissions, and global warming, as well as the supply shocks are translated efficiently to the research funding in this field.

47.4.5   The Most Prolific Institutions in the Sugar Feedstock-based Bioethanol Fuels

The most prolific 11 institutions publishing papers on the sugar feedstock-based bioethanol fuels with at least 2.8% of the sample papers each given in Table 47.3 have shaped the development of the research in this field.

The most prolific institutions are the State University of Campinas, University of Sao Paulo and to a lesser extent Biorenewables National Laboratory, Khon Kaen University, and University of Utrecht. Similarly, the top countries for these most prolific institutions are Brazil and the USA. In total, only four countries house these top institutions.

On the other hand, the institutions with the most citation impact are the State University of Campinas and to a lesser extent Biorenewables National Laboratory, University of Utrecht, Fermentec Inc., and University of Illinois Urbana Champaign. These findings confirm the dominance of Brazilian and to a lesser extent of US, Thai, and Dutch institutions for these HCPs. This is not surprising as Brazil is the major producer of sugarcane-based bioethanol fuels.

47.4.6   The Most Prolific Funding Bodies in the Sugar Feedstock-based Bioethanol Fuels

The most prolific eight funding bodies funding at least 2.8% of the sample papers each is given in Table 47.4. It is notable that only 46% and 36% of the sample and population papers were funded, respectively.

The most prolific funding bodies are Sao Paulo Research Foundation and to a lesser extent the National Council for Scientific and Technological Development and European Commission. The most prolific countries for these top funding bodies are Brazil and, to a lesser extent, China. This is not surprising as Brazil is the major producer of sugarcane-based bioethanol fuels.

These findings on the funding of the research in this field suggest that the level of the funding, mostly since 2008, is not highly intensive and it has been nevertheless largely instrumental in enhancing the research in this field (Ebadi and Schiffauerova, 2016) in the light of North's institutional framework (North, 1991). It is also notable that the funding rate in this field is inferior to those in the other research fronts of the bioethanol fuels such as algal bioethanol fuels. Further, it is expected that this high funding rate would improve in the light of the recent supply shocks. Further, it emerges that Brazil has heavily funded the research on the sugar cane-based bioethanol fuels as the major producer of sugarcane-based bioethanol fuels.

47.4.7   The Most Prolific Source Titles in the Sugar Feedstock-based Bioethanol Fuels

The most prolific 17 source titles publishing at least 1.8% of the sample papers each in the sugar feedstock-based bioethanol fuels have shaped the development of the research in this field (Table 47.5).

The most prolific source titles are the Biomass and Bioenergy, Bioresource Technology and to a lesser extent Energy Policy, Industrial Crops and Products, Renewable and Sustainable Energy Reviews, and Applied Energy.

On the other hand, the source titles with the most citation impact are the Biomass and Bioenergy, Bioresource Technology and to a lesser extent Applied Energy, Energy Policy, and International Journal of Life Cycle Assessment.

It is notable that these top source titles are primarily related to the bioresources, energy, and life cycle assessment. This finding suggests that Bioresource Technology, Biomass and Bioenergy, and the other prolific journals in these fields have significantly shaped the development of the research in this field as they focus primarily on the sugar feedstock-based bioethanol fuels with a high yield. In this context, the influence of the top two journals is quite extraordinary with 24% of the sample papers, in total. It is also notable that the energy-related journals have also published papers in the areas of techno-economics environmental impact, land use change, economics, and labor relations as the social science-related journals.

47.4.8 THE MOST PROLIFIC COUNTRIES IN THE SUGAR FEEDSTOCK-BASED BIOETHANOL FUELS

The most prolific 12 countries publishing at least 3.7% of the sample papers each have significantly shaped the development of the research in this field (Table 47.6).

The most prolific countries are Brazil, the USA and, to a lesser extent, China, Netherlands, India, Thailand, Australia, and France. Further, five European countries listed in Table 47.6 produce 22% and 11% of the sample and population papers, respectively.

On the other hand, the countries with the most citation impact are the USA and, to a lesser extent, Netherlands, Switzerland, Brazil, and France. Similarly, the countries with the least impact are the UK, India, and Thailand.

The close examination of these findings suggests that Brazil, the USA, Europe, and, to a lesser extent, China, Thailand, and Australia are the major producers of the research in this field. It is a fact that the USA has been a major player in science (Leydesdorff and Wagner, 2009). The USA has further developed a strong research infrastructure to support its corn and grass-based bioethanol industry (Gillon, 2010).

However, China has been a rising mega star in scientific research in competition with the USA and Europe (Leydesdorff and Zhou, 2005). China is also a major player in this field as a major producer of bioethanol (Fang et al., 2010).

Next, Europe has been a persistent player in the scientific research in competition with both the USA and China (Leydesdorff, 2000). Europe has also been a persistent producer of bioethanol along with the USA and Brazil (Gnansounou, 2010).

Finally, Brazil has been the major producer and user of sugar cane-based bioethanol fuels since the 1970s as a world lead in this respect (Goldemberg et al., 2008; Martinelli and Filoso, 2008).

47.4.9 THE MOST PROLIFIC SCOPUS SUBJECT CATEGORIES IN THE SUGAR FEEDSTOCK-BASED BIOETHANOL FUELS

The most prolific eight Scopus subject categories indexing at least 6.4% of the sample papers each, respectively, given in Table 47.7 have shaped the development of the research in this field.

The most prolific Scopus subject categories in the sugar feedstock-based bioethanol fuels are Environmental Science, Energy and to a lesser extent Agricultural and Biological Sciences and Chemical Engineering. It is also notable that Social Sciences including Economics and Business account have a strong presence in both sample and population studies.

On the other hand, the Scopus subject categories with the most citation impact are Environmental Science and Energy. Similarly, the least influential subject categories are Agricultural and Biological Sciences, Engineering, Biochemistry, Genetics and Molecular Biology, and Social Sciences.

These findings are thought-provoking suggesting that the primary subject categories are related to energy, environmental science, engineering, genetics, and social sciences as the core of the research in this field concerns with the sugar feedstock-based bioethanol fuels. The other finding is that social sciences are unusually well represented in both the sample and population papers contrary to the most fields in bioethanol fuels. The social and economic studies account for the field of social sciences (Gnansounou et al., 2005; Goldemberg et al., 2008; Martinelli and Filoso, 2008).

47.4.10   The Most Prolific Keywords in the Sugar Feedstock-based Bioethanol Fuels

A limited number of keywords have shaped the development of the research in this field as shown in Table 47.8 and the Appendix. These keywords are grouped under the five headings: biomass, fermentation, hydrolysis and hydrolysates, products, and evaluation.

The most prolific keywords across all of the research fronts are ethanol, biofuels, sugar, fermentation, Brazil, sugar cane, sugarcane, alcohol, bioethanol, sweet sorghum, ethanol production, yeast, saccharomyces, and greenhouse gases. Similarly, the most influential keywords are sugar, ethanol, biofuels, Brazil, sugar cane, fermentation, sugarcane, alcohol, yeast, ethanol production, greenhouse gases, life cycle assessment, saccharomyces, and sweet sorghum.

These findings suggest that it is necessary to determine the keyword set carefully to locate the relevant research in each of these research fronts. Additionally, the size of the samples for each keyword highlights the intensity of the research in the relevant research areas. These findings also highlight different spelling of some strategic keywords: sugar cane v. sugarcane v. saccharum, bioethanol v. bio-ethanol, sweet sorghum v. sorghum bicolor, sugar beet v. sugarbeet v. beta vulgaris.

47.4.11   The Most Prolific Research Fronts in the Sugar Feedstock-based Bioethanol Fuels

Information about the research fronts for the sample papers in the sugar feedstock-based bioethanol fuels with regard to the sugar feedstocks used for the bioethanol production is given in Table 47.9.

As this table shows, there are two primary research fronts for this field: sugarcane and sweet sorghum while the other sugar feedstocks used in these studies are sugar beet and energy cane. It is notable that energy cane a genetically engineered version of sugar cane (Matsuoka et al., 2014).

Information about the thematic research fronts for the sample papers in the sugar feedstock-based bioethanol fuels is given in Table 47.10. As this table shows, there are five primary research fronts: biomass pretreatments and hydrolysis, hydrolysate fermentation, and production and evaluation sugar feedstock-based bioethanol fuels.

For the research front of sugar feedstock pretreatments, the primary research fronts are the chemical and enzymatic pretreatments. Further, the most prolific feedstock used for the fermentation and bioethanol production is sweet sorghum. Similarly, the most prolific feedstock for the bioethanol evaluation is sugar cane.

These findings are thought-provoking in seeking ways to increase sugar feedstock-based bioethanol yield at the global scale. It is clear that all of these research fronts have public importance and merit substantial funding and other incentives. Further, it is notable that sugar feedstock-based bioethanol fuels have become a core unit of the bioethanol research to make it more competitive with the crude oil-based gasoline and petrodiesel fuels, especially for the USA and the other countries with vast farmlands.

In comparison with the other feedstock-based research fronts, it is notable that evaluation of the bioethanol fuel emerges as a primary research front for this field with 75% of the sample papers. This suggests that the primary stakeholders have been primarily interested in the evaluation of the sugar feedstock-based bioethanol fuels such as technoeconomics, life cycle, economics, social science, land use, labor, and environmental impact-related studies as a case study for the bioethanol fuels together with algal and corn feedstocks in this field. In this context, Brazil has been the global leader in the production and use of the sugar cane-based bioethanol fuels since the 1970s in the aftermath of the global crude oil crisis in the early 1970s.

In the end, these most cited papers in this field hint that the efficiency of sugar feedstock-based bioethanol fuels could be optimized using the structure, processing, and property relationships of these sugar feedstocks such as sugar cane, sweet sorghum, sugar beet, and energy cane in the fronts of the feedstock pretreatment and hydrolysis, and hydrolysate fermentation (Formela et al., 2016; Konur, 2018a, 2020b, 2021a,b,c,d; Konur and Matthews, 1989).

47.5   CONCLUSION AND FUTURE RESEARCH

The research on the sugar feedstock-based bioethanol fuels has been mapped through a scientometric study of both sample (100 papers) and population (1,093 papers) datasets.

The critical issue in this study has been to obtain a representative sample of the research as in any other scientometric study. Therefore, the keyword set has been carefully devised and optimized after a number of runs in the Scopus database. It is a representative sample of the wider population studies. This keyword set was provided in the Appendix, and the relevant keywords are presented in Table 47.8. However, it should be noted that it has been very difficult to compile a representative keyword set since this research field has been connected closely with many other fields. Therefore, it has been necessary to compile a keyword list to exclude papers concerned with the other research fields.

It is notable in this context that the research on the production of bioethanol fuels from wastes of the sugar feedstocks such as sugarcane bagasse and sugarcane straw is closely related to the research on the bioethanol production from the feedstocks themselves, such as sugar cane or sweet sorghum juice. Therefore, it is crucial to collect data on these two interconnected research fronts separately. Hence, the studies on the production and evaluation of the wastes of the sugar feedstocks for the bioethanol production were presented separately in another section in the fourth volume.

The other issue has been the selection of a multidisciplinary database to carry out the scientometric study of the research in this field. For this purpose. Scopus database has been selected. The journal coverage of this database has been notably wider than that of Web of Science and other multi-subject databases.

The key scientometric properties of the research in this field have been determined and discussed in this book chapter. It is evident that a limited number of documents, authors, institutions, publication years, institutions, funding bodies, source titles, countries, Scopus subject categories, Scopus keywords, and research fronts have shaped the development of the research in this field.

There is ample scope to increase the efficiency of the scientometric studies in this field in the author and document domains by developing consistent policies and practices in both domains across all the academic databases. In this respect, it seems that authors, journals, and academic databases have a lot to do. Furthermore, the significant gender deficit as in most scientific fields emerges as a public policy issue. The potential deficits on the basis of age, race, disability, and sexuality need also to be explored in this field as in other scientific fields.

The research in this field has boomed since 2008, possibly promoted by the public concerns on global warming, greenhouse gas emissions, and climate change. Furthermore, the recent COVID-19 pandemic and Russian invasion of Ukraine have resulted in a global supply shocks shifting the focus of the stakeholders from the crude oil-based fuels to biomass-based fuels such as bioethanol fuels. It is expected that there would be further incentives for the key stakeholders to carry out the research for the sugar feedstock-based bioethanol fuels to increase the ethanol yield and to make it more competitive with the crude oil-based gasoline and petrodiesel fuels. This might be truer for the crude oil- and foreign exchange-deficient countries to maintain the energy security at the face of the global supply shocks.

The relatively modest funding rate of 46% and 36% for the sample and population papers, respectively, suggests that funding in this field nevertheless enhanced the research in this field primarily since 2008, possibly more than doubling in the current decade. However, it is evident that there is ample room for more funding and other incentives to enhance the research in this field further considering the decline the research output for the population papers after 2012.

The institutions from Brazil and to a lesser extent of the USA, Thailand, and Netherlands have mostly shaped the research in this field. Further, Brazil, the USA, Europe, and, to a lesser extent, China, India, Thailand, and Australia have been the major producers of the research in this field as the major producers and users of bioethanol fuels from different types of biomass such as corn, sugarcane, and grass as well as other types of biomass. It is evident that these countries have well-developed research infrastructure in bioethanol fuels and their derivatives. It is also notable all of these major countries have access to the large farmlands.

It emerges that ethanol is more popular than bioethanol as a keyword with strong implications for the search strategy. In other words, the search strategy using only bioethanol keyword would not be much helpful. The Scopus keywords are grouped under the eight headings: biomass, fermentation, bacteria, hydrolysis, hydrolysates, pretreatments, other processes, and products of the fermentation. It is also notable to consider the Latin keywords for the sugar feedstocks in the search strategy as it has been done in this book chapter.

As Table 47.9 shows, there are two primary research fronts for this field with regard to the feedstocks: sugar cane and sweet sorghum while the other sugar feedstocks used in these studies are sugar beet and energy cane. On the other hand, Table 47.10 shows that there are five primary thematic research fronts: biomass pretreatments and hydrolysis, hydrolysate fermentation, and production and evaluation of sugar feedstock-based bioethanol fuels.

These findings are thought-provoking in seeking ways to increase bioethanol yield through the sugar feedstock-based bioethanol fuels at the global scale. It is clear that all of these research fronts have public importance and merit substantial funding and other incentives. Further, it is notable that sugar feedstock-based bioethanol fuels, as a first generation biofuels, have become a core unit of the bioethanol research to make it more competitive with the crude oil-based gasoline and diesel fuels, especially for the countries with large access to the farmlands.

In comparison with the other feedstock-based research fronts, it is notable that evaluation of the bioethanol fuel emerges as a primary research front for this field with 75% of the sample papers. This suggests that the primary stakeholders have been primarily interested in the evaluation of the sugar feedstock-based bioethanol fuels such as technoeconomics, life cycle, economics, social sciences, land use, labor, and environmental impact-related studies in this field. These studies show that residual sugar feedstocks such as sugarcane bagasse are more suitable for the bioethanol fuel production due to the public concerns about the food security.

In this context, Brazil has been the global leader in the production and use of the sugar cane-based bioethanol fuels since the 1970s in the aftermath of the global crude oil crisis in the early 1970s.

Thus, the scientometric analysis has a great potential to gain valuable insights into the evolution of the research in this field as in other scientific fields especially in the aftermath of the significant global supply shocks such as COVID-19 pandemic and the Russian invasion of Ukraine.

It is recommended that further scientometric studies are carried out for the primary research fronts. It is further recommended that reviews of the most cited papers are carried out for each primary research front to complement these scientometric studies. Next, the scientometric studies of the hot papers in these primary fields are carried out.

ACKNOWLEDGMENTS

The contribution of the highly cited researchers in the field of the sugar feedstock-based bioethanol fuels has been gratefully acknowledged.

APPENDIX: THE KEYWORD SET FOR SUGAR FEEDSTOCK-BASED BIOETHANOL FUELS

(((TITLE (ethanol OR bioethanol) AND TITLE (*diesel OR *hydrogen OR h2 OR *butanol OR biogas OR *methane OR oil OR jet OR protein) AND TITLE (sugarcane OR saccharum OR {sugar beet} OR sugarbeet OR {sugar cane} OR {sweet sorghum} OR {sorghum bicolor} OR {energy cane} OR energycane OR cane OR beet OR {sugar juice} OR {beta vulgaris} OR {sugar mills} OR {sugar juice} OR brazil* OR paulo)) AND NOT TITLE (bagasse OR baggase OR straw OR leaves OR tops OR pulp OR {by-products} OR *cellulosic OR *cellulose OR trash OR "second generation" OR residu* OR {co-products} OR silage OR molasses OR pomaces OR vinasse OR {distillers' grains} OR ddgs OR solubles OR 2g OR mud OR byproduct OR biomass OR cake OR *electricity OR *power OR biorefinery OR cogeneration OR clones OR chp OR gasification OR algae)) OR ((((TITLE (sugarcane OR saccharum OR {sugar beet} OR sugarbeet OR {sugar cane} OR {sweet sorghum} OR {sorghum bicolor} OR {energy cane} OR energycane OR cane OR beet OR {beta vulgaris} OR {sugar juice} OR {sugar mills}) AND TITLE (ethanol OR bioethanol OR saccharification OR *hydrolysis OR digestibili* OR recalcitrance OR hydrolysate* OR hydrolyzate* OR ferment* OR coferment* OR delignification OR depolymerization OR milling OR grinding OR pretreat* OR "pre-treat*" OR "juice extraction" OR {sugar yield})) OR (TITLE (ethanol OR bioethanol) AND TITLE (brazil* OR paulo))) AND NOT (TITLE (bagasse OR biochar OR tailings OR leaves OR nematode OR anaerobic OR domestic* OR waste* OR flower* OR *alkanoate OR residu* OR straw OR race OR uptake OR "Second Generation" OR *butyric OR tops OR trash OR *cellulose OR pulp OR *cellulosic OR {by-products} OR oil* OR {co-products} OR *hydrogen OR silage OR molasses OR *diesel OR pomaces OR *jet OR vinasse OR lactic OR *butanol OR acer OR maple OR {distillers' grains} OR ddgs OR solubles OR 2g OR gut OR nano* OR baggase OR vitro OR pigment OR *methane OR mud OR aguardente OR byproduct OR roots OR borer OR *malic OR fatty OR biomass OR cake OR *electricity OR *power OR hyperion OR biorefinery OR cogeneration OR algae OR clones OR chp OR gasification OR protein OR maize OR cachaça) OR SRCTITLE (food* OR botan* OR polymer* OR protein* OR soil* OR materials OR plant* OR chromat* OR macromol* OR carbohydrate* OR breed* OR adsorp* OR organic OR euphytica OR inorganic OR hydrogen OR agron* OR phyto* OR dairy OR lwt OR nutrit* OR cellulose OR composites OR animal OR data OR analyst OR cereal) OR SUBJAREA (medi OR vete OR nurs OR phar OR heal OR neur OR dent OR psyc)))) AND (LIMIT-TO (SRCTYPE, "j") OR LIMIT-TO (SRCTYPE, "b") OR LIMIT-TO (SRCTYPE, "k")) AND (LIMIT-TO (DOCTYPE, "ar") OR LIMIT-TO (DOCTYPE, "cp") OR LIMIT-TO (DOCTYPE, "ch") OR LIMIT-TO (DOCTYPE, "re") OR LIMIT-TO (DOCTYPE, "no") OR LIMIT-TO (DOCTYPE, "le") OR LIMIT-TO (DOCTYPE, "bk") OR LIMIT-TO (DOCTYPE, "ed") OR LIMIT-TO (DOCTYPE, "sh")) AND (LIMIT-TO (LANGUAGE, "English"))

REFERENCES

Aman, V. 2018. Does the Scopus author ID suffice to track scientific international mobility? A case study based on Leibniz laureates. *Scientometrics* 117:705–720.

Angelici, C., B. M. Weckhuysen and P. C. A. Bruijnincx. 2013. Chemocatalytic conversion of ethanol into butadiene and other bulk chemicals. *ChemSusChem* 6:1595–1614.

Antolini, E. 2007. Catalysts for direct ethanol fuel cells. *Journal of Power Sources* 170:1–12.

Antolini, E. 2009. Palladium in fuel cell catalysis. *Energy and Environmental Science* 2:915–931.

Beaudry, C. and V. Lariviere. 2016. Which gender gap? Factors affecting researchers' scientific impact in science and medicine. *Research Policy* 45:1790–1817.

Blankenship, K. M. 1993. Bringing gender and race in: US employment discrimination policy. *Gender & Society* 7:204–226.

Burnham, J. F. 2006. Scopus database: A review. *Biomedical Digital Libraries* 3:1–8.

Carlson, K. M., J. S. Gerber and D. Mueller, et al. 2017. Greenhouse gas emissions intensity of global croplands. *Nature Climate Change* 7:63–68.

Chandel, A. K., M. L. Narasu, M., G. Chandrasekhar, A. Manikyam and L. V. Rao. 2009. Use of *Saccharum spontaneum* (wild sugarcane) as biomaterial for cell immobilization and modulated ethanol production by thermotolerant *Saccharomyces cerevisiae* VS3. *Bioresource Technology* 100:2404–2410.

Chandel, A. K., O. V. Singh, R. L. Venkateswar, G. Chandrasekhar and M. L. Narasu. 2011. Bioconversion of novel substrate *Saccharum spontaneum*, a weedy material, into ethanol by *Pichia stipitis* NCIM3498. *Bioresource Technology* 102:1709–1714.

Change, C. 2007. Climate change impacts, adaptation and vulnerability. *Science of the Total Environment* 326:95–112.

De Carvalho Macedo, I. C., J. E.A. Seabra, and J. E. A. R. Silva. 2008. Green house gases emissions in the production and use of ethanol from sugarcane in Brazil: The 2005/2006 averages and a prediction for 2020. *Biomass and Bioenergy* 32:582–595.

Dirth, T. P. and N. R. Branscombe. 2017. Disability models affect disability policy support through awareness of structural discrimination. *Journal of Social Issues* 73:413–442.

Ebadi, A. and A. Schiffauerova. 2016. How to boost scientific production? A statistical analysis of research funding and other influencing factors. *Scientometrics* 106:1093–1116.

Fang, X., Y. Shen, J. Zhao, X. Bao and Y. Qu. 2010. Status and prospect of lignocellulosic bioethanol production in China. *Bioresource Technology* 101:4814–4819.

Fauci, A. S., H. C. Lane and R. R. Redfield. 2020. Covid-19-navigating the uncharted. *New England Journal of Medicine* 382:1268–1269.

Fernando, S., S. Adhikari, C. Chandrapal and M. Murali. 2006. Biorefineries: Current status, challenges, and future direction. *Energy & Fuels* 20:1727–1737.

Formela, K., A. Hejna, L. Piszczyk, M. R. Saeb and X. Colom. 2016. Processing and structure-property relationships of natural rubber/wheat bran biocomposites. *Cellulose* 23:3157–3175.

Garfield, E. 1955. Citation indexes for science. *Science* 122:108–111.

Gillon, S. 2010. Fields of dreams: Negotiating an ethanol agenda in the Midwest United States. *Journal of Peasant Studies* 37:723–748.

Gnansounou, E. 2010. Production and use of lignocellulosic bioethanol in Europe: Current situation and perspectives. *Bioresource Technology* 101:4842–4850.

Gnansounou, E., A. Dauriat and C. E. Wyman. 2005. Refining sweet sorghum to ethanol and sugar: Economic trade-offs in the context of North China. *Bioresource Technology* 96:985–1002.

Goldemberg, J., S. T. Coelho and P. Guardabassi. 2008. The sustainability of ethanol production from sugarcane. *Energy Policy* 36:2086–2097.

Hahn-Hagerdal, B., M. Galbe, M. F. Gorwa-Grauslund, G. Liden and G. Zacchi. 2006. Bio-ethanol - The fuel of tomorrow from the residues of today. *Trends in Biotechnology* 24:549–556.

Halleux, H., S. Lassaux, R. Renzoni and A. Germain. 2008. Comparative life cycle assessment of two biofuels: Ethanol from sugar beet and rapeseed methyl ester. *International Journal of Life Cycle Assessment* 13:184–190.

Hamilton, J. D. 1983. Oil and the macroeconomy since World War II. *Journal of Political Economy* 91:228–248.

Hamilton, J. D. 2003. What is an oil shock? *Journal of Econometrics* 113:363–398.

Hamilton, J. D. 2009. Causes and consequences of the oil shock of 2007-08. *Brookings Papers on Economic Activity* 2009:215–261.

Hill, J., E. Nelson, D. Tilman, S. Polasky and D. Tiffany. 2006. Environmental, economic, and energetic costs and benefits of biodiesel and ethanol biofuels. *Proceedings of the National Academy of Sciences of the United States of America* 103:11206–11210.

Hill, J., S. Polasky and E. Nelson, et al. 2009. Climate change and health costs of air emissions from biofuels and gasoline. *Proceedings of the National Academy of Sciences of the United States of America* 106:2077–2082.

Hsieh, W. D., R. H. Chen, T. L. Wu and T. H. Lin. 2002. Engine performance and pollutant emission of an SI engine using ethanol-gasoline blended fuels. *Atmospheric Environment* 36:403–410.

Huang, H. J., S. Ramaswamy, U. W. Tschirner and B. V. Ramarao. 2008. A review of separation technologies in current and future biorefineries. *Separation and Purification Technology* 62:1–21.

Jones, T. C. 2012. America, oil, and war in the Middle East. *Journal of American History* 99:208–218.

Kerr, R. A. 2007. Global warming is changing the world. *Science* 316:188–190.

Kilian, L. 2008. Exogenous oil supply shocks: How big are they and how much do they matter for the US economy? *Review of Economics and Statistics* 90:216–240.

Kilian, L. 2009. Not all oil price shocks are alike: Disentangling demand and supply shocks in the crude oil market. *American Economic Review*, 99:1053–69.

Kim, J. 2018. Evaluating author name disambiguation for digital libraries: A case of DBLP. *Scientometrics* 116:1867–1886.

Kim, M. and D. F. Day. 2011. Composition of sugar cane, energy cane, and sweet sorghum suitable for ethanol production at Louisiana sugar mills. *Journal of Industrial Microbiology and Biotechnology* 38:803–807.

Konur, O. 2000. Creating enforceable civil rights for disabled students in higher education: An institutional theory perspective. *Disability & Society* 15:1041–1063.

Konur, O. 2002a. Access to nursing education by disabled students: Rights and duties of nursing programs. *Nurse Education Today* 22:364–374.

Konur, O. 2002b. Assessment of disabled students in higher education: Current public policy issues. *Assessment and Evaluation in Higher Education* 27:131–152.

Konur, O. 2002c. Access to employment by disabled people in the UK: Is the Disability Discrimination Act working? *International Journal of Discrimination and the Law* 5:247–279.

Konur, O. 2006a. Participation of children with dyslexia in compulsory education: Current public policy issues. *Dyslexia* 12:51–67.

Konur, O. 2006b. Teaching disabled students in higher education. *Teaching in Higher Education* 11:351–363.

Konur, O. 2007a. A judicial outcome analysis of the *Disability Discrimination Act*: A windfall for the employers? *Disability & Society* 22:187–204.

Konur, O. 2007b. Computer-assisted teaching and assessment of disabled students in higher education: The interface between academic standards and disability rights. *Journal of Computer Assisted Learning* 23:207–219.

Konur, O. 2011. The scientometric evaluation of the research on the algae and bio-energy. *Applied Energy* 88:3532–3540.

Konur, O. 2012a. Prof. Dr. Ayhan Demirbas' scientometric biography. *Energy Education Science and Technology Part A: Energy Science and Research* 28:727–738.

Konur, O. 2012b. The evaluation of the biogas research: A scientometric approach. *Energy Education Science and Technology Part A: Energy Science and Research* 29:1277–1292.

Konur, O. 2012c. The evaluation of the global energy and fuels research: A scientometric approach. *Energy Education Science and Technology Part A: Energy Science and Research* 30:613–628.

Konur, O. 2012d. The evaluation of the research on the biodiesel: A scientometric approach. *Energy Education Science and Technology Part A: Energy Science and Research* 28:1003–1014.

Konur, O. 2012e. The evaluation of the research on the bioethanol: A scientometric approach. *Energy Education Science and Technology Part A: Energy Science and Research* 28:1051–1064.

Konur, O. 2012f. The evaluation of the research on the biofuels: A scientometric approach. *Energy Education Science and Technology Part A: Energy Science and Research* 28:903–916.

Konur, O. 2012g. The evaluation of the research on the biohydrogen: A scientometric approach. *Energy Education Science and Technology Part A: Energy Science and Research* 29:323–338.

Konur, O. 2012h. The evaluation of the research on the microbial fuel cells: A scientometric approach. *Energy Education Science and Technology Part A: Energy Science and Research* 29:309–322.

Konur, O. 2012i. The scientometric evaluation of the research on the production of bioenergy from biomass. *Biomass and Bioenergy* 47:504–515.

Konur, O. 2015. Current state of research on algal bioethanol. In *Marine Bioenergy: Trends and Developments*, Ed. S. K. Kim and C. G. Lee, pp. 217–244. Boca Raton, FL: CRC Press.

Konur, O., Ed. 2018a. *Bioenergy and Biofuels*. Boca Raton, FL: CRC Press.

Konur, O. 2018b. Bioenergy and biofuels science and technology: Scientometric overview and citation classics. In *Bioenergy and Biofuels*, Ed. O. Konur, pp. 3–63. Boca Raton: CRC Press.

Konur, O. 2019. Cyanobacterial bioenergy and biofuels science and technology: A scientometric overview. In *Cyanobacteria: From Basic Science to Applications*, Ed. A. K. Mishra, D. N. Tiwari and A. N. Rai, pp. 419–442. Amsterdam: Elsevier.

Konur, O. 2020a. The scientometric analysis of the research on the bioethanol production from green macroalgae. In *Handbook of Algal Science, Technology and Medicine*, Ed. O. Konur, pp. 385–401. London: Academic Press.

Konur, O., Ed. 2020b. *Handbook of Algal Science, Technology and Medicine*. London: Academic Press.

Konur, O., Ed. 2021a. *Handbook of Biodiesel and Petrodiesel Fuels: Science, Technology, Health, and Environment*. Boca Raton, FL: CRC Press.

Konur, O., Ed. 2021b. *Handbook of Biodiesel and Petrodiesel Fuels: Science, Technology, Health, and Environment. Volume 1. Biodiesel Fuels: Science, Technology, Health, and Environment*. Boca Raton, FL: CRC Press.

Konur, O., Ed. 2021c. *Handbook of Biodiesel and Petrodiesel Fuels: Science, Technology, Health, and Environment. Volume 2. Biodiesel Fuels based on the Edible and Nonedible Feedstocks, Wastes, and Algae: Science, Technology, Health, and Environment*. Boca Raton, FL: CRC Press.

Konur, O., Ed. 2021d. *Handbook of Biodiesel and Petrodiesel Fuels: Science, Technology, Health, and Environment. Volume 3. Petrodiesel Fuels: Science, Technology, Health, and Environment*. Boca Raton, FL: CRC Press.

Konur, O. and F. L. Matthews. 1989. Effect of the properties of the constituents on the fatigue performance of composites: A review. *Composites* 20:317–328.

Kruyt, B., D. P. van Vuuren, H. J. de Vries and H. Groenenberg. 2009. Indicators for energy security. *Energy Policy* 37:2166–2181.

Laopaiboon, L., S. Nuanpeng, P. Srinophakun, P. Klanrit and P. Laopaiboon. 2009. Ethanol production from sweet sorghum juice using very high gravity technology: Effects of carbon and nitrogen supplementations. *Bioresource Technology* 100:4176–4182.

Laopaiboon, L., P. Thanonkeo, P. Jaisil and P. Laopaiboon. 2007. Ethanol production from sweet sorghum juice in batch and fed-batch fermentations by *Saccharomyces cerevisiae*. *World Journal of Microbiology and Biotechnology* 23:1497–1501.

Leydesdorff, L. 2000. Is the European Union becoming a single publication system? *Scientometrics* 47:265–280.

Leydesdorff, L. and C. Wagner. 2009. Is the United States losing ground in science? A global perspective on the world science system. *Scientometrics* 78:23–36.

Leydesdorff, L. and P. Zhou. 2005. Are the contributions of China and Korea upsetting the world system of science? *Scientometrics* 63:617–630.

Li, H., S. M. Liu, X. H. Yu, S. L. Tang and C. K. Tang. 2020. Coronavirus disease 2019 (COVID-19): Current status and future perspectives. *International Journal of Antimicrobial Agents* 55:105951.

Lin, Y. and S. Tanaka. 2006. Ethanol fermentation from biomass resources: Current state and prospects. *Applied Microbiology and Biotechnology* 69:627–642.

Ma, X., L. Sun and C. Song. 2002. A new approach to deep desulfurization of gasoline, diesel fuel and jet fuel by selective adsorption for ultra-clean fuels and for fuel cell applications. *Catalysis Today* 77:107–116.

Martinelli, L. A. and S. Filoso. 2008. Expansion of sugarcane ethanol production in Brazil: Environmental and social challenges. *Ecological Applications* 18:885–898.

Martinelli, L. A., R. Garrett, S. Ferraz and R. Naylor. 2011. Sugar and ethanol production as a rural development strategy in Brazil: Evidence from the state of São Paulo. *Agricultural Systems* 104:419–428.

Matsuoka, S., A. J. Kennedy, E. G. D. dos Santos, A. L. Tomazela and L. C. Rubio. 2014. Energy cane: Its concept, development, characteristics, and prospects. *Advances in Botany* 2014:597275.

Morschbacker, A. 2009. Bio-ethanol based ethylene. *Polymer Reviews* 49:79–84.

Najafi, G., B. Ghobadian and T. Tavakoli, et al. 2009. Performance and exhaust emissions of a gasoline engine with ethanol blended gasoline fuels using artificial neural network. *Applied Energy* 86:630–639.

Newman, P. W. G. and J. R. Kenworthy. 1989. Gasoline consumption and cities: A comparison of U.S. cities with a global survey. *Journal of the American Planning Association* 55:24–37.

North, D. C. 1991. Institutions. *Journal of Economic Perspectives* 5:97–112.

Ogbonna, J. C., H. Mashima and H. Tanaka. 2001. Scale up of fuel ethanol production from sugar beet juice using loofa sponge immobilized bioreactor. *Bioresource Technology* 76:1–8.

Olsson, L. and B. Hahn-Hagerdal. 1996. Fermentation of lignocellulosic hydrolysates for ethanol production. *Enzyme and Microbial Technology* 18:312–331.

Reeves, S. 2014. To Russia with love: How moral arguments for a humanitarian intervention in Syria opened the door for an invasion of the Ukraine. *Michigan State University International Law Review* 23:199.

Renouf, M.A., M. K. Wegener and L. K. Nielsen. 2008. An environmental life cycle assessment comparing Australian sugarcane with US corn and UK sugar beet as producers of sugars for fermentation. Biomass and Bioenergy 32:1144–1155.

Sanchez, O. J. and C. A. Cardona. 2008. Trends in biotechnological production of fuel ethanol from different feedstocks. *Bioresource Technology* 99:5270–5295.

Siqueira, G., A. M. F. Milagres, W. Carvalho, G. Koch and A. Ferraz. 2011. Topochemical distribution of lignin and hydroxycinnamic acids in sugar-cane cell walls and its correlation with the enzymatic hydrolysis of polysaccharides. *Biotechnology for Biofuels* 4:7.

Sun, Y. and J. Cheng. 2002. Hydrolysis of lignocellulosic materials for ethanol production: A review. *Bioresource Technology* 83:1–11.

Taherzadeh, M. J. and K. Karimi. 2007. Enzyme-based hydrolysis processes for ethanol from lignocellulosic materials: A review. *Bioresources* 2:707–738.

Taherzadeh, M. J. and K. Karimi. 2008. Pretreatment of lignocellulosic wastes to improve ethanol and biogas production: A review. *International Journal of Molecular Sciences* 9:1621–1651.

Winzer, C. 2012. Conceptualizing energy security. *Energy Policy* 46:36–48.

Yang, B. and C. E. Wyman. 2008. Pretreatment: The key to unlocking low-cost cellulosic ethanol. *Biofuels, Bioproducts and Biorefining* 2:26–40.

48 First Generation Sugar Feedstock-based Bioethanol Fuels

Review

Ozcan Konur
(Formerly) Ankara Yildirim Beyazit University

48.1 INTRODUCTION

Crude oil-based gasoline fuels (Ma et al., 2002; Newman and Kenworthy, 1989) have been widely used in the transportation sector since the 1920s. However, there have been great public concerns over the adverse environmental and human impact of these fuels (Hill et al., 2006, 2009). Hence, biomass-based bioethanol fuels (Hill et al., 2006; Konur, 2012, 2015, 2019, 2020) have increasingly been used in blending gasoline fuels (Hsieh et al., 2002; Najafi et al., 2009), in fuel cells (Antolini, 2007, 2009), and in biochemical production (Angelici et al., 2013; Morschbacker, 2009) in a biorefinery context (Fernando et al., 2006; Huang et al., 2008).

However, it is necessary to pretreat the biomass (Alvira et al., 2010; Taherzadeh and Karimi, 2008) to enhance the yield of bioethanol (Hahn-Hagerdal et al., 2006; Sanchez and Cardona, 2008) prior to bioethanol fuel production from sugar feedstocks through hydrolysis (Sun and Cheng, 2002; Taherzadeh and Karimi, 2007) and fermentation (Lin and Tanaka, 2006; Olsson and Hahn-Hagerdal, 1996) of the biomass and hydrolysates, respectively.

One of the most-studied feedstocks for bioethanol fuels has been sugar feedstocks. Research in the field of sugar feedstock-based bioethanol fuels has intensified in this context in the key research fronts of pretreatment of sugar feedstocks (Chandel et al., 2009, 2011; Siqueira et al., 2011), hydrolysis of sugar feedstocks (Chandel et al., 2009, 2011; Siqueira et al., 2011), fermentation of sugar feedstock-based hydrolysates (Chandel et al., 2009; Laopaiboon et al., 2007; Renouf et al., 2008), and production (Kim and Day, 2011; Laopaiboon et al., 2009; Gnansounou et al., 2005) and evaluation (Macedo et al., 2008; Goldemberg et al., 2008; Martinelli and Filoso, 2008) of sugar feedstock-based bioethanol fuels. Further, sugarcane (Goldemberg et al., 2008; Macedo et al., 2008; Martinelli and Filoso, 2008), sweet sorghum (Gnansounou et al., 2005; Kim and Day, 2011; Laopaiboon et al., 2007) and to a lesser extent sugar beet (Halleux et al., 2008; Ogbonna et al., 2001; Renouf et al., 2008) and energy cane (Kim and Day, 2011) have been studied intensively in this context.

However, it is essential to develop efficient incentive structures (North, 1991) for the primary stakeholders to enhance research in this field (Konur, 2000, 2002a,b,c, 2006a,b, 2007a,b). Although there have been a limited number of review papers on bioethanol fuel production from sugar feedstocks (Almodares and Hadi, 2009; Ayodele et al., 2020; Zabed et al., 2020), there has been no review of the most cited 25 papers in this field.

Thus, this chapter presents a review of the most-cited 25 articles in the field of bioethanol fuel production from sugar feedstocks. Then, it discusses the key findings of these highly influential papers and comments on future research priorities in this field.

DOI: 10.1201/9781003226451-62

48.2 MATERIALS AND METHODS

Search for this study was carried out using the Scopus database (Burnham, 2006) in July 2022.

As a first step for search of the relevant literature, keywords were selected using the most-cited first 200 population papers. The selected keyword list was then optimized to obtain a representative sample of papers for the searched research field. This final keyword set was provided in the Appendix of Konur (2023) for future replication studies.

As a second step, a sample dataset was used for this study. The first 25 articles with at least 91 citations each were selected for the review study. Key findings from each paper were taken from the abstracts of these papers and were discussed. Additionally, a number of brief conclusions were drawn and a number of relevant recommendations were made to enhance the future research landscape.

48.3 RESULTS

Brief information about the 25 most-cited papers with at least 91 citations each on sugar feedstock-based bioethanol fuels is given below. The primary research fronts are the production and evaluation of sugar feedstock-based bioethanol fuels with 9 and 16 highly cited papers (HCPs).

48.3.1 PRODUCTION OF SUGAR FEEDSTOCK-BASED BIOETHANOL FUELS

There are nine HCPs for the production of sugar feedstock-based bioethanol fuels (Table 48.1).

Gnansounou et al. (2005) evaluated bioethanol, sugar, and bioelectricity production from sweet sorghum juice and hemicellulose and cellulose in sorghum bagasse in China in a paper with 293 citations. They found that the production of bioethanol from hemicellulose and cellulose in bagasse was more favorable than burning it to make electricity, but the relative merits of making bioethanol or sugar from the juice were very sensitive to the price of sugar in China. Thus, they recommended a flexible plant capable of making both sugar and bioethanol from the juice. Overall, bioethanol production from bagasse was very favorable, but other agricultural residues such as corn stover and rice hulls would likely provide a more attractive feedstock for making bioethanol in the medium and long term due to their extensive availability in China and their independence from other markets. Furthermore, the process for bagasse conversion was based on particular design assumptions, and other technologies could enhance competitiveness while considerations such as perceived risk could impede applications.

Kim and Day (2011) determined the composition and ethanol yield of sugarcane, energy cane, and sweet sorghum suitable for bioethanol production in the US sugar biorefineries in a paper with 283 citations. They noted that the sugar biorefineries in Louisiana operated only 3 months of the year. For year-round operation, they recommended that energy cane and sweet sorghum must be used as additional feedstocks as they had different harvest times compared to sugarcane. They determined that the juice of energy cane and sweet sorghum contained 9.8% and 11.8% fermentable sugars, respectively. The chemical composition of sugarcane bagasse was 42% cellulose, 25% hemicellulose, and 20% lignin; of energy cane was 43% cellulose, 24% hemicellulose, and 22% lignin; and of sweet sorghum was 45% cellulose, 27% hemicellulose, and 21% lignin. The theoretical ethanol yields were 3,609, 12,938, and 5,804 kg/ha from sugarcane, energy cane, and sweet sorghum, respectively.

Laopaiboon et al. (2009) produced bioethanol from sweet sorghum juice using very high gravity (VHG) fermentation using *Saccharomyces cerevisiae* NP01 in a paper with 228 citations. They used sucrose, sugarcane molasses, and ammonium sulfate as carbon and nitrogen supplementations. When sucrose was used as an additive, they found that the sweet sorghum juice containing total sugar of 280 g/L, 3 g yeast extract/L, and 5 g peptone/L resulted in maximum bioethanol production efficiency with concentration, productivity, and yield of 120.68, 2.01, and 0.51 g/g,

TABLE 48.1

Production of Sugar Feedstock-based Bioethanol Fuels

No.	Papers	Biomass/Hydrolysate	Prts.	Yeasts	Parameters	Keywords	Lead Author	Affil.	Cits
3	Gnansounou et al. (2005)	Sweet sorghum juice and bagasse	Na	Na	Bioethanol, sugar, and electricity production from sweet sorghum juice and bagasse	Sweet sorghum, ethanol	Gnansounou, Edgard 6508334495	Swiss Fed. Inst. Technol. Switzerland	293
4	Kim and Day (2011)	Sugarcane, energy cane, sweet sorghum	Na	Na	Composition and ethanol yield of sugarcane, energy cane, and sweet sorghum	Sugarcane, energy cane, sweet sorghum, ethanol	Kim, Misook 55686300000	Dankook Univ. S. Korea	283
8	Laopaiboon et al. (2009)	Sweet sorghum juice	Na	S. cerevisiae	Bioethanol production from sweet sorghum juice, additives, bioethanol efficiency, productivity, yield, Thailand	Sweet sorghum, ethanol	Laopaiboon, Lakkana* 6506483310	Khon Kaen Univ. Thailand	228
9	Limtong et al. (2007)	Sugarcane juice	Na	K. marxianus	Bioethanol production from sugarcane juice, optimization, ethanol concentration, productivity, yield, Thailand	Sugarcane, ethanol	Limtong, Savitree 6507260821	Kasetsart Univ. Thailand	201
10	Dien et al. (2009)	Sweet sorghum	Acid, cellulases	Yeasts	Bioethanol production, feedstock engineering, sugar and ethanol yield	Sorghum bicolor, ethanol	Dien, Bruce S. 6603685796	USDA Agric. Serv. USA	184

(Continued)

TABLE 48.1 (*Continued*)
Production of Sugar Feedstock-based Bioethanol Fuels

No.	Papers	Biomass/ Hydrolysate	Prts.	Yeasts	Parameters	Keywords	Lead Author	Affil.	Cits
14	Laopaiboon et al. (2007)	Sweet sorghum juice	Na	S. cerevisiae	Bioethanol from sweet sorghum juice in batch and fed-batch fermentations, bioethanol efficiency, productivity, yield, Thailand	Sweet sorghum, ethanol, fermentation	Laopaiboon, Lakkana* 6506483310	Khon Kaen Univ. Thailand	131
18	Chandel et al. (2009)	Sugarcane	Na	S. cerevisiae	bioethanol from wild sugarcane, batch and repeated batch fermentation, free and immobilized yeasts, bioethanol yield	Saccharum, ethanol	Rao, L. Venkateswar 36793766300	Osmania Univ. India	112
20	Ogbonna et al. (2001)	Sugar beet juice	Na	S. cerevisiae	Bioethanol production from sugar beet juice, immobilized yeasts, external loop bioreactor, system scaling up	Sugar beet, ethanol	Ogbonna, James C. 7003690993	Univ. Nigeria Nigeria	102
25	Dodic et al. (2009)	Sugar beet juice	Na	S. cerevisiae	Bioethanol from sugar beet juice in batch culture, sugar concentration, ethanol yield, CO_2 emissions, Serbia	Sugar beet, bioethanol	Dodic, Sinisa 6603034801	Univ. Novi Sad Serbia	91

*: Female; Cits., Number of citations received for each paper; Na, non-available; Prt, Biomass pretreatments.

respectively. Further, when sugarcane molasses were used as an additive, the juice under the same conditions resulted in maximum bioethanol concentration, productivity, and yield with the values of 109.34, 1.5, and 0.45 g/g, respectively. In addition, ammonium sulfate was not suitable for use as a nitrogen supplement in sweet sorghum juice for bioethanol production since it caused a reduction in bioethanol concentration and yield of approximately 14% when compared to those of un-supplemented juices.

Limtong et al. (2007) produced bioethanol at a high temperature from sugarcane juice using *Kluyveromyces marxianus* in a paper with 201 citations. They found that this strain supplemented with 4% (w/v) of ethanol at 35°C produced high concentrations of ethanol at both 40°C and 45°C. Ethanol production by this strain in shaking flask cultivation in sugarcane juice media at 37°C was highest in a medium containing 22% of total sugars and some chemicals and having a pH of 5.0; the ethanol concentration reached 8.7% (w/v), productivity of 1.45 g/L/h, and yield 77.5% of theoretical yield. Further, at 40°C, a maximal ethanol concentration of 6.78% (w/v), a productivity of 1.13, and a yield of 60.4% of theoretical yield were obtained from the same medium, except that the pH was adjusted to 5.5. In a study on ethanol production in a 5 L jar fermenter with an agitation speed of 300 rpm and an aeration rate of 0.2 vvm throughout fermentation, this strain resulted in a final ethanol concentration of 6.43% (w/v), a productivity of 1.3 g/L/h, and a yield of 57.1% of theoretical yield.

Dien et al. (2009) improved sugar conversion and ethanol yield for forage sorghum (*Sorghum bicolor* L. Moench) lines with reduced lignin contents in a paper with 184 citations. They evaluated forage sorghum plants carrying *brown midrib* (*bmr*) mutations, which reduced lignin contents. They found that the near-isogenic lines evaluated were wild type, *bmr*-6, *bmr*-12, and *bmr*-6 *bmr*-12 double mutant. Bmr-6 and *bmr*-12 mutations were equally efficient at reducing lignin contents by 13% and 15%, respectively, and the effects were additive (27%) for the double mutant. They pretreated sorghum biomass samples with dilute acid and recovered solids washed and hydrolyzed with cellulase. Glucose yields for the sorghum biomass were improved by 27%, 23%, and 34% for *bmr*-6, *bmr*-12, and the double mutant, respectively, compared to the wild type. Conversion of cellulose to ethanol for dilute-acid-pretreated sorghum biomass was improved by 22%, 21%, and 43% for *bmr*-6, *bmr*-12, and the double mutant compared to the wild type, respectively. They observed an increased number of lignin globules in double-mutant tissues as compared to the wild type, suggesting that the lignin had become more pliable. Further, the mutations were also effective in improving ethanol yields when the degrained sorghum was pretreated with dilute alkali instead of dilute acid. Following pretreatment with dilute ammonium hydroxide and simultaneous saccharification and fermentation, ethanol conversion yields were 116 and 130 mg ethanol/g dry biomass for the double-mutant samples and 98 and 113 mg/g for the wild-type samples.

Laopaiboon et al. (2007) produced bioethanol from sweet sorghum juice in batch and fed-batch fermentations in a paper with 131 citations. They used the sweet sorghum juice supplemented with 0.5% ammonium sulfate for bioethanol production by *S. cerevisiae* TISTR 5048. They found that in batch fermentation, kinetic parameters for bioethanol production were dependent on initial cell and sugar concentrations. The optimum initial cell and sugar concentrations in batch fermentation were 1×10^8 cells/mL and 24°Bx respectively. During these conditions, ethanol concentration (P), yield (Y_{ps}), and productivity (Q_p) were 100 g/L, 0.42 g/g, and 1.67 g/L/h respectively. In fed-batch fermentation, the optimum substrate feeding strategy for ethanol production at the initial sugar concentration of 24°Bx was one-time substrate feeding, where P, Y_{ps}, and Q_p were 120 g/L, 0.48 g/g and 1.11 g/L/h respectively. They concluded that fed-batch fermentation improved the efficiency of ethanol production in terms of ethanol concentration and product yield.

Chandel et al. (2009) produced bioethanol from wild sugarcane (*Saccharum spontaneum*) in a paper with 112 citations. This biomass consisted of 45.10% cellulose and 22.75% of hemicellulose on a dry solid basis. They found that aqueous ammonia delignified *S. spontaneum* yielded total reducing sugars, 53.91 g/L (539.10 mg/g of substrate) with a hydrolytic efficiency of 77.85%. They tested the enzymatic hydrolysate of *S. spontaneum* for ethanol production under batch and repeated batch production systems using in situ entrapped *S. cerevisiae* VS3 cells in *S. spontaneum* stalks. Batch fermentation of VS3 free cells and immobilized cells resulted in ethanol production, 19.45 g/L (yield, 0.410 g/g) and 21.66 g/L (yield, 0.434 g/g), respectively. Immobilized VS3 cells showed maximum ethanol production (22.85 g/L, yield, 0.45 g/g) up to eighth cycle during repeated batch fermentation followed by a gradual reduction in subsequent cycles of fermentation.

Ogbonna et al. (2001) scaled up bioethanol production from sugar beet juice in a paper with 102 citations. They used cells immobilized on a loofa sponge and *S. cerevisiae* IR2. When compared with a 2 L bubble column bioreactor, they observed that mixing was not sufficient in an 8 L bioreactor containing a bed of sliced loofa sponges, and consequently, the immobilized cells were not uniformly distributed within the bed. Most of the cells were immobilized in the lower part of the bed and this resulted in decreased ethanol productivity. By using an external loop bioreactor, constructing the fixed bed with cylindrical loofa sponges, dividing the bed into upper, middle, and lower sections with approximately 1 cm spaces between them, and circulating the broth through the loop during the immobilization, they obtained uniform cell distribution within the bed. Using this method, they scaled up the system to 50 L, and when compared with the 2 L bubble column bioreactor, there were no significant differences in ethanol productivity and yield. In conclusion, by using an external loop bioreactor to immobilize the cells uniformly on loofa sponge beds, efficient large-scale ethanol production systems could be constructed.

Dodic et al. (2009) produced bioethanol from sugar beet juice in batch culture by free *S. cerevisiae* cells in a paper with 91 citations. They diluted thick juice and molasses of sugar beet with distilled water to give a total sugar concentration of 5%, 10%, 15%, 20%, and 25% (w/w). The optimal initial concentration of fermentable sugars was 20% (w/w) in the culture medium while the maximal ethanol yield (68%) and maximal CO_2 evolution rate were realized, amounting to more than 90 g/L/h. The optimal concentration of fermentable sugar from thick juice for bioethanol production by free *S. cerevisiae* cells was 20% (w/w) at 30°C, pH 5 and agitation rate 200 rpm gave maximum ethanol concentration of 12% (v/v).

48.3.2 Evaluation of Sugar Feedstock-based Bioethanol Fuels

There are 16 HCPs for the evaluation of the sugar feedstock-based bioethanol fuels (Table 48.2).

Macedo et al. (2008) evaluated the energy balance and greenhouse gas (GHG) emissions in the production and use of sugarcane-based bioethanol in Brazil presenting the 2005/2006 data with a prediction for 2020 in a paper with 630 citations. They used a sample of mills processing up to 100 million tons of sugarcane per year for 2005/2006. They found that the fossil energy ratio was 9.3 for 2005/2006 and might reach 11.6 in 2020 with technologies already commercial. For anhydrous ethanol production, the total GHG emission was 436 kg CO_2 eq/m³ ethanol for 2005/2006, decreasing to 345 kg CO_2 eq/m³ in the 2020 scenario. Avoided GHG emissions depending on the final use for E100 use in Brazil in 2005/2006 were 2,181 kg CO_2 eq/m³ ethanol and for E25 were 2,323 kg CO_2 eq/m³ ethanol. Both values would increase about 26% for the conditions assumed for 2020 mostly due to the large increase in sales of bioelectricity surpluses. Further, they found the high impact of sugarcane productivity and ethanol yield variation on these balances (and the impacts of average cane transportation distances, level of soil cultivation, and some others) and of sugarcane bagasse and bioelectricity surpluses on GHG emission avoidance.

Goldemberg et al. (2008) discussed the sustainability of sugarcane-based bioethanol production in a paper with 431 citations. The positive impact of bioethanol production was the elimination of lead compounds from gasoline and the reduction of noxious and CO_2 emissions. These positive impacts were particularly noticeable in the air quality improvement of metropolitan areas but also in rural areas where mechanized harvesting of green cane was introduced, eliminating the burning of sugarcane. Negative impacts such as future large-scale bioethanol production might lead to the destruction or damage of high biodiversity areas, deforestation, degradation, or damaging of soils through the use of chemicals and soil decarbonization, water resource contamination or depletion,

TABLE 48.2
Evaluation of Sugar Feedstock-based Bioethanol Fuels

No.	Papers	Biomass/ Hydrolysate	Parameters	Keywords	Lead Author	Affil.	Cits
1	Macedo et al. (2008)	Sugarcane	Brazil energy balance and GHG emissions, sugarcane productivity, ethanol yield	Sugarcane, ethanol	Macedo, Isaias C. 6603142465	State Univ. Campinas Brazil	631
2	Goldemberg et al. (2008)	Sugarcane	Impact of sugar-based bioethanol production in Brazil, GHG emissions, biodiversity, deforestation, soil, water, food security, labor	Sugarcane, ethanol	Goldemberg, Jose 7003862006	Univ. Sao Paulo Brazil	431
3	Martinelli and Filoso (2008)	Sugarcane	Environmental and social impacts associated with bioethanol production in Brazil	Sugarcane, ethanol	Martinelli, Luiz A. 7102366222	Univ. Sao Paulo Brazil	274
4	Quintero et al. (2008)	Sugarcane	Economic and environmental performance of the bioethanol production process from sugarcane and corn in Colombia	Sugarcane, ethanol	Cardona, Carlos A. 57214443163	Natl. Univ. Colombia Colombia	251
5	Luo et al. (2009)	Sugarcane	Comparative life cycle assessment (LCA) and costing (LCC) of sugarcane-based bioethanol and gasoline in Brazil	Sugarcane, bioethanol	Luo, Lin 35307615000	Leiden Univ. Netherlands	236
6	Smeets et al. (2008)	Sugarcane	Environmental and socioeconomic impact of sugarcane-based bioethanol production in Brazil, sugarcane production	Ethanol, Brazilian	Smeets, Edward 15078634500	Utrecht Univ. Netherlands	166
7	Renouf et al. (2008)	Sugarcane, sugar beet	Environmental LCA of sugarcane production and processing in Australia, energy input, GHG emissions, acidification, N_2O emissions, eutrophication	Sugarcane, sugar beet, fermentation	Renouf, Marguerite A.* 23989436100	Univ. Queensland Australia	163
8	Van den Wall Bake et al. (2009)	Sugarcane	Reductions in feedstock and industrial production cost of Brazilian sugarcane-based ethanol, progress ratio, experience curve	Ethanol, Brazilian	Junginger, Martin 57195636600	Utrecht Univ. Netherlands	144

(Continued)

TABLE 48.2 (*Continued*)
Evaluation of Sugar Feedstock-based Bioethanol Fuels

No.	Papers	Biomass/ Hydrolysate	Parameters	Keywords	Lead Author	Affil.	Cits
9	Cavalett et al. (2013)	Sugarcane	Comparative LCA of sugarcane-based bioethanol and gasoline in Brazil using different LCA methods	Ethanol, Brazil	Cavalett, Otavio 9846419000	CTBE Brazil	129
10	De Carvalho Macedo (1998)	Sugarcane	GHG emissions and energy balances in sugarcane-based bioethanol production and evaluation in Brazil	Bioethanol, Brazil	De Carvalho Macedo, Isaias 6603142465	State Univ. Campinas Brazil	118
11	Walter et al. (2011)	Sugarcane	Sustainability assessment of sugarcane-based bioethanol production in Brazil considering land use change, GHG emissions, and socioeconomic aspects	Bioethanol, Brazil	Walter, Arnaldo 7102205549	State Univ. Campinas Brazil	115
12	Ometto et al. (2009)	Sugarcane	LCA of sugarcane-based bioethanol in Brazil, global warming, ozone formation, acidification, nutrient enrichment, ecotoxicity, human toxicity	Sugarcane, ethanol, Brazil	Aldo R. Ometto 14066545600	Univ. Sao Paulo Brazil	107
13	Zhang et al. (2010)	Sweet sorghum	Productive potentials of sweet sorghum-based bioethanol in China	Sweet sorghum, ethanol	Xie, Gaodi 7202981418	Chinese Acad. Sci. China	99
14	Pereira and Ortega (2010)	Sugarcane	Sustainability assessment of large-scale sugarcane-based ethanol production in Brazil	Sugarcane, Ethanol	Ortega, Enrique 9839187400	State Univ. Campinas, Brazil	99
15	Halleux et al. (2008)	Sugar beet	Comparative LCA of sugar beet-based bioethanol and rapeseed methyl ester (RME) in Belgium	Sugar beet, ethanol	Halleux, Hubert 24381733300	Univ. Liege Belgium	98
16	Tsiropoulos et al. (2015)	Sugarcane	LCA of bioplastics from sugarcane-based bioethanol from India and Brazil, polyethylene, GHG emissions	Sugarcane, ethanol	Tsiropoulos, I 35516560700	Utrecht Univ. Netherlands	95

*, Female; Cits., Number of citations received for each paper; Na, non-available; Prt, Biomass pretreatments..

competition between food and fuel production decreasing food security and a worsening of labor conditions on the fields. They clarified the sustainability aspects of bioethanol production mainly in the Sao Paulo State, where more than 60% of Brazil's sugarcane plantations was located and responsible for 62% of ethanol production.

Martinelli and Filoso (2008) discussed the environmental and social impacts associated with bioethanol production in Brazil in a paper with 274 citations. They cautioned that environmental and social impacts associated with bioethanol production in Brazil could become important obstacles to sustainable bioethanol fuel production worldwide. Atmospheric pollution from the burning of sugarcane for harvesting, degradation of soils and aquatic systems, and exploitation of sugarcane cutters were among the issues that deserved immediate attention from the Brazilian government and international organizations. Expansion of sugarcane crops to the areas presently cultivated for soybeans also represented an environmental threat because it might increase the deforestation pressure from soybean crops in the Amazon region. They made recommendations to help policymakers and the Brazilian government establish new initiatives to produce a code for bioethanol production that was environmentally sustainable and economically fair. Recommendations included proper planning and environmental risk assessments for the expansion of sugarcane to new regions such as Central Brazil, improvement of land use practices to reduce soil erosion and nitrogen pollution, proper protection of streams and riparian ecosystems, banning of sugarcane burning practices, and fair working conditions for sugarcane cutters. They also supported the creation of a more constructive approach for international stakeholders and trade organizations to promote sustainable development for bioethanol production in developing countries such as Brazil. Finally, they supported the inclusion of environmental values in the price of bioethanol fuels in order to discourage excessive replacement of natural ecosystems such as forests, wetlands, and pastures by bioenergy crops.

Quintero et al. (2008) compared the economic and environmental performance of the bioethanol production process from sugarcane and corn in Colombia in a paper with 251 citations. They integrated the net present value and total output rate of potential environmental impact into one index using the analytical hierarchy process (AHP) approach. They determined the sugarcane-based bioethanol process as the best choice for Colombian bioethanol production facilities. AHP scores obtained in this study for sugarcane and corn ethanol were 0.571 and 0.429, respectively. However, they further found that starch feedstocks like corn, cassava, or potatoes could potentially cause a higher impact on rural communities and boost their economies if social matters were considered.

Luo et al. (2009) performed the comparative life cycle assessment (LCA) and costing (LCC) of sugarcane-based bioethanol and gasoline in Brazil in a paper with 236 citations. The base case was the bioethanol production from sugarcane and bioelectricity generation from sugarcane bagasse whereas the future case was bioethanol production from both sugarcane and bagasse and bioelectricity generation from sugarcane wastes. In both cases, sugar was co-produced. The life cycles of fuels included gasoline production, agricultural production of sugarcane, bioethanol production, sugar and bioelectricity co-production, blending ethanol with gasoline to produce E10 (10% of ethanol) and E85 (85%), and finally the use of gasoline, E10, E85, and pure ethanol (E100). Furthermore, they conducted LCC to give an indication of fuel economy in both cases. They found that in base cases, less GHG was emitted while the overall evaluation of these fuel options depended on the importance attached to different impacts. The future case was certainly more economically attractive, which had been the driving force for the development in the bioethanol industry in Brazil. Nevertheless, the outcomes depended very much on the assumed price for crude oil. In real market, the prices of fuels were very much dependent on taxes and subsidies. Technological developments could help in lowering both the environmental impact and the prices of bioethanol fuels.

Smeets et al. (2008) evaluated the environmental and socioeconomic impact of sugarcane-based bioethanol production in the state of Sao Paulo of Brazil in a paper with 166 citations. They classified 14 areas of concern as a minor or medium bottleneck. For seven areas of concern, they calculated the additional costs to avoid or reduce undesirable effects at less than 10% for each area of concern. Due to higher yields and overlapping costs, the total additional production cost of compliance with various environmental and socioeconomic criteria were about 36%. Further, the energy input-to-output ratio could be increased and GHG emissions be reduced by increasing bioethanol production per ton of sugarcane and by increasing the use of sugarcane waste for electricity production. A major bottleneck for sustainable and certified production was the increase in sugarcane production and the possible impact on biodiversity and competition with food production. Genetically modified sugarcane was presently being developed, but at this moment has not yet been applied. Both a ban on and the allowance of the use of genetically modified sugarcane could become a major bottleneck considering the potentially large benefits and disadvantages, which were both highly uncertain at this moment.

Renouf et al. (2008) performed an environmental LCA of sugarcane production and processing for bioethanol production in Australia in a paper with 163 citations. They compared Australian sugarcane with the US corn and UK sugar beet. They found that sugarcane had an advantage in respect of energy input, GHG emissions, and possibly acidification potential due to its high saccharide yield and the displacement of fossil fuels with surplus renewable energy from sugarcane bagasse. However, Australian sugarcane could exhibit high nitrous oxide (N_2O) emissions, which would reduce GHG advantages in some regions. For eutrophication, sugar beet provided advantages due to the avoided production of other agricultural crops displaced by the use of beet pulp as an animal feed. The three factors that had the most influence on the environmental impact of these agro-industrial systems are the commodities displaced by by-products, agricultural yields, and nitrogen use efficiency.

Van den Wall Bake et al. (2009) evaluated the reductions in feedstock and industrial production costs of Brazilian sugarcane-based ethanol in a paper with 144 citations. They found that the progress ratio (PR) for feedstock and industrial production costs was 0.68 and 0.81, respectively. Further, the experience curve of total production costs resulted in a PR of 0.80. Cost breakdowns of sugarcane production showed that all subprocesses contributed to the total, but that increasing yields had been the main driving force. Industrial costs mainly decreased because of the increasing scales of ethanol plants. The total production cost at present was approximately 340 US\$/$m_{ethanol}^3$ (16 US\$/GJ). Based on the experience curves for feedstock and industrial costs, they estimated the total ethanol production costs in 2020 between US\$ 200 and 260/m^3 (9.4–12.2 US\$/GJ). They concluded that using disaggregated experience curves for feedstock and industrial processing costs provided more insights into the factors that lowered costs in the past and allowed more accurate estimations for future cost developments.

Cavalett et al. (2013) performed a comparative LCA of sugarcane-based bioethanol and gasoline in Brazil using different LCA methods in a paper with 129 citations. They considered the LCA methods of CML 2001, Impact 2002+, EDIP 2003, Eco-indicator 99, TRACI 2, ReCiPe, and Ecological Scarcity 2006. They used energy allocation to split the environmental burdens between ethanol and surplus bioelectricity generated at the sugarcane biorefinery. Finally, they considered the phases of feedstock and bioethanol production, distribution, and use in system boundaries. At the midpoint level, they found that a comparison of different LCA methods showed that ethanol presented lower impacts than gasoline in important categories such as global warming, fossil depletion, and ozone layer depletion. However, ethanol presented a higher impact in acidification, eutrophication, photochemical oxidation, and agricultural land use categories. Further, regarding single-score indicators, ethanol presented better performance than gasoline using the ReCiPe Endpoint LCA method. Using IMPACT 2002+, Eco-indicator 99, and Ecological Scarcity 2006, they verified higher scores for ethanol, mainly due to the impact related to particulate emissions and land use impacts. In conclusion, although there was a relative agreement on the results regarding equivalent environmental impact categories using different LCA methods at a midpoint level, when single-score indicators were considered, the use of different LCA methods led to different conclusions. Single-score results also limited the interpretability at an endpoint level, as a consequence of small contributions of relevant environmental impact categories weighted in a single-score indicator.

De Carvalho Macedo (1998) evaluated the GHG emissions and energy balances in sugarcane-based bioethanol production and utilization in Brazil in 1996 in a paper with 118 citations. They found that the production of sugarcane in 1996 was 273 million t (harvested wet wt)/year, leading to 13.7 million m^3 of ethanol and 13.5 million t of sugar. They evaluated GHG emissions for the agronomic and industrial production processes and product utilization including N_2O and methane. Updating the energy balance from 1985 to 1995 showed the effect of the main technological trends as fossil fuel consumption due to increasing agricultural mechanization was largely offset by technological advances in transportation and overall conversion efficiencies (both agricultural and industrial). On the other hand, the output/input energy ratio in bioethanol grew to 9.2 (average) and 11.2 (best values). Net savings in CO_2 (equivalent) emissions, due to bioethanol and sugarcane bagasse substitution for fossil fuels, corresponded to 46.7×10^6 t CO_2 (equivalent)/year, nearly 20% of all CO_2 emissions from fuels in Brazil. Further, bioethanol alone was responsible for 64% of the net avoided emissions.

Walter et al. (2011) performed a sustainability assessment of sugarcane-based bioethanol production in Brazil considering land use change, GHG emissions, and socioeconomic aspects in a paper with 115 citations. They found that the recent expansion of sugarcane had mostly occurred at the expense of pasture lands and other temporary crops and that the hypothesis of induced deforestation was not confirmed. Avoided GHG emissions due to the use of anhydrous bioethanol blended with gasoline in Brazil (E25) were 78%, while this figure would be 70% in the case of its use in Europe (E10). Conversely, considering the direct impacts of land use change, the avoided emissions (e.g., bioethanol consumed in Europe) would vary from −2.2 (i.e., emissions slightly higher than gasoline) to 164.8% (a remarkable carbon capture effect) depending on the management practices employed previous to land use change and also along sugarcane cropping. In addition, they found that where the bulk of sugarcane production took place, in the state of Sao Paulo, there were positive socioeconomic aspects. In conclusion, a significant share of bioethanol production in Brazil could be considered sustainable, in particular regarding the three aspects assessed.

Ometto et al. (2009) performed LCA of sugarcane-based bioethanol as E100 vehicle fuel in Brazil in a paper with 107 citations. The product system included agricultural and industrial activities, distribution, cogeneration of bioelectricity and steam, ethanol use during car driving, and industrial by-product recycling to irrigate sugarcane fields. LCA covered the emission-related impact categories: global warming, ozone formation, acidification, nutrient enrichment, ecotoxicity, and human toxicity (HT). The ethanol lifecycle used a high quantity and diversity of non-renewable resources. The input of renewable resources was also high mainly because of water consumption in the industrial phases. During the lifecycle of ethanol, there was a surplus of electric energy due to the cogeneration activity. Another focus point was the quantity of emissions to the atmosphere and the diversity of the substances emitted. Harvesting was the unit process that contributed most to global warming. For photochemical ozone formation, harvesting was also the activity with the strongest contributions due to the burning in harvesting and the emissions from diesel fuel. The acidification impact potential was mostly due to the NO_x emitted by the combustion of ethanol during use, on account of the sulfuric acid use in the industrial process and because of the NO_x emitted by burning in harvesting. The main consequence of the intensive use of fertilizers in the field was the high nutrient enrichment impact potential associated with this activity. The main contributions to the ecotoxicity impact potential came from chemical applications during crop growth. The activity that presented the highest impact potential for HT via air and soil was harvesting. Via water, the HT potential was high in harvesting due to lubricant use on the machines. Nutrient enrichment, acidification, and human toxicity via air and water were the most significant impact potentials for the lifecycle of fuel ethanol. In conclusion, the ethanol lifecycle contributed negatively to all the impact potentials of global warming, ozone formation, acidification, nutrient enrichment, ecotoxicity, and HT. Concerning energy consumption, it consumed less energy than its own production largely because of the bioelectricity cogeneration system, but this process was highly dependent on water. The main causes for the biggest impact potential indicated by normalization were the nutrient application, burning in harvesting, and use of diesel fuel. Recommendations for ethanol lifecycle were harvesting sugarcane without burning, doing more environmentally benign agricultural practices, using renewable fuel rather than petrodiesel, not washing sugarcane and implementing water recycling systems during the industrial processing, and improving the system of gaseous emission control during the use of ethanol in cars, mainly for NO_x.

Zhang et al. (2010) evaluated the productive potentials of sweet sorghum-based bioethanol in China in a paper with 99 citations. They found that although sweet sorghum could be planted in the majority of lands in China, the suitable unused lands for large-scale planting were only as much as 78.6×10^4 hm^2, and the productive potentials of ethanol from these lands were 157.1×10^4–294.6×10^4 t/year, which could only meet 24.8%–46.4% of the current demand for E10 fuel in China. On the other hand, if all the common grain sorghum at present were replaced by sweet sorghum, the average ethanol yield of 244×10^4 t/year could be added, and thus the productive potentials of sweet sorghum ethanol could satisfy 63.2%–84.9% of the current demand for E10 fuel of China. In general, Heilongjiang, Jilin, Inner Mongolia, and Liaoning ranked the highest in the productive potentials of sweet sorghum ethanol, followed by Hebei, Shanxi, Sichuan, and some other provinces.

Pereira and Ortega (2010) performed a sustainability assessment of large-scale sugarcane-based ethanol production in Brazil in a paper with 99 citations. It covered both farm and industrial production phases. They found that about 1.82 kg of topsoil was eroded, and 18.4 L of water and 1.52 m^2 of land were needed to produce 1 L of ethanol from sugarcane. Further, 0.28 kg of CO_2 was released per liter of ethanol produced. The energy content of ethanol was 8.2 times greater than the fossil-based energy required to produce it. The transformity of ethanol was about the same as those calculated for fossil fuels. The renewability of ethanol was 30%. The other energy indices indicated important environmental impact as well as natural resource consumption. In conclusion, sugarcane and bioethanol production presented low renewability when a large-scale system was adopted.

Halleux et al. (2008) performed a comparative LCA of sugar beet-based bioethanol and rapeseed methyl ester (RME) in Belgium in a paper with 98 citations. The assessment included cultivation, extraction, processing, and final use of biofuels as well as valorization of by-products in substitution of animal feed or biochemicals. They found that without valorization of the by-products, the impacts of the two biofuels over their whole life cycle were not significantly lower than those of the related fossil fuels. However, credit-accorded thanks to valorization of by-products lead to improvements in the environmental performances in comparison to fossil fuels, especially for RME. The impacts of agriculture and production of fertilizers were more important in the case of RME because the yield per hectare was lower than in the production of sugar beet-based ethanol. However, the extraction of sugar from beet, fermentation, purification of ethanol, and drying of pulps and slops consumed much more energy than the extraction and esterification of rapeseed oil. Moreover, the by-products of RME were more valuable than those of ethanol from sugar beet. This induced a more important environmental benefit in the production of RME. These parameters led to a global environmental impact that was higher for the production of ethanol than for RME. In conclusion, RME allowed a considerable improvement in environmental performances compared to fossil diesel, while bioethanol offered a more limited benefit compared to petrol.

Tsiropoulos et al. (2015) performed LCA of bioplastics from sugarcane-based bioethanol from India and Brazil in a paper with 95 citations. They compared fully high-density biopolyethylene and partial biopolyethylene terephthalate with their petrochemical counterparts in Europe. They found that biopolyethylene resulted in GHG emissions of around -0.75 kg $CO^2_{eq}/kg_{polyethylene}$, i.e., 140% lower than petrochemical polyethylene, while savings on nonrenewable energy use were approximately 65%. GHG emissions of partial biopolyethylene terephthalate were similar to petrochemical production ($\pm10\%$), and nonrenewable energy use was lower by up to 10%. Assuming that the process energy provided by combined heat and power reduced GHG emissions of partial biopolyethylene terephthalate production to a range from -4% (higher) to 15% (lower) compared to petrochemical polyethylene terephthalate. Internal technical improvements such as fuel switch, new plants, and best available technology offered savings of up to 30% in GHG emissions compared to the current production of petrochemical polyethylene terephthalate. The combination of some of these measures and the use of biomass for the supply of process steam could reduce GHG emissions even further. In human health and ecosystem quality, the impact of biopolymers was up to two orders of magnitude higher, primarily due to pesticide use, preharvesting burning practices in Brazil, and land occupation. When improvements were assumed across the supply chain, such as pesticide control and elimination of burning practices, the impact of biopolymers could be significantly reduced.

48.4 DISCUSSION

48.4.1 INTRODUCTION

Crude oil-based gasoline fuels have been widely used in the transportation sector since the 1920s. However, there have been great public concerns over the adverse environmental and human impact of these fuels. Hence, biomass-based bioethanol fuels have increasingly been used in blending gasoline and petrodiesel fuels, in fuel cells, and in biochemical production in a biorefinery context.

However, it is necessary to pretreat the biomass to enhance the yield of bioethanol prior to bioethanol fuel production from sugar feedstocks through hydrolysis and fermentation of the biomass and hydrolysates, respectively. One of the most-studied feedstocks for bioethanol fuels has been sugar feedstocks, such as sugarcane, sugar beet, sweet sorghum, and energy cane. Research in the field of sugar feedstock-based bioethanol fuels has intensified in this context in the key research fronts of pretreatment and hydrolysis of sugar feedstocks, fermentation of sugar feedstock-based hydrolysates, and production and evaluation of sugar feedstock-based bioethanol fuels. Further, sugarcane and sweet sorghum and to a lesser extent sugar beet and energy cane have been studied intensively in this context.

However, it is essential to develop efficient incentive structures for the primary stakeholders to enhance research in this field. Although there have been a number of review papers for this field, there has been no review of the most-cited 25 articles in this field.

Thus, this chapter presents a review of the most-cited 25 articles on bioethanol fuel production from sugar feedstocks. Then, it discusses the key findings of these highly influential papers and comments on future research priorities in this field.

As a first step for the search of the relevant literature, keywords were selected using the most-cited first 200 population papers. The selected keyword list was then optimized to obtain a representative sample of papers for the searched research field. This keyword list was provided in the Appendix of Konur (2023) for future replicative studies.

As a second step, a sample dataset was used for this study. The first 25 articles with at least 91 citations each were selected for the review study. Key findings from each paper were taken from the abstracts of these papers and were discussed. Additionally, a number of brief conclusions were drawn and a number of relevant recommendations were made to enhance the future research landscape.

Information about the research fronts for the sample papers in bioethanol fuel production from sugar feedstocks with regard to sugar feedstocks used in these processes is given in Table 48.3. As this table shows, there are four primary research fronts for this field with regard to the feedstocks: sugarcane, sweet sorghum, sugar beet, and energy canes with 69%, 24%, 16%, and 4% of the HCPs, respectively.

On the other hand, sugar beet is the most influential research front with an 9% surplus while sweet sorghum and sugarcane are the least influential research fronts with 6% and 5% deficits, respectively.

Information about the thematic research fronts for the sample papers in bioethanol fuel production from sugar feedstocks is given in Table 48.4. As this table, shows there are two primary

TABLE 48.3
The Most Prolific Research Fronts for Sugar Feedstock-based Bioethanol Fuels

No.	Research Fronts	N Paper (%) Review	N Paper (%) Sample	Surplus (%)
1	Sugarcane	68.0	73.4	−5.4
2	Sweet sorghum	24.0	30.3	−6.3
3	Sugar beet	16.0	7.3	8.7
4	Energy cane	4.0	3.7	0.3

N Paper (%) review, The number of papers in the sample of 25 reviewed papers; N paper (%) sample, The number of papers in the population sample of 109 papers.

TABLE 48.4
The Most Prolific Thematic Research Fronts for Sugar Feedstock-based Bioethanol Fuel Fuels

No.	Research Fronts	N Paper (%) Review	N Paper (%) Sample	Surplus (%)
1	Biomass Pretreatments	4.0	12.8	−8.8
	Chemical pretreatments	4.0	8.3	−4.3
	Acid pretreatment	4.0	4.6	−0.6
	Sweet sorghum	4.0	0.9	3.1
	Enzymatic pretreatments	4.0	10.1	−6.1
	Feedstock genetic engineering	4.0	1.8	2.2
2	Biomass Hydrolysis	4.0	10.1	−6.1
	Enzymatic hydrolysis	4.0	10.1	−6.1
	Sweet sorghum	4.0	5.5	−1.5
3	Hydrolysate Fermentation	12.0	28.5	−16.5
	Sweet sorghum	8.0	19.3	−11.3
	Sugarcane	4.0	6.4	−2.4
4	Bioethanol Production	36.0	35.7	0.3
	Sweet sorghum	20.0	22.9	−2.9
	Sugarcane	12.0	7.3	4.7
	Sugar beet	8.0	3.7	4.3
	Energy cane	4.0	1.8	2.2
5	Bioethanol Fuel Evaluation	64.0	75.2	−11.2
	Sugarcane	56.0	64.2	−8.2
	Sweet sorghum	4.0	6.4	−2.4
	Sugar beet	8.0	2.8	5.2

N Paper (%) review, The number of papers in the sample of 25 reviewed papers; N paper (%) sample, The number of papers in the population sample of 109 papers.

research fronts for this field: the evaluation and production of bioethanol fuels with 64% and 36% of these HCPs, respectively. The other research fronts are hydrolysate fermentation and pretreatments and hydrolysis of sugar feedstocks with 12%, 4%, and 4% of HCPs, respectively.

Further, sugarcane and sweet sorghum feedstocks dominate the evaluation and production of bioethanol fuels with 56% and 20% of HCPs, respectively. Further, hydrolysate fermentation, bioethanol fuel evaluation, and biomass pretreatments and hydrolysis are the least influential fronts with 16%, 11%, 8%, and 6% deficits, respectively.

48.4.2 PRODUCTION OF SUGAR FEEDSTOCK-BASED BIOETHANOL FUELS

There are nine HCPs for the production of sugar feedstock-based bioethanol fuels (Table 48.1).

These HCPs show a sample of research on the production of sugar-based bioethanol fuels. These studies hint that the required steps in bioethanol production are the pretreatments and enzymatic hydrolysis of sugar feedstocks, followed by fermentation of hydrolysates (Alvira et al., 2010; Taherzadeh and Karimi, 2008).

Gnansounou et al. (2005) evaluated bioethanol, sugar, and bioelectricity production from sweet sorghum juice and hemicellulose and cellulose in sorghum bagasse in China and found that the relative merits of making bioethanol or sugar from the juice were very sensitive to the price of sugar in China. Further, Kim and Day (2011) determined the composition and ethanol yield of sugarcane, energy cane, and sweet sorghum suitable for bioethanol production in the US sugar biorefineries and recommended that energy cane and sweet sorghum must be used as additional feedstocks as they had different harvest times compared to sugarcane.

Laopaiboon et al. (2009) produced bioethanol from sweet sorghum juice using VHG fermentation using *S. cerevisiae* NP01 and found that the sweet sorghum juice resulted in maximum bioethanol production efficiency with concentration, productivity, and yield. Further, Limtong et al. (2007) produced bioethanol at a high temperature from sugarcane juice using *K. marxianus* and found that this strain supplemented with 4% (w/v) ethanol at 35°C produced high concentrations of ethanol at both 40°C and 45°C.

Dien et al. (2009) improved sugar conversion and ethanol yield for forage sorghum lines with reduced lignin contents and found that the conversion of cellulose to ethanol for dilute acid-pretreated sorghum biomass was improved by 22%, 21%, and 43% for *bmr*-6, *bmr*-12, and the double mutants compared to wild type, respectively. Further, Laopaiboon et al. (2007) produced bioethanol from sweet sorghum juice in batch and fed-batch fermentations and found that in batch fermentation, the kinetic parameters for bioethanol production were dependent on initial cell and sugar concentrations.

Chandel et al. (2009) produced bioethanol from wild sugarcane and found that batch fermentation of VS3 free cells and immobilized cells resulted in ethanol production, 19.45 g/L (yield, 0.410 g/g) and 21.66 g/L (yield, 0.434 g/g), respectively. Further, Ogbonna et al. (2001) scaled up bioethanol production from sugar beet juice and found that there were no significant differences in ethanol productivity and yield. Finally, Dodic et al. (2009) produced bioethanol from sugar beet juice in a batch culture by free *S. cerevisiae* cells and found that the maximal ethanol yield (68%) and maximal CO_2 evolution rate were realized, amounting to more than 90 g/L/h.

48.4.3 EVALUATION OF SUGAR FEEDSTOCK-BASED BIOETHANOL FUELS

There are 16 HCPs for the evaluation of the sugar feedstock-based bioethanol fuels (Table 48.2). These HCPs show a sample of research on the evaluation of sugar-based bioethanol fuels. These studies hint that a large number of variables are considered for the evaluation of these bioethanol fuels: energy balance, GHG emissions, biodiversity, deforestation, soil, water, food security, labor, environmental and social impacts, economic and environmental performance, acidification, N_2O emissions, eutrophication, sustainability, and land use change. It is notable that sugar feedstock-based bioethanol fuels have positive evaluations in most of these variables compared to gasoline fuels and some other biofuels. Thus, Brazil has become a global leader in both the production and utilization of sugarcane-based bioethanol fuels since the 1970s in the aftermath of the global oil crisis in the early 1970s.

Macedo et al. (2008) evaluated the energy balance and GHG emissions in the production and use of sugarcane-based bioethanol in Brazil presenting the 2005/2006 data with a prediction for 2020 and found that the fossil energy ratio was 9.3 for 2005/2006 and might reach 11.6 in 2020 with technologies that are already commercial. Further, Goldemberg et al. (2008) discussed the sustainability of sugarcane-based bioethanol production and found that the positive impact of bioethanol production was the elimination of lead compounds from gasoline and the reduction of noxious and CO_2 emissions.

Martinelli and Filoso (2008) discussed the environmental and social impacts associated with bioethanol production in Brazil and cautioned that environmental and social impacts associated with bioethanol production in Brazil could become important obstacles to sustainable bioethanol fuel production worldwide. Quintero et al. (2008) compared the economic and environmental performance of the bioethanol production process from sugarcane and corn under Colombian conditions and determined the sugarcane-based bioethanol process as the best choice for Colombian bioethanol production facilities.

Luo et al. (2009) performed the comparative LCA and LCC of sugarcane-based bioethanol and gasoline in Brazil and found that in the base case, less GHG was emitted while the overall evaluation of these fuel options was dependent on the importance attached to different impacts. Further, Smeets et al. (2008) evaluated the environmental and socioeconomic impact of sugarcane-based bioethanol production in the state of Sao Paulo of Brazil and classified 14 areas of concern as a minor or medium bottleneck.

Renouf et al. (2008) performed an environmental LCA of sugarcane production and processing of bioethanol production in Australia and found that sugarcane had an advantage in respect of energy input, GHG emissions, and possibly acidification potential. Van den Wall Bake et al. (2009) evaluated the reductions in feedstock and industrial production costs of Brazilian sugarcane-based ethanol and found that the PR for feedstock and industrial production cost was 0.68 and 0.81, respectively.

Cavalett et al. (2013) performed the comparative LCA of sugarcane-based bioethanol and gasoline in Brazil using different LCA methods and found that comparison of different LCA methods showed that ethanol presented lower impacts than gasoline in important categories such as global warming, fossil depletion, and ozone layer depletion. Further, De Carvalho Macedo (1998) evaluated GHG emissions and energy balances in sugarcane-based bioethanol production and utilization and found that the production of sugarcane in 1996 was 273 million t (harvested wet wt)/year, leading to 13.7 million m^3 ethanol and 13.5 million t of sugar.

Walter et al. (2011) performed a sustainability assessment of sugarcane-based bioethanol production in Brazil considering land use change, GHG emissions, and socioeconomic aspects and found that the recent expansion of sugarcane had mostly occurred at the expense of pasture lands and other temporary crops and that the hypothesis of induced deforestation was not confirmed.

Ometto et al. (2009) performed the LCA of sugarcane-based bioethanol as E100 vehicle fuel in Brazil and found that during the lifecycle of ethanol, there was a surplus of bioelectric energy due to the cogeneration activity. Zhang et al. (2010) evaluated the productive potentials of sweet sorghum-based bioethanol in China and found that although sweet sorghum could be planted in the majority of lands in China, the suitable unused lands for large-scale planting were only as much as 78.6×10^4 hm^2 and the productive potentials of ethanol from these lands were $157.1 \times 10^4 - 294.6 \times 10^4$ t/year.

Pereira and Ortega (2010) performed a sustainability assessment of large-scale sugarcane-based ethanol production in Brazil and found that about 1.82 kg of topsoil was eroded, and 18.4 L of water and 1.52 m^2 of land were needed to produce 1 L of ethanol from sugarcane. Halleux et al. (2008) performed a comparative LCA of sugar beet-based bioethanol and RME in Belgium and found that without valorization of the by-products, the impacts of the two biofuels over their whole life cycle were not significantly lower than those of the related fossil fuels. Finally, Tsiropoulos et al. (2015) performed LCA of bioplastics from sugarcane-based bioethanol from India and Brazil and found that biopolyethylene resulted in GHG emissions of 140% lower than petrochemical polyethylene while savings on nonrenewable energy use were approximately 65%.

48.5 CONCLUSION AND FUTURE RESEARCH

Brief information about the key research fronts covered by the 25 most-cited papers with at least 91 citations each is given under two primary headings: production and evaluation of bioethanol fuels.

The usual characteristics of these HCPs are that pretreatments and hydrolysis of sugar feedstocks and fermentation of the resultant hydrolysates are the primary processes for bioethanol fuel production from sugar feedstocks to improve ethanol yield as the sugar feedstocks are one of the most studied feedstocks for bioethanol production, especially for countries with large farmlands such as Brazil.

The key findings on these research fronts should be read in light of the increasing public concerns about climate change, GHG emissions, and global warming as these concerns have been certainly behind the boom in research on bioethanol fuel production from sugar feedstocks as an alternative to crude oil-based gasoline and diesel fuels in the last decades. It is also a sustainable alternative to crude oil-based gasoline or petrodiesel fuels. Recent supply shocks caused by the COVID-19 pandemic and the Russian invasion of Ukraine also highlight the importance of the production and utilization of bioethanol fuels as an alternative to crude oil-based gasoline and diesel fuels.

As Table 48.3 shows, there are three primary research fronts for this field with regard to feedstocks: sugarcane and to a lesser extent sweet sorghum and sugar beet. The other front is energy cane which is a genetically engineered version of sugarcane (Matsuoka et al., 2009).

Similarly, as Table 48.4 shows, there are two primary thematic research fronts for this field: the evaluation and production of bioethanol fuels. The other research fronts are the fermentation of hydrolysates and pretreatments and hydrolysis of sugar feedstocks. Further, sugarcane and sweet sorghum feedstocks dominate the evaluation and production of bioethanol fuels, respectively. Further, hydrolysate fermentation, bioethanol fuel evaluation, and biomass pretreatments and hydrolysis are the least influential fronts.

These studies emphasize the importance of proper incentive structures for efficient bioethanol fuel production from sugar feedstocks in light of North's institutional framework (North, 1991). In this context, the major producers and users of bioethanol fuels such as the United States and Canada with vast farmlands have developed strong incentive structures for efficient bioethanol fuel production from sugar feedstocks. In light of supply shocks caused primarily by the COVID-19 pandemic and the Russian invasion of Ukraine, it is expected that the incentive structures such as public funding would be enhanced to increase the share of bioethanol fuels in the global fuel portfolio as a strong alternative to crude oil-based gasoline and petrodiesel fuels.

In this context, it is expected that the most prolific researchers, institutions, countries, funding bodies, and journals in this field would have a first-mover advantage to benefit from such potential incentives. This is especially true for the Brazilian stakeholders as Brazil has become the global leader in both the production and utilization of sugar feedstock-based bioethanol fuels.

It is recommended that such review studies are performed for the primary research fronts of bioethanol fuel production from sugar feedstocks.

ACKNOWLEDGMENTS

The contribution of highly cited researchers in the field of bioethanol fuel production from sugar feedstocks has been gratefully acknowledged.

REFERENCES

Almodares, A. and M. R. Hadi. 2009. Production of bioethanol from sweet sorghum: A review. *African Journal of Agricultural Research* 4:772–780.

Alvira, P., E. Tomas-Pejo, M. Ballesteros and M. J. Negro. 2010. Pretreatment technologies for an efficient bioethanol production process based on enzymatic hydrolysis: A review. *Bioresource Technology* 101:4851–4861.

Angelici, C., B. M. Weckhuysen and P. C. A. Bruijnincx. 2013. Chemocatalytic conversion of ethanol into butadiene and other bulk chemicals. *ChemSusChem* 6:1595–1614.

Antolini, E. 2007. Catalysts for direct ethanol fuel cells. *Journal of Power Sources* 170:1–12.

Antolini, E. 2009. Palladium in fuel cell catalysis. *Energy and Environmental Science* 2:915–931.

Ayodele, B. V., M. A. Alsaffar and S. I. Mustapa. 2020. An overview of integration opportunities for sustainable bioethanol production from first- and second-generation sugar-based feedstocks. *Journal of Cleaner Production* 245:118857.

Burnham, J. F. 2006. Scopus database: A review. *Biomedical Digital Libraries* 3:1–8.

Cavalett, O., M. F. Chagas, J. E. Seabra and A. Bonomi. 2013. Comparative LCA of ethanol versus gasoline in Brazil using different LCIA methods. *International Journal of Life Cycle Assessment* 18:647–658.

Chandel, A. K., M. L. Narasu, M., G. Chandrasekhar, A. Manikyam and L. V. Rao. 2009. Use of *Saccharum spontaneum* (wild sugarcane) as biomaterial for cell immobilization and modulated ethanol production by thermotolerant *Saccharomyces cerevisiae* VS3. *Bioresource Technology* 100:2404–2410.

Chandel, A. K., O. V. Singh, R. L. Venkateswar, G. Chandrasekhar and M. L. Narasu. 2011. Bioconversion of novel substrate *Saccharum spontaneum*, a weedy material, into ethanol by *Pichia stipitis* NCIM3498. *Bioresource Technology* 102:1709–1714.

De Carvalho Macedo, I. 1998. Greenhouse gas emissions and energy balances in bio-ethanol production and utilization in Brazil (1996). *Biomass and Bioenergy* 14:77–81.

Dien, B. S., G. Sarath, and J. F. Pedersen, et al. 2009. Improved sugar conversion and ethanol yield for forage sorghum (*Sorghum bicolor* L. Moench) lines with reduced lignin contents. *Bioenergy Research* 2:153–164.

Dodic, S., S. Popov and J. Dodi, et al. 2009. Bioethanol production from thick juice as intermediate of sugar beet processing. *Biomass and Bioenergy* 33:822–827.

Fernando, S., S. Adhikari, C. Chandrapal and M. Murali. 2006. Biorefineries: Current status, challenges, and future direction. *Energy & Fuels* 20:1727–1737.

Gnansounou, E., A. Dauriat and C. E. Wyman. 2005. Refining sweet sorghum to ethanol and sugar: Economic trade-offs in the context of North China. *Bioresource Technology* 96:985–1002.

Goldemberg, J., S. T. Coelho and P. Guardabassi. 2008. The sustainability of ethanol production from sugarcane. *Energy Policy* 36:2086–2097.

Hahn-Hagerdal, B., M. Galbe, M. F. Gorwa-Grauslund, G. Liden and G. Zacchi. 2006. Bio-ethanol - The fuel of tomorrow from the residues of today. *Trends in Biotechnology* 24:549–556.

Halleux, H., S. Lassaux, R. Renzoni and A. Germain. 2008. Comparative life cycle assessment of two biofuels: Ethanol from sugar beet and rapeseed methyl ester. *International Journal of Life Cycle Assessment* 13:184–190.

Hill, J., E. Nelson, D. Tilman, S. Polasky and D. Tiffany. 2006. Environmental, economic, and energetic costs and benefits of biodiesel and ethanol biofuels. *Proceedings of the National Academy of Sciences of the United States of America* 103:11206–11210.

Hill, J., S. Polasky and E. Nelson, et al. 2009. Climate change and health costs of air emissions from biofuels and gasoline. *Proceedings of the National Academy of Sciences of the United States of America* 106:2077–2082.

Hsieh, W. D., R. H. Chen, T. L. Wu and T. H. Lin. 2002. Engine performance and pollutant emission of an SI engine using ethanol-gasoline blended fuels. *Atmospheric Environment* 36:403–410.

Huang, H. J., S. Ramaswamy, U. W. Tschirner and B. V. Ramarao. 2008. A review of separation technologies in current and future biorefineries. *Separation and Purification Technology* 62:1–21.

Kim, M. and D. F. Day. 2011. Composition of sugar cane, energy cane, and sweet sorghum suitable for ethanol production at Louisiana sugar mills. *Journal of Industrial Microbiology and Biotechnology* 38:803–807.

Konur, O. 2000. Creating enforceable civil rights for disabled students in higher education: An institutional theory perspective. *Disability & Society* 15:1041–1063.

Konur, O. 2002a. Access to nursing education by disabled students: Rights and duties of nursing programs. *Nurse Education Today* 22:364–374.

Konur, O. 2002b. Assessment of disabled students in higher education: Current public policy issues. *Assessment and Evaluation in Higher Education* 27:131–152.

Konur, O. 2002c. Access to employment by disabled people in the UK: Is the Disability Discrimination Act working? *International Journal of Discrimination and the Law* 5:247–279.

Konur, O. 2006a. Participation of children with dyslexia in compulsory education: Current public policy issues. *Dyslexia* 12:51–67.

Konur, O. 2006b. Teaching disabled students in higher education. *Teaching in Higher Education* 11:351–363.

Konur, O. 2007a. A judicial outcome analysis of the *Disability Discrimination Act*: A windfall for the employers? *Disability & Society* 22:187–204.

Konur, O. 2007b. Computer-assisted teaching and assessment of disabled students in higher education: The interface between academic standards and disability rights. *Journal of Computer Assisted Learning* 23:207–219.

Konur, O. 2012. The evaluation of the research on the bioethanol: A scientometric approach. *Energy Education Science and Technology Part A: Energy Science and Research* 28:1051–1064.

Konur, O. 2015. Current state of research on algal bioethanol. In *Marine Bioenergy: Trends and Developments*, Ed. S. K. Kim and C. G. Lee, pp. 217–244. Boca Raton, FL: CRC Press.

Konur, O. 2019. Cyanobacterial bioenergy and biofuels science and technology: A scientometric overview. In *Cyanobacteria: From Basic Science to Applications*, Ed. A. K. Mishra, D. N. Tiwari and A. N. Rai, pp. 419–442. Amsterdam: Elsevier.

Konur, O. 2020. The scientometric analysis of the research on the bioethanol production from green macroalgae. In *Handbook of Algal Science, Technology and Medicine*, Ed. O. Konur, pp. 385–401. London: Academic Press.

Konur, O. 2023. First generation sugar feedstock-based bioethanol fuels: Scientometric study. In *Feedstock-based Bioethanol Fuels. I. Non-Waste Feedstocks: Starch, Sugar, Grass, Wood, Cellulose, Algae, and Biosyngas-based Bioethanol Fuels. Handbook of Bioethanol Fuels Volume 3*, Ed. O. Konur. Boca Raton, FL: CRC Press.

Laopaiboon, L., S. Nuanpeng, P. Srinophakun, P. Klanrit and P. Laopaiboon. 2009. Ethanol production from sweet sorghum juice using very high gravity technology: Effects of carbon and nitrogen supplementations. *Bioresource Technology* 100:4176–4182.

Laopaiboon, L., P. Thanonkeo, P. Jaisil and P. Laopaiboon. 2007. Ethanol production from sweet sorghum juice in batch and fed-batch fermentations by *Saccharomyces cerevisiae*. *World Journal of Microbiology and Biotechnology* 23:1497–1501.

Limtong, S., C. Sringiew and W. Yongmanitchai. 2007. Production of fuel ethanol at high temperature from sugar cane juice by a newly isolated *Kluyveromyces marxianus*. *Bioresource Technology* 98:3367–3374.

Lin, Y. and S. Tanaka. 2006. Ethanol fermentation from biomass resources: Current state and prospects. *Applied Microbiology and Biotechnology* 69:627–642.

Luo, L., E. van der Voet and G. Huppes. 2009. Life cycle assessment and life cycle costing of bioethanol from sugarcane in Brazil. *Renewable and Sustainable Energy Reviews* 13:1613–1619.

Ma, X., L. Sun and C. Song. 2002. A new approach to deep desulfurization of gasoline, diesel fuel and jet fuel by selective adsorption for ultra-clean fuels and for fuel cell applications. *Catalysis Today* 77:107–116.

Macedo, I. C., J. E. A. Seabra and J. E. A. R. Silva. 2008. Green house gases emissions in the production and use of ethanol from sugarcane in Brazil: The 2005/2006 averages and a prediction for 2020. *Biomass and Bioenergy* 32:582–595.

Martinelli, L. A. and S. Filoso. 2008. Expansion of sugarcane ethanol production in Brazil: Environmental and social challenges. *Ecological Applications* 18:885–898.

Matsuoka, S., J. Ferro and P. Arruda. 2009. The Brazilian experience of sugarcane ethanol industry. *In Vitro Cellular & Developmental Biology-Plant* 45:372–381.

Morschbacker, A. 2009. Bio-ethanol based ethylene. *Polymer Reviews* 49:79–84.

Najafi, G., B. Ghobadian and T. Tavakoli, et al. 2009. Performance and exhaust emissions of a gasoline engine with ethanol blended gasoline fuels using artificial neural network. *Applied Energy* 86:630–639.

Newman, P. W. G. and J. R. Kenworthy. 1989. Gasoline consumption and cities: A comparison of U.S. cities with a global survey. *Journal of the American Planning Association* 55:24–37.

North, D. C. 1991. Institutions. *Journal of Economic Perspectives* 5:97–112.

Ogbonna, J. C., H. Mashima and H. Tanaka. 2001. Scale up of fuel ethanol production from sugar beet juice using loofa sponge immobilized bioreactor. *Bioresource Technology* 76:1–8.

Olsson, L. and B. Hahn-Hagerdal. 1996. Fermentation of lignocellulosic hydrolysates for ethanol production. *Enzyme and Microbial Technology* 18:312–331.

Ometto, A. R., M. Z. Hauschild and W. N. L. Roma. 2009. Lifecycle assessment of fuel ethanol from sugarcane in Brazil. *International Journal of Life Cycle Assessment* 14:236–247.

Pereira, C. L. F. and E. Ortega. 2010. Sustainability assessment of large-scale ethanol production from sugarcane. *Journal of Cleaner Production* 18:77–82.

Quintero, J. A., M. I. Montoya, O. J. Sanchez, O. H. Giraldo and C. A. Cardona. 2008. Fuel ethanol production from sugarcane and corn: Comparative analysis for a Colombian case. *Energy* 33:385–399.

Renouf, M. A., M. K. Wegener and L. K. Nielsen. 2008. An environmental life cycle assessment comparing Australian sugarcane with US corn and UK sugar beet as producers of sugars for fermentation. *Biomass and Bioenergy* 32:1144–1155.

Sanchez, O. J. and C. A. Cardona. 2008. Trends in biotechnological production of fuel ethanol from different feedstocks. *Bioresource Technology* 99:5270–5295.

Siqueira, G., A. M. F. Milagres, W. Carvalho, G. Koch and A. Ferraz. 2011. Topochemical distribution of lignin and hydroxycinnamic acids in sugar-cane cell walls and its correlation with the enzymatic hydrolysis of polysaccharides. *Biotechnology for Biofuels* 4:7.

Smeets, E., M. Junginger and A. Faaij, et al. 2008. The sustainability of Brazilian ethanol-An assessment of the possibilities of certified production. *Biomass and Bioenergy* 32:781–813.

Sun, Y. and J. Cheng. 2002. Hydrolysis of lignocellulosic materials for ethanol production: A review. *Bioresource Technology* 83:1–11.

Taherzadeh, M. J. and K. Karimi. 2007. Enzyme-based hydrolysis processes for ethanol from lignocellulosic materials: A review. *Bioresources* 2:707–738.

Taherzadeh, M. J. and K. Karimi. 2008. Pretreatment of lignocellulosic wastes to improve ethanol and biogas production: A review. *International Journal of Molecular Sciences* 9:1621–1651.

Tsiropoulos, I., A. P. C. Faaij and L. Lundquist, L., et al. 2015. Life cycle impact assessment of bio-based plastics from sugarcane ethanol. *Journal of Cleaner Production* 90:114–127.

Van den Wall Bake, J. D., M. Junginger, A. Faaij, T. Poot and A. Walter. 2009. Explaining the experience curve: Cost reductions of Brazilian ethanol from sugarcane. *Biomass and Bioenergy* 33:644–658.

Walter, A., P. Dolzan and O. Quilodran, et al. 2011. Sustainability assessment of bio-ethanol production in Brazil considering land use change, GHG emissions and socio-economic aspects. *Energy Policy* 39:5703–5716.

Zabed, H., G. Faruq and J. N. Sahu, et al. 2014. Bioethanol production from fermentable sugar juice. *Scientific World Journal* 2014:957102.

Zhang, C., G. Xie, S. Li, L. Ge and T. He. 2010. The productive potentials of sweet sorghum ethanol in China. *Applied Energy* 87:2360–2368.

Part 11

Grass-based Bioethanol Fuels

Part II

49 Grass-based Bioethanol Fuels
Scientometric Study

Ozcan Konur

(Formerly) Ankara Yildirim Beyazit University

49.1 INTRODUCTION

Crude oil-based gasoline fuels (Ma et al., 2002; Newman and Kenworthy, 1989) have been widely used in the transportation sector since the 1920s. However, there have been great public concerns over the adverse environmental and human impact of these fuels (Hill et al., 2006, 2009). Hence, biomass-based bioethanol fuels (Hill et al., 2006; Konur, 2012e, 2015, 2019, 2020a) have increasingly been used in blending gasoline fuels (Hsieh et al., 2002; Najafi et al., 2009), in fuel cells (Antolini, 2007, 2009), and in biochemical production (Angelici et al., 2013; Morschbacker, 2009) in a biorefinery context (Fernando et al., 2006; Huang et al., 2008).

Bioethanol fuels also play a critical role in maintaining energy security (Kruyt et al., 2009; Winzer, 2012) in the supply shocks (Kilian, 2008, 2009) related to oil price shocks (Hamilton, 2003, 2009), COVID-19 pandemic (Fauci et al., 2020; Li et al., 2020), or wars (Hamilton, 1983; Jones, 2012) in the aftermath of the Russian invasion of Ukraine (Reeves, 2014).

However, it is necessary to pretreat the biomass (Taherzadeh and Karimi, 2008; Yang and Wyman, 2008) to enhance the yield of bioethanol (Hahn-Hagerdal et al., 2006; Sanchez and Cardona, 2008) prior to bioethanol production through hydrolysis (Sun and Cheng, 2002; Taherzadeh and Karimi, 2007) and fermentation (Lin and Tanaka, 2006; Olsson and Hahn-Hagerdal, 1996) of the biomass and hydrolysates, respectively.

One of the most-studied biomass for bioethanol production and use has been grass biomass. Research in the field of grass-based bioethanol fuels has intensified in this context in the key research fronts of pretreatment of grass biomass (Li et al., 2010, Mani et al., 2004; Sun and Cheng, 2005), hydrolysis of grass biomass (Hu and Wen, 2008; Li et al., 2010; Sun and Cheng, 2005), fermentation of grass hydrolysates (Faga et al., 2010; Fu et al., 2011; Jin et al., 2010), and grass-based bioethanol fuels in general (Fu et al., 2011; Pimentel and Patzek, 2005; Schmer et al., 2008). Further, switchgrass (Li et al., 2010; Mani et al., 2004; Pimentel and Patzek, 2005) and to a lesser extent miscanthus (Murnen et al., 2007; Xu et al., 2012; Yoshida et al., 2008), Bermuda grass (Sun and Cheng, 2005; Wang et al., 2010), giant reed (Zhang et al., 2011), and alfalfa (Dien et al., 2006; Sreenath et al., 1999) have been studied intensively at the expense of other grass biomasses in this context.

However, it is essential to develop efficient incentive structures (North, 1991) for the primary stakeholders to enhance research in this field (Konur, 2000, 2002a,b,c, 2006a,b, 2007a,b). Scientometric analysis has been used in this context to inform the primary stakeholders about the current state of research in a selected research field (Garfield, 1955; Konur, 2011, 2012a,b,c,d,e,f,g,h,i, 2015, 2018b, 2019, 2020a).

As there have been no published scientometric studies in this field, this chapter presents a scientometric study of research in grass-based bioethanol fuels. It examines the scientometric characteristics of both the sample and population data presenting the scientometric characteristics of these both datasets in the order of documents, authors, publication years, institutions, funding bodies, source titles, countries, Scopus subject categories, Scopus keywords, and research fronts.

DOI: 10.1201/9781003226451-64

201

49.2 MATERIALS AND METHODS

Search for this study was carried out using the Scopus database (Burnham, 2006) in June 2022.

As a first step for the search of the relevant literature, keywords were selected using the first most-cited 200 population papers. The selected keyword list was then optimized to obtain a representative sample of papers for the searched research field. This keyword list was provided in the appendix for future replicative studies.

As a second step, two sets of data were used for this study. First, a population sample of 1,006 papers was used to examine the scientometric characteristics of the population data. Secondly, a sample of 100 most-cited papers, corresponding to 10% of population papers, was used to examine the scientometric characteristics of these citation classics.

The scientometric characteristics of these both sample and population datasets were presented in the order of documents, authors, publication years, institutions, funding bodies, source titles, countries, Scopus subject categories, Scopus keywords, and research fronts.

Lastly, the key scientometric findings for both datasets were discussed to highlight the research landscape for grass-based bioethanol fuels. Additionally, a number of brief conclusions were drawn and a number of relevant recommendations were made to enhance the future research landscape.

49.3 RESULTS

49.3.1 THE MOST PROLIFIC DOCUMENTS IN GRASS-BASED BIOETHANOL FUELS

Information on the types of documents for both datasets is given in Table 49.1. The articles and conference papers, published in journals, dominate both the sample (98%) and population (95%) papers as they are over-represented in the sample papers by 3%. Further, review papers and short surveys have a 0.6% surplus as they are over-represented in the sample papers by 0.6% as they constitute 2% and 1.4% of sample and population papers, respectively.

It is further noted that 99% of population papers was published in journals while 1% of them was published in book series and books. On the contrary, all sample papers were published in journals.

49.3.2 THE MOST PROLIFIC AUTHORS IN GRASS-BASED BIOETHANOL FUELS

Information about the most prolific 18 authors with at least 4% of sample papers each is given in Table 49.2. The most prolific authors are Arthur J. Ragauskas and Blake A. Simmons with 9% of

TABLE 49.1

Documents in Grass-based Bioethanol Fuels

Documents	Sample Dataset (%)	Population Dataset (%)	Surplus (%)
Article	95.0	93.2	1.8
Conference paper	3.0	2.3	0.7
Review	2.0	1.2	0.8
Book chapter	0.0	0.6	−0.6
Letter	0.0	0.2	−0.2
Note	0.0	0.2	−0.2
Short survey	0.0	0.2	−0.2
Editorial	0.0	0.1	−0.1
Book	0.0	0.0	0.0
Sample size	100	1,006	

Population dataset, The number of papers (%) in the set of 1,006 population papers; Sample dataset, The number of papers (%) in the set of 100 highly cited papers.

TABLE 49.2

The Most Prolific Authors in Grass-based Bioethanol Fuels

No.	Author Name	Author Code	Sample Papers (%)	Population Papers (%)	Surplus	Institution	Country	HI	N	Res. Front
1	Ragauskas, Arthur J.	7006265204	9.0	2.5	6.5	Univ. Tennessee Knoxville	USA	91	747	PT
2	Simmons, Blake, A.	7102183263	9.0	2.1	6.9	Lawrence Berkeley Natl. Lab.	USA	74	436	PT
3	Dale, Bruce E.	7201511969	8.0	1.9	6.1	Michigan State Univ.	USA	90	429	PT
4	Cheng, Jay J.	15046539600	8.0	1.8	6.2	NC State Univ.	USA	14	55	PT
5	Singh, Seema*	35264950300	8.0	1.8	6.2	Sandia Natl. Labs.	USA	57	179	PT
6	Balan, Venkatesh	15757087100	7.0	1.5	5.5	Univ. Houston	USA	55	211	PT
7	Shi, Jian	55537446300	6.0	1.3	4.7	Univ. Kentucky	USA	39	80	PT
8	Lee, Yoon Y.	8948274900	6.0	1.0	5.0	Auburn Univ.	USA	53	115	PT
9	Dien, Bruce S.	6603685796	5.0	2.2	2.8	USDA Agr. Res. Serv.	USA	61	186	PT
10	Wyman, Charles E.	7004396809	5.0	1.8	3.2	Univ. Calif. Riverside	USA	80	286	PT
11	Peng, Liangcai	36802605200	5.0	1.4	3.6	Huazhong Agr. Univ.	USA	32	108	PT
12	Holtzapple, Mark T.	7004167004	5.0	1.1	3.9	Texas A&M Univ.	USA	46	195	PT
13	Li, Chenlin	35760453700	5.0	0.5	4.5	Idaho Natl. Lab.	USA	23	53	PT
14	Tu, Yuanyuan	55324399900	4.0	1.2	2.8	Huazhong Agr. Univ.	China	23	36	PT
15	Brosse, Nicolas	6603382062	4.0	0.8	3.2	Univ. Lorraine	France	36	149	PT
16	Elander, Richard T.	6603931116	4.0	0.8	3.2	NREL	USA	31	61	PT
17	Ladisch, Michael R.	7005670397	4.0	0.8	3.2	Purdue Univ.	USA	59	290	PT
18	Vogel, Kenneth P.	7102498441	4.0	0.6	3.4	Univ. Nebraska Lincoln	USA	63	146	PT

*, Female researchers; Author code, the unique code given by Scopus to authors; Population papers, the number of papers authored in the population dataset; PT, pretreatments; Sample papers, the number of papers authored in the sample dataset.

sample papers each, followed by Bruce E. Dale, Jay J. Cheng, and Seema Singh with 8% of sample papers each. The other prolific authors are Venkatesh Balan, Jian Shi, and Yoon Y. Lee with 6% of sample papers each.

On the other hand, the most influential authors are Blake A. Simmons and Arthur J. Ragauskas with 6.9% and 6.5% surplus, respectively, followed by Jay J. Cheng, Seema Singh, and Bruce E. Dale with 6.1%–6.2% surplus each.

The most prolific institution for the sample dataset is the Huazhong Agricultural University with two authors each. In total, 17 institutions house these top authors. On the other hand, the most prolific country for the sample dataset is the United States with 16 authors. In total, only three countries house these top authors.

There is only one primary research front for these top authors: Pretreatments for the grass biomass. On the other hand, there is a significant gender deficit (Beaudry and Lariviere, 2016) for the sample dataset as surprisingly only one of these top researchers is female with a representation rate of 6%.

Additionally, there are other authors with a relatively low citation impact and with 0.8%–1.2% of population papers each: Marli Cammasola, Ratna R. Sharma-Shivappa, Kasiviswanathan Muthukumarappan, Yunqiao Pu, Aldo J. P. Dillon, Jiele Xu, Joseph C. Burns, Richard A. Dixon, Bin Yang, Junhua Zhang, Vera E. Budaeva, Michael D. Casler, Iwona Cybulska, Bryon S. Donohue, Mirvat E. Ebrik, Chunxiang Fu, Praveen Kolar, John M. Prausnitz, Miguel Rodriguez, Gautum Sarath, Shahab Sokhansanj, and Zeng-Yu Wang.

49.3.3 THE MOST PROLIFIC RESEARCH OUTPUT BY YEARS IN GRASS-BASED BIOETHANOL FUELS

Information about papers published between 1970 and 2022 is given in Figure 49.1. This figure clearly shows that the bulk of research papers in the population dataset was published primarily

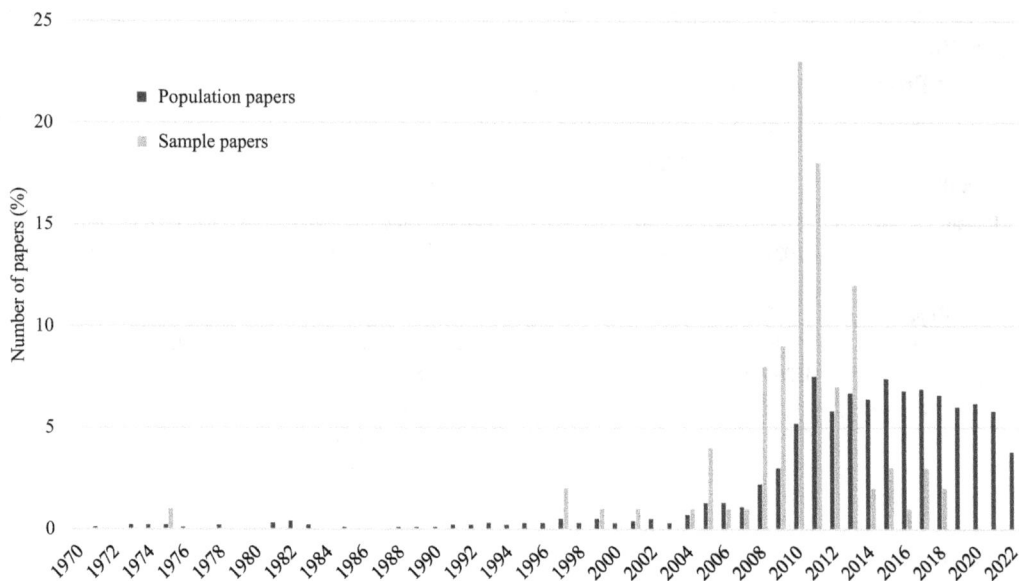

FIGURE 49.1   Research output by years regarding grass-based bioethanol fuels.

in the 2010s and the early 2020s with 65 and 16% of population dataset, respectively. Similarly, publication rates for the 2000s, 1990s, 1980s, and 1970s were 11%, 3%, 1%, and 1% respectively. Additionally, 0.9% of population papers was published in the pre-1970s.

Similarly, the bulk of research papers in the sample dataset was published in the 2000s and 2010s with 25 and 71% of sample dataset, respectively. Similarly, publication rates for the 1990s, 1980s, and 1970s were 3%, 0%, and 1% of sample papers, respectively.

The most prolific publication years for the population dataset were 2017 and 2016 with 6.9% and 6.8% of the dataset, respectively. Further, 81% of population papers was published between 2010 and 2022. Similarly, 88% of sample papers was published between 2008 and 2017 while the most prolific publication year was 2010 with 23% of sample papers. The other prolific years were 2011 and 2013 with 18% and 13% of sample papers, respectively. The number of population papers rose from 2008 to 2111; thereafter, it steadied above 5% of population papers each year, losing its momentum.

49.3.4   The Most Prolific Institutions in Grass-based Bioethanol Fuels

Information about the most prolific 19 institutions publishing papers on grass-based bioethanol fuels with at least 4% of sample papers each is given in Table 49.3.

The most prolific institution is the NC State University with 11% of sample papers, followed by the USDA Agricultural Research Service, Georgia Institute of Technology, and Joint Bioenergy Institute with 10%, 10%, and 9% of sample papers, respectively. The other prolific institutions are Michigan State University, Sandia National Laboratories, and University of Nebraska Lincoln with 8% of sample papers each.

Similarly, the top country for these most prolific institutions is the United States with 16 institutions. In total, only three countries house these top institutions.

On the other hand, the institution with the most citation impact is the Georgia Institute of Technology with an 8% surplus, followed by the Joint BioEnergy Institute, NC State University, University of Nebraska Lincoln, Sandia National Laboratories, and Michigan State University with 5.6%–6.6% of sample papers each.

TABLE 49.3
The Most Prolific Institutions in Grass-based Bioethanol Fuels

No.	Institutions	Country	Sample Papers (%)	Population Papers (%)	Surplus (%)
1	NC State Univ.	USA	11.0	4.6	6.4
2	USDA Agric. Res. Serv.	USA	10.0	5.0	5.0
3	Georgia Inst. Technol.	USA	10.0	2.0	8.0
4	Joint BioEnergy Inst.	USA	9.0	2.4	6.6
5	Michigan State Univ.	USA	8.0	2.4	5.6
6	Sandia Natl. Labs.	USA	8.0	2.1	5.9
7	Univ. Nebraska Lincoln	USA	8.0	1.7	6.3
8	Oak Ridge Natl. Lab.	USA	6.0	3.6	2.4
9	Lawrence Berkeley Natl. Lab.	USA	6.0	1.9	4.1
10	Auburn Univ.	USA	6.0	1.4	4.6
11	Univ. Calif. Riverside	USA	5.0	2.0	3.0
12	NREL	USA	5.0	1.6	3.4
13	Texas A&M Univ.	USA	5.0	1.4	3.6
14	Univ. Wisconsin Madison	USA	4.0	1.9	2.1
15	Univ. Calif. Berkeley	USA	4.0	1.7	2.3
16	Huazhong Agr. Univ.	China	4.0	1.5	2.5
17	Purdue Univ.	USA	4.0	1.2	2.8
18	Genencor Inc.	USA	4.0	0.8	3.2
19	Univ. Lorraine	France	4.0	0.8	3.2

Additionally, there are other institutions with a relatively low citation impact and with 0.9%–2.7% of population papers each: University of Tennessee, Knoxville, Chinese Academy of Sciences, South Dakota State University, University of Georgia, DOE Bioenergy Research Centers, University of Illinois Urbana-Champaign, Oklahoma State University, Northwest A&F University, Aberystwyth University, Chulalongkorn University, Iowa State University, Beijing Forestry University, University of Caxias do Sul, Samuel Roberts Noble Foundation, University of Minnesota Twin Cities, Wageningen University & Research, Oregon State University, University of California Davis, Washington State University Tri-Cities, Michigan Technological University, Russian Academy of Sciences, South China University of Technology, Virginia Polytechnic Institute and State University, Aberystwyth University, Indian Institute of Technology Roorkee, and Rural Development Administration.

49.3.5 THE MOST PROLIFIC FUNDING BODIES IN GRASS-BASED BIOETHANOL FUELS

Information about the most prolific 12 funding bodies funding at least 3% of sample papers each is given in Table 49.4. Further, only 51% and 52% of sample and population papers were funded, respectively.

The most prolific funding body is the U.S. Department of Energy with 15% of sample papers, followed by the Office of Science, Biological and Environmental Research, Higher Education Discipline Innovation Project, and Lawrence Berkeley National Laboratory with 5%–7% of sample papers each. On the other hand, the most prolific country for these top funding bodies is the United States with seven funding bodies, followed by the European Union (EU) and South Korea with two funding bodies each. In total, only three countries and the EU house these top funding bodies.

The funding body with the most citation impact is the U.S. Department of Energy with a 7.1% surplus while the other influential funding bodies are the Higher Education Discipline Innovation

TABLE 49.4

The Most Prolific Funding Bodies in Grass-based Bioethanol Fuels

No.	Funding Bodies	Country	Sample Paper No. (%)	Population Paper No. (%)	Surplus (%)
1	U.S. Department of Energy	USA	15.0	7.9	7.1
2	Office of Science	USA	7.0	4.1	2.9
3	Biological and Environmental Research	USA	6.0	2.7	3.3
4	Higher Education Discipline Innovation Project	China	5.0	1.0	4.0
5	Lawrence Berkeley National Laboratory	USA	5.0	1.0	4.0
6	National Science Foundation	USA	3.0	2.8	0.2
7	European Commission	EU	3.0	2.2	0.8
8	European Regional Development Fund	EU	3.0	1.2	1.8
9	Ministry of Education, Science and Technology	South Korea	3.0	0.6	2.4
10	National Research Foundation of Korea	South Korea	3.0	0.6	2.4
11	North Carolina State University	USA	3.0	0.5	2.5
12	North Carolina Biotechnology Center	USA	3.0	0.3	2.7

Project, Lawrence Berkeley National Laboratory, Biological and Environmental Research, and Office of Science with 2.9%–4.0% surplus each.

The other funding bodies with a relatively low citation impact and with 0.8%–6.4% of population papers each are the National Natural Science Foundation of China, U.S. Department of Agriculture, National Institute of Food and Agriculture, National Council for Scientific and Technological Development, National Key Research and Development Program of China, Ministry of Science and Technology of China, CAPES Foundation, Ministry of Education, Culture, Sports, Science and Technology, Biotechnology and Biological Sciences Research Council, Sao Paulo Research Foundation, Ministry of Education of China, South Dakota State University, Ministry of Science, Technology, and Innovation, Natural Sciences and Engineering Research Council of Canada, National Research Foundation of Korea, Oak Ridge National Laboratory, and Seventh Framework Program of EU.

49.3.6 THE MOST PROLIFIC SOURCE TITLES IN GRASS-BASED BIOETHANOL FUELS

Information about the most prolific 14 source titles publishing at least 2% of sample papers each in grass-based bioethanol fuels is given in Table 49.5.

The most prolific source title is Bioresource Technology with 39% of sample papers. The other prolific titles are Biomass and Bioenergy, Biotechnology for Biofuels, Bioenergy Research, and Proceedings of the National Academy of Sciences of the United States of America with 4%–8% of sample papers each.

On the other hand, the source title with the most citation impact is Bioresource Technology with a 24% surplus. The other influential titles are Biomass and Bioenergy, Proceedings of the National Academy of Sciences of the United States of America, Polymer Degradation and Stability, and Biotechnology Progress with 2.5%–3.8% surplus each. Similarly, the source title with the least impact is Applied Biochemistry and Biotechnology with a 0.3% deficit.

The other source titles with a relatively low citation impact with 0.5%–5.3% of population papers each are the Industrial Crops and Products, Bioresources, ACS Sustainable Chemistry and Engineering, Crop Science, Biomass Conversion And Biorefinery, Cellulose, Energies, Journal of Environmental Quality, Renewable Energy, Process Biochemistry, Transactions of the ASABE, Nature, Waste and Biomass Valorization, Biofuels, Green Chemistry, Fuel, Journal of

TABLE 49.5
The Most Prolific Source Titles in Grass-based Bioethanol Fuels

No.	Source Titles	Sample Papers (%)	Population Papers (%)	Surplus (%)
1	Bioresource Technology	39.0	14.9	24.1
2	Biomass and Bioenergy	8.0	4.2	3.8
3	Biotechnology for Biofuels	5.0	4.0	1.0
4	Bioenergy Research	5.0	3.3	1.7
5	Proceedings of the National Academy of Sciences of the United States of America	4.0	0.4	3.6
6	Biotechnology and Bioengineering	3.0	0.8	2.2
7	Applied Biochemistry and Biotechnology Part A Enzyme Engineering and Biotechnology	3.0	0.7	2.3
8	Biotechnology Progress	3.0	0.5	2.5
9	Polymer Degradation and Stability	3.0	0.4	2.6
10	Applied Biochemistry and Biotechnology	2.0	2.3	−0.3
11	Energy and Fuels	2.0	1.5	0.5
12	Plos One	2.0	1.2	0.8
13	Industrial and Engineering Chemistry Research	2.0	1.1	0.9
14	International Journal of Life Cycle Assessment	2.0	0.3	1.7

Cleaner Production, GCB Bioenergy, Scientific Reports, Asian Journal of Chemistry, Cellulose Chemistry and Technology, Environmental Technology United Kingdom, Processes, and Science of the Total Environment.

49.3.7 THE MOST PROLIFIC COUNTRIES IN GRASS-BASED BIOETHANOL FUELS

Information about the most prolific 12 countries publishing at least 2% of sample papers each in grass-based bioethanol fuels is given in Table 49.6.

The most prolific country is the United States with 75% of sample papers, followed by China, France, and Colombia with 4 to 8% of sample papers each. Further, four European countries listed in 49.6 produce 12% and 4% of sample and population papers, respectively.

On the other hand, the country with the most citation impact is the United States with a 34% surplus, followed by Colombia and France with 2.9% and 2.4% surplus, respectively. Similarly, the country with the least citation impact is China with a 7.7% deficit while India, Canada, the United Kingdom, Japan, and South Korea are the other countries with the least influence with 0.3% and 2.1% deficit each.

Additionally, the other countries with a relatively low citation impact and with 0.5%–3.5% of sample papers each are Brazil, Thailand, Spain, Denmark, Netherlands, Poland, Russia, Sweden, Germany, Malaysia, Nigeria, Ireland, Portugal, Belgium, Uruguay, Iran, South Africa, Argentina, Australia, Austria, Greece, Mexico, Norway, Pakistan, and Turkey.

49.3.8 THE MOST PROLIFIC SCOPUS SUBJECT CATEGORIES IN GRASS-BASED BIOETHANOL FUELS

Information about the most prolific eight Scopus subject categories indexing at least 7% of sample papers each is given in Table 49.7.

TABLE 49.6
The Most Prolific Countries in Grass-based Bioethanol Fuels

No.	Countries	Sample Papers (%)	Population Papers (%)	Surplus (%)
1	USA	75.0	40.9	34.1
2	China	8.0	15.7	−7.7
3	France	5.0	2.6	2.4
4	Colombia	4.0	1.1	2.9
5	India	3.0	5.1	−2.1
6	UK	3.0	4.0	−1.0
7	Japan	3.0	3.3	−0.3
8	South Korea	3.0	3.3	−0.3
9	Canada	2.0	3.9	−1.9
10	Taiwan	2.0	1.2	0.8
11	Finland	2.0	0.8	1.2
12	Italy	2.0	0.8	1.2

TABLE 49.7
The Most Prolific Scopus Subject Categories in Grass-based Bioethanol Fuels

No.	Scopus Subject Categories	Sample Papers (%)	Population Papers (%)	Surplus (%)
1	Energy	67.0	42.4	24.6
2	Environmental Science	63.0	41.5	21.5
3	Chemical Engineering	56.0	39.1	16.9
4	Biochemistry and Genetics and Molecular Biology	22.0	20.7	1.3
5	Immunology and Microbiology	16.0	13.6	2.4
6	Agricultural and Biological Sciences	15.0	24.5	−9.5
7	Engineering	8.0	9.3	−1.3
8	Chemistry	7.0	12.4	−5.4

The most prolific Scopus subject category in grass-based bioethanol fuels is Energy with 67% of sample papers, followed by Environmental Science and Chemical Engineering with 63% and 56% of sample papers, respectively. The other prolific subject categories are Biochemistry, Genetics and Molecular Biology, Immunology and Microbiology, and Agricultural and Biological Sciences with 15%–22% of sample papers each. It is noted that Social Sciences including Economics and Business account only for 2.9% of population studies.

On the other hand, the Scopus subject category with the most citation impact is Energy with a 25% surplus, followed by Environmental Science and Chemical Engineering with a 22% and 17% surplus, respectively. Similarly, the least influential subject categories are Chemistry and Engineering with a 5% and 1% deficit, respectively.

49.3.9 THE MOST PROLIFIC KEYWORDS IN GRASS-BASED BIOETHANOL FUELS

Information about the Scopus keywords used with at least 10% or 3.5% of sample or population papers, respectively, is given in Table 49.8. For this purpose, keywords related to the keyword set given in the appendix are selected from a list of the most prolific keyword set provided by the Scopus database.

TABLE 49.8
The Most Prolific Keywords in Grass-based Bioethanol Fuels

No.	Keywords	Sample Papers (%)	Population Papers (%)	Surplus (%)
1	Biomass and biomass constituents			
	Lignin	83.0	37.2	45.8
	Cellulose	83.0	32.8	50.2
	Panicum	74.0	25.3	48.7
	Biomass	62.0	38.0	24.0
	Grass	60.0	32.4	27.6
	Switchgrass	46.0	17.6	28.4
	Poaceae	28.0	15.4	12.6
	Lignocellulose	27.0	12.0	15.0
	Carbohydrates	24.0	12.7	11.3
	Hemicellulose	15.0	7.3	7.7
	Miscanthus	14.0	19.9	−5.9
	Xylan	13.0	5.3	7.7
	Lignocellulosic biomass	11.0	11.0	0.0
	Energy crops	11.0	4.8	6.2
	Feedstocks	6.0	7.2	−1.2
	Pennisetum	5.0	4.6	0.4
	Arundo	5.0	3.5	1.5
	Alfalfa	0.0	3.6	−3.6
2	Fermentation	0.0	0.0	0.0
	Fermentation	30.0	21.3	8.7
	SSF		4.5	−4.5
3	Bacteria			
	Yeast	10.0	7.2	2.8
	Saccharomyces	7.0	5.6	1.4
4	Hydrolysis			
	Hydrolysis	55.0	33.9	21.1
	Enzymatic hydrolysis	41.0	27.4	13.6
	Saccharification	28.0	20.2	8.0
	Digestibility	14.0	5.2	9.0
	Enzymatic digestibility	13.0	5.0	8.0
	Sugar yield	10.0	2.3	8.0
	Enzymatic saccharification	7.0	5.3	2.0
5	Hydrolysates			
	Sugar	76.0	24.8	51.0
	Glucose	34.0	19.1	15.0
	Xylose	21.0	8.0	13.0
	Fermentable sugars	8.0	4.6	3.0
6	Biomass pretreatment			
	Pre-treatment	56.0	24.5	32.0
	Enzyme activity	54.0	22.8	31.0
	Pretreatment	41.0	21.0	20.0

(Continued)

TABLE 49.8 (*Continued*)
The Most Prolific Keywords in Grass-based Bioethanol Fuels

No.	Keywords	Sample Papers (%)	Population Papers (%)	Surplus (%)
	Cellulases	36.0	24.0	12.0
	Temperature	32.0	12.5	20.0
	Enzymes	24.0	13.2	11.0
	Sulfuric acid	18.0	7.3	11.0
	Ionic liquids	16.0	8.1	8.0
	Ammonia	13.0	7.1	6.0
	Sodium hydroxide	11.0	6.6	4.0
	Enzymolysis	11.0	6.5	5.0
	Delignification	9.0	9.5	−1.0
	Alkalinity	5.0	3.8	1.0
7	Other processes			
	Biotechnology	22.0	8.3	14.0
	Fractionation	5.0	3.8	1.0
8	Products			
	Ethanol	48.0	28.6	19.4
	Biofuels	38.0	24.1	13.9
	Alcohol	26.0	10.0	16.0
	Bioenergy	13.0	5.0	8.0
	Bioethanol	12.0	15.1	−3.1
	Alcohol production	11.0	4.5	6.5
	Ethanol production	10.0	5.1	4.9
	Cellulosic ethanol	8.0	4.9	3.1
	Bioethanol production		7.6	−7.6

These keywords are grouped into eight headings: biomass, fermentation, bacteria, hydrolysis, hydrolysates, pretreatments, other processes, and products.

The most prolific keywords related to biomass and biomass constituents are lignin and cellulose with 83% of sample papers each, followed by panicum, biomass, grass, and switchgrass with 46%–74% of sample papers each. Further, the most prolific keywords related to bacteria are yeasts and saccharomyces with 10% and 7% of sample papers, respectively.

The most prolific keyword related to hydrolysis is hydrolysis with 55% of sample papers, followed by enzymatic hydrolysis and saccharification with 41% and 28% of sample papers, respectively. Further, the most prolific keyword related to fermentation is fermentation with 30% of sample papers.

The most prolific keyword related to the hydrolysates is sugar with 76% of sample papers, followed by glucose and xylose with 34% and 21% of sample papers, respectively. Further, the most prolific keywords related to biomass pretreatments are pre-treatment and enzyme activity with 56% and 54% of sample papers, respectively, followed by pretreatments, cellulases, and temperature with 32%–41% of sample papers each.

The most prolific keyword related to other processes is biotechnology with 22% of sample papers. Finally, the most prolific keyword related to products is ethanol with 48% of sample papers, followed by biofuels and alcohol with 38% and 26% of sample papers, respectively.

On the other hand, the most influential keywords are sugars, cellulose, panicum, lignin, pretreatment, enzyme activity, switchgrass, grass, and biomass with 24%–51% surplus each. Similarly, the most prolific keywords across all research fronts are cellulose, sugar, panicum, biomass, grass, pretreatments, hydrolysis, enzyme activity, ethanol, switchgrass, enzymatic hydrolysis, pretreatments, biofuels, cellulases, glucose, temperature, fermentation, saccharification, Poaceae, lignocellulose, and alcohol with 26%–83% of sample papers each.

49.3.10   THE MOST PROLIFIC RESEARCH FRONTS IN GRASS-BASED BIOETHANOL FUELS

Information about the research fronts for sample papers in grass-based bioethanol fuels with regard to the grass biomass used for bioethanol fuels is given in Table 49.9.

As this table shows, there are two primary research fronts for this field – switchgrass and miscanthus – with 62% and 17% of sample papers, respectively. The other grass biomasses used in these studies are giant reed, Bermuda grass, grass, ryegrass, common reed, alfalfa, napier grass, canary grass, and other grass with at least 2% of sample papers each.

Information about the thematic research fronts for sample papers in grass-based bioethanol fuels is given in Table 49.10. As this table shows, there are four primary research fronts – grass biomass pretreatments and hydrolysis, grass hydrolysate fermentation, and grass-based bioethanol fuels in general – with 87%, 78%, 18%, and 35% of sample papers, respectively.

For the research front of grass biomass pretreatments, the primary research fronts are chemical, enzymatic, and hydrothermal pretreatments with 70%, 79%, and 20% of sample papers, respectively. Additionally, there are four further research fronts – mechanical and microbial pretreatments, pretreatments in general, and genetic engineering of the grass biomass – with 7%, 3%, 2%, and 9% of sample papers, respectively. Further, the most prolific chemical pretreatments are alkaline, acid, ammonia, and ionic liquid pretreatments with 29%, 25%, 14%, and 9% of sample papers, respectively.

TABLE 49.9
The Most Prolific Research Fronts for Grass-based Bioethanol Fuels

No.	Research Fronts	N Paper (%) Sample
1	Switchgrass	62
2	Miscanthus	17
3	Giant reed	6
4	Bermuda grass	6
5	Grass	5
6	Ryegrass	4
7	Common reed	3
8	Alfalfa	3
9	Napier grass	2
10	Canary grass	2
11	Other grasses	3
	Silvergrass	1
	Elephant grass	1
	Cordgrass	1

N paper (%) sample, The number of papers in the population sample of 100 papers.

TABLE 49.10
The Most Prolific Thematic Research Fronts for Grass-based Bioethanol Fuels

No.	Research Fronts	N Paper (%) Sample
1	Grass Biomass Pretreatments	87
	Grass biomass pretreatments in general	2
	Switchgrass	2
	Mechanical pretreatments	7
	Milling pretreatments	4
	Switchgrass	3
	Miscanthus	1
	Microwave pretreatments	3
	Switchgrass	2
	Bermuda grass	1
	Ultrasound pretreatments	0
	Hydrothermal pretreatments	20
	Steam explosion pretreatment	4
	Switchgrass	2
	Other grasses	2
	Liquid hot water pretreatment	12
	Switchgrass	9
	Other grasses	3
	Autohydrolysis pretreatment	3
	Miscanthus	1
	Bermuda grass	2
	Hot compressed water pretreatment	0
	Wet oxidation pretreatment	1
	Miscanthus	1
	Chemical pretreatments	70
	Acid pretreatment	25
	Switchgrass	15
	Bermuda grass	2
	Canary grass	2
	Alfalfa	2
	Ryegrass	2
	Other grasses	2
	Solvent pretreatment	14
	Switchgrass	6
	Miscanthus	6
	Other grasses	2
	Alkaline pretreatment	29
	Switchgrass	14
	Other grasses	4
	Miscanthus	4
	Bermuda grass	3
	Grass	2

(Continued)

TABLE 49.10 (*Continued*)
The Most Prolific Thematic Research Fronts for Grass-based Bioethanol Fuels

No.	Research Fronts	N Paper (%) Sample
	Common reed	2
	Sulfur dioxide pretreatment	3
	Switchgrass	3
	Ionic liquid pretreatment	9
	Switchgrass	8
	Miscanthus	1
	Sulfite pretreatment	0
	Ammonia pretreatment	14
	Switchgrass	10
	Miscanthus	2
	Other grasses	2
	Surfactant pretreatment	2
	Other chemical pretreatments	3
	Enzymatic pretreatments	79
	Switchgrass	39
	Miscanthus	11
	Bermuda grass	7
	Grass	5
	Giant reed	4
	Ryegrass	3
	Alfalfa	2
	Canary grass	2
	Common reed	3
	Other grasses	3
	Microbial pretreatments	3
	Switchgrass	3
	Grass genetic engineering	9
	Switchgrass	9
2	Grass Biomass Hydrolysis	78
	Enzymatic hydrolysis	78
	Switchgrass	39
	Miscanthus	11
	Bermuda grass	7
	Grass	5
	Giant reed	4
	Ryegrass	4
	Alfalfa	2
	Canary grass	2
	Common reed	3
	Other grasses	3
	Autohydrolysis	0

(*Continued*)

TABLE 49.10 (*Continued*)
The Most Prolific Thematic Research Fronts for Grass-based Bioethanol Fuels

No.	Research Fronts	N Paper (%) Sample
	Acid hydrolysis	0
	Hydrolysis in general	0
3	Grass Hydrolysate Fermentation	18
	Switchgrass	6
	Giant reed	3
	Bermuda grass	2
	Miscanthus	2
	Other grasses	5
4	Grass-based Bioethanol Production in General	35
	Switchgrass	23
	Napier grass	3
	Other grasses	9

N paper (%) sample, The number of papers in the population sample of 100 papers.

49.4 DISCUSSION

49.4.1 INTRODUCTION

Crude oil-based gasoline fuels have been widely used in the transportation sector since the 1920s. However, there have been great public concerns over the adverse environmental and human impact of these fuels. Hence, biomass-based bioethanol fuels have increasingly been used in blending gasoline fuels, in fuel cells, and in biochemical production in a biorefinery context.

However, it is necessary to pretreat the biomass to enhance the yield of bioethanol prior to bioethanol production through hydrolysis and fermentation. One of the most-studied biomasses for bioethanol production and use has been grass biomass. Research in the field of grass-based bioethanol fuels has intensified in this context in the key research fronts of pretreatment and hydrolysis of grass biomass, fermentation of grass hydrolysates, and grass-based bioethanol fuels in general. Further, switchgrass and to a lesser extent miscanthus, Bermuda grass, giant reed, and alfalfa have been studied intensively at the expense of other grass biomasses in this context.

However, it is essential to develop efficient incentive structures for the primary stakeholders to enhance research in this field. This is especially important to maintain energy security in the cases of supply shocks such as oil price shocks and war-related shocks as in the case of the Russian invasion of Ukraine or COVID-19 shocks starting.

Scientometric analysis has been used in this context to inform the primary stakeholders about the current state of research in a selected research field. As there has been no scientometric study in this field, this chapter presents a scientometric study of research in grass-based bioethanol fuels. It examines the scientometric characteristics of both the sample and population data presenting the scientometric characteristics of these both datasets in the order of documents, authors, publication years, institutions, funding bodies, source titles, countries, Scopus subject categories, Scopus keywords, and research fronts.

As a first step for the search of the relevant literature, keywords were selected using the first most-cited 200 papers. The selected keyword list was then optimized to obtain a representative sample of papers for the searched research field. A copy of this extended keyword list was provided in the appendix for future replicative studies. Further, a selected list of keywords was presented in Table 49.8.

As a second step, two sets of data were used for this study. First, a population sample of 1,006 papers was used to examine the scientometric characteristics of the population data. Secondly, a sample of 100 most-cited papers, corresponding to 10% of population dataset was used to examine the scientometric characteristics of these citation classics.

The scientometric characteristics of these sample and population datasets were presented in the order of documents, authors, publication years, institutions, funding bodies, source titles, countries, Scopus subject categories, Scopus keywords, and research fronts.

Lastly, the key scientometric findings for both datasets were discussed to highlight the research landscape for grass-based bioethanol fuels. Additionally, a number of brief conclusions were drawn and a number of relevant recommendations were made to enhance the future research landscape.

49.4.2   The Most Prolific Documents in Grass-based Bioethanol Fuels

Articles (together with conference papers) dominate both the sample (98%) and population (95%) papers (Table 49.1). Further, review papers and articles have a surplus (1%) and deficit (3%), respectively. The representation of reviews in sample papers is relatively insignificant (2%).

Scopus differs from the Web of Science database in differentiating and showing articles (95%) and conference papers (3%) published in journals separately. However, it should be noted that these conference papers are also published in journals as articles, compared to those published only in conference proceedings. Hence, the total number of articles and review papers in the sample dataset is 98 and 2%, respectively.

It is observed during the search process that there has been inconsistency in the classification of documents in Scopus as well as in other databases such as the Web of Science. This is especially relevant for the classification of papers as reviews or articles as the papers not involving a literature review may be erroneously classified as a review paper. There is also a case of review papers being classified as articles. For example, the total number of reviews in the sample dataset was manually found as nearly 3% compared to 2% as indexed by Scopus, reducing the number of articles and conference papers to 97% for the sample dataset.

In this context, it would be helpful to provide a classification note for the published papers in books and journals in the first instance. It would also be helpful to use the document types listed in Table 49.1 for this purpose. Book chapters may also be classified as articles or reviews as an additional classification to differentiate review chapters from experimental chapters as it is done by the Web of Science. It would be further helpful to additionally classify the conference papers as articles or review papers as well as it is done in the Web of Science database.

49.4.3   The Most Prolific Authors in Grass-based Bioethanol Fuels

There have been the most prolific 18 authors with at least 4% of sample papers each as given in Table 49.2. These authors have shaped the development of research in this field.

The most prolific authors are Arthur J. Ragauskas, Blake A. Simmons, Bruce E. Dale, Jay J. Cheng, and Seema Singh and to a lesser extent Venkatesh Balan, Jian Shi, and Yoon Y. Lee.

It is important to note the inconsistencies in indexing the author names in Scopus and other databases. It is especially an issue for names with more than two components such as 'Blake Sam de Hyun Dale'. The probable outcomes are 'Dale, B.S.D.H.', 'de Hyun Dale, B.S.' or 'Hyun Dale, B.S.D.'. The first choice is the gold standard of the publishing sector as the last word in the name is taken as the last name. In most of the academic databases such as PUBMED and EBSCO databases, this version is used predominantly. The second choice is a strong alternative while the last choice is an undesired outcome as two last words are taken as the last name. It is a good practice to combine the words of the last name with a hyphen: 'Hyun-Dale, B.S.D.'. It is noted that inconsistent indexing of author names may cause substantial inefficiencies in the search process for the papers as well as allocating credit to the authors as there are different author entries for each outcome in the databases.

There are also inconsistencies in shortening Chinese names, for example, 'YangYing Cheng' is often shortened as 'Cheng, Y.', 'Cheng, Y.-Y.', and 'Cheng, Y.Y.' as it is done in the Web of Science database as well. However, the gold standard in this case is 'Cheng, Y' where the last word is taken as the last name and the first word is taken as a single forename. In most of the academic databases such as PUBMED and EBSCO, this first version is used predominantly. However, it makes sense to use the third option to differentiate Chinese names efficiently: 'Cheng, Y.Y.' Therefore, there have been difficulties in locating papers for the Chinese authors. In such cases, the use of unique author codes provided for each author by the Scopus database has been helpful.

There is also a difficulty in allowing credit for authors especially for authors with common names such as 'Cheng, X.' in conducting scientometric studies. These difficulties strongly influence the efficiency of scientometric studies as well as allocating credit to authors as there are same author entries for different authors with the same name, e.g., 'Cheng, X.' in the databases.

In this context, the coding of authors in the Scopus database is a welcome innovation compared to other databases such as the Web of Science. In this process, Scopus allocates a unique number to each author in the database (Aman, 2018). However, there might still be substantial inefficiencies in this coding system, especially for common names. For example, some of the papers for a certain author may be allocated to another researcher with a different author code. It is possible that Scopus uses a number of software programs to differentiate author names and the program may not be false-proof (Kim, 2018).

In this context, it does not help that author names are not given in full in some journals and books. This makes it difficult to differentiate authors with common names and makes the scientometric studies further difficult in the author domain. Therefore, author names should be given in all books and journals in the first instance. There is also a cultural issue where some authors do not use their full names in their papers. Instead, they use initials for their forenames: 'Cheng, H.J.', 'Cheng', 'Cheng, H.', or 'Cheng, J.' instead of 'Cheng, Hyun Jae'.

There are also inconsistencies in naming the authors with more than two components by the authors themselves in journal papers and book chapters, for example, 'Dale, A.P.C.' might be given as 'Dale, A.' or 'Dale, A.C.' or 'Dale, A.P.' or 'Dale, C.' in the journals and books. This also makes the scientometric studies difficult in the author domain. Hence, contributing authors should use their names consistently in their publications.

The other critical issue regarding author names is the inconsistencies in the spelling of author names in national spellings (e.g. Göğüşçöl, Gökçe) rather than in English spellings (e.g. Goguscol, Gokce) in the Scopus database. Scopus differs from the Web of Science database and many other databases in this respect where the author names are given only in English spellings. It is observed that national spellings of author names do not help much in conducting scientometric studies as well as in allocating credits to authors as sometimes there are different author entries for English and national spellings in the Scopus database.

The most prolific institution for the sample dataset is Huazhong Agricultural University. Further, the most prolific country for the sample dataset is the United States. These findings confirm the dominance of the United States in this field. The only research front is pretreatments of the grass biomass.

It is also noted that there is a significant gender deficit for the sample dataset surprisingly with a representation rate of 6%. This finding is the most thought-provoking with strong public policy implications. Hence, institutions, funding bodies, and policymakers should take efficient measures to reduce the gender deficit in this field as well as other scientific fields with a strong gender deficit. In this context, it is worth noting the level of representation of researchers from minority groups in science on the basis of race, sexuality, age, and disability, besides gender (Blankenship, 1993; Dirth and Branscombe, 2017; Konur, 2000, 2002a,b,c, 2006a,b, 2007a,b).

49.4.4 THE MOST PROLIFIC RESEARCH OUTPUT BY YEARS IN GRASS-BASED BIOETHANOL FUELS

The research output observed between 1970 and 2022 is illustrated in Figure 49.1. This figure clearly shows that the bulk of research papers in the population dataset was published primarily in the 2010s and to a lesser extent the early 2020s. Similarly, the bulk of research papers in the sample dataset was published in the 2010s and to a lesser extent the 2000s. Further, the number of population papers rose from 2008 to 2011; thereafter, it steadied above 5% of population papers each year, losing its momentum.

These findings suggest that the most prolific sample and population papers were primarily published in the 2010s. These are thought-provoking findings as there has been a significant research boom since 2008. In this context, the increasing public concerns about climate change (Change, 2007), greenhouse gas emissions (Carlson et al., 2017), and global warming (Kerr, 2007) have been certainly behind the boom in research in this field in the last two decades. Furthermore, the supply shocks experienced due to the COVID-19 pandemic might also be behind the research boom in this field since 2019. However, there has been no sharp rise due to this shock in 2020 and 2021 yet.

Based on these findings, the size of population papers is likely to be more than double in the current decade, provided that the public concerns about climate change, greenhouse gas emissions, global warming, as well as supply shocks are translated efficiently into research funding in this field.

49.4.5 THE MOST PROLIFIC INSTITUTIONS IN GRASS-BASED BIOETHANOL FUELS

The most prolific 19 institutions publishing papers on grass-based bioethanol fuels with at least 4% of sample papers each given in Table 49.3 have shaped the development of research in this field.

The most prolific institutions are the NC State University, USDA Agricultural Research Service, and Georgia Institute of Technology and to a lesser extent Joint Bioenergy Institute, Michigan State University, Sandia National Laboratories, and University of Nebraska Lincoln. Similarly, the top country for these most prolific institutions is the United States. In total, only three countries house these top institutions.

On the other hand, the institutions with the most citation impact are the Georgia Institute of Technology and to a lesser extent the Joint BioEnergy Institute, NC State University, University of Nebraska Lincoln, Sandia National Laboratories, and Michigan State University. These findings confirm the dominance of the US institutions for these highly cited papers.

49.4.6 THE MOST PROLIFIC FUNDING BODIES IN GRASS-BASED BIOETHANOL FUELS

The most prolific 12 funding bodies funding at least 3% of sample papers each are given in Table 49.4. It is noted that only 52% and 51% of sample and population papers were funded, respectively.

The most prolific funding bodies are the U.S. Department of Energy and to a lesser extent the Office of Science, Biological and Environmental Research, Higher Education Discipline Innovation Project, and Lawrence Berkeley National Laboratory. The most prolific country for these top funding bodies is the United States.

These findings on the funding of research in this field suggest that the level of funding, mostly since 2010, is relatively intensive, and it has been largely instrumental in enhancing research in this field (Ebadi and Schiffauerova, 2016) in light of North's institutional framework (North, 1991). It is also noted that the funding rate in this field is superior to those in the other research fronts of bioethanol fuels such as wood-based grass-based bioethanol fuels. Further, it is expected that this high funding rate would improve in light of recent supply shocks.

49.4.7 THE MOST PROLIFIC SOURCE TITLES IN GRASS-BASED BIOETHANOL FUELS

The most prolific 14 source titles publishing at least 2% of sample papers each in grass-based bioethanol fuels have shaped the development of research in this field (Table 49.5).

The most prolific source titles are Bioresource Technology and to a lesser extent Biomass and Bioenergy, Biotechnology for Biofuels, Bioenergy Research, and Proceedings of the National Academy of Sciences of the United States of America.

On the other hand, the source titles with the most citation impact are Bioresource Technology and to a lesser extent the Biomass and Bioenergy, Proceedings of the National Academy of Sciences of the United States of America, Polymer Degradation and Stability, and Biotechnology Progress. Similarly, the source title with the least impact is Applied Biochemistry and Biotechnology.

It is notable that these top source titles are primarily related to bioresources, energy, and biotechnology. This finding suggests that Bioresource Technology and the other prolific journals in these fields have significantly shaped the development of research in this field as they focus primarily on grass-based bioethanol fuels with a high yield. In this context, the influence of Bioresource Technology is quite extraordinary with 38% of sample papers.

49.4.8   The Most Prolific Countries in Grass-based Bioethanol Fuels

The most prolific 12 countries publishing at least 2% of sample papers each have significantly shaped the development of research in this field (Table 49.6).

The most prolific countries are the United States and to a lesser extent China, France, and Colombia. On the other hand, the countries with the most citation impact are the United States and to a lesser extent Colombia and France. Similarly, the countries with the least impact are China and to a lesser extent India, Canada, the United Kingdom, Japan, and South Korea.

Close examination of these findings suggests that the United States and to a lesser extent, Europe, China, and Colombia are the major producers of research in this field. It is a fact that the United States has been a major player in science (Leydesdorff and Wagner, 2009). The United States has further developed a strong research infrastructure to support its corn- and grass-based bioethanol industry (Gillon, 2010).

However, China has been a rising megastar in scientific research in competition with the United States and Europe (Leydesdorff and Zhou, 2005). China is also a major player in this field as a major producer of bioethanol (Fang et al., 2010).

Next, Europe has been a persistent player in scientific research in competition with both the United States and China (Leydesdorff, 2000). Europe has also been a persistent producer of bioethanol along with the United States and Brazil (Gnansounou, 2010).

49.4.9   The Most Prolific Scopus Subject Categories in Grass-based Bioethanol Fuels

The most prolific eight Scopus subject categories indexing at least 7% of sample papers each, respectively, as given in Table 49.7 have shaped the development of research in this field.

The most prolific Scopus subject categories in grass-based bioethanol fuels are Energy and to a lesser extent Environmental Science, Chemical Engineering, Biochemistry, Genetics and Molecular Biology, Immunology and Microbiology, and Agricultural and Biological Sciences. On the other hand, the Scopus subject categories with the most citation impact are Energy and to a lesser extent Environmental Science and Chemical Engineering. Similarly, the least influential subject categories are Chemistry and Engineering.

These findings are thought provoking suggesting that the primary subject categories are related to energy, environmental science, genetics, microbiology, and chemical engineering as the core of research in this field concerns grass-based bioethanol fuels. The other finding is that social sciences are not well represented in both the sample and population papers as in most fields in bioethanol fuels.

49.4.10 THE MOST PROLIFIC KEYWORDS IN GRASS-BASED BIOETHANOL FUELS

A limited number of keywords have shaped the development of research in this field as shown in Table 49.8 and the Appendix. These keywords are grouped into eight headings: biomass, fermentation, bacteria, hydrolysis, hydrolysates, pretreatments, other processes, and products of fermentation.

The most prolific keywords across all research fronts are cellulose, sugar, panicum, biomass, grass, pretreatments, hydrolysis, enzyme activity, ethanol, switchgrass, enzymatic hydrolysis, pretreatments, biofuels, cellulases, glucose, temperature, fermentation, saccharification, Poaceae, lignocellulose, and alcohol. Similarly, the most influential keywords are sugars, cellulose, panicum, lignin, pre-treatment, enzyme activity, switchgrass, grass, and biomass.

These findings suggest that it is necessary to determine the keyword set carefully to locate the relevant research in each of these research fronts. Additionally, the size of samples for each keyword highlights the intensity of research in the relevant research areas.

49.4.11 THE MOST PROLIFIC RESEARCH FRONTS IN GRASS-BASED BIOETHANOL FUELS

Information about the research fronts for the sample papers in grass-based bioethanol fuels with regard to the grass biomass used for bioethanol production is given in Table 49.9. As this table shows, there are two primary research fronts in this field: switchgrass and to a lesser extent miscanthus. The other grass biomasses used in these studies are giant reed, Bermuda grass, grass in general, ryegrass, common reed, alfalfa, napier grass, canary grass, and other grasses.

Information about the thematic research fronts for the sample papers in grass-based bioethanol fuels is given in Table 49.10. As this table shows, there are four primary research fronts: grass biomass pretreatments and hydrolysis, and to a lesser extent grass hydrolysate fermentation, and grass-based bioethanol fuels in general.

For the research front of grass biomass pretreatments, the primary research fronts are chemical, enzymatic, and hydrothermal pretreatments. Additionally, there are four further research fronts: mechanical and microbial pretreatments, pretreatments in general, and genetic engineering of the grass biomass. Further, the most prolific chemical pretreatments are alkaline, acid, ammonia, and ionic liquid pretreatments.

These findings are thought provoking in seeking ways to increase grass-based bioethanol yield at the global scale. It is clear that all these research fronts have public importance and merit substantial funding and other incentives. Further, it is noted that grass-based bioethanol fuels have become a core of bioethanol research to make it more competitive with crude oil-based gasoline and diesel fuels, especially in the United States and other countries with vast farmlands.

In the end, these most-cited papers in this field hint that the efficiency of grass-based bioethanol production could be optimized using the structure, processing, and property relationships of these grass species in the fronts of grass biomass pretreatments and hydrolysis, and hydrolysate fermentation (Formela et al., 2016; Konur, 2018a, 2020b, 2021a,b,c,d; Konur and Matthews, 1989).

49.5 CONCLUSION AND FUTURE RESEARCH

Research on grass-based bioethanol fuels has been mapped through a scientometric study of both sample (100 papers) and population (1,006 papers) datasets.

The critical issue in this study has been to obtain a representative sample of research as in any other scientometric studies. Therefore, the keyword set has been carefully devised and optimized after a number of runs in the Scopus database. It is a representative sample of wider population studies. This keyword set was provided in the Appendix and the relevant keywords are presented in Table 49.8. However, it should be noted that it has been very difficult to compile a representative keyword set since this research field has been connected closely with many other fields. Therefore, it has been necessary to compile a keyword list to exclude papers concerned with other research fields.

The other issue has been the selection of a multidisciplinary database to carry out the sciento-metric study of research in this field. For this purpose, the Scopus database has been selected. The journal coverage of this database has been notably wider than that of the Web of Science and other multisubject databases.

The key scientometric properties of research in this field have been determined and discussed in this chapter. It is evident that a limited number of documents, authors, institutions, publication years, institutions, funding bodies, source titles, countries, Scopus subject categories, Scopus key-words, and research fronts have shaped the development of research in this field.

There is ample scope to increase the efficiency of scientometric studies in this field in the author and document domains by developing consistent policies and practices in both domains across all academic databases. In this respect, it seems that authors, journals, and academic databases have a lot to do. Furthermore, a significant gender deficit as in most scientific fields emerges as a public policy issue. The potential deficits on the basis of age, race, disability, and sexuality need also to be explored in this field as in other scientific fields.

Research in this field has boomed since 2008, possibly promoted by public concerns about global warming, greenhouse gas emissions, and climate change. Furthermore, the recent COVID-19 pan-demic and the Russian invasion of Ukraine have resulted in global supply shocks shifting the focus of the stakeholders from crude oil-based fuels to biomass-based fuels such as bioethanol fuels. It is expected that there would be further incentives for the key stakeholders to carry out research for grass-based bioethanol production to increase ethanol yield and to make it more competitive with crude oil-based gasoline and petrodiesel fuels. This might be truer for the crude oil- and foreign exchange-deficient countries to maintain energy security in the face of global supply shocks.

The relatively significant funding rate of 52% and 51% for the sample and population papers, respectively, suggests that funding in this field significantly enhanced research in this field primar-ily since 2008, possibly more than doubling in the current decade. However, it is evident that there is ample room for more funding and other incentives to enhance research in this field further bearing in mind that the research output for the population papers lost its momentum after 2011.

Institutions from the United States have mostly shaped research in this field. Further, the United States and to a lesser extent China, Europe, and Colombia have been the major producers of research in this field as the major producers and users of bioethanol fuels from different types of biomasses such as corn, sugarcane, grass as well as other types of biomass. It is evident that these countries have a well-developed research infrastructure for bioethanol fuels and their derivatives. It is also noted that all these major countries have access to large farmlands.

It emerges that ethanol is more popular than bioethanol as a keyword with strong implications for the search strategy. In other words, the search strategy using only the bioethanol keyword would not be much helpful. Scopus keywords are grouped into eight headings: biomass, fermentation, bacte-ria, hydrolysis, hydrolysates, pretreatments, other processes, and products of fermentation. It is also noted that there is a need to use Latin keywords for grass in the search strategy.

As Table 49.9 shows, there are two primary research fronts for this field with respect to grass biomass: switchgrass and to a lesser extent miscanthus. The other grass biomasses used in these studies are giant reed, Bermuda grass, grass, ryegrass, common reed, alfalfa, napier grass, canary grass, and other grasses.

On the other hand, Table 49.10 shows that there are four primary thematic research fronts: grass biomass pretreatments and hydrolysis, and to a lesser extent grass hydrolysate fermentation, and grass-based bioethanol production in general. For the research front of grass biomass pretreatments, the primary research fronts are chemical, enzymatic, and hydrothermal pretreatments. Additionally, there are four further research fronts: mechanical and microbial pretreatments, pretreatments in general, and genetic engineering of the grass biomass. Further, the most prolific chemical pretreat-ments are alkaline, acid, ammonia, and ionic liquid pretreatments.

These findings are thought provoking in seeking ways to increase bioethanol yield through grass-based bioethanol fuels at the global scale. It is clear that all these research fronts have public importance and merit substantial fundings and other incentives. Further, it is noted that grass-based bioethanol fuels, as a second generation biofuel, have become a core unit of bioethanol research to make it more competitive with crude oil-based gasoline and petrodiesel fuels, especially for countries with large access to farmlands.

Thus, scientometric analysis has a great potential to gain valuable insights into the evolution of research in this field as in other scientific fields, especially in the aftermath of significant global supply shocks such as the COVID-19 pandemic and the Russian invasion of Ukraine.

It is recommended that further scientometric studies are carried out for the primary research fronts. It is further recommended that reviews of the most-cited papers are carried out for each primary research front to complement these scientometric studies. Next, scientometric studies of hot papers in these primary fields are carried out.

ACKNOWLEDGMENTS

The contribution of highly cited researchers in the field of grass-based bioethanol fuels has been gratefully acknowledged.

APPENDIX: THE KEYWORD SET FOR GRASS-BASED BIOETHANOL FUELS

((TITLE (*grass OR miscanthus OR alfalfa OR reed OR arundo OR bode OR taro OR panicum OR lolium OR medicago OR pennisetum OR phalaris OR phragmites OR cynodon OR spartina OR fescue OR festuca OR zostera) AND TITLE (saccharification OR *hydrolysis OR digestibili* OR ssf OR shf OR ethanol OR bioethanol OR recalcitrance OR hydrolysate* OR hydrolyzate* OR ferment* OR coferment* OR cellulase* OR fractionation OR delignification OR depolymerization OR microwave* OR ultrasound OR sonicat* OR extrusion OR milling OR grinding OR pretreat* OR "pre-treat*" OR bioorganosolve OR sugar OR "size reduction" OR steam* OR "hot water" OR "hot compressed" OR "sulfuric acid*" OR sulfite* OR sporl OR lime OR "sulfur dioxide" OR so2 OR alkali* OR naoh* OR extruder OR "sodium hydroxide*" OR "ionic liquid*" OR solvent* OR organosolv* OR ammonia OR "wet oxidation" OR "wet explosion" OR flowthrough OR hydrothermolysis OR "supercritical co2" OR "supercritical carbon dioxide" OR surfactant* OR afex* OR "supercritical water" OR tween* OR milled OR hydrothermal OR pelleting OR pelletizing)) AND NOT (TITLE (seagrass* OR anti* OR combustion OR fatty OR *hydrogen OR damage OR *butanol OR oils OR biogas OR activated OR rumen OR carp OR seed* OR root* OR diet* OR wilt* OR *methane OR rhisoz* OR horse OR levulinic OR soil* OR *butyric OR sprout* OR virus OR zea OR ruminal OR depression OR amino* OR sheep OR rhiz* OR lactic OR cytochrome OR biochar OR coal OR ocean OR drought OR herbicid* OR fertil* OR fiberboard OR indole OR pollut* OR pyrolysis OR mud OR vitro OR *dimers OR lab OR snow OR mine OR pasture OR absor* OR protein* OR drying OR catalysis OR bioremed* OR caproic OR ssr OR h2 OR furfural OR digestion OR bleach* OR activated OR gasifier OR biocarbon OR syngas OR liquor OR protease OR lactobacillus OR clover OR torref* OR infection OR loop OR silage OR ensil* OR soybean* OR carbonate OR biosorption OR gasification OR hydrochar OR dye OR purif* OR battery) OR SRCTITLE (animal* OR nutrit* OR botan* OR livestock* OR ocean* OR ecol* OR dairy OR materials OR meat* OR geo* OR marine OR soil* OR aqua* OR biogeo* OR mosq* OR phyto* OR ruminant OR pollut* OR zoo* OR fish* OR physiol* OR silage OR atmos* OR hydrogen* OR poult* OR plant* OR composite* OR membrane OR lipid* OR pest* OR hydro* OR macromol* OR estuar* OR hort* OR agron* OR biomacro* OR grass* OR weed* OR pyrolysis OR food* OR materials OR insect* OR agric* OR small OR nutrient* OR agro* OR {crop sci*} OR toxic* OR carbohydrate* OR cereal* OR agrar* OR forage OR water* OR euphytica OR entomol* OR {crop prot*} OR fertilizer* OR manag* OR biomed* OR field OR oeco* OR zemdirbyste OR allelopathy OR ekol* OR organic) OR SUBJAREA (vete OR medi OR nurs OR phar OR heal OR neur OR dent))) OR ((TITLE (ethanol OR bioethanol) AND TITLE (*diesel OR *butanol OR biogas OR *hydrogen OR *methane)) AND TITLE (*grass OR miscanthus OR alfalfa OR reed

OR arundo OR bode OR taro OR panicum OR lolium OR medicago OR pennisetum OR phalaris OR phragmites OR cynodon OR spartina OR fescue OR festuca OR zostera)) AND (LIMIT-TO (SRCTYPE, "j") OR LIMIT-TO (SRCTYPE, "b") OR LIMIT-TO (SRCTYPE, "k")) AND (LIMIT-TO (DOCTYPE, "ar") OR LIMIT-TO (DOCTYPE, "cp") OR LIMIT-TO (DOCTYPE, "re") OR LIMIT-TO (DOCTYPE, "ch") OR LIMIT-TO (DOCTYPE, "le") OR LIMIT-TO (DOCTYPE, "no") OR LIMIT-TO (DOCTYPE, "sh") OR LIMIT-TO (DOCTYPE, "ed")) AND (LIMIT-TO (LANGUAGE, "English"))

REFERENCES

Aman, V. 2018. Does the Scopus author ID suffice to track scientific international mobility? A case study based on Leibniz laureates. *Scientometrics* 117:705–720.

Angelici, C., B. M. Weckhuysen and P. C. A. Bruijnincx. 2013. Chemocatalytic conversion of ethanol into butadiene and other bulk chemicals. *ChemSusChem* 6:1595–1614.

Antolini, E. 2007. Catalysts for direct ethanol fuel cells. *Journal of Power Sources* 170:1–12.

Antolini, E. 2009. Palladium in fuel cell catalysis. *Energy and Environmental Science* 2:915–931.

Beaudry, C. and V. Lariviere. 2016. Which gender gap? Factors affecting researchers' scientific impact in science and medicine. *Research Policy* 45:1790–1817.

Blankenship, K. M. 1993. Bringing gender and race in: US employment discrimination policy. *Gender & Society* 7:204–226.

Burnham, J. F. 2006. Scopus database: A review. *Biomedical Digital Libraries* 3:1–8.

Carlson, K. M., J. S. Gerber and D. Mueller, et al. 2017. Greenhouse gas emissions intensity of global croplands. *Nature Climate Change* 7:63–68.

Change, C. 2007. Climate change impacts, adaptation and vulnerability. *Science of the Total Environment* 326:95–112.

Dien, B.S., J. H. G. Jung and K. P. Vogel, et al. 2006. Chemical composition and response to dilute-acid pretreatment and enzymatic saccharification of alfalfa, reed canarygrass, and switchgrass. *Biomass and Bioenergy* 30:880–891.

Dirth, T. P. and N. R. Branscombe. 2017. Disability models affect disability policy support through awareness of structural discrimination. *Journal of Social Issues* 73:413–442.

Ebadi, A. and A. Schiffauerova. 2016. How to boost scientific production? A statistical analysis of research funding and other influencing factors. *Scientometrics* 106:1093–1116.

Faga, B. A., M. R. Wilkins and I. M. Banat. 2010. Ethanol production through simultaneous saccharification and fermentation of switchgrass using Saccharomyces cerevisiae D5A and thermotolerant *Kluyveromyces marxianus* IMB strains. *Bioresource Technology* 101:2273–2279.

Fang, X., Y. Shen, J. Zhao, X. Bao and Y. Qu. 2010. Status and prospect of lignocellulosic bioethanol production in China. *Bioresource Technology* 101:4814–4819.

Fauci, A. S., H. C. Lane and R. R. Redfield. 2020. Covid-19-navigating the uncharted. *New England Journal of Medicine* 382:1268–1269.

Fernando, S., S. Adhikari, C. Chandrapal and M. Murali. 2006. Biorefineries: Current status, challenges, and future direction. *Energy & Fuels* 20:1727–1737.

Formela, K., A. Hejna, L. Piszczyk, M. R. Saeb and X. Colom. 2016. Processing and structure-property relationships of natural rubber/wheat bran biocomposites. *Cellulose* 23:3157–3175.

Fu, C., J. R. Mielenz and X. Xiao, et al. 2011. Genetic manipulation of lignin reduces recalcitrance and improves ethanol production from switchgrass. *Proceedings of the National Academy of Sciences of the United States of America* 108:3803–3808.

Garfield, E. 1955. Citation indexes for science. *Science* 122:108–111.

Gillon, S. 2010. Fields of dreams: Negotiating an ethanol agenda in the Midwest United States. *Journal of Peasant Studies* 37:723–748.

Gnansounou, E. 2010. Production and use of lignocellulosic bioethanol in Europe: Current situation and perspectives. *Bioresource Technology* 101:4842–4850.

Hahn-Hagerdal, B., M. Galbe, M. F. Gorwa-Grauslund, G. Liden and G. Zacchi. 2006. Bio-ethanol - The fuel of tomorrow from the residues of today. *Trends in Biotechnology* 24:549–556.

Hamilton, J. D. 1983. Oil and the macroeconomy since World War II. *Journal of Political Economy* 91:228–248.

Hamilton, J. D. 2003. What is an oil shock? *Journal of Econometrics* 113:363–398.

Hamilton, J. D. 2009. Causes and consequences of the oil shock of 2007-08. *Brookings Papers on Economic Activity* 2009:215–261.

Hill, J., E. Nelson, D. Tilman, S. Polasky and D. Tiffany. 2006. Environmental, economic, and energetic costs and benefits of biodiesel and ethanol biofuels. *Proceedings of the National Academy of Sciences of the United States of America* 103:11206–11210.

Hill, J., S. Polasky and E. Nelson, et al. 2009. Climate change and health costs of air emissions from biofuels and gasoline. *Proceedings of the National Academy of Sciences of the United States of America* 106:2077–2082.

Hsieh, W. D., R. H. Chen, T. L. Wu and T. H. Lin. 2002. Engine performance and pollutant emission of an SI engine using ethanol-gasoline blended fuels. *Atmospheric Environment* 36:403–410.

Hu, Z. and Z. Wen. 2008. Enhancing enzymatic digestibility of switchgrass by microwave-assisted alkali pretreatment. *Biochemical Engineering Journal* 38:369–378.

Huang, H. J., S. Ramaswamy, U. W. Tschirner and B. V. Ramarao. 2008. A review of separation technologies in current and future biorefineries. *Separation and Purification Technology* 62:1–21.

Jin, M., M. W. Lau, V. Balan, and B. E. Dale. 2010. Two-step SSCF to convert AFEX-treated switchgrass to ethanol using commercial enzymes and *Saccharomyces cerevisiae* 424A(LNH-ST). *Bioresource Technology* 101:8171–8178.

Jones, T. C. 2012. America, oil, and war in the Middle East. *Journal of American History* 99:208–218.

Kerr, R. A. 2007. Global warming is changing the world. *Science* 316:188–190.

Kilian, L. 2008. Exogenous oil supply shocks: How big are they and how much do they matter for the US economy? *Review of Economics and Statistics* 90:216–240.

Kilian, L. 2009. Not all oil price shocks are alike: Disentangling demand and supply shocks in the crude oil market. *American Economic Review*, 99:1053–69.

Kim, J. 2018. Evaluating author name disambiguation for digital libraries: A case of DBLP. *Scientometrics* 116:1867–1886.

Konur, O. 2000. Creating enforceable civil rights for disabled students in higher education: An institutional theory perspective. *Disability & Society* 15:1041–1063.

Konur, O. 2002a. Access to nursing education by disabled students: Rights and duties of nursing programs. *Nurse Education Today* 22:364–374.

Konur, O. 2002b. Assessment of disabled students in higher education: Current public policy issues. *Assessment and Evaluation in Higher Education* 27:131–152.

Konur, O. 2002c. Access to employment by disabled people in the UK: Is the Disability Discrimination Act working? *International Journal of Discrimination and the Law* 5:247–279.

Konur, O. 2006a. Participation of children with dyslexia in compulsory education: Current public policy issues. *Dyslexia* 12:51–67.

Konur, O. 2006b. Teaching disabled students in higher education. *Teaching in Higher Education* 11:351–363.

Konur, O. 2007a. A judicial outcome analysis of the *Disability Discrimination Act*: A windfall for the employers? *Disability & Society* 22:187–204.

Konur, O. 2007b. Computer-assisted teaching and assessment of disabled students in higher education: The interface between academic standards and disability rights. *Journal of Computer Assisted Learning* 23:207–219.

Konur, O. 2011. The scientometric evaluation of the research on the algae and bio-energy. *Applied Energy* 88:3532–3540.

Konur, O. 2012a. The evaluation of the biogas research: A scientometric approach. *Energy Education Science and Technology Part A: Energy Science and Research* 29:1277–1292.

Konur, O. 2012b. The evaluation of the educational research: A scientometric approach. *Energy Education Science and Technology Part B: Social and Educational Studies* 4:1935–1948.

Konur, O. 2012c. The evaluation of the global energy and fuels research: A scientometric approach. *Energy Education Science and Technology Part A: Energy Science and Research* 30:613–628.

Konur, O. 2012d. The evaluation of the research on the biodiesel: A scientometric approach. *Energy Education Science and Technology Part A: Energy Science and Research* 28:1003–1014.

Konur, O. 2012e. The evaluation of the research on the bioethanol: A scientometric approach. *Energy Education Science and Technology Part A: Energy Science and Research* 28:1051–1064.

Konur, O. 2012f. The evaluation of the research on the biofuels: A scientometric approach. *Energy Education Science and Technology Part A: Energy Science and Research* 28:903–916.

Konur, O. 2012g. The evaluation of the research on the biohydrogen: A scientometric approach. *Energy Education Science and Technology Part A: Energy Science and Research* 29:323–338.

Konur, O. 2012h. The evaluation of the research on the microbial fuel cells: A scientometric approach. *Energy Education Science and Technology Part A: Energy Science and Research* 29:309–322.

Konur, O. 2012i. The scientometric evaluation of the research on the production of bioenergy from biomass. *Biomass and Bioenergy* 47:504–515.

Konur, O. 2015. Current state of research on algal bioethanol. In *Marine Bioenergy: Trends and Developments*, Ed. S. K. Kim and C. G. Lee, pp. 217–244. Boca Raton, FL: CRC Press.

Konur, O., Ed. 2018a. *Bioenergy and Biofuels*. Boca Raton, FL: CRC Press.

Konur, O. 2018b. Bioenergy and biofuels science and technology: Scientometric overview and citation classics. In *Bioenergy and Biofuels*, Ed. O. Konur, pp. 3–63. Boca Raton: CRC Press.

Konur, O. 2019. Cyanobacterial bioenergy and biofuels science and technology: A scientometric overview. In *Cyanobacteria: From Basic Science to Applications*, Ed. A. K. Mishra, D. N. Tiwari and A. N. Rai, pp. 419–442. Amsterdam: Elsevier.

Konur, O. 2020a. The scientometric analysis of the research on the bioethanol production from green macroalgae. In *Handbook of Algal Science, Technology and Medicine*, Ed. O. Konur, pp. 385–401. London: Academic Press.

Konur, O., Ed. 2020b. *Handbook of Algal Science, Technology and Medicine*. London: Academic Press.

Konur, O., Ed. 2021a. *Handbook of Biodiesel and Petrodiesel Fuels: Science, Technology, Health, and Environment*. Boca Raton, FL: CRC Press.

Konur, O., Ed. 2021b. *Handbook of Biodiesel and Petrodiesel Fuels: Science, Technology, Health, and Environment. Volume 1. Biodiesel Fuels: Science, Technology, Health, and Environment.* Boca Raton, FL: CRC Press.

Konur, O., Ed. 2021c. *Handbook of Biodiesel and Petrodiesel Fuels: Science, Technology, Health, and Environment. Volume 2. Biodiesel Fuels based on the Edible and Nonedible Feedstocks, Wastes, and Algae: Science, Technology, Health, and Environment.* Boca Raton, FL: CRC Press.

Konur, O., Ed. 2021d. *Handbook of Biodiesel and Petrodiesel Fuels: Science, Technology, Health, and Environment. Volume 3. Petrodiesel Fuels: Science, Technology, Health, and Environment.* Boca Raton, FL: CRC Press.

Konur, O. and F. L. Matthews. 1989. Effect of the properties of the constituents on the fatigue performance of composites: A review. *Composites* 20:317–328.

Kruyt, B., D. P. van Vuuren, H. J. de Vries and H. Groenenberg. 2009. Indicators for energy security. *Energy Policy* 37:2166–2181.

Leydesdorff, L. 2000. Is the European Union becoming a single publication system? *Scientometrics* 47:265–280.

Leydesdorff, L. and C. Wagner. 2009. Is the United States losing ground in science? A global perspective on the world science system. *Scientometrics* 78:23–36.

Leydesdorff, L. and P. Zhou. 2005. Are the contributions of China and Korea upsetting the world system of science? *Scientometrics* 63:617–630.

Li, C., B. Knierim and C. Manisseri, et al. 2010. Comparison of dilute acid and ionic liquid pretreatment of switchgrass: Biomass recalcitrance, delignification and enzymatic saccharification. *Bioresource Technology* 101:4900–4906.

Li, H., S. M. Liu, X. H. Yu, S. L. Tang and C. K. Tang. 2020. Coronavirus disease 2019 (COVID-19): Current status and future perspectives. *International Journal of Antimicrobial Agents* 55:105951.

Lin, Y. and S. Tanaka. 2006. Ethanol fermentation from biomass resources: Current state and prospects. *Applied Microbiology and Biotechnology* 69:627–642.

Ma, X., L. Sun and C. Song. 2002. A new approach to deep desulfurization of gasoline, diesel fuel and jet fuel by selective adsorption for ultra-clean fuels and for fuel cell applications. *Catalysis Today* 77:107–116.

Mani, S., L.G. Tabil and S. Sokhansanj. 2004. Grinding performance and physical properties of wheat and barley straws, corn stover and switchgrass. *Biomass and Bioenergy* 27:339–352

Morschbacker, A. 2009. Bio-ethanol based ethylene. *Polymer Reviews* 49:79–84.

Murnen, H. K., V. Balan and S. P. S. Chundawat, et al. 2007. Optimization of Ammonia Fiber Expansion (AFEX) pretreatment and enzymatic hydrolysis of *Miscanthus x giganteus* to fermentable sugars. *Biotechnology Progress* 23:846–850.

Najafi, G., B. Ghobadian and T. Tavakoli, et al. 2009. Performance and exhaust emissions of a gasoline engine with ethanol blended gasoline fuels using artificial neural network. *Applied Energy* 86:630–639.

Newman, P. W. G. and J. R. Kenworthy. 1989. Gasoline consumption and cities: A comparison of U.S. cities with a global survey. *Journal of the American Planning Association* 55:24–37.

North, D. C. 1991. Institutions. *Journal of Economic Perspectives* 5:97–112.

Olsson, L. and B. Hahn-Hagerdal. 1996. Fermentation of lignocellulosic hydrolysates for ethanol production. *Enzyme and Microbial Technology* 18:312–331.

Pimentel, D. and T. W. Patzek. 2005. Ethanol production using corn, switchgrass, and wood; biodiesel production using soybean and sunflower. *Natural Resources Research* 14:65–76.

Reeves, S. 2014. To Russia with love: How moral arguments for a humanitarian intervention in Syria opened the door for an invasion of the Ukraine. *Michigan State University International Law Review* 23:199.

Sanchez, O. J. and C. A. Cardona. 2008. Trends in biotechnological production of fuel ethanol from different feedstocks. *Bioresource Technology* 99:5270–5295.

Schmer, M. R., K. P. Vogel, M. R. Mitchell and R. K. Perrin. 2008. Net energy of cellulosic ethanol from switchgrass. *Proceedings of the National Academy of Sciences of the United States of America* 105:464–469.

Sreenath, H. K., R. G. Koegel, A. B. Moldes, T. W. Jeffries and R. J. Straub. 1999. Enzymic saccharification of alfalfa fibre after liquid hot water pretreatment. *Process Biochemistry* 35:33–41.

Sun, Y. and J. Cheng. 2002. Hydrolysis of lignocellulosic materials for ethanol production: A review. *Bioresource Technology* 83:1–11.

Sun, Y. and J. J. Cheng. 2005. Dilute acid pretreatment of rye straw and bermudagrass for ethanol production. *Bioresource Technology* 96:1599–1606.

Taherzadeh, M. J. and K. Karimi. 2007. Enzyme-based hydrolysis processes for ethanol from lignocellulosic materials: A review. *Bioresources* 2:707–738.

Taherzadeh, M. J. and K. Karimi. 2008. Pretreatment of lignocellulosic wastes to improve ethanol and biogas production: A review. *International Journal of Molecular Sciences* 9:1621–1651.

Wang, Z., D. R. Keshwani, A. P. Redding and J. J. Cheng. 2010. Sodium hydroxide pretreatment and enzymatic hydrolysis of coastal Bermuda grass. *Bioresource Technology* 101:3583–3585.

Winzer, C. 2012. Conceptualizing energy security. *Energy Policy* 46:36–48.

Xu, N., W. Zhang and S. Ren, et al. 2012. Hemicelluloses negatively affect lignocellulose crystallinity for high biomass digestibility under NaOH and H_2SO_4 pretreatments in Miscanthus. *Biotechnology for Biofuels* 5:58.

Yang, B. and C. E. Wyman. 2008. Pretreatment: The key to unlocking low-cost cellulosic ethanol. *Biofuels, Bioproducts and Biorefining* 2:26–40.

Yoshida, M., Y. Liu and S. Uchida, et al. 2008. Effects of cellulose crystallinity, hemicellulose, and lignin on the enzymatic hydrolysis of *Miscanthus sinensis* to monosaccharides. *Bioscience, Biotechnology and Biochemistry* 72:805–810.

Zhang, J., M. Siika-Aho, M. Tenkanen and L. Viikari. 2011. The role of acetyl xylan esterase in the solubilization of xylan and enzymatic hydrolysis of wheat straw and giant reed. *Biotechnology for Biofuels* 4:60.

50 Grass-based Bioethanol Fuels
Review

Ozcan Konur
(Formerly) Ankara Yildirim Beyazit University

50.1 INTRODUCTION

Crude oil-based gasoline fuels (Ma et al., 2002; Newman and Kenworthy, 1989) have been widely used in the transportation sector since the 1920s. However, there have been great public concerns over the adverse environmental and human impact of these fuels (Hill et al., 2006, 2009). Hence, biomass-based bioethanol fuels (Hill et al., 2006; Konur, 2012, 2015, 2019, 2020) have increasingly been used in blending gasoline fuels (Hsieh et al., 2002; Najafi et al., 2009), in fuel cells (Antolini, 2007, 2009), and in biochemical production (Angelici et al., 2013; Morschbacker, 2009) in a biorefinery context (Fernando et al., 2006; Huang et al., 2008).

However, it is necessary to pretreat the biomass (Alvira et al., 2010; Taherzadeh and Karimi, 2008) to enhance the yield of the bioethanol (Hahn-Hagerdal et al., 2006; Sanchez and Cardona, 2008) before the grass-based bioethanol production through the hydrolysis (Sun and Cheng, 2002; Taherzadeh and Karimi, 2007) and fermentation (Lin and Tanaka, 2006; Olsson and Hahn-Hagerdal, 1996) of the biomass and hydrolysates, respectively.

One of the most-studied biomass for the bioethanol fuels and use has been grass biomass. Research in the field of grass-based bioethanol fuels has intensified in this context in the key research fronts of the pretreatment of the grass biomass (Li et al., 2010, Mani et al., 2004; Sun and Cheng, 2005), hydrolysis of grass biomass (Hu and Wen, 2008; Li et al., 2010; Sun and Cheng, 2005), fermentation of grass hydrolysates (Faga et al., 2010; Fu et al., 2011; Jin et al., 2010), and grass-based bioethanol fuels in general (Fu et al., 2011; Pimentel and Patzek, 2005; Schmer et al., 2008). Furthermore, the switchgrass (Li et al., 2010; Mani et al., 2004; Pimentel and Patzek, 2005) and to a lesser extent miscanthus (Murnen et al., 2007; Xu et al., 2012; Yoshida et al., 2008), Bermudagrass (Sun and Cheng, 2005; Wang et al., 2010), giant reed (Scordia et al., 2011; Zhang et al., 2011), and alfalfa (Dien et al., 2006; Sreenath et al., 1997) have been studied intensively at the expense of the other grass biomass in this context.

However, it is essential to develop efficient incentive structures (North, 1991) for the primary stakeholders to enhance research in this field (Konur, 2000, 2002a,b,c, 2006a,b, 2007a,b). Although there have been many review papers on grass-based bioethanol fuels (Akin et al., 2007; de Oliveira et al., 2015; Keshwani and Cheng, 2009), there has been no review of the 25 most-cited papers in this field.

Thus, this book chapter presents a review of the 25 most-cited articles in the field of grass-based bioethanol fuels. Then, it discusses the key findings of these highly influential papers and comments on future research priorities in this field.

50.2 MATERIALS AND METHODS

The search for this study was carried out using the Scopus database (Burnham, 2006) in June 2022.

As a first step for the search of the relevant literature, the keywords were selected using the first 200 most-cited population papers. The selected keyword list was then optimized to obtain a

DOI: 10.1201/9781003226451-65

representative sample of papers for the searched research field. This final keyword set was provided in the appendix of Konur (2023) for future replication studies.

As a second step, a sample dataset was used for this study. The first 25 articles with at least 155 citations each were selected for the review study. Key findings from each paper were taken from the abstracts of these papers and were discussed. Additionally, many brief conclusions were drawn and several relevant recommendations were made to enhance future research landscape.

50.3 RESULTS

The brief information about the 25 most-cited papers with at least 155 citations each on the grass-based bioethanol fuels is given as follows. The primary research fronts are the pretreatments of grass biomass and the grass-based bioethanol fuels in general with 18 and 7 highly cited papers (HCPs), respectively. Further, the most prolific pretreatments are chemical pretreatments with 13 HCPs. The other research fronts for the biomass pretreatments are the multiple, enzymatic, and mechanical pretreatments with one, three, and one HCPs, respectively. There are five research fronts for the chemical pretreatments: multiple chemical, acid, alkali, ammonia, and ionic liquid (IL) pretreatments with two, five, three, two, and one HCPs, respectively.

50.3.1 THE GRASS BIOMASS PRETREATMENTS FOR THE GRASS-BASED BIOETHANOL PRODUCTION

There are 18 HCPs for the grass biomass pretreatments (Table 50.1). The most prolific pretreatments for the grass-based bioethanol production are the chemical pretreatments with 13 HCPs. The other research fronts for the biomass pretreatments are the multiple, enzymatic, and mechanical pretreatments with one, three, and one HCPs, respectively. There are also five research fronts for the chemical pretreatments: multiple chemical, acid, alkali, ammonia, and IL pretreatments with two, five, three, two, and one HCPs, respectively.

50.3.1.1 Multiple Pretreatments

Hu and Wen (2008) enhanced the enzymatic digestibility of switchgrass by microwave-assisted alkaline pretreatment in a paper with 344 citations. When switchgrass was soaked in water and treated by microwave, they observed that the total sugar yield from the combined pretreatment and hydrolysis was 34.5 g/100 g biomass, equivalent to 58.5% of the maximum potential sugars released. This yield was 53% higher than that obtained from conventional heating of switchgrass. To further improve the sugar yield, they presoaked switchgrass in different concentrations of alkaline solutions and then treated by microwave or conventional heating. With alkaline loading from 0.05 to 0.3 g alkali/g biomass, microwave pretreatment resulted in a higher sugar yield than conventional heating, with the highest yield (90% of maximum potential sugars) being achieved at 0.1 g/g of alkaline loading due to the disruption of recalcitrant structures. At optimal conditions of 190°C, 50 g/L solid content, and 30-min treatment time, the sugar yield from the combined pretreatment and hydrolysis was 58.7 g/100 g biomass, equivalent to 99% of potential maximum sugars. They concluded that microwave-assisted alkali pretreatment was an efficient way to improve the enzymatic digestibility of switchgrass.

50.3.1.2 Chemical Pretreatments

There are five research fronts for chemical pretreatments for the grass-based bioethanol production: multiple chemical, acid, alkaline, ammonia, and IL pretreatments with two, five, three, two, and one HCPs, respectively.

50.3.1.2.1 Multiple Chemical Pretreatments

Li et al. (2010) compared the dilute acid and IL pretreatments of switchgrass in terms of delignification, hydrolysis efficiency, and sugar yields in a paper with 840 citations. When subjected to IL

TABLE 50.1

Grass Biomass Pretreatments for the Grass-based Bioethanol Production

No.	Papers	Biomass/ Hydrolysate	Prts.	Parameters	Keywords	Lead Author	Affil.	Cits
1	Li et al. (2010)	Switchgrass	Acid, ionic liquids, enzymes	Switchgrass acid and ionic liquid pretreatments, delignification, hydrolysis efficiency, sugar yields	Switchgrass, acid, ionic liquid, pretreatment, recalcitrance, delignification, saccharification	Singh, Seema* 35264950300	Sandia Natl. Labs. USA	840
2	Mani et al. (2004)	Switchgrass	Milling	Switchgrass milling pretreatment, specific energy consumption, calorific value, ash content	Switchgrass, grinding	Tabil, Lope G. 6701349307	Univ. Saskatchewan Canada	517
3	Sun and Cheng (2005)	Bermudagrass	H_2SO_4, T. reesei	Bermudagrass acid and enzymatic pretreatment, acid content and residence time, hydrolysis efficiency	Bermudagrass, ethanol	Cheng, Jay J. 15046539600	N.C. State Univ. USA	408
4	Dien et al. (2006)	Alfalfa, canary grass, switchgrass	H_2SO_4, cellulases	Acid pretreatment, hydrolysis, maturity effect, carbohydrate and cellulose content, sugar recovery	Alfalfa, canary grass, switchgrass, acid, saccharification, pretreatment	Dien, Bruce S. 6603685796	USDA Agr. Res. Serv. USA	403
5	Hu and Wen (2008)	Switchgrass	Microwave, alkali, cellulases	Microwave and alkali pretreatment, hydrolysis, sugar yield, optimization	Switchgrass, digestibility, alkali, pretreatment, microwave	Wen, Zhiyou 14012373200	Iowa State Univ. USA	344
6	Singh et al. (2009)	Switchgrass	[EMIM][OAc], cellulases	Ionic liquid pretreatment visualization, biomass solubilization and cellulose regeneration, hydrolysis	Switchgrass, ionic liquids, pretreatment	Simmons, Blake A. 7102183263	Lawrence Berkeley Natl. Lab. USA	327
7	Alzadeh et al. (2005)	Switchgrass	AFEX	AFEX pretreatment, optimization, hydrolysis efficiency, bioethanol yield	Switchgrass, ammonia, AFEX, pretreatment	Dale, Bruce E. 7201511969	Michigan State Univ. USA	314
8	Yoshida et al. (2008)	Miscanthus	Milling, cellulase, β-glucosidase, xylanase, sodium chlorite	Enzymatic pretreatment, hydrolysis, sugar yield, cellulose crystallinity, delignification	Miscanthus, hydrolysis	Yoshida, Makoto 57080400900	Tokyo Univ. Agr. Technol. Japan	299
9	Guo et al. (2008)	Silver grass	H_2SO_4, enzymes	Acid pretreatment, hydrolysis, sugar yield, grass-specific surface area	Silver grass, ethanol, acid, pretreatment	Hwang, Wen-Song	Inst. Nucl. Ener. Res. Taiwan	225

(Continued)

TABLE 50.1 (Continued)
Grass Biomass Pretreatments for the Grass-based Bioethanol Production

No.	Papers	Biomass/ Hydrolysate	Prts.	Parameters	Keywords	Lead Author	Affil.	Cits
10	Chang et al. (1997)	Switchgrass	Lime, milling	Lime pretreatment, optimization, sugar yield, solubilization	Switchgrass, lime, pretreatment	Holtzapple, Mark T. 7004167004	Texas A&M Univ. USA	221
11	Samuel et al. (2010)	Switchgrass	H_2SO_4, milling	Acid pretreatment, lignin structure and type, S/G ratio	Switchgrass, acid, pretreatment	Ragauskas, Arthur J. 7006265204	Univ. Tennessee Knoxville USA	213
12	Xu et al. (2012)	Miscanthus	NaOH, H_2SO_4	Alkali and acid pretreatments, hemicelluloses, CrI, biomass digestibility, sugar yield	Miscanthus, digestibility, NaOH, H_2SO_4	Peng, Liangcai 36802605200	Huazhong Agr. Univ. China	206
13	Demartini et al. (2013)	Switchgrass	Enzymes	Enzymatic digestibility, lignin, hemicellulose, xylan, hydrolysis, sugar yield, recalcitrance	Switchgrass, recalcitrance	Wyman, Charles E. 7004396809	Univ. Calif. Riverside USA	196
14	Xu et al. (2010)	Switchgrass	NaOH, cellulase, cellobiase	Alkali pretreatment, hydrolysis, optimization, sugar yield, delignification	Switchgrass, sodium hydroxide, pretreatment, ethanol	Cheng, Jay J. 15046539600	N.C. State Univ. USA	194
15	Xu et al. (2011)	Switchgrass	Acid, Pv4CL1, Pv4CL2	Switchgrass engineering, acid pretreatment, hydrolysis efficiency, Pv4CL1 and Pv4CL2 genes, transgenic biomass	Switchgrass, fermentable, sugar,	Zhao, Bingyu 5526097390 0	Virginia Polytech. Inst. State Univ. USA	181
16	Wang et al. (2010)	Bermudagrass	NaOH, enzymes	Alkali pretreatment, hydrolysis efficiency, optimization, sugar yield	Bermuda grass, sodium hydroxide pretreatment, hydrolysis	Cheng, Jay J. 15046539600	N.C. State Univ. USA	166
17	Murnen et al. (2007)	Miscanthus	AFEX, cellulase, β-glucosidase, washing, xylanase, Tween-80	AFEX pretreatment, hydrolysis efficiency, optimization, sugar yield	Miscanthus, ammonia, AFEX, pretreatment, hydrolysis, fermentable, sugars	Balan, Venkatesh 15757087100	Univ. Houston USA	163
18	Chuck et al. (2011)	Switchgrass	H_2SO_4, accellerase, amyloglucosidase	Acid pretreatment, switchgrass engineering, hydrolysis efficiency, transgenic biomass, Cg1 genes, flowering, starch content	Switchgrass, digestibility	Chuck, George S. 57203407102	Univ. Calif. Berkeley USA	155

*, female; Cits., number of citations received for each paper; Na, nonavailable; Prt, biomass pretreatments.

pretreatment, they observed that switchgrass had reduced cellulose crystallinity, increased surface area, and decreased lignin content compared with dilute acid pretreatment. Furthermore, IL pretreatment enabled a significant enhancement in the rate of enzymatic hydrolysis of the cellulose component of switchgrass, with a rate increase of 16.7-fold and a glucan yield of 96.0% obtained in 24 h. They concluded that IL pretreatment was superior to the dilute acid pretreatment process for switchgrass.

Xu et al. (2012) studied the effect of the hemicelluloses on the lignocellulose crystallinity for high biomass digestibility under sodium hydroxide (NaOH) and sulfuric acid (H_2SO_4) pretreatments in miscanthus in a paper with 206 citations. They selected six typical pairs of miscanthus samples with different cell wall compositions and then compared their cellulose crystallinity and biomass digestibility after various chemical pretreatments. They observed that a miscanthus sample with a high hemicellulose level had a relatively low cellulose crystallinity index (CI) and enhanced biomass digestibility at similar rates after pretreatments of NaOH and H_2SO_4 with three concentrations. By contrast, a miscanthus sample with a high cellulose or lignin level showed increased CI and low biomass saccharification, particularly after H_2SO_4 pretreatment. The cellulose CI negatively affected biomass digestion. Furthermore, increased hemicellulose level by 25% or decreased cellulose and lignin contents by 31% and 37% resulted in increased hexose yields by 1.3 times to 2.2 times released from enzymatic hydrolysis after NaOH or H_2SO_4 pretreatments. They concluded that hemicelluloses were the dominant factor that positively determined biomass digestibility after pretreatments with NaOH or H_2SO_4 by negatively affecting cellulose crystallinity.

50.3.1.2.2 Acid Pretreatments

Sun and Cheng (2005) studied the dilute H_2SO_4 pretreatment of Bermudagrass for bioethanol production in a paper with 408 citations. They pretreated this feedstock at a solid loading rate of 10% at 121°C with different H_2SO_4 concentrations (0.6%, 0.9%, 1.2%, and 1.5%, w/w) and residence times (30, 60, and 90 min). They then hydrolyzed the solid residues from the prehydrolysate by cellulases. With the increasing acid concentration and residence time, they observed that the amount of arabinose and galactose increased while the glucose concentration in the prehydrolysate of Bermudagrass increased with the increase in pretreatment severity. The xylose concentration increased with the increase in H_2SO_4 concentration and residence time. Most of the arabinan, galactan, and xylan in the biomass were hydrolyzed during acid pretreatment. Cellulose remaining in the pretreated feedstock was highly digestible by cellulases from *Trichoderma reesei*.

Dien et al. (2006) studied the dilute H_2SO_4 pretreatment and enzymatic hydrolysis of alfalfa, canary grass, and switchgrass using a commercial cellulase in a paper with 403 citations. They harvested each feedstock at two or three maturity stages. They found that the total carbohydrate and cellulose content of the feedstocks varied from 518 to 655 g/kg and from 209 to 322 g/kg dry matter (DM), respectively. Carbohydrate and lignin contents were lower for samples from early maturity samples compared with samples from late maturity harvests. They observed that a significant amount of the available carbohydrates were in the form of soluble sugars and storage carbohydrates (4.3% − 16.3% wt/wt). Recovery of soluble sugars following dilute acid pretreatment was problematic, especially that of fructose. Fructose was extremely tolerant to the dilute acid pretreatments. Next, the efficiency at which available glucose was recovered was inversely correlated with maturity and lignin content. However, total glucose yields were higher for the later maturities because of higher cellulose contents compared with the earlier maturity samples.

Guo et al. (2008) characterized dilute H_2SO_4 pretreatment of silver grass for bioethanol production in a paper with 225 citations. They observed that the highest yield of xylose from silver grass was between 70% and 75%, which was similar to bagasse. However, silver grass gave a higher level of fermentability than bagasse using the hydrolysate because less acetic acid was formed. The release of sugars resulted in an about 2.0-fold increase in the specific surface area of the pretreated silver grass. Increasing the specific surface area did not obviously enhance enzymatic digestibility.

They asserted that the increase in hydrophilicity might enhance enzymatic adsorption onto lignin and increase the accumulation of cellobiose for enzymatic hydrolysis as pretreatment severity increases.

Samuel et al. (2010) characterized and compared switchgrass ball-milled lignin before and after dilute H_2SO_4 pretreatment in a paper with 213 citations. They isolated ball-milled lignin from switchgrass before and after pretreatment. They observed that ball-milled switchgrass lignin was of coumaryl-guaiacyl-syringyl (HGS) type with a considerable amount of p-coumarate and felurate esters of lignin. The major ball-milled lignin interunit was the β-O-4 linkage, and a minor amount of phenylcoumarin, resinol, and spirodienone units were also present. As a result of acid pretreatment, there was 36% decrease in β-O-4 linkage observed. In addition to these changes, the syringyl/guaiacyl (S/G) ratio decreased from 0.80 to 0.53.

Chuck et al. (2011) overexpressed the maize *Corngrass1* (*Cg1*) micro ribonucleic acid (microRNA) to prevent flowering, improve digestibility, and increase the starch content of switchgrass in a paper with 155 citations. The maize *Cg1* gene encoded a microRNA that promoted juvenile cell wall identities and morphology. To test the hypothesis that juvenile biomass had superior qualities as a potential bioethanol feedstock, they transferred the *Cg1* gene into switchgrass. They observed that the transgenic biomass had up to 250% more starch, resulting in higher glucose release from saccharification assays with or without acid pretreatment. In addition, they observed a complete inhibition of flowering in both greenhouse and field-grown switchgrass.

50.3.1.2.3   Alkaline Pretreatments

Chang et al. (1997) carried out the lime ($Ca(OH)_2$) pretreatment of switchgrass in a paper with 219 citations. After studying many conditions, the optimal pretreatment conditions were 2 h, 100°C and 120°C, 0.1 g $Ca(OH)_2$/g dry biomass, and 9 mL/g water dry biomass. There was little benefit of grinding below 20 mesh as even coarse particles (4–10 mesh) digested well. Using the optimal conditions, they observed that the 3-d reducing sugar yield was five times that of untreated switchgrass, the 3-d total sugar (glucose + xylose) yield was seven times, the 3-d glucose yield was five times, and the 3-d xylose yield was 21 times. Finally, little glucan (10%) was solubilized as a result of the $Ca(OH)_2$ pretreatment, whereas about 26% of xylan and 29% of lignin became solubilized.

Xu et al. (2010) carried out the NaOH pretreatment of switchgrass for ethanol production in a paper with 194 citations. They pretreated the biomass at 121°C, 50°C, and 21°C, and a solid/liquid ratio of 0.1 g/mL, respectively, for 0.25–1, 1–48, and 1–96 h at different NaOH concentrations (0.5%, 1.0%, and 2.0%, w/v). At the best pretreatment conditions (50°C, 12 h, and 1.0% NaOH), they found that the yield of total reducing sugars was 453.4 mg/g raw biomass, which was 3.78 times that of untreated biomass, and the glucan and xylan conversions reached 74.4 and 62.8%, respectively. Lignin reduction was closely related to the degree of pretreatment. The maximum lignin reductions were 85.8% at 121°C, 77.8% at 50°C, and 62.9% at 21°C, all of which were obtained at the combinations of the longest residence times and the greatest NaOH concentration. Cellulase and cellobiase loadings of 15 filter paper unit (FPU)/g dry biomass and 20 cellobiase unit (CBU)/g dry biomass were sufficient to maximize sugar production.

Wang et al. (2010) studied the NaOH pretreatment and enzymatic hydrolysis of coastal Bermudagrass in a paper with 166 citations. They pretreated the biomass with NaOH at concentrations from 0.5% to 3% (w/v) for a residence time from 15 to 90 min at 121°C. They observed up to 86% lignin removal. The optimal NaOH pretreatment conditions at 121°C for total reducing sugar production and glucose and xylose yields were 15 min and 0.75% NaOH. Under these optimal pretreatment conditions, they found that the total reducing sugar yield was about 71% of the theoretical maximum, and the overall conversion efficiencies for glucan and xylan were 90.43% and 65.11%, respectively.

50.3.1.2.4   Ionic Liquid Pretreatments

Singh et al. (2009) studied biomass solubilization and cellulose regeneration during IL pretreatment of switchgrass in a paper with 327 citations. They observed that the addition of ground switchgrass to 1-n-ethyl-3-methylimidazolium acetate ([EMIM]OAc) resulted in the disruption and solubilization of the plant cell wall at mild temperatures. They further observed the swelling of the plant cell wall, attributed to disruption of inter- and intramolecular hydrogen bonding between cellulose fibrils and lignin, followed by complete dissolution of biomass. They next observed the subsequent cellulose regeneration *in situ* and provided direct evidence of significant rejection of lignin from the recovered polysaccharides. In comparison with untreated biomass, IL-pretreated biomass produced cellulose that was efficiently hydrolyzed with commercial cellulase with high sugar yields over a relatively short time interval.

50.3.1.2.5 Ammonia Pretreatments

Alizadeh et al. (2005) pretreated switchgrass by ammonia fiber explosion (AFEX) in a paper with 14 citations. They observed that the optimal pretreatment conditions for switchgrass were near 100°C reactor temperature and ammonia loading of 1:1 kg of ammonia: kg of DM with 80% moisture content (dry weight basis [dwb]) at 5-min residence time. Hydrolysis results of AFEX-treated and AFEX-untreated samples showed 93% vs 16% glucan conversion, respectively. Finally, the ethanol yield of optimized AFEX-treated switchgrass was about 0.2 g ethanol/g dry biomass, which was 2.5 times more than that of the untreated sample.

Murnen et al. (2007) optimized AFEX pretreatment and enzymatic hydrolysis of *Miscanthus x giganteus* to fermentable sugars in a paper with 163 citations. They varied the pretreatment conditions including temperature, moisture, ammonia loading, residence time, and enzyme loadings to maximize hydrolysis yields. In addition, they performed the soaking of the biomass before AFEX and washing the pretreated material to improve sugar yields. The optimal AFEX conditions determined were 160°C, 2:1 (w/w) ammonia to biomass loading, 233% moisture (dry weight basis), and 5-min reaction time for water-soaked miscanthus. They obtained 96% glucan and 81% xylan conversions after 168-h enzymatic hydrolysis at 1% glucan loading using 15 FPU/(g of glucan) of cellulase and 64 p-normal-phase glucose unit (p-NPGU)/(g of glucan) of β-glucosidase along with xylanase and Tween-80 supplementation.

50.3.1.3 Enzymatic Pretreatments

There are three HCPs for the enzymatic pretreatment of the grass biomass.

Yoshida et al. (2008) studied the effects of cellulose crystallinity, hemicellulose, and lignin on the enzymatic hydrolysis of *Miscanthus sinensis* to sugars in a paper with 299 citations. They ground an air-dried biomass by ball milling and separated the powder into four fractions by passage through a series of sieves with mesh sizes of 250–355, 150–250, 63–150, and less than 63 μm. They then hydrolyzed each fraction with commercially available cellulase and β-glucosidase. They observed that the yield of sugars increased as the crystallinity of the substrate decreased. The addition of xylanase increased the yield of both pentoses and glucose. Delignification by the sodium chlorite method improved the initial rate of hydrolysis by cellulolytic enzymes significantly, resulting in a higher yield of sugars as compared with that for untreated samples. Finally, they found that when delignified *M. sinensis* was hydrolyzed with cellulase, β-glucosidase, and xylanase, hemicellulose was hydrolyzed completely into sugars, and the conversion rate of glucan to glucose was 90.6%.

Demartini et al. (2013) studied the plant cell wall components that affected biomass recalcitrance in poplar and switchgrass in a paper with 196 citations. They found that lignin and hemicellulose influenced the enzymatic digestibility of both poplar and switchgrass, but the degree of influence varied significantly. Xylan removal from switchgrass resulted in materials that achieved nearly 100% glucose yields at high enzyme loading in subsequent enzymatic hydrolysis, whereas chlorite extractions that reduced the lignin content had the most beneficial effect in poplar. They identified subsets of hemicellulose that were key recalcitrance-causing factors in switchgrass. They recommended that different strategies should be adopted when trying to engineer poplar and switchgrass for reduced recalcitrance or when designing processing conditions to efficiently convert a specific biomass feedstock into sugars.

Xu et al. (2011) silenced 4-coumarate, coenzyme A ligase (4CL), in switchgrass leading to reduced lignin content and improved fermentable sugar yields for bioethanol production in a paper with 181 citations. They identified two homologous *4CL* genes, *Pv4CL1* and *Pv4CL2*, in switchgrass through phylogenetic analysis. They found that Pv4CL1 was involved in monolignol biosynthesis. They obtained stable transgenic plants with *Pv4CL1* downregulated. RNA interference of *Pv4CL1* reduced extractable 4CL activity by 80%, leading to a reduction in lignin content with decreased guaiacyl unit composition. They observed altered lignification patterns in the stems of ribonucleic acid interference (RNAi) transgenic plants. The transgenic plants also had uncompromised biomass yields. After dilute acid pretreatment, the low lignin transgenic biomass had significantly increased cellulose hydrolysis efficiency. They concluded that Pv4CL1, but not Pv4CL2, was the key 4CL isozyme involved in lignin biosynthesis, and reducing lignin content in switchgrass biomass by silencing *Pv4CL1* remarkably increased the efficiency of fermentable sugar release for bioethanol production.

50.3.1.4 Mechanical Pretreatments
There is only one HCP for the mechanical pretreatment of the grass biomass.

50.3.1.4.1 Milling
Mani et al. (2004) studied the milling performance and physical properties of wheat and barley straws, corn stover, and switchgrass in a paper with 517 citations. They ground these feedstocks using a hammer mill with three different screen sizes (3.2, 1.6, and 0.8 mm). Among the four materials, they observed that switchgrass had the highest specific energy consumption (27.6 kW h/t), and corn stover had the least specific energy consumption (11.0 kW h/t) at 3.2 mm screen size. They developed second- or third-order polynomial models relating bulk and particle densities of grinds to geometric mean diameter within the range of 0.18–1.43 mm. Furthermore, they found that switchgrass had the highest calorific value and the lowest ash content among these feedstocks.

50.3.2 GRASS-BASED BIOETHANOL FUELS

There are two and five HCPs for the production and evaluation of grass-based bioethanol fuels (Table 50.2). All of these HCPS are related to switchgrass-based bioethanol fuels.

50.3.2.1 Production of Switchgrass-based Bioethanol Fuels
Fu et al. (2011) studied the effect of the genetic manipulation of lignin on the recalcitrance of switchgrass and switchgrass-based bioethanol production in a paper with 515 citations. They observed that genetic modification of switchgrass produced phenotypically normal plants that had reduced thermal–chemical (≤180 °C), enzymatic, and microbial recalcitrance. Downregulation of the switchgrass caffeic acid O-methyltransferase (COMT) gene decreased lignin content modestly, reduced the syringyl:guaiacyl (S/G) lignin monomer ratio, improved forage quality, and, most importantly, increased the ethanol yield by up to 38% using conventional biomass fermentation processes. The downregulated lines required less severe pretreatment and 300%–400% lower cellulase dosages for equivalent product yields using simultaneous saccharification and fermentation (SSF) with yeast. Furthermore, fermentation of diluted acid-pretreated transgenic switchgrass using *Clostridium thermocellum* with no added enzymes showed better product yields than those obtained with unmodified switchgrass. Therefore, this apparent reduction in the recalcitrance of transgenic switchgrass had the potential to lower processing costs for switchgrass-based bioethanol significantly.

Bokinsky et al. (2011) synthesized three advanced biofuels from IL-pretreated switchgrass using engineered *Escherichia coli* in a consolidated bioprocessing process in a paper with 278 citations. They engineered a native *E. coli* to grow using both the cellulose and hemicellulose fractions of several types of switchgrass biomass pretreated with ILs. These engineered strains expressed cellulase, xylanase, β-glucosidase, and xylobiosidase enzymes under the control of native *E. coli* promoters

TABLE 50.2
Switchgrass-based Bioethanol Production in General

No.	Papers	Biomass/ Hydrolysate	Prts.	Yeasts	Parameters	Keywords	Lead Author	Affil.	Cits
1	Pimentel and Patzek (2005)	Switchgrass	Na	Na	Net energy of bioethanol production from switchgrass, corn grains, and wood	Switchgrass, ethanol	Pimentel, David 7005471319	Cornell Univ. USA	1012
2	Schmer et al. (2008)	Switchgrass	Na	Na	Bioethanol production, biomass and net energy yield, GHG emissions, forecast	Switchgrass, ethanol	Vogel, Kenneth P. 7102498441	Univ. Nebraska Lincoln USA	824
3	Fu et al. (2011)	Switchgrass	Genetic eng., acid	C. thermocellum	Switchgrass genetic engineering, recalcitrance, bioethanol yield, delignification, SSF, costs	Switchgrass, ethanol, recalcitrance	Dixon, Richard A. 7402020530	Univ. N. Texas USA	515
4	Bokinsky et al. (2011)	Switchgrass	Ionic liquids	E. coli	Bioethanol production, yeast engineering, consolidated bioprocessing, enzymes, ionic liquid pretreatment	Switchgrass, ionic liquids	Keasling, Jay D. 7005564120	Univ. Calif. Berkeley USA	278
5	Tao et al. (2011)	Switchgrass	AFEX, acid, lime, LHW, SAA, SO₂, steam explosion	Na	Pretreatment type, technoeconomic and process analysis, sugar, and bioethanol yield	Switchgrass, ethanol, pretreatment	Tao, Ling 34769458000	NREL USA	234
6	Spatari et al. (2005)	Switchgrass	Na	Na	Bioethanol use, LCA, GHG emissions	Switchgrass, ethanol	Maclean, Heather L.* 7005613727	Univ. Toronto Canada	233
7	Cherubini and Jungmeier (2010)	Switchgrass	Na	Na	Bioethanol production, LCA, GHG emissions, acidification, eutrophication, C sequestration	Switchgrass, ethanol	Cherubini, Francesco 23468479700	Norwegian Univ. Sci. Technol. Norway	216

*, female; Cits., number of citations received for each paper; Na, nonavailable; Prt, biomass pretreatments.

selected to optimize growth on model cellulosic and hemicellulosic substrates. Furthermore, these strains grew using either the cellulose or hemicellulose components of IL-pretreated biomass or on both components when combined as a coculture. Next, they further engineered both cellulolytic and hemicellulolytic strains with three biofuel synthesis pathways for the production of fuel substitutes suitable for gasoline engines directly from IL-pretreated switchgrass without externally supplied hydrolase enzymes. With improvements in both biofuel synthesis pathways and biomass digestion capabilities, the consolidated bioprocessing would provide an economical route to the production of advanced biofuels.

50.3.2.1 Evaluation of the Switchgrass-based Bioethanol Fuels

Pimentel and Patzek (2005) determined the net energy of bioethanol production using corn, switchgrass, and wood in a paper with 1,012 citations. They found that the energy outputs from ethanol produced using corn, switchgrass, and wood biomass were each less than the respective fossil energy inputs. Ethanol production using corn grain, switchgrass, and wood biomass required 29%, 50%, and 57% more fossil energy than the ethanol fuel produced, respectively.

Schmer et al. (2008) determined the agricultural energy input costs, biomass yield, estimated ethanol output, greenhouse gas emissions (GHGs), and net energy yield (NEY) of switchgrass-based bioethanol based on known farm inputs and harvested yields in a paper with 824 citations. They managed switchgrass in field trials of 3–9 ha on marginal cropland on ten farms across a wide precipitation and temperature gradient in Nebraska. They found that annual biomass yields of established fields averaged 5.2–11.1 Mg/ha with a NEY of 60 GJ/ha/y. Switchgrass produced 540% more renewable than nonrenewable energy consumed. Switchgrass monocultures managed for high yield produced 93% more biomass yield and an equivalent estimated NEY than previous estimates from human-made prairies that received low agricultural inputs. The estimated average GHG emissions from switchgrass-based bioethanol were 94% lower than the estimated GHG from gasoline. This was a baseline study that represented the genetic material and agronomic technology available for switchgrass production in 2000 and 2001, when the fields were planted. They forecasted that the improved genetics and agronomics would further enhance the energy sustainability and biofuel yield of switchgrass.

Tao et al. (2011) carried out a process and technoeconomic analysis of the leading six pretreatments for switchgrass-based bioethanol production in a paper with 234 citations. These were AFEX, dilute acid, $Ca(OH)_2$, liquid hot water (LHW), soaking in aqueous ammonia (SAA), and sulfur dioxide-impregnated steam explosion pretreatments. They determined the overall bioethanol production, total capital investment, and minimum ethanol price along with selected sensitivity analysis. They found limited differentiation between the projected economic performances of the pretreatment options, except for processes that exhibited significantly lower sugar and resulting bioethanol yields.

Spatari et al. (2005) carried out the life cycle assessment of switchgrass- and corn stover-based bioethanol-fueled cars to examine the environmental implications of the production and use of bioethanol fuels in cars in Ontario in a paper with 233 citations. They compared the results to those of low-sulfur reformulated gasoline (RFG) in a functionally equivalent car using two time frames, one near-term (2010) and one midterm (2020), which assumed technological improvements in the switchgrass-based bioethanol life cycle. Near-term results showed that, compared to a RFG car, life cycle GHG emissions were 57% and 65% lower for an E85-fueled car produced from switchgrass and corn stover, respectively, on grams of CO_2 equivalent per kilometer basis. Corn stover bioethanol exhibited slightly lower life cycle GHG emissions, primarily due to sharing emissions with grain production. Through projected improvements in crop and ethanol yields, results for the midterm scenario showed that GHG emissions could be 25%–35% lower than those in 2010 and that, even with anticipated improvements in RFG cars, E85 cars could still achieve up to 70% lower GHG emissions across the life cycle.

Cherubini and Jungmeier (2010) carried out a life cycle analysis (LCA) of a biorefinery concept producing bioethanol, bioenergy, and chemicals from switchgrass in a paper with 215 citations.

They found that the use of switchgrass in a biorefinery offset GHG emissions and reduced fossil energy demand as GHG emissions were decreased by 79% and about 80% of nonrenewable energy was saved. Soil C sequestration was responsible for a large GHG benefit (65 kt CO_2-eq/a, for the first 20 years), while switchgrass production was the most important contributor to total GHG emissions of the system. If compared with the fossil reference system, the biorefinery system released more nitrous oxide (N_2O) emissions, while both CO_2 and methane (CH_4) emissions were reduced. Furthermore, the biorefinery had higher impacts in the categories of acidification and eutrophication. They discussed that these results were mainly affected by the switchgrass production and land use change effects. Steps that mainly influenced the production of switchgrass were soil N_2O emissions, manufacture of fertilizers, processing (i.e., pelletizing and drying), and transport. Even if the biorefinery chain had higher primary energy demand than the fossil reference system, it was mainly based on renewable energy as the provision of biomass with sustainable practices was then a crucial point to ensure a renewable energy supply to biorefineries. They argued that the determination of the real GHG and energy balance was complex, and a certain degree of uncertainty was always present in the final results. Furthermore, ranges in final results could be even more widened by applying different combinations of biomass feedstocks, conversion routes, fuels, end-use applications, and methodological assumptions. They showed that the switchgrass enhanced carbon sequestration in soils if established on set-aside land, thus considerably increasing the GHG savings of the system for the first 20 years after crop establishment. However, high biomass yields were extremely important in achieving high GHG emission savings, although the use of chemical fertilizers to enhance plant growth could reduce these savings. Hence, some strategies, aiming at simultaneously maintaining crop yield and reducing N fertilization application through alternative management, could be adopted. However, even if a reduction in GHG emissions was achieved, it should not be disregarded that additional environmental impacts (like acidification and eutrophication) might be caused. This aspect could not be ignored by policymakers, even if they had climate change mitigation objectives as the main goal.

50.4 DISCUSSION

50.4.1 INTRODUCTION

Crude oil-based gasoline fuels have been widely used in the transportation sector since the 1920s. However, there have been great public concerns over the adverse environmental and human impact of these fuels. Hence, biomass-based bioethanol fuels have increasingly been used in blending gasoline and petrodiesel fuels, in fuel cells, and in biochemical production in a biorefinery context.

However, it is necessary to pretreat the biomass to enhance the yield of the bioethanol fuels before the grass-based bioethanol production through the hydrolysis and fermentation of the biomass and hydrolysates, respectively. One of the most-studied biomass for bioethanol production and use has been grass biomass. Research in the field of grass-based bioethanol fuels has intensified in this context in the key research fronts of the pretreatment and hydrolysis of grass biomass, fermentation of grass hydrolysates, and grass-based bioethanol fuels in general. Furthermore, the switchgrass and to a lesser extent miscanthus, Bermudagrass, giant reed, and alfalfa have been studied intensively at the expense of the other grass biomass in this context.

However, it is essential to develop efficient incentive structures for the primary stakeholders to enhance research in this field. Although there have been many review papers for this field, there has been no review of the 25 most-cited articles in this field.

Thus, this book chapter presents a review of the 25 most-cited articles on the grass-based bioethanol fuels. Then, it discusses the key findings of these highly influential papers and comments on future research priorities in this field.

As a first step for the search of the relevant literature, the keywords were selected using the most-cited first 200 population papers. The selected keyword list was then optimized to obtain a representative sample of papers for the searched research field. This keyword list was provided in the appendix of Konur (2023) for future replicative studies.

As a second step, a sample dataset was used for this study. The first 25 articles with at least 155 citations each were selected for the review study. Key findings from each paper were taken from the abstracts of these papers and were discussed. Additionally, many brief conclusions were drawn and several relevant recommendations were made to enhance future research landscape.

Information about research fronts for the sample papers in grass-based bioethanol fuels with regard to the grass biomass used in these processes is given in Table 50.3. As this table shows, there are four primary research fronts for this field: switchgrass, miscanthus, Bermudagrass, and other grass biomass with 76%, 12%, 8%, and 12% of the HCPs, respectively.

However, the switchgrass is the most influential research front with 14% surplus, while miscanthus and grass are the least influential research fronts with 5% deficit each.

Information about the thematic research fronts for the sample papers in the grass-based bioethanol production is given in Table 50.4. As this table shows, there are two primary research fronts for this field: the pretreatments and hydrolysis of grass biomass with 84% and 60% of these HCPs, respectively. The other prolific research fronts are bioethanol production and evaluation and grass hydrolysate fermentation with 28% and 8% of the HCPs, respectively.

Furthermore, the most prolific pretreatments are chemical and enzymatic pretreatments with 76% and 60% of these HCPs, respectively. The other research fronts are the mechanical, hydrothermal, and microbial pretreatments of grass biomass together with the grass engineering with 20%, 4%, 4%, and 12% of the HCPs, respectively.

The most prolific chemical pretreatment of the grass biomass is acid pretreatment with 40% of the HCPs, followed by alkaline pretreatments with 24% of the HCPs. The other prolific pretreatments are the IL and ammonia pretreatments with 12% of the HCPs each. Similarly, the most prolific mechanical pretreatment is the milling pretreatment with 16% of the HCPs.

As this table shows, some research fronts are not represented at all or barely represented in these HCPs: ultrasound, microwave, steam, LHW, autohydrolysis, hot compressed water pretreatment, wet oxidation, solvent, SO_2, sulfite, surfactants, and microbial pretreatments.

TABLE 50.3
Most Prolific Research Fronts for the Grass-based Bioethanol Fuels

No.	Research Fronts	N Paper (%) Review	N Paper (%) Sample	Surplus (%)
1	Switchgrass	76	62	14
2	Miscanthus	12	17	−5
3	Bermudagrass	8	6	2
4	Other grass biomass	12		12
	Alfalfa	4	3	1
	Canary grass	4		4
	Silver grass	4		4
	Grass	0	5	−5
	Ryegrass	0	4	−4
	Common reed	0	3	−3
	Other grass	0	3	−3
	Canary grass	0	2	−2
	Napier grass	0	2	−2

N Paper (%) review, the number of papers in the sample of 25 reviewed papers; N paper (%) sample, the number of papers in the population sample of 100 papers.

TABLE 50.4

Most Prolific Thematic Research Fronts for the Grass-based Bioethanol Fuels

No.	Research Fronts	N Paper (%) Review	N Paper (%) Sample	Surplus (%)
1	Grass Biomass Pretreatments	84	87	−3
	Biomass pretreatments in general	0	2	−2
	Mechanical pretreatments	20	7	13
	Milling pretreatments	16	4	12
	Microwave pretreatments	4	3	1
	Ultrasound pretreatments	0	0	0
	Hydrothermal pretreatments	4	20	−16
	Liquid hot water pretreatment	4	12	−8
	Steam explosion pretreatment	4	4	0
	Autohydrolysis pretreatment	0	3	−3
	Wet oxidation pretreatment	0	1	−1
	Hot compressed water pretreatment	0	0	0
	Chemical pretreatments	76	70	6
	Acid pretreatment	40	25	15
	Alkaline pretreatment	24	29	−5
	Ammonia pretreatment	12	14	−2
	Ionic liquid pretreatment	12	9	3
	Other chemical pretreatments	8	3	5
	Sulfur dioxide pretreatment	4	3	1
	Solvent pretreatment	0	14	−14
	Sulfite pretreatment	0	0	0
	Enzymatic pretreatments	60	79	−19
	Microbial pretreatments	4	3	1
	Grass engineering	12	9	3
2	Grass Biomass Hydrolysis	60	78	−18
	Enzymatic hydrolysis	60	78	−18
	Autohydrolysis	0	0	0
	Acid hydrolysis	0	0	0
	Hydrolysis in general	0	0	0
3	Grass Hydrolysate Fermentation	8	18	−10
4	Grass-based Bioethanol Production In General	28	35	−7
	Switchgrass bioethanol production and evaluation	28	23	5

N Paper (%) review, the number of papers in the sample of 25 reviewed papers; N paper (%) sample, the number of papers in the population sample of 100 papers.

Further, the grass biomass hydrolysis is the least influential front with 18% deficit, followed by grass hydrolysate fermentation and grass-based bioethanol production, and grass biomass pretreatments with 10%, 7%, and 3% deficit, respectively. However, on an individual basis, acid and milling pretreatments are the most prolific research fronts with 13% and 12% surplus, respectively. Similarly, enzymatic pretreatments, enzymatic hydrolysis, hydrothermal pretreatments, solvent pretreatment, and LHW pretreatment are the least influential research fronts with 8%–19% deficits ach.

50.4.2   THE GRASS BIOMASS PRETREATMENTS FOR THE GRASS-BASED BIOETHANOL PRODUCTION

There are 18 HCPs for the grass biomass pretreatments (Table 50.1). The most prolific pretreatments for the grass-based bioethanol production are the chemical pretreatments with 13 HCPs. The other research fronts for the biomass pretreatments are the multiple, enzymatic, and mechanical pretreatments with one, three, and one HCPs, respectively. There are five research fronts for the chemical pretreatments: multiple chemical, acid, alkali, ammonia, and IL pretreatments with two, five, three, two, and one HCPs, respectively.

These HCPs show a sample of research on the pretreatments of grass biomass for the grass-based bioethanol production. These studies hint that the chemical and enzymatic pretreatments and to a lesser extent mechanical pretreatments enhance both the sugar and ethanol yield during the grass-based bioethanol production. In other words, the pretreatment stage is one of the most important phases of the grass-based bioethanol production as in the case of the other types of biomass (Alvira et al., 2010; Taherzadeh and Karimi, 20008).

It is notable that the most prolific chemical pretreatment is acid pretreatment of grass biomass. Furthermore, it is notable that there is no dedicated HCP in the research fronts of the hydrothermal pretreatments of grass biomass such as steam explosion pretreatments and other mechanical pretreatments such as ultrasound pretreatments. It is interesting to note that most of these biomass studies for these pretreatments are related to switchgrass.

50.4.1.1   Multiple Pretreatments

Hu and Wen (2008) enhanced the enzymatic digestibility of switchgrass by microwave-assisted alkaline pretreatment and observed that the total sugar yield from the combined pretreatment and hydrolysis was 34.5 g/100 g biomass, equivalent to 58.5% of the maximum potential sugars released. This study shows that pretreatments are often used in combination with other pretreatments.

50.4.1.2   Chemical Pretreatments

There are five research fronts for the chemical pretreatments for the grass-based bioethanol production: multiple chemical, acid, alkaline, ammonia, and IL pretreatments with two, five, three, two, and one HCPs, respectively.

50.4.1.2.1   Multiple Chemical Pretreatments

Li et al. (2010) compared the dilute acid and IL pretreatments of switchgrass in terms of delignification, hydrolysis efficiency, and sugar yields and observed that switchgrass had reduced cellulose crystallinity, increased surface area, and decreased lignin content compared with dilute acid pretreatment. Furthermore, Xu et al. (2012) studied the effect of the hemicelluloses on the lignocellulose crystallinity for high biomass digestibility under NaOH and H_2SO_4 pretreatments in miscanthus and observed that a miscanthus sample with a high hemicellulose level had a relatively low cellulose CI and enhanced biomass digestibility at similar rates after pretreatments of NaOH and H_2SO_4 with three concentrations.

50.4.1.2.2   Acid Pretreatments

Sun and Cheng (2005) studied the dilute H_2SO_4 pretreatment of Bermudagrass for ethanol production, and with the increasing acid concentration and residence time, they observed that the amount of arabinose and galactose increased, while the glucose concentration in the prehydrolysate of Bermudagrass increased with the increase in pretreatment severity. Furthermore, Dien et al. (2006) studied the dilute H_2SO_4 pretreatment and enzymatic hydrolysis of alfalfa, canary grass, and switchgrass using a commercial cellulase and found that the recovery of soluble sugars following dilute acid pretreatment was problematic, especially that of fructose.

Guo et al. (2008) characterized dilute H_2SO_4 pretreatment of silver grass for bioethanol production and observed that the highest yield of xylose from silver grass was between 70% and 75%, which was similar to bagasse. Furthermore, Samuel et al. (2010) characterized and compared switchgrass ball-milled lignin before and after dilute H_2SO_4 pretreatment and observed that ball-milled switchgrass lignin was of HGS type with a considerable amount of p-coumarate and felurate esters of lignin.

Finally, Chuck et al. (2011) overexpressed the maize *Cgl* microRNA to prevent flowering, improve digestibility, and increase the starch content of switchgrass and observed that the transgenic biomass had up to 250% more starch, resulting in higher glucose release from saccharification assays with or without acid pretreatment.

50.4.1.2.3 Alkaline Pretreatments

Chang et al. (1997) carried out the lime pretreatment of switchgrass and observed that the reducing sugar yield was five times that of untreated switchgrass, the total sugar yield was seven times, the glucose yield was five times, and the xylose yield was 21 times. Furthermore, Xu et al. (2010) carried out the NaOH pretreatment of switchgrass for ethanol production and found that the yield of total reducing sugars was 453.4 mg/g raw biomass, which was 3.78 times that of untreated biomass, and the glucan and xylan conversions reached 74.4 and 62.8%, respectively. Finally, Wang et al. (2010) studied the NaOH pretreatment and enzymatic hydrolysis of coastal Bermudagrass and found that the total reducing sugar yield was about 71% of the theoretical maximum, and the overall conversion efficiencies for glucan and xylan were 90.43% and 65.11%, respectively.

50.4.1.2.4 Ionic Liquid Pretreatments

Singh et al. (2009) studied biomass solubilization and cellulose regeneration during IL pretreatment of switchgrass and observed that the addition of ground switchgrass to the [EMIM]OAc resulted in the disruption and solubilization of the plant cell wall at mild temperatures.

50.4.1.2.5 Ammonia Pretreatments

Alizadeh et al. (2005) pretreated switchgrass by the AFEX process and observed for the AFEX-treated and AFEX-untreated samples 93% vs 16% glucan conversion, respectively. Furthermore, Murnen et al. (2007) optimized the AFEX pretreatment and enzymatic hydrolysis of miscanthus to fermentable sugars and obtained 96% glucan and 81% xylan conversions after 168-h enzymatic hydrolysis at 1% glucan loading using 15 FPU/(g of glucan) of cellulase and 64 p-NPGU/(g of glucan) of β-glucosidase along with xylanase and Tween-80 supplementation.

50.4.1.3 Enzymatic Pretreatments

There are three HCPs for the enzymatic pretreatment of the grass biomass.

Yoshida et al. (2008) studied the effects of cellulose crystallinity, hemicellulose, and lignin on the enzymatic hydrolysis of *Miscanthus sinensis* to sugars and observed that the yield of sugars increased as the crystallinity of the substrate decreased. Furthermore, Demartini et al. (2013) studied the plant cell wall components that affected biomass recalcitrance in poplar and switchgrass and found that lignin and hemicellulose influenced the enzymatic digestibility of both poplar and switchgrass, but the degree of influence varied significantly. Finally, Xu et al. (2011) silenced 4CL, in switchgrass, leading to reduced lignin content and improved fermentable sugar yields for bioethanol production, and concluded that Pv4CL1, but not Pv4CL2, was the key 4CL isozyme involved in lignin biosynthesis, and reducing lignin content in switchgrass biomass by silencing *Pv4CL1* remarkably increased the efficiency of fermentable sugar release for bioethanol production.

50.4.1.4   Mechanical Pretreatments

There is only one HCP for the mechanical pretreatment of the grass biomass.

50.4.1.4.1   Milling

Mani et al. (2004) studied the milling performance and physical properties of wheat and barley straws, corn stover, and switchgrass and observed that switchgrass had the highest specific energy consumption and corn stover had the least specific energy consumption at 3.2 mm screen size.

50.4.3   Grass-based Bioethanol Fuels in General

There are seven HCPs for the grass-based bioethanol fuels in general (Table 50.2). All of these HCPS are related to switchgrass-based bioethanol fuels.

These HCPs show a sample of research on the grass-based bioethanol production. They cover the technoeconomics and life cycle studies together with the conventional bioethanol production studies involving the pretreatment and hydrolysis of grass mass and the fermentation of the resulting hydrolysates.

50.4.3.1   Switchgrass-based Bioethanol Fuel Production

Fu et al. (2011) studied the effect of the genetic manipulation of lignin on the recalcitrance of switchgrass and switchgrass-based bioethanol production and observed that genetic modification of switchgrass increased the ethanol yield by up to 38% using conventional biomass fermentation processes. Furthermore, Bokinsky et al. (2011) synthesized three advanced biofuels from IL-pretreated switchgrass using engineered *Escherichia coli* in a consolidated bioprocessing process and further engineered both cellulolytic and hemicellulolytic strains with three biofuel synthesis pathways for the production of fuel substitutes suitable for gasoline engines directly from IL-pretreated switchgrass without externally supplied hydrolase enzymes.

50.4.3.2   Switchgrass-based Bioethanol Fuel Evaluation

Pimentel and Patzek (2005) determined the net energy of bioethanol production using corn, switchgrass, and wood in a paper with 1,012 citations. They found that the energy outputs from ethanol produced using this biomass were each less than the respective fossil energy inputs. Ethanol production using corn grain, switchgrass, and wood biomass required 29, 50, and 57% more fossil energy than the ethanol fuel produced, respectively. Furthermore, Schmer et al. (2008) determined the agricultural energy input costs, biomass yield, estimated ethanol output, GHGs, and NEY of switchgrass-based bioethanol based on known farm inputs and harvested yields, and in contrast to Pimentel and Patzek (2005), they found that switchgrass produced 540% more renewable than nonrenewable energy consumed.

Tao et al. (2011) carried out a process and technoeconomic analysis of the leading six pretreatments for switchgrass-based bioethanol production and found limited differentiation between the projected economic performances of the pretreatment options, except for processes that exhibited significantly lower sugar and resulting bioethanol yields. Furthermore, Spatari et al. (2005) carried out the life cycle assessment of switchgrass- and corn stover-based bioethanol-fueled cars to examine the environmental implications of the production and use of bioethanol fuels in cars in Ontario and found that the midterm scenario showed that GHG emissions could be 25%–35% lower than those in 2010 and that, even with anticipated improvements in RFG cars, E85 cars could still achieve up to 70% lower GHG emissions across the life cycle.

Finally, Cherubini and Jungmeier (2010) carried out a LCA of a biorefinery concept producing bioethanol, bioenergy, and chemicals from switchgrass and found that the use of switchgrass in a biorefinery offset GHG emissions and reduced fossil energy demand as GHG emissions were decreased by 79% and about 80% of nonrenewable energy was saved.

50.5   CONCLUSION AND FUTURE RESEARCH

The brief information about the key research fronts covered by the 25 most-cited papers with at least 155 citations each is given under two primary headings: the pretreatments of grass biomass and grass-based bioethanol fuels in general.

The usual characteristics of these HCPs are that the pretreatments and hydrolysis of grass biomass and fermentation of grass hydrolysates are the primary processes for the grass-based bioethanol production to improve the ethanol yield as the grass biomass is one of the most studied feedstocks for bioethanol production, especially for the countries with the large farmlands such as the USA and Canada.

The key findings on these research fronts should be read in light of the increasing public concerns about climate change, GHG emissions, and global warming as these concerns have been certainly behind the boom in research on the grass-based bioethanol production as an alternative to crude oil-based gasoline and petrodiesel fuels in the last decades. It is also a sustainable alternative to food crop- and waste-based bioethanol fuels. The recent supply shocks caused by the coronavirus disease 2019 (COVID-19) pandemic and the Russian invasion of Ukraine also highlight the importance of the production and utilization of the bioethanol fuels as an alternative to crude oil-based gasoline and petrodiesel fuels.

As Table 50.3 shows, there are four primary research fronts for this field: switchgrass and to a lesser extent miscanthus, Bermudagrass, and other grass biomass at the expense of the other biomass listed in this table. However, the switchgrass is the most influential research front, while miscanthus and grass in general are the least influential research fronts.

Similarly, as Table 50.4 shows there are two primary thematic research fronts for this field: the pretreatments and hydrolysis of grass biomass. The other research fronts are bioethanol production and evaluation of bioethanol fuels and grass hydrolysate fermentation.

Furthermore, the most prolific pretreatments are chemical and enzymatic pretreatments. The other research fronts are the mechanical, hydrothermal, and microbial pretreatments of grass biomass together with the grass engineering.

The most prolific chemical pretreatment of the grass biomass is acid pretreatment, followed by alkaline pretreatments. The other prolific pretreatments are the IL and ammonia pretreatments. Similarly, the most prolific mechanical pretreatment is the milling pretreatment.

These studies emphasize the importance of proper incentive structures for efficient grass-based bioethanol production in light of North's institutional framework (North, 1991). In this context, the major producers and users of bioethanol fuels such as the USA and Canada with vast farmlands have developed strong incentive structures for efficient grass-based bioethanol production. In light of the supply shocks caused primarily by the COVID-19 pandemic and the Russian invasion of Ukraine, it is expected that the incentive structures such as public funding would be enhanced to increase the share of bioethanol fuels in the global fuel portfolio as a strong alternative to crude oil-based gasoline and petrodiesel fuels.

In this context, it is expected that the most prolific researchers, institution countries, funding bodies, and journals in this field would have a first-mover advantage to benefit from such potential incentives.

It is recommended that such review studies are performed for the primary research fronts of the grass-based bioethanol production.

ACKNOWLEDGMENTS

The contribution of the highly cited researchers in the field of grass-based bioethanol production has been gratefully acknowledged.

REFERENCES

Akin, D. E. 2007. Grass lignocellulose: Strategies to overcome recalcitrance. *Applied Biochemistry and Biotechnology* 137–140:3–15.

Alizadeh, H., F. Teymouri, T. I. Gilbert and B. E. Dale. 2005. Pretreatment of switchgrass by ammonia fiber explosion (AFEX). *Applied Biochemistry and Biotechnology - Part A Enzyme Engineering and Biotechnology* 124:1133–1141.

Alvira, P., E. Tomas-Pejo, M. Ballesteros and M. J. Negro. 2010. Pretreatment technologies for an efficient bioethanol production process based on enzymatic hydrolysis: A review. *Bioresource Technology* 101:4851–4861.

Angelici, C., B. M. Weckhuysen and P. C. A. Bruijnincx. 2013. Chemocatalytic conversion of ethanol into butadiene and other bulk chemicals. *ChemSusChem* 6:1595–1614.

Antolini, E. 2007. Catalysts for direct ethanol fuel cells. *Journal of Power Sources* 170:1–12.

Antolini, E. 2009. Palladium in fuel cell catalysis. *Energy and Environmental Science* 2:915–931.

Bokinsky, G., Y. P. Peralta-Yahya and A. George, et al. 2011. Synthesis of three advanced biofuels from ionic liquid-pretreated switchgrass using engineered *Escherichia coli*. *Proceedings of the National Academy of Sciences of the United States of America*, 108:19949–19954.

Burnham, J. F. 2006. Scopus database: A review. *Biomedical Digital Libraries* 3:1–8.

Chang, V. S., B. Burr and M. T. Holtzapple. 1997. Lime pretreatment of switchgrass. *Applied Biochemistry and Biotechnology - Part A Enzyme Engineering and Biotechnology* 63–65:3–19.

Cherubini, F. and G. Jungmeier. 2010. LCA of a biorefinery concept producing bioethanol, bioenergy, and chemicals from switchgrass. *International Journal of Life Cycle Assessment* 15:53–66.

Chuck, G. S., C. Tobias and L. Sun, et al. 2011. Overexpression of the maize Corngrass1 microRNA prevents flowering, improves digestibility, and increases starch content of switchgrass. *Proceedings of the National Academy of Sciences of the United States of America* 108:17550–17555.

De Oliveira, D. M., A. Finger-Teixeira, T. R. Mota, et al. 2015. Ferulic acid: A key component in grass lignocellulose recalcitrance to hydrolysis. *Plant Biotechnology Journal* 13:1224–1232.

Demartini, J. D., S. Pattathil and J. S. Miller, et al. 2013. Investigating plant cell wall components that affect biomass recalcitrance in poplar and switchgrass. *Energy and Environmental Science* 6:898–909.

Dien, B.S., J. H. G. Jung and K. P. Vogel, et al. 2006. Chemical composition and response to dilute-acid pretreatment and enzymatic saccharification of alfalfa, reed canarygrass, and switchgrass. *Biomass and Bioenergy* 30:880–891.

Faga, B. A., M. R. Wilkins and I. M. Banat. 2010. Ethanol production through simultaneous saccharification and fermentation of switchgrass using Saccharomyces cerevisiae D5A and thermotolerant *Kluyveromyces marxianus* IMB strains. *Bioresource Technology* 101:2273–2279.

Fernando, S., S. Adhikari, C. Chandrapal and M. Murali. 2006. Biorefineries: Current status, challenges, and future direction. *Energy & Fuels* 20:1727–1737.

Fu, C., J. R. Mielenz and X. Xiao, et al. 2011. Genetic manipulation of lignin reduces recalcitrance and improves ethanol production from switchgrass. *Proceedings of the National Academy of Sciences of the United States of America* 108:3803–3808.

Guo, G. L., W. H. Chen, W. H. Chen, L. C. Men and W. S. Hwang. 2008. Characterization of dilute acid pretreatment of silvergrass for ethanol production. *Bioresource Technology* 99:6046–6053.

Hahn-Hagerdal, B., M. Galbe, M. F. Gorwa-Grauslund, G. Liden and G. Zacchi. 2006. Bio-ethanol - The fuel of tomorrow from the residues of today. *Trends in Biotechnology* 24:549–556.

Hill, J., E. Nelson, D. Tilman, S. Polasky and D. Tiffany. 2006. Environmental, economic, and energetic costs and benefits of biodiesel and ethanol biofuels. *Proceedings of the National Academy of Sciences of the United States of America* 103:11206–11210.

Hill, J., S. Polasky and E. Nelson, et al. 2009. Climate change and health costs of air emissions from biofuels and gasoline. *Proceedings of the National Academy of Sciences of the United States of America* 106:2077–2082.

Hsieh, W. D., R. H. Chen, T. L. Wu and T. H. Lin. 2002. Engine performance and pollutant emission of an SI engine using ethanol-gasoline blended fuels. *Atmospheric Environment* 36:403–410.

Hu, Z. and Z. Wen. 2008. Enhancing enzymatic digestibility of switchgrass by microwave-assisted alkali pretreatment. *Biochemical Engineering Journal* 38:369–378.

Huang, H. J., S. Ramaswamy, U. W. Tschirner and B. V. Ramarao. 2008. A review of separation technologies in current and future biorefineries. *Separation and Purification Technology* 62:1–21.

Jin, M., M. W. Lau, V. Balan and B. E. Dale. 2010. Two-step SSCF to convert AFEX-treated switchgrass to ethanol using commercial enzymes and *Saccharomyces cerevisiae* 424A(LNH-ST). *Bioresource Technology* 101:8171–8178.

Keshwani, D. R. and J. J. Cheng. 2009. Switchgrass for bioethanol and other value-added applications: A review. *Bioresource Technology* 100:1515–1523.

Konur, O. 2000. Creating enforceable civil rights for disabled students in higher education: An institutional theory perspective. *Disability & Society* 15:1041–1063.

Konur, O. 2002a. Access to nursing education by disabled students: Rights and duties of nursing programs. *Nurse Education Today* 22:364–374.

Konur, O. 2002b. Assessment of disabled students in higher education: Current public policy issues. *Assessment and Evaluation in Higher Education* 27:131–152.

Konur, O. 2002c. Access to employment by disabled people in the UK: Is the Disability Discrimination Act working? *International Journal of Discrimination and the Law* 5:247–279.

Konur, O. 2006a. Participation of children with dyslexia in compulsory education: Current public policy issues. *Dyslexia* 12:51–67.

Konur, O. 2006b. Teaching disabled students in higher education. *Teaching in Higher Education* 11:351–363.

Konur, O. 2007a. A judicial outcome analysis of the *Disability Discrimination Act*: A windfall for the employers? *Disability & Society* 22:187–204.

Konur, O. 2007b. Computer-assisted teaching and assessment of disabled students in higher education: The interface between academic standards and disability rights. *Journal of Computer Assisted Learning* 23:207–219.

Konur, O. 2012. The evaluation of the research on the bioethanol: A scientometric approach. *Energy Education Science and Technology Part A: Energy Science and Research* 28:1051–1064.

Konur, O. 2015. Current state of research on algal bioethanol. In *Marine Bioenergy: Trends and Developments*, Ed. S. K. Kim and C. G. Lee, pp. 217–244. Boca Raton, FL: CRC Press.

Konur, O. 2019. Cyanobacterial bioenergy and biofuels science and technology: A scientometric overview. In *Cyanobacteria: From Basic Science to Applications*, Ed. A. K. Mishra, D. N. Tiwari and A. N. Rai, pp. 419–442. Amsterdam: Elsevier.

Konur, O. 2020. The scientometric analysis of the research on the bioethanol production from green macroalgae. In *Handbook of Algal Science, Technology and Medicine*, Ed. O. Konur, pp. 385–401. London: Academic Press.

Konur, O. 2023. Grass-based bioethanol fuel production: Scientometric study. In *Feedstock-based Bioethanol Fuels. I. Non-Waste Feedstocks: Starch, Sugar, Grass, Wood, Cellulose, Algae, and Biosyngas-based Bioethanol Fuels. Handbook of Bioethanol Fuels Volume 3*, Ed. O. Konur. Boca Raton, FL: CRC Press.

Li, C., B. Knierim and C. Manisseri, et al. 2010. Comparison of dilute acid and ionic liquid pretreatment of switchgrass: Biomass recalcitrance, delignification and enzymatic saccharification. *Bioresource Technology* 101:4900–4906.

Lin, Y. and S. Tanaka. 2006. Ethanol fermentation from biomass resources: Current state and prospects. *Applied Microbiology and Biotechnology* 69:627–642.

Ma, X., L. Sun and C. Song. 2002. A new approach to deep desulfurization of gasoline, diesel fuel and jet fuel by selective adsorption for ultra-clean fuels and for fuel cell applications. *Catalysis Today* 77:107–116.

Mani, S., L.G. Tabil and S. Sokhansanj. 2004. Grinding performance and physical properties of wheat and barley straws, corn stover and switchgrass. *Biomass and Bioenergy* 27:339–352.

Morschbacker, A. 2009. Bio-ethanol based ethylene. *Polymer Reviews* 49:79–84.

Murnen, H. K., V. Balan and S. P. S. Chundawat, et al. 2007. Optimization of Ammonia Fiber Expansion (AFEX) pretreatment and enzymatic hydrolysis of *Miscanthus x giganteus* to fermentable sugars. *Biotechnology Progress* 23:846–850.

Najafi, G., B. Ghobadian and T. Tavakoli, et al. 2009. Performance and exhaust emissions of a gasoline engine with ethanol blended gasoline fuels using artificial neural network. *Applied Energy* 86:630–639.

Newman, P. W. G. and J. R. Kenworthy. 1989. Gasoline consumption and cities: A comparison of U.S. cities with a global survey. *Journal of the American Planning Association* 55:24–37.

North, D. C. 1991. Institutions. *Journal of Economic Perspectives* 5:97–112.

Olsson, L. and B. Hahn-Hagerdal. 1996. Fermentation of lignocellulosic hydrolysates for ethanol production. *Enzyme and Microbial Technology* 18:312–331.

Pimentel, D. and T. W. Patzek. 2005. Ethanol production using corn, switchgrass, and wood; Biodiesel production using soybean and sunflower. *Natural Resources Research* 14:65–76.

Samuel, R., Y. Pu, B. Raman and A. K. Ragauskas. 2010. Structural characterization and comparison of switchgrass ball-milled lignin before and after dilute acid pretreatment. *Applied Biochemistry and Biotechnology* 162:62–74.

Sanchez, O. J. and C. A. Cardona. 2008. Trends in biotechnological production of fuel ethanol from different feedstocks. *Bioresource Technology* 99:5270–5295.

Schmer, M. R., K. P. Vogel, M. R. Mitchell and R. K. Perrin. 2008. Net energy of cellulosic ethanol from switch-grass. *Proceedings of the National Academy of Sciences of the United States of America* 105:464–469.

Scordia, D., S. L. Cosentino, J. W. Lee and T. W. Jeffries. 2011. Dilute oxalic acid pretreatment for biorefining giant reed (*Arundo donax* L.). *Biomass and Bioenergy* 35:3018–3024.

Singh, S., B. A. Simmons and K. P. Vogel. 2009. Visualization of biomass solubilization and cellulose regeneration during ionic liquid pretreatment of switchgrass. *Biotechnology and Bioengineering* 104:68–75.

Spatari, S., Y. Zhang and H. L. Maclean. 2005. Life cycle assessment of switchgrass- and corn stover-derived ethanol-fueled automobiles. *Environmental Science and Technology* 39:9750–9758.

Sreenath, H. K., R. G. Koegel, A. B. Moldes, T. W. Jeffries and R. J. Straub. 1999. Enzymic saccharification of alfalfa fibre after liquid hot water pretreatment. *Process Biochemistry* 35:33–41.

Sun, Y. and J. Cheng. 2002. Hydrolysis of lignocellulosic materials for ethanol production: A review. *Bioresource Technology* 83:1–11.

Sun, Y. and J. J. Cheng. 2005. Dilute acid pretreatment of rye straw and bermudagrass for ethanol production. *Bioresource Technology* 96:1599–1606.

Taherzadeh, M. J. and K. Karimi. 2007. Enzyme-based hydrolysis processes for ethanol from lignocellulosic materials: A review. *Bioresources* 2:707–738.

Taherzadeh, M. J. and K. Karimi. 2008. Pretreatment of lignocellulosic wastes to improve ethanol and biogas production: A review. *International Journal of Molecular Sciences* 9:1621–1651.

Tao, L., A. Aden and R. T. Elander, et al. 2011. Process and technoeconomic analysis of leading pretreatment technologies for lignocellulosic ethanol production using switchgrass. *Bioresource Technology* 102:11105–11114.

Wang, Z., D. R. Keshwani, A. P. Redding and J. J. Cheng. 2010. Sodium hydroxide pretreatment and enzymatic hydrolysis of coastal Bermuda grass. *Bioresource Technology* 101:3583–3585.

Xu, B., L. L. Escamilla-Trevino and N. Sathitsuksanoh, et al. 2011. Silencing of 4-coumarate: Coenzyme A ligase in switchgrass leads to reduced lignin content and improved fermentable sugar yields for biofuel production. *New Phytologist* 192:611–625.

Xu, J., J. J. Cheng, R. R. Sharma-Shivappa and J. C. Burns. 2010. Sodium hydroxide pretreatment of switch-grass for ethanol production. *Energy and Fuels* 24:2113–2119.

Xu, N., W. Zhang and S. Ren, et al. 2012. Hemicelluloses negatively affect lignocellulose crystallinity for high biomass digestibility under NaOH and H_2SO_4 pretreatments in Miscanthus. *Biotechnology for Biofuels* 5:58.

Yoshida, M., Y. Liu and S. Uchida, et al. 2008. Effects of cellulose crystallinity, hemicellulose, and lignin on the enzymatic hydrolysis of *Miscanthus sinensis* to monosaccharides. *Bioscience, Biotechnology and Biochemistry* 72:805–810.

Zhang, J., M. Siika-Aho, M. Tenkanen and L. Viikari. 2011. The role of acetyl xylan esterase in the solubilization of xylan and enzymatic hydrolysis of wheat straw and giant reed. *Biotechnology for Biofuels* 4:60.

Part 12

Wood-based Bioethanol Fuels

51 Wood-based Bioethanol Fuels
Scientometric Study

Ozcan Konur
(Formerly) Ankara Yildirim Beyazit University

51.1 INTRODUCTION

Crude oil-based gasoline fuels (Ma et al., 2002; Newman and Kenworthy, 1989) have been widely used in the transportation sector since the 1920s. However, there have been great public concerns over the adverse environmental and human impact of these fuels (Hill et al., 2006, 2009). Hence, biomass-based bioethanol fuels (Hill et al., 2006; Konur, 2012e, 2015, 2019, 2020a) have increasingly been used in blending gasoline fuels (Hsieh et al., 2002; Najafi et al., 2009), in fuel cells (Antolini, 2007, 2009), and in biochemical production (Angelici et al., 2013; Morschbacker, 2009) in a biorefinery context (Fernando et al., 2006; Huang et al., 2008).

Bioethanol fuels also play a critical role in maintaining energy security (Kruyt et al., 2009; Winzer, 2012) in the supply shocks (Kilian, 2008, 2009) related to oil price shocks (Hamilton, 2003, 2009), coronavirus disease 2019 (COVID-19) pandemic (Fauci et al., 2020; Li et al., 2020), or wars (Hamilton, 1983; Jones, 2012) in the aftermath of the Russian invasion of Ukraine (Reeves, 2014).

However, it is necessary to pretreat the biomass (Taherzadeh and Karimi, 2008; Yang and Wyman, 2008) to enhance the yield of the bioethanol (Hahn-Hagerdal et al., 2006; Sanchez and Cardona, 2008) before the bioethanol production through hydrolysis (Sun and Cheng, 2002; Taherzadeh and Karimi, 2007) and fermentation (Lin and Tanaka, 2006; Olsson and Hahn-Hagerdal, 1996) of the biomass and hydrolysates, respectively.

One of the most-studied biomass for the bioethanol production has been wood. Research in the field of wood-based bioethanol fuels has intensified in this context in the key research fronts of the pretreatment of wood (Kilpelainen et al., 2007; Schwanninger et al., 2004; Sun et al., 2009), hydrolysis of the wood (Mooney et al., 1998; Palonen et al., 2004; Studer et al., 2011), fermentation of the wood hydrolysates (Jonsson et al., 1998; Larsson et al., 1999a,b), and bioethanol production from the wood in general (Pan et al., 2005; Pimentel and Patzek, 2005; Wingren et al., 2003). Furthermore, both the softwood such as pine (Kilpelainen et al., 2007; Sun et al., 2009; Zhu et al., 2009) and hardwood such as poplar (Kumar et al., 2009; Pan et al., 2006; Studer et al., 2011) have been studied intensively.

However, it is essential to develop efficient incentive structures (North, 1991) for the primary stakeholders to enhance research in this field (Konur, 2000, 2002a,b,c, 2006a,b, 2007a,b). The scientometric analysis has been used in this context to inform the primary stakeholders about the current state of research in a selected research field (Garfield, 1955; Konur, 2011, 2012a,b,c,d,e,f,g,h,i, 2015, 2018b, 2019, 2020a).

As there have been no published scientometric studies in this field, this book chapter presents a scientometric study of research in wood-based bioethanol fuels. It examines the scientometric characteristics of both the sample and population data presenting the scientometric characteristics of these both datasets in the order of documents, authors, publication years, institutions, funding bodies, source titles, countries, Scopus subject categories, Scopus keywords, and research fronts.

51.2 MATERIALS AND METHODS

The search for this study was carried out using the Scopus database (Burnham, 2006) in June 2022.

DOI: 10.1201/9781003226451-67

As a first step for the search of the relevant literature, the keywords were selected using the first 200 most-cited population papers. The selected keyword list was then optimized to obtain a representative sample of papers for the searched research field. This keyword list was provided in the Appendix for future replicative studies.

As a second step, two sets of data were used for this study. First, a population sample of 5,460 papers was used to examine the scientometric characteristics of the population data. Second, a sample of 200 most-cited papers, corresponding to 3.7% of the population papers, was used to examine the scientometric characteristics of these citation classics.

The scientometric characteristics of these both sample and population datasets were presented in the order of documents, authors, publication years, institutions, funding bodies, source titles, countries, Scopus subject categories, Scopus keywords, and research fronts.

Lastly, the key scientometric findings for both datasets were discussed to highlight research landscape for wood-based bioethanol fuels. Additionally, a number of brief conclusions were drawn and a number of relevant recommendations were made to enhance the future research landscape.

51.3 RESULTS

51.3.1 THE MOST PROLIFIC DOCUMENTS IN THE WOOD-BASED BIOETHANOL FUELS

The information on the types of documents for both datasets is given in Table 51.1. The articles and conference papers, published in journals, dominate both the sample (96%) and population (97%) papers as they are underrepresented in the sample papers by 1%. Furthermore, review papers and short surveys have a surplus as they are overrepresented in the sample papers by 3% as they constitute 4% and 1% of the sample and population papers, respectively.

It is further notable that 98.8% of the population papers were published in journals, while 0.8 and 0.4% of them were published in book series and books, respectively. On the contrary, all of the sample papers were published in journals.

51.3.2 THE MOST PROLIFIC AUTHORS IN THE WOOD-BASED BIOETHANOL FUELS

The information about the 18 most prolific authors with at least 2.5% of sample papers each is given in Table 51.2. The most prolific author is Jack N. Saddler of the University of British Columbia with 12.5% of the sample papers, followed by Guido Zacchi and Xuejun Pan with 9.5% and 8.0% of

TABLE 51.1
Documents in Wood-based Bioethanol Fuels

Documents	Sample Dataset (%)	Population Dataset (%)	Surplus (%)
Article	88.5	94.7	−6.2
Conference paper	7.5	2.5	5.0
Review	2.5	1.1	1.4
Short survey	1.5	0.2	1.3
Letter	0.0	0.5	−0.5
Note	0.0	0.5	−0.5
Book chapter	0.0	0.4	−0.4
Editorial	0.0	0.0	0.0
Book	0.0	0.0	0.0
Sample size	200	5,460	

Population dataset, the number of papers (%) in the set of 5,460 population papers; Sample dataset, the number of papers (%) in the set of 200 highly cited papers.

TABLE 51.2
Most Prolific Authors in Wood-based Bioethanol Fuels

No.	Author Name	Author Code	Sample Papers (%)	Population Papers (%)	Surplus	Institution	Country	HI	N	Res. Front
1	Saddler, Jack N.	7005297559	12.5	1.8	10.7	Univ. British Columbia	Canada	97	405	H, P
2	Zacchi, Guido	7006727748	9.5	1.2	8.3	Lund Univ.	Sweden	67	204	E, F, H, P
3	Pan, Xuejun	57203296000	8.0	0.5	7.5	Univ. Wisconsin Madison	USA	44	117	H, P
4	Galbe, Mats	7003788758	6.0	0.9	5.1	Lund Univ.	Sweden	50	131	E, F, H, P
5	Zhu, Junyong	7405692678	6.0	0.7	5.3	USDA Forest Serv.	USA	62	307	H, P
6	Gregg, David J.	7005324246	5.0	0.4	4.6	Univ. British Columbia	Canada	21	25	H, P
7	Sun, Runchang	5566152560	4.0	1.5	2.5	Dalian Polytech. Univ.	China	110	1057	H, P
8	Wyman, Charles E.	7004396809	4.0	0.4	3.6	Univ. Calif. Riverside	USA	80	286	H, P
9	Ragauskas, Arthur J.	7006265204	3.5	0.8	2.7	Univ. Tennessee Knoxville	USA	91	746	E, H, P
10	Gleisner, Roland	6603256087 55928749700	3.5	0.4	3.1	USDA Forest Serv.	USA	28	69	H, P
11	Ladisch, Michael R.	7005670397	3.5	0.2	3.3	Purdue Univ.	USA	59	290	H, P
12	Parajo, Juan C.	7005594129	3.0	1.3	1.7	Univ. Vigo	Spain	62	300	E, F, H, P
13	Viikari, Liisa*	7006720604	3.0	0.2	2.8	Univ. Helsinki	Finland	55	194	H, P
14	Gilkes, Neil	35493889000	3.0	0.2	2.8	Univ. British Columbia	Canada	51	87	H, P
15	Garrote, Gil	6603849654	2.5	0.3	2.2	Univ. Vigo	Spain	43	98	E, F, H, P
16	Ballesteros, Ignacio	6602732963	2.5	0.3	2.2	CSIC	Spain	37	70	E, F, H, P
17	Negro, Maria J.*	6701512649	2.5	0.2	2.3	CIEMAT	Spain	37	72	E, F, H, P
18	Kim, Youngmi*	5699576800	2.5	0.2	2.3	Univ. Wisconsin River Falls	USA	26	35	H, P

*, Female researchers; Author code, the unique code given by Scopus to the authors; E, ethanol production. F, fermentation of the feedstocks; H, hydrolysis of the biomass; P, pretreatments of the biomass; Population papers: the number of papers authored in the population dataset; Sample papers, the number of papers authored in the sample dataset.

the sample papers, respectively. The other prolific authors are Mats Galbe, Junyong Zhu, David J. Gregg, Runchang Sun, and Charles E. Wyman with 4%–6% of the sample papers each.

Furthermore, the most influential author is Jack N. Saddler with 10.7% surplus, followed by Guido Zacchi and Xuejun Pan with 8.3% and 7.5% surplus, respectively. The other influential authors are Junyong Zhu, Mats Galbe, David J. Gregg, Charles E. Wyman, Michael R. Ladisch, and Roland Gleisner with 3.1%–5.3% surplus each.

The most prolific institution for the sample dataset is the University of British Columbia with three authors, while Lund University, University of Vigo, and USDA Forest Service have two authors each. In total, 13 institutions house these top authors.

However, the most prolific country for the sample dataset is the USA with seven authors, followed by Spain, Canada, and Sweden with four, three, and two authors, respectively. In total, six countries house these top authors.

There are four primary research fronts for these top authors: ethanol production in general (E), wood hydrolysate fermentation (F), wood biomass hydrolysis (H), and wood biomass pretreatments (P) with 8, 7, 18, and 18 authors, respectively.

However, there is a significant gender deficit (Beaudry and Lariviere, 2016) for the sample dataset as surprisingly only three of these top researchers are female with a representation rate of 17%.

Additionally, there are other authors with a relatively low citation impact and with 0.3%–0.9% of the population papers each: Feng Xu, Valentin Santos, Shiro Saka, Qiang Yong, Hisashi Miyafuji, Herbert Sixta, Eulogio Castro, Juanita Freer, Jia-Long Wen, Yunqiao Pu, Jose Luis Alonso, Caoxing Huang, Hasan Jameel, Xianzhi Meng, Junhua Zhang, Qingxi Hou, Boris N. Kuznetsov, Leif J. Jonsson, Stefan M. Willfor, Chenhuan Lai, Aloia Romani, Shao-Ni Sun, Andrey Pranovich, Songlin Yi, Jaime Baeza, In-Gyu Choi, Shijie Liu, Encarnacion Ruiz, Blake A. Simmons, Chang Geung Yoo, and Tong-Qi Yuan.

51.3.3 THE MOST PROLIFIC RESEARCH OUTPUT BY YEARS IN THE WOOD-BASED BIOETHANOL FUELS

Information about papers published between 1970 and 2022 is given in Figure 51.1. This figure clearly shows that the bulk of research papers in the population dataset were published primarily

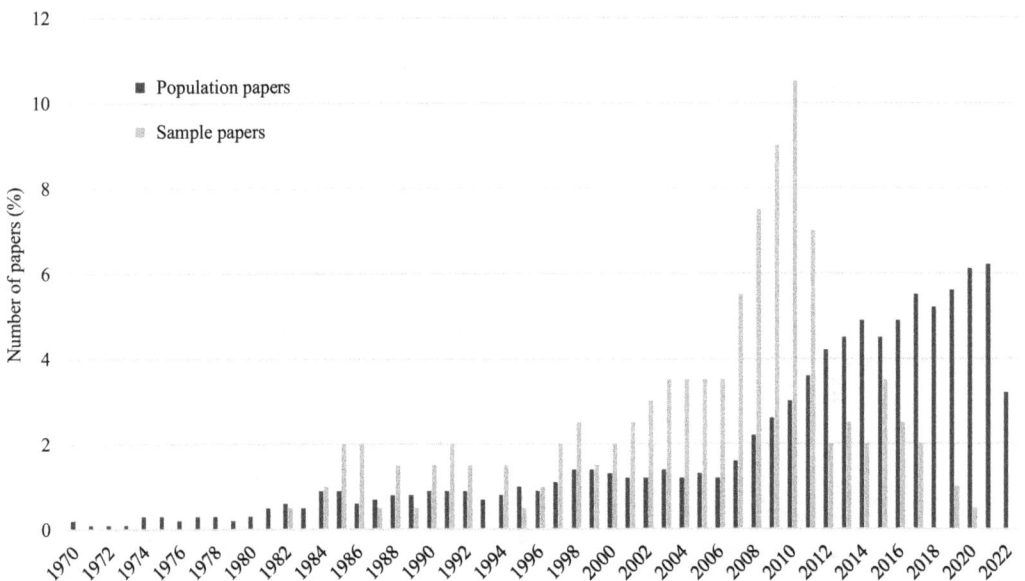

FIGURE 51.1 Research output by years regarding the wood-based bioethanol fuels.

in the 2010s with 46% of the population datasets. The publication rates for the early 2020s, 2000s, 1990s, 1980s, and 1970s were 16%, 15%, 10%, 7%, and 2% respectively. Additionally, 5% of the population papers were published in the pre-1970s.

Similarly, the bulk of research papers in the sample dataset were published in the 2000s and 2010s with 44% and 33% of the sample datasets, respectively. The publication rates for the 1990s, 1980s, and 1970s were 14%, 8%, and 0% of the sample papers, respectively. Additionally, 1% of the sample papers were published in the pre-1970s.

The most prolific publication years for the population dataset were 2020 and 2021 with 6.1% and 6.2% of the dataset, respectively. Furthermore, 66% of the population papers were published between 2008 and 2022. Similarly, 78% of the sample papers were published between 1998 and 2016, while the most prolific publication year was 2010 with 10.5% of the sample papers. The other prolific years were 2008 and 2011 with 7.5% and 7.0% of the sample papers, respectively.

51.3.4 THE MOST PROLIFIC INSTITUTIONS IN THE WOOD-BASED BIOETHANOL FUELS

Information about the 18 most prolific institutions publishing papers on wood-based bioethanol fuels with at least 2.5% of the sample papers each is given in Table 51.3.

The most prolific institution is the University of British Columbia with 11.5% of the sample papers, followed by Lund University with 10.5% of the sample papers. The other prolific institutions are the University of Wisconsin Madison, USDA Forest Service, South China University of Technology, National Renewable Energy Laboratory (NREL), and Georgia Institute of Technology with 4.5%–8.0% of the sample papers each.

The top country for these most prolific institutions is the USA with 10 institutions, while China, Finland, and Spain house two institutions each. In total, six countries house these top institutions.

However, the institution with the most citation impact is the University of British Columbia with 9.2% surplus, followed by Lund University with 8.8% surplus. The other influential institutions are University of Wisconsin Madison, USDA Forest Service, and NREL with 4.5%–6.9% surplus each. Similarly, the institution with the least citation impact is the Beijing Forestry University with 0.1% deficit.

Additionally, there are other institutions with a relatively low citation impact and with 0.5%–2.7% of the population papers each: Nanjing Forestry University, Kyoto University, Chinese Academy of Forestry, Abo Akademi University, State University of New York (SUNY), Aalto University, Technical University Zvolen, Chinese Academy of Sciences, University of Sao Paulo, University of Laval, Qilu University of Technology, Chalmers University of Technology, Tianjin University of Science & Technology, University of Tennessee Knoxville, Northwest A&F University, Royal Institute of Technology, Lulea Technical University, Forestry and Forest Products Research Institute, SCION, Northeast Forestry University, University of Concepcion, University of Jaen, Kyoto Prefectural University, University of Tokyo, University of Toronto, Russian Academy of Sciences, Umea University, CNRS, Washington State University Pullman, Guangxi University, Michigan State University, and Seoul National University.

51.3.5 THE MOST PROLIFIC FUNDING BODIES IN THE WOOD-BASED BIOETHANOL FUELS

Information about the 15 most prolific funding bodies funding at least 1.5% of the sample papers each is given in Table 51.4. Only 34% and 41% of the sample and population papers were funded, respectively.

The most prolific funding body is the National Natural Science Foundation of China with 4% of the sample papers, followed by the European Commission and the US Department of Energy with 3.5% of the sample papers each. However, the most prolific country for these top funding bodies is the USA with five funding bodies, followed by Canada and the European Union (EU) with three funding bodies each. The other prolific countries are China and Sweden with two funding bodies each. In total, only four countries and the EU house these top funding bodies.

TABLE 51.3
Most Prolific Institutions in Wood-based Bioethanol Fuels

No.	Institutions	Country	Sample Papers (%)	Population Papers (%)	Surplus (%)
1	Univ. Brit. Columbia	Canada	11.5	2.3	9.2
2	Lund Univ.	Sweden	10.5	1.7	8.8
3	Univ. Wisconsin Madison	USA	8.0	1.1	6.9
4	USDA Forest Serv.	USA	7.5	1.9	5.6
5	S. China Univ. Technol.	China	5.5	2.8	2.7
6	NREL	USA	5.5	1.0	4.5
7	Georgia Inst. Technol.	USA	4.5	0.7	3.8
8	Helsinki Univ.	Finland	4.0	0.8	3.2
9	Univ. Calif. Riverside	USA	4.0	0.5	3.5
10	Dartmouth Coll.	USA	4.0	0.3	3.7
11	Beijing Forestry Univ.	China	3.5	3.6	−0.1
12	NC State Univ.	USA	3.5	1.4	2.1
13	VTT Tech. Res. Ctr.	Finland	3.5	0.8	2.7
14	CIEMAT	Spain	3.5	0.5	3.0
15	Purdue Univ.	USA	3.5	0.3	3.2
16	Univ. Vigo	Spain	3.0	1.4	1.6
17	Oak Ridge Natl. Lab.	USA	3.0	1.4	1.6
18	Auburn Univ.	USA	2.5	0.5	2.0

TABLE 51.4
Most Prolific Funding Bodies in Wood-based Bioethanol Fuels

No.	Funding Bodies	Country	Sample Paper No. (%)	Population Paper No. (%)	Surplus (%)
1	Natl. Natr. Sci. Found.	China	4.0	8.7	−4.7
2	US Dept. Energy	USA	3.5	2.3	1.2
3	Eur. Comm.	EU	3.5	1.6	1.9
4	Natr. Sci. Eng. Res. Counc.	Canada	3.0	1.7	1.3
5	Natr. Resourc. Canada	Canada	3.0	0.4	2.6
6	US Dept. Agric.	USA	2.0	1.2	0.8
7	Govnt. Canada	Canada	2.0	1.0	1.0
8	Swedish Energ. Agcy.	Sweden	2.0	0.6	1.4
9	Seventh Frame. Prog.	EU	2.0	0.3	1.7
10	Swedish Natl. Brd. Ind. Tech. Devnt.	Sweden	2.0	0.1	1.9
11	Biol. Env. Res.	USA	1.5	0.9	0.6
12	Eur. Reg. Devnt. Fund	EU	1.5	1.3	0.2
13	Natl. Sci. Found.	USA	1.5	1.2	0.3
14	Natl. Sci. Found. Jiangsu Prov.	China	1.5	0.6	0.9
15	US Forest Serv.	USA	1.5	0.4	1.1

The funding body with the most citation impact is Natural Resources Canada with 2.6% surplus, while the other influential funding bodies are Natural Resources Canada, Swedish National Board for Industrial and Technical Development, and Seventh Framework Program of the EU with 1.7% – 1.9% surplus each. Similarly, the funding body with the least citation impact is the National Natural Science Foundation of China with 4.7% deficit.

The other funding bodies with a relatively low citation impact and with 0.5%–1.5% of the population papers each are the Ministry of Education, Culture, Sports, Science and Technology, Japan Society for the Promotion of Science, National Key Research and Development Program of China, Fundamental Research Funds for the Central Universities, Ministry of Education of China, Ministry of Science and Technology of China, Office of Science, Brazilian National Council for Scientific and Technological Development, China Scholarship Council, National Research Foundation of Korea, Ministry of Economics and competitiveness, Coordination for the Improvement of Higher Education Personnel (CAPES), National Institute of Food and Agriculture, Priority Academic Program Development of Jiangsu Higher Education Institutions, Nanjing Forestry University, Ministry of Finance, Sao Paulo Research Foundation, Foundation for Science and Technology, State Key Laboratory of Pulp and Paper Engineering, and China Postdoctoral Science Foundation.

51.3.6 THE MOST PROLIFIC SOURCE TITLES IN THE WOOD-BASED BIOETHANOL FUELS

Information about the 19 most prolific source titles publishing at least 1.5% of the sample papers each in wood-based bioethanol fuels is given in Table 51.5.

The most prolific source title is Bioresource Technology with 17% of the sample papers, followed by Biotechnology and Bioengineering with 12% of the sample papers. The other prolific titles are Enzyme and Microbial Technology, Applied Biochemistry and Biotechnology Part A Enzyme Engineering and Biotechnology, Biotechnology Progress, and Green Chemistry with 5.5% – 7.0% of the sample papers each.

However, the source title with the most citation impact is Biotechnology and Bioengineering with 10.8% surplus, followed by Bioresource Technology with 9.7% surplus. The other influential

TABLE 51.5
Most Prolific Source Titles in Wood-based Bioethanol Fuels

No.	Source Titles	Sample Papers (%)	Population Papers (%)	Surplus (%)
1	Bioresource Technology	17.0	7.3	9.7
2	Biotechnology and Bioengineering	12.0	1.2	10.8
3	Enzyme and Microbial Technology	7.0	0.8	6.2
4	Applied Biochemistry and Biotechnology Part A Enzyme Engineering and Biotechnology	6.0	1.1	4.9
5	Biotechnology Progress	6.0	0.7	5.3
6	Green Chemistry	5.5	0.9	4.6
7	Industrial and Engineering Chemistry Research	4.0	1.2	2.8
8	Biotechnology for Biofuels	3.0	1.6	1.4
9	Biomass and Bioenergy	3.0	1.4	1.6
10	Applied Biochemistry and Biotechnology	3.0	1.4	1.6
11	Applied Microbiology and Biotechnology	2.5	0.3	2.2
12	Industrial Crops and Products	2.0	2.6	−0.6
13	Bioresources	1.5	4.9	−3.4
14	Holzforschung	1.5	3.8	−2.3
15	Wood Science and Technology	1.5	2.5	−1.0
16	Journal of Agricultural and Food Chemistry	1.5	0.6	0.9
17	Process Biochemistry	1.5	0.5	1.0
18	Chemsuschem	1.5	0.3	1.2
19	Journal of Biotechnology	1.5	0.2	1.3

titles are Enzyme and Microbial Technology, Biotechnology Progress, Applied Biochemistry and Biotechnology Part A Enzyme Engineering and Biotechnology, and Green Chemistry with 4.6%–6.2% surplus each. Similarly, the source title with the least impact is Bioresources with 3.5% deficit, followed by Holzforschung, Wood Science and Technology, and Industrial Crops and Products with 0.6%–2.3% deficit each.

The other source titles with a relatively low citation impact with 0.5%–1.9% of the population paper each are the Journal of Wood Chemistry and Technology, Journal of Wood Science, ACS Sustainable Chemistry and Engineering, Cellulose, European Journal of Wood and Wood Products, Wood Research, Journal of Agricultural and Food Chemistry, Carbohydrate Polymers, Cellulose Chemistry and Technology, Journal of the American Chemical Society, Biomass Conversion and Biorefinery, Chemistry of Natural Compounds, RSC Advances, Agricultural and Biological Chemistry, Journal of Applied Polymer Science, Journal of Chemical Technology and Biotechnology, Nature, Wood and Fiber Science, Fuel, Industrial and Engineering Chemistry, International Journal of Biological Macromolecules, Scientific Reports, Science of the Total Environment, Energy and Fuels, Polymers, Biotechnology Letters, and Renewable Energy.

51.3.7 THE MOST PROLIFIC COUNTRIES IN THE WOOD-BASED BIOETHANOL FUELS

Information about the most prolific 12 countries publishing at least 2% of sample papers each in wood-based bioethanol fuels is given in Table 51.6.

The most prolific country is the USA with 37.5% of the sample papers, followed by Sweden, Canada, and China with 15.5%, 15%, and 12.5% of the sample papers, respectively. Finland and Spain are the other prolific countries with 8 and 7.5% of the sample papers, respectively. Further, six European countries listed in Table 51.6 produce 38% and 21% of the sample and population papers, respectively.

However, the country with the most citation impact is the USA with 18.3% surplus, followed by Sweden and Canada with 10% and 7.1% surplus, respectively. The other influential countries are Finland and Spain with 3.3% and 2.7% surplus, respectively. Similarly, the country with the least citation impact is China with 7.1% deficit, while China and Japan are countries with the least influence with 5.7% and 4.8% deficit, respectively. The other least influential countries are Brazil, Germany, and the UK with 0.4%–1.4% deficit each.

Additionally, there are other countries with relatively low citation impact and with 0.5%–3.2% of the sample papers each: South Korea, France, India, Russia, Italy, Australia, Slovakia, Turkey,

TABLE 51.6
Most Prolific Countries in Wood-based Bioethanol Fuels

No.	Countries	Sample Papers (%)	Population Papers (%)	Surplus (%)
1	USA	37.5	19.2	18.3
2	Sweden	15.5	5.5	10.0
3	Canada	15.0	7.9	7.1
4	China	12.5	18.2	−5.7
5	Finland	8.0	4.7	3.3
6	Spain	7.5	4.8	2.7
7	Japan	4.0	8.8	−4.8
8	Germany	2.5	2.9	−0.4
9	Austria	2.5	0.9	1.6
10	Brazil	2.0	3.4	−1.4
11	UK	2.0	2.4	−0.4
12	New Zealand	2.0	1.2	0.8

Poland, Chile, Portugal, Czech Republic, Malaysia, Indonesia, Switzerland, Romania, Netherlands, Belgium, Denmark, Taiwan, Thailand, Norway, Iran, Slovenia, Argentina, Mexico, and Hungary.

51.3.8 THE MOST PROLIFIC SCOPUS SUBJECT CATEGORIES IN THE WOOD-BASED BIOETHANOL FUELS

Information about the most prolific nine Scopus subject categories indexing at least 7% of the sample papers each is given in Table 51.7.

The most prolific Scopus subject category in wood-based bioethanol fuels is Chemical Engineering with 64% of the sample papers, followed by Biochemistry, Genetics and Molecular Biology, Immunology and Microbiology, Environmental Science, and Energy with 29%–49% of the sample papers each. It is notable that Social Sciences including Economics and Business account only for 2.9% of the population studies.

However, the Scopus subject category with the most citation impact is Immunology and Microbiology with 29.3% surplus each, followed by Biochemistry, Genetics and Molecular Biology and Chemical Engineering with 28.3% and 27.7% surplus, respectively. Similarly, the least influential subject categories are Materials Science and Agricultural and Biological Sciences with 16.5% and 15.3% deficit, respectively. The other least influential subject categories are Chemistry and Engineering with 12.2% and 4.1% deficit, respectively.

51.3.9 THE MOST PROLIFIC KEYWORDS IN THE WOOD-BASED BIOETHANOL FUELS

Information about the Scopus keywords used with at least 6% or 2.1% of the sample or population papers, respectively, is given in Table 51.8. For this purpose, keywords related to the keyword set given in the appendix are selected from a list of the most prolific keyword set provided by the Scopus database.

These keywords are grouped under eight headings: biomass, fermentation, bacteria, hydrolysis, hydrolysates, pretreatments, other processes, and products.

The most prolific keyword related to the biomass and biomass constituents is lignin with 88% of the sample papers, followed by cellulose, wood, biomass, hemicellulose, lignocellulose, and softwoods with 23%–55% of the sample papers each. Furthermore, the most prolific keywords related to the bacteria are saccharomyces and fungi with 12 and 10% of the sample papers, respectively.

The most prolific keyword related to hydrolysis is hydrolysis with 60% of the sample papers, followed by enzymatic hydrolysis with 31% of the sample papers. Furthermore, the most prolific keyword related to fermentation is fermentation with 27% of the sample papers.

TABLE 51.7
Most Prolific Scopus Subject Categories in Wood-based Bioethanol Fuels

No.	Scopus Subject Categories	Sample Papers (%)	Population Papers (%)	Surplus (%)
1	Biochemistry, Genetics and Molecular Biology	48.5	20.0	28.5
2	Immunology and Microbiology	40.5	11.2	29.3
3	Chemical Engineering	64.0	36.3	27.7
4	Environmental Science	36.5	27.7	8.8
5	Agricultural and Biological Sciences	9.0	24.3	−15.3
6	Energy	29.0	20.0	9.0
7	Chemistry	12.0	24.2	−12.2
8	Engineering	8.0	12.1	−4.1
9	Materials Science	7.0	23.5	−16.5

TABLE 51.8

The Most Prolific Keywords in Wood-based Bioethanol Fuels

No.	Keywords	Sample Papers (%)	Population Papers (%)	Surplus (%)
1	Biomass and biomass constituents			
	Lignin	80.0	28.7	51.3
	Cellulose	54.5	22.9	31.6
	Wood	47.0	29.0	18.0
	Biomass	33.5	14.9	18.6
	Hemicellulose	27.5	9.2	18.3
	Lignocellulose	25.5	5.3	20.2
	Softwoods	23.0	6.9	16.1
	Eucalyptus	14.0	6.8	7.2
	Spruce	14.0	4.6	9.4
	Pine	12.5	5.3	7.2
	Poplar	11.0	7.6	3.4
	Xylan	9.5	2.9	6.6
	Hardwoods	8.0	4.4	3.6
	Bamboo	3.5	5.7	−2.2
	Wood products		5.2	−5.2
2	Fermentation			
	Fermentation	26.5	10.2	16.3
	Furfural	6.0	2.6	3.4
3	Bacteria			
	Fungi	12.0	3.2	8.8
	S. cerevisiae	10.5	2.4	8.1
	Yeast	6.5	3.0	3.5
4	Hydrolysis			
	Hydrolysis	59.5	19.4	40.1
	Enzymatic hydrolysis	30.5	12.9	17.6
	Saccharification	15.5	8.1	7.4
5	Hydrolysates			
	Sugar	31.5	11.4	20.1
	Glucose	26.5	10.1	16.4
	Xylose	12.5	4.3	8.2
	Polysaccharides	7.5	2.9	4.6
6	Biomass pretreatment			
	Enzyme activity	30.5	10.0	20.5
	Enzymes	26.5	6.2	20.3
	Cellulases	26.5	5.2	21.3
	Steam	21.5	6.8	14.7
	Pre-treatment	17.5	8.9	8.6
	Pretreatment	17.0	7.7	9.3
	Sulfuric acid	15.5	3.6	11.9
	Temperature	12.5	4.8	7.7
	Water vapor	12.5	2.5	10.0
	Water	9.0	5.3	3.7
	Acetic acid	9.0	4.2	4.8

(Continued)

TABLE 51.8 (*Continued*)
The Most Prolific Keywords in Wood-based Bioethanol Fuels

No.	Keywords	Sample Papers (%)	Population Papers (%)	Surplus (%)
	Acids	8.0	1.7	6.3
	Trichoderma	7.5	0.0	7.5
	Solvent	7.0	5.5	1.5
	β-glucosidase	6.5	0.0	6.5
	Hypocrea	6.5	0.0	6.5
	Ionic liquids	5.0	6.1	−1.1
	Enzymolysis	5.0	3.8	1.2
	pH	4.5	3.9	0.6
	Microwaves		2.2	−2.2
	Hydrogen peroxide		2.1	−2.1
7	Other processes			
	Biotechnology	20.5	4.0	16.5
	Adsorption	8.5	2.1	6.4
	Fractionation	8.0	3.7	4.3
	Dissolution	8.0	2.9	5.1
	Delignification	7.5	7.1	0.4
	Degradation	6.5	3.8	2.7
	Dissolving	6.5	1.6	4.9
	Extraction	5.5	5.7	−0.2
8	Products			
	Ethanol	36.0	12.1	23.9
	Alcohol	22.0	5.2	16.8
	Bioethanol		4.8	−4.8

The most prolific keyword related to hydrolysates is sugar with 32% of the sample papers, followed by glucose with 27% of the sample papers. Further, the most prolific keyword related to biomass pretreatments is enzyme activity with 31% of the sample papers, followed by enzymes and cellulases with 27% of the sample papers each. The other prolific keywords are steam, pre-treatment, pretreatment, and sulfuric acid with 16%–27% of the Sample papers each.

The most prolific keyword related to the other processes is biotechnology with 21% of the sample papers, followed by biotechnology with 12% of the sample papers. Finally, the most prolific keyword related to the products is ethanol with 36% of the sample papers, followed by alcohol with 22% of the sample papers. It is notable that bioethanol accounts only for 5% of the population papers.

However, the most influential keywords are lignin, hydrolysis, cellulose, ethanol, cellulases, enzyme activity, enzymes, lignocellulose, and sugar with 20%–51% surplus each. Similarly, the least influential keywords are wood products and bioethanol with 5% deficit.

Similarly, the most prolific keywords across all of research fronts are lignin, hydrolysis, cellulose, wood, ethanol, biomass, sugar, enzymatic hydrolysis, enzyme activity, hemicellulose, fermentation, glucose, enzymes, cellulases, and lignocellulose.

51.3.10 THE MOST PROLIFIC RESEARCH FRONTS IN THE WOOD-BASED BIOETHANOL FUELS

Information about research fronts for the sample papers in wood-based bioethanol fuels with regard to the wood biomass used for the bioethanol fuels is given in Table 51.9.

TABLE 51.9

Most Prolific Research Fronts for the Wood-based Bioethanol Fuels

No.	Research Fronts	N Paper (%) Sample
1	Hardwood	49.0
	Poplar	18.0
	Eucalyptus	9.5
	Aspen	5.0
	Hardwood in general	4.5
	Birch	4.0
	Bamboo	3.5
	Willow	3.5
	Other hardwoods	1.0
2	Softwood	35.5
	Pine	13.0
	Spruce	9.5
	Softwood in general	8.5
	Douglas fir	4.5
3	Wood in general	20.0

N paper (%) sample, the number of papers in the population sample of 200 papers.

As this table shows, there are three primary research fronts for this field with respect to wood biomass: hardwood, softwood, and wood in general with 49%, 36%, and 20% of the sample papers, respectively. Furthermore, the most prolific hardwood is popular with 18% of the sample papers, followed by eucalyptus with 10% of the sample papers. However, the most prolific softwood is pine with 13% of the sample papers, followed by spruce with 10% of the sample papers.

Information about the thematic research fronts for the sample papers in wood-based bioethanol fuels is given in Table 51.10. As this table shows, the most prolific research front is the wood biomass pretreatments with 87.5% of the sample papers, followed by wood biomass hydrolysis with 62.5% of the sample papers. The other prolific research fronts are fermentation of the wood hydrolysates and bioethanol production in general with 18% of the sample papers each.

Furthermore, the most prolific pretreatments are chemical and hydrothermal pretreatments and to a lesser extent mechanical and hydrothermal pretreatments of the wood, while the most prolific hydrolysis is the enzymatic pretreatment of the wood. Furthermore, on individual terms, the most prolific pretreatments are steam explosion, acid, solvent, alkaline, sulfur dioxide, liquid hot water, ionic liquid, milling, autohydrolysis, sulfite, hot compressed water, and ammonia pretreatments.

51.4 DISCUSSION

51.4.1 Introduction

Crude oil-based gasoline fuels have been widely used in the transportation sector since the 1920s. However, there have been great public concerns over the adverse environmental and human impact of these fuels. Hence, biomass-based bioethanol fuels have increasingly been used in blending gasoline fuels, in fuel cells, and in biochemical production in a biorefinery context.

However, it is necessary to pretreat the biomass to enhance the yield of the bioethanol before the bioethanol production through hydrolysis and fermentation. One of the most-studied biomass for the bioethanol production has been wood. Research in the field of wood-based bioethanol fuels has intensified

TABLE 51.10

Most Prolific Thematic Research Fronts for the Wood-based Bioethanol Fuels

No.	Research Fronts	N Paper (%) Sample
1	Wood Biomass Pretreatments	87.5
	Mechanical pretreatments	8.0
	Milling pretreatments	6.0
	Other pretreatments	2.0
	Hydrothermal pretreatments	42.0
	Steam explosion pretreatment	27.0
	Liquid hot water pretreatment	7.0
	Autohydrolysis pretreatment	5.0
	Hot compressed water pretreatment	2.5
	Wet oxidation pretreatment	0.5
	Chemical pretreatments	60.0
	Acid pretreatment	24.5
	Solvent pretreatment	12.0
	Alkaline pretreatment	9.5
	Sulfur dioxide pretreatment	7.5
	Ionic liquid pretreatment	6.5
	Sulfite pretreatment	4.0
	Ammonia pretreatment	2.5
	Other chemical pretreatments	2.5
	Enzymatic pretreatments	7.0
2	Wood Biomass Hydrolysis	62.5
	Enzymatic hydrolysis	53.0
	Autohydrolysis	5.0
	Acid hydrolysis	4.5
3	Wood Hydrolysate Fermentation	18.0
	Fermentation inhibitors	7.0
	SSF	6.5
	Hydrolysate detoxification	2.5
	Fermentation in general	2.0
4	Ethanol Production in General	18.0

N paper (%) sample, the number of papers in the population sample of 200 papers.

in this context in the key research fronts of the pretreatment and hydrolysis of the wood, fermentation of the wood hydrolysates, and the bioethanol production from the wood in general. Furthermore, both the softwood such as pine and hardwood such as poplar have been studied intensively.

However, it is essential to develop efficient incentive structures for the primary stakeholders to enhance research in this field. This is especially important to maintain energy security in the cases of supply shocks such as oil price shocks, war-related shocks as in the case of the Russian invasion of Ukraine starting, or COVID-19 shocks.

The scientometric analysis has been used in this context to inform the primary stakeholders about the current state of research in a selected research field. As there has been no scientometric study in this field, this book chapter presents a scientometric study of research in wood-based bioethanol fuels. It examines the scientometric characteristics of both the sample and population data presenting the scientometric characteristics of these both datasets in the order of documents, authors, publication years, institutions, funding bodies, source titles, countries, Scopus subject categories, Scopus keywords, and research fronts.

As a first step for the search of the relevant literature, the keywords were selected using the 200 first most-cited papers. The selected keyword list was then optimized to obtain a representative sample of papers for the searched research field. A copy of this extended keyword list was provided in the appendix for future replicative studies. Furthermore, a selected list of the keywords was presented in Table 51.8.

As a second step, two sets of data were used for this study. First, a population sample of 5,460 papers was used to examine the scientometric characteristics of the population data. Second, a sample of 200 most-cited papers, corresponding to 3.7% of the population dataset, was used to examine the scientometric characteristics of these citation classics.

The scientometric characteristics of these sample and population datasets were presented in the order of documents, authors, publication years, institutions, funding bodies, source titles, countries, Scopus subject categories, Scopus keywords, and research fronts.

Lastly, the key scientometric findings for both datasets were discussed to highlight research landscape for wood-based bioethanol fuels. Additionally, a number of brief conclusions were drawn and a number of relevant recommendations were made to enhance the future research landscape.

51.4.2 THE MOST PROLIFIC DOCUMENTS IN THE WOOD-BASED BIOETHANOL FUELS

Articles (together with conference papers) dominate both the sample (96%) and population (97%) papers (Table 51.1). Furthermore, review papers and articles have a surplus (3%) and deficit (1%), respectively. The representation of reviews in the sample papers is relatively modest (4%).

Scopus differs from the Web of Science database in differentiating and showing articles (89%) and conference papers (8%) published in journals separately. However, it should be noted that these conference papers are also published in journals as articles, compared with those published only in the conference proceedings. Hence, the total number of articles and review papers in the sample dataset is 96% and 4%, respectively.

It is observed during the search process that there has been inconsistency in the classification of the documents in Scopus and in other databases such as Web of Science. This is especially relevant for the classification of papers as reviews or articles as the papers not involving a literature review may be erroneously classified as a review paper. There is also a case of review papers being classified as articles. For example, the total number of reviews in the sample dataset was manually found as nearly 3% compared with 4% as indexed by Scopus, increasing the number of articles and conference papers to 97% for the sample datasets.

In this context, it would be helpful to provide a classification note for the published papers in the books and journals at the first instance. It would also be helpful to use the document types listed in Table 51.1 for this purpose. Book chapters may also be classified as articles or reviews as an additional classification to differentiate review chapters from experimental chapters as it is done by the Web of Science. It would be further helpful to additionally classify the conference papers as articles or review papers and it is done in the Web of Science database.

51.4.3 THE MOST PROLIFIC AUTHORS IN THE WOOD-BASED BIOETHANOL FUELS

There have been 18 most prolific authors with at least 2.5% of the sample papers each as given in Table 51.2. These authors have shaped the development of research in this field.

The most prolific authors are Guido Zacchi, Xuejun Pan, and to a lesser extent Mats Galbe, Junyong Zhu, David J. Gregg, Runchang Sun, and Charles E. Wyman.

It is important to note the inconsistencies in indexing of the author names in Scopus and other databases. It is especially an issue for names with more than two components such as 'Blake Sam de

Hyun Pan'. The probable outcomes are 'Pan, B.S.D.H.', 'de Hyun Pan, B.S.', or 'Hyun Pan, B.S.D.'. The first choice is the gold standard of the publishing sector as the last word in the name is taken as the last name. In most of the academic databases such as PubMed and EBSCO, this version is used predominantly. The second choice is a strong alternative, while the last choice is an undesired outcome as two last words are taken as the last name. It is good practice to combine the words of the last name by a hyphen: 'Hyun-Pan, B.S.D.'. It is notable that inconsistent indexing of the author names may cause substantial inefficiencies in the search process for the papers and allocating credit to the authors as there are different author entries for each outcome in the databases.

There are also inconsistencies in the shortening of Chinese names. For example, 'YangYing Pan' is often shortened as 'Pan, Y.', 'Pan, Y.-Y.', and 'Pan, Y.Y.' as it is done in the Web of Science database as well. However, the gold standard in this case is 'Pan, Y' where the last word is taken as the last name and the first word is taken as a single forename. In most of the academic databases such as PubMed and EBSCO, this first version is used predominantly. However, it makes sense to use the third option to differentiate Chinese names efficiently: 'Pan, Y.Y.'. Therefore, there have been difficulties in locating papers for Chinese authors. In such cases, the use of the unique author codes provided for each author by the Scopus database has been helpful.

There is also a difficulty in allowing credit for the authors, especially for the authors with common names such as 'Pan, X.' in conducting scientometric studies. These difficulties strongly influence the efficiency of the scientometric studies and allocating credit to the authors as there are the same author entries for different authors with the same name, e.g., 'Pan, X.' in the databases.

In this context, the coding of authors in the Scopus database is a welcome innovation compared with other databases such as Web of Science. In this process, Scopus allocates a unique number to each author in the database (Aman, 2018). However, there might still be substantial inefficiencies in this coding system, especially for common names. For example, some of the papers for a certain author may be allocated to another researcher with a different author code. It is possible that Scopus uses a number of software programs to differentiate the author names and the program may not be false-proof (Kim, 2018).

In this context, it does not help that author names are not given in full in some journals and books. This makes it difficult to differentiate authors with common names and makes scientometric studies further difficult in the author domain. Therefore, the author names should be given in all books and journals at the first instance. There is also a cultural issue where some authors do not use their full names in their papers. Instead, they use initials for their forenames: 'Pan, H.J.', 'Pan', 'Pan, H.', or 'Pan, J.' instead of 'Pan, Hyun Jae'.

There are also inconsistencies in the naming of the authors with more than two components by the authors themselves in journal papers and book chapters. For example, 'Pan, APC' might be given as 'Pan, A' or 'Pan, A.C.' or 'Pan, A.P.' or 'Pan, C.' in journals and books. This also makes scientometric studies difficult in the author domain. Hence, contributing authors should use their name consistently in their publications.

The other critical issue regarding the author names is the inconsistencies in the spelling of the author names in the national spellings (e.g., Göğüşçöl, Gökçe) rather than in English spellings (e.g., Goguscol, Gokce) in Scopus database. Scopus differs from the Web of Science database and many other databases in this respect where the author names are given only in English spellings. It is observed that national spellings of the author names do not help much in conducting scientometric studies and in allocating credits to the authors as sometimes there are different author entries for the English and National spellings in the Scopus database.

The most prolific institutions for the sample dataset are the University of British Columbia and to a lesser extent Lund University, University of Vigo, and The U.S. Department of Agriculture (USDA) Forest Service.

The most prolific countries for the sample dataset are the USA and to a lesser extent Spain, Canada, and Sweden. These findings confirm the dominance of the USA, Europe, and to a lesser extent of Canada in this field. The most prolific research fronts are the bioethanol production, biomass hydrolysis, hydrolysate fermentation, and biomass pretreatments.

It is also notable that there is a significant gender deficit for the sample dataset as surprisingly with a representation rate of 17%. This finding is the most thought-provoking with strong public policy implications. Hence, institutions, funding bodies, and policymakers should take efficient measures to reduce the gender deficit in this field and other scientific fields with strong gender deficit. In this context, it is worth to note the level of representation of the researchers from minority groups in science based on race, sexuality, age, and disability, besides gender (Blankenship, 1993; Dirth and Branscombe. 2017; Konur, 2000, 2002a,b,c, 2006a,b, 2007a,b).

51.4.4   THE MOST PROLIFIC RESEARCH OUTPUT BY YEARS IN THE WOOD-BASED BIOETHANOL FUELS

Research output observed between 1970 and 2022 is illustrated in Figure 51.1. This figure clearly shows that the bulk of research papers in the population dataset were published primarily in the 2010s. Similarly, the bulk of research papers in the sample dataset were published in the last two decades.

These findings suggest that the most prolific sample and population papers were primarily published in the last two decades. These are the thought-provoking findings as there has been a significant research boom in the last two decades. In this context, the increasing public concerns about climate change (Change, 2007), greenhouse gas emissions (Carlson et al., 2017), and global warming (Kerr, 2007) have been certainly behind the boom in research in this field in the last two decades. Furthermore, the supply shocks experienced due to the COVID-19 pandemic might also be behind research boom in this field since 2019.

Based on these findings, the size of the population papers is likely to more than double in the current decade, provided that the public concerns about climate change, greenhouse gas emissions, and global warming, as well as the supply shocks, are translated efficiently to research funding in this field.

51.4.5   THE MOST PROLIFIC INSTITUTIONS IN THE WOOD-BASED BIOETHANOL FUELS

The 18 most prolific institutions publishing papers on wood-based bioethanol fuels with at least 2.5% of the sample papers each given in Table 51.3 have shaped the development of research in this field.

The most prolific institutions are the University of British Columbia, Lund University, and to a lesser extent University of Wisconsin Madison, USDA Forest Service, South China University of Technology, NREL, and Georgia Institute of Technology.

Similarly, the top countries for these most prolific institutions are the USA and to a lesser extent China, Finland, and Spain. In total, six countries house these top institutions.

However, the institutions with the most citation impact are University of British Columbia, Lund University, and to a lesser extent University of Wisconsin Madison, USDA Forest Service, and NREL. These findings confirm the dominance of the US, European, and to a lesser extent of Canadian and Chinese institutions.

51.4.6   THE MOST PROLIFIC FUNDING BODIES IN THE WOOD-BASED BIOETHANOL FUELS

The 15 most prolific funding bodies funding at least 1.3% of the sample papers each are given in Table 51.4. It is notable that only 34% and 41% of the sample and population papers each were funded.

The most prolific funding bodies are the National Natural Science Foundation of China, European Commission, and US Department of Energy. The most prolific countries for these top funding bodies are the USA and to a lesser extent Canada, Europe, China, and Sweden. In total, only four countries and the EU house these top funding bodies.

These findings on the funding of research in this field suggest that the level of funding, mostly in the last two decades, is highly modest, and it has been largely instrumental in enhancing research in this field (Ebadi and Schiffauerova, 2016) in light of North's institutional framework (North, 1991). However, considering the relatively low levels of funding there is ample room to enhance funding in this field.

51.4.7   The Most Prolific Source Titles in the Wood-based Bioethanol Fuels

The 19 most prolific source titles publishing at least 1.5% of the sample papers each in wood-based bioethanol fuels have shaped the development of research in this field (Table 51.5).

The most prolific source titles are Bioresource Technology, Biotechnology and Bioengineering, and to a lesser extent Enzyme and Microbial Technology, Applied Biochemistry and Biotechnology Part A Enzyme Engineering and Biotechnology, Biotechnology Progress, and Green Chemistry.

The source titles with the most citation impact are Biotechnology and Bioengineering, Bioresource Technology and to a lesser extent Enzyme and Microbial Technology, Biotechnology Progress, Applied Biochemistry and Biotechnology Part A Enzyme Engineering and Biotechnology, and Green Chemistry. Similarly, the source titles with the least impact are Bioresources and to a lesser extent Holzforschung, Wood Science and Technology, and Industrial Crops and Products.

It is notable that these top source titles are primarily related to bioresources, bioengineering, enzymes, and biotechnology. This finding suggests that Bioresource Technology, Biotechnology and Bioengineering, and the other prolific journals in these fields have significantly shaped the development of research in this field as they focus primarily on wood-based bioethanol fuels with a high yield.

51.4.8   The Most Prolific Countries in the Wood-based Bioethanol Fuels

The 12 most prolific countries publishing at least 2% of the sample papers each have significantly shaped the development of research in this field (Table 51.6).

The most prolific countries are the USA and to a lesser extent Sweden, Canada, and China, Finland, and Spain. Furthermore, six European countries listed in Table 51.6 produce 38% and 21% of the sample and population papers, respectively.

However, countries with the most citation impact are the USA and to a lesser extent Sweden and Canada, Finland, and Spain. Similarly, countries with the least impact are China, Japan, and to a lesser extent Brazil, Germany, and the UK.

A close examination of these findings suggests that the USA, Europe, and to a lesser extent China and Canada are the major producers of research in this field. It is a fact that the USA has been a major player in science (Leydesdorff and Wagner, 2009). The USA has further developed a strong research infrastructure to support its corn and grass-based bioethanol industry (Gillon, 2010).

However, China has been a rising mega star in scientific research in competition with the USA and Europe (Leydesdorff and Zhou, 2005). China is also a major player in this field as a major producer of bioethanol (Fang et al., 2010).

Next, Europe has been a persistent player in scientific research in competition with both the USA and China (Leydesdorff, 2000). Europe has also been a persistent producer of bioethanol along with the USA and Brazil (Gnansounou, 2010).

51.4.9   The Most Prolific Scopus Subject Categories in
the Wood-based Bioethanol Fuels

The nine most prolific Scopus subject categories indexing at least 7% of the sample papers each, respectively, given in Table 51.7 have shaped the development of research in this field.

The most prolific Scopus subject categories in wood-based bioethanol fuels are Chemical Engineering and to a lesser extent Biochemistry, Genetics and Molecular Biology, Immunology and Microbiology, Environmental Science, and Energy.

The Scopus subject categories with the most citation impact are Immunology and Microbiology and to a lesser extent Biochemistry, Genetics and Molecular Biology, and Chemical Engineering. Similarly, the least influential subject categories are Materials Science and Agricultural and Biological Sciences and to a lesser extent Chemistry and Engineering.

These findings are thought-provoking suggesting that the primary subject categories are related to chemical engineering, genetics, biochemistry, environmental science, and energy as the core of research in this field concerns wood-based ethanol production to increase the ethanol yield. The other finding is that social sciences are not well represented in both the sample and population papers as in most fields in bioethanol fuels.

51.4.10 THE MOST PROLIFIC KEYWORDS IN THE WOOD-BASED BIOETHANOL FUELS

A limited number of keywords have shaped the development of research in this field as shown in Table 51.8 and the Appendix. These keywords are grouped under eight headings: biomass, fermentation, bacteria, hydrolysis, hydrolysates, pretreatments, other processes, and products of fermentation.

Similarly, the most prolific keywords across all of research fronts are lignin, hydrolysis, cellulose, wood, ethanol, biomass, sugar, enzymatic hydrolysis, enzyme activity, hemicellulose, fermentation, glucose, enzymes, cellulases, and lignocellulose.

However, the most influential keywords are lignin, hydrolysis, cellulose, ethanol, cellulases, enzyme activity, enzymes, lignocellulose, and sugar. Similarly, the least influential keywords are wood products and bioethanol.

These findings suggest that it is necessary to determine the keyword set carefully to locate the relevant research in each of these research fronts. Additionally, the size of the samples for each keyword highlights the intensity of research in the relevant research areas.

51.4.11 THE MOST PROLIFIC RESEARCH FRONTS IN THE WOOD-BASED BIOETHANOL FUELS

Table 51.9 shows that there are three primary research fronts for this field: hardwood, softwood, and wood in general. Furthermore, the most prolific hardwood is poplar, followed by eucalyptus. However, the most prolific softwood is pine, followed by spruce.

Information about the thematic research fronts for the sample papers in wood-based bioethanol fuels is given in Table 51.10. As this table shows, the most prolific research front is biomass pretreatments, followed by biomass hydrolysis. The other prolific research fronts are fermentation of hydrolysates and bioethanol production.

These findings are thought-provoking in seeking ways to increase bioethanol yield through the wood-based bioethanol production at the global scale. It is clear that all of these research fronts have public importance and merit substantial funding and other incentives. Furthermore, it is notable that wood-based bioethanol fuels have become a core of bioethanol research to increase the ethanol yield and to make them more competitive with crude oil-based gasoline and diesel fuels, especially for countries with large forests.

In the end, these most-cited papers in this field hint that the efficiency of wood-based bioethanol production could be optimized using the structure, processing, and property relationships of these bioethanol production processes in the fronts of biomass pretreatments, biomass hydrolysis, and hydrolysate fermentation (Formela et al., 2016; Konur, 2018a, 2020b, 2021a,b,c,d; Konur and Matthews, 1989).

51.5 CONCLUSION AND FUTURE RESEARCH

Research on wood-based bioethanol fuels has been mapped through a scientometric study of both sample (200 papers) and population (5,460 papers) datasets.

The critical issue in this study has been to obtain a representative sample of research as in any other scientometric study. Therefore, the keyword set has been carefully devised and optimized after a number of runs in the Scopus database. It is a representative sample of the wider population studies. This keyword set was provided in the Appendix, and the relevant keywords are presented in Table 51.8. However, it should be noted that it has been very difficult to compile a representative keyword set since this research field has been connected closely with many other fields. Therefore, it has been necessary to compile a keyword list to exclude papers concerned with the other research fields.

The other issue has been the selection of a multidisciplinary database to carry out the scientometric study of research in this field. For this purpose, the Scopus database has been selected. The journal coverage of this database has been notably wider than that of the Web of Science and other multi-subject databases.

The key scientometric properties of research in this field have been determined and discussed in this book chapter. It is evident that a limited number of documents, authors, institutions, publication years, institutions, funding bodies, source titles, countries, Scopus subject categories, Scopus keywords, and research fronts have shaped the development of research in this field.

There is ample scope to increase the efficiency of the scientometric studies in this field in the author and document domains by developing consistent policies and practices in both domains across all academic databases. In this respect, it seems that authors, journals, and academic databases have a lot to do. Furthermore, the significant gender deficit as in most scientific fields emerges as a public policy issue. The potential deficits based on age, race, disability, and sexuality need also to be explored in this field as in other scientific fields.

Research in this field has boomed in the last two decades possibly promoted by the public concerns on global warming, greenhouse gas emissions, and climate change. Furthermore, the recent COVID-19 pandemic and the Russian invasion of Ukraine have resulted in global supply shocks shifting the focus of the stakeholders from crude oil-based fuels to biomass-based fuels such as bioethanol fuels. It is expected that there would be further incentives for the key stakeholders to carry out research for the wood-based bioethanol production to increase the ethanol yield and to make them more competitive with crude oil-based gasoline and diesel fuels. This might be truer for crude oil- and foreign exchange-deficient countries to maintain energy security in the face of the global supply shocks.

A relatively modest funding rate of 41% for the population papers suggests that funding in this field significantly enhanced research in this field primarily in the last two decades, possibly more than doubling in the current decade. However, it is evident that there is ample room for more funding and other incentives to enhance research in this field further.

The institutions from the USA, Europe, and to a lesser extent China have mostly shaped research in this field. Furthermore, the USA, Europe, and to a lesser extent China have been the major producers of research in this field as the major producers and users of bioethanol fuels from different types of biomass such as corn, sugarcane, and grass, as well as other types of biomass. It is evident that these countries have well-developed research infrastructure in bioethanol fuels and their derivatives.

It emerges that ethanol is more popular than bioethanol as a keyword with strong implications for the search strategy. In other words, the search strategy using only the bioethanol keyword would not be much helpful. The Scopus keywords are grouped under eight headings: biomass, fermentation, bacteria, hydrolysis, hydrolysates, pretreatments, other processes, and products of fermentation.

There are three primary research fronts regarding the wood biomass used for the bioethanol production: hardwood, softwood, and wood in general. Furthermore, the most prolific hardwood is poplar, followed by eucalyptus, while the most prolific softwood is pine, followed by spruce.

Similarly, the most prolific thematic research fronts are the wood biomass pretreatments, followed by wood biomass hydrolysis and to a lesser extent fermentation of the wood hydrolysates and wood-based bioethanol production in general, including the technoeconomic studies of the bioethanol production.

These findings are thought-provoking in seeking ways to increase bioethanol yield from the wood-based bioethanol production. It is clear that all of these research fronts have public importance and merit substantial funding and other incentives. Furthermore, it is notable that the wood-based bioethanol fuels have become a core of bioethanol research to make them more competitive with crude oil-based gasoline and diesel fuels.

Thus, the scientometric analysis has a great potential to gain valuable insights into the evolution of research in this field as in other scientific fields, especially in the aftermath of the significant global supply shocks such as COVID-19 pandemic and the Russian invasion of Ukraine.

It is recommended that further scientometric studies are carried out for the primary research fronts. It is further recommended that reviews of the most-cited papers are carried out for each primary research front to complement these scientometric studies. Next, the scientometric studies of the hot papers in these primary fields are carried out.

ACKNOWLEDGMENTS

The contribution of the highly cited researchers in the field of wood-based bioethanol fuels has been gratefully acknowledged.

APPENDIX: THE KEYWORD SET FOR WOOD-BASED BIOETHANOL FUELS

(TITLE (saccharif* OR hydrolyze* OR prehydroly* OR posthydroly* OR *hydrolysis OR digestibili* OR ssf OR shf OR sscf OR accessibility OR ethanol OR bioethanol OR detoxif* OR recalcitrance OR hydrolysate* OR hydrolyzate* OR inhibitor* OR ferment* OR coferment* OR cellulase* OR fractionation OR delignification OR depolymerization OR microwave* OR ultrasound OR sonicat* OR extrusion OR milling OR grinding OR pretreatment* OR pretreated OR "pre-treat*" OR bioorganosolve OR sugar OR "size reduction" OR steam* OR "hot water" OR "hot compressed" OR acid* OR h2so4 OR "hydrogen peroxide*" OR h2o2 OR hcl OR sulfite* OR sporl OR lime* OR "sulphur dioxide" OR "sulfur dioxide" OR so2 OR alkali* OR naoh* OR "sodium hydroxide*" OR "ionic liquid*" OR solvent* OR organosolv* OR ammonia OR ozon* OR "wet oxidation" OR flowthrough OR "supercritical co2" OR "supercritical carbon dioxide" OR surfactant* OR afex* OR "supercritical water" OR tween*) AND TITLE (wood OR softwood* OR hardwood* OR woody OR cottonwood OR "olive tree*" OR eucalyptus OR pine OR pinus OR araucaria OR poplar* OR aspen* OR populus OR cedar OR cedrus OR spruce OR sweetgum OR picea OR beech OR fagus OR oak OR quercus OR willow* OR salix OR cypress OR cupressus OR prosopis OR birch OR betula OR alder OR alnus OR bamboo* OR phyllostachys OR pseudotsuga OR castanea OR maple* OR acer OR yew OR *taxus OR tsuga OR larch OR larix OR fir OR ulmus OR hickory OR carya OR mahogany OR swietenia OR "palm tree" OR sycamore OR "black locust" OR tamarix OR catalpa)) AND NOT (SUBJAREA (medi OR phar OR vete OR nurs OR neur OR dent OR heal OR psyc) OR SRCTITLE (ecol* OR soil* OR plant* OR botan* OR tree* OR phyto* OR geo* OR forest* OR remote OR oecol* OR coal* OR climat* OR conserv* OR system* OR pyrolysis OR ecog* OR biogeo* OR hazard* OR genetic* OR genome* OR biomed* OR atmosph* OR breed* OR molecular OR composite* OR mycol* OR material* OR chromat* OR pulp OR drying OR appita OR biodeterioration OR "organic chem*" OR feed* OR pollut* OR hort* OR building* OR carbon OR corrosion OR hydrol* OR pollut*) OR TITLE (nanofib* OR litter* OR adsorb* OR diels OR spectral OR antioxidant* OR hydrochars OR algorithm* OR taxol OR *dna OR adhesive* OR pcr OR methanolysis OR fibers OR miscanthus OR frog OR papermaking OR activated OR soil* OR nanotube* OR mushroom* OR nanopart* OR phyto* OR lignification OR forest* OR liquefaction OR diode* OR hydrocracking OR fruit OR filter* OR leaves OR allerg* OR pyrolysis OR nanocrystal* OR spectro* OR resins OR decolor* OR bark OR *synthesis OR tissue* OR biosorp* OR *bleaching OR leaf OR aryls OR biochar* OR fatty OR biosorb* OR fibre* OR metagen* OR solar OR ptr OR heating OR gasification OR genes OR hardness OR effluent* OR endophyte* OR nematode* OR pathogen OR plus OR chromato* OR termite* OR genom* OR torref* OR {milled wood} OR wastewater* OR dye* OR succinic OR decay* OR stems OR veneer* OR channel* OR kraft OR *coal OR insect* OR {thermal pretreat*} OR liquor OR composites OR fire OR ecol* OR combustion OR biogas OR *alkanoates OR nanowhisk* OR panels OR {non-wood} OR capture OR aerogel OR embryos OR anhydrides OR methane OR esterif* OR wine OR butanol OR derivatives)) AND (LIMIT-TO (SRCTYPE, "j") OR LIMIT-TO (SRCTYPE, "k") OR LIMIT-TO (SRCTYPE, "b")) AND (LIMIT-TO (DOCTYPE, "ar") OR LIMIT-TO (DOCTYPE, "cp") OR LIMIT-TO (DOCTYPE, "re") OR LIMIT-TO (DOCTYPE, "no") OR LIMIT-TO (DOCTYPE, "ch") OR LIMIT-TO (DOCTYPE, "le") OR LIMIT-TO (DOCTYPE, "sh") OR LIMIT-TO (DOCTYPE, "bk") OR LIMIT-TO (DOCTYPE, "ed") OR LIMIT-TO (DOCTYPE, "cr")) AND (LIMIT-TO (LANGUAGE, "English"))

REFERENCES

Aman, V. 2018. Does the Scopus author ID suffice to track scientific international mobility? A case study based on Leibniz laureates. *Scientometrics* 117:705–720.

Angelici, C., B. M. Weckhuysen and P. C. A. Bruijnincx. 2013. Chemocatalytic conversion of ethanol into butadiene and other bulk chemicals. *ChemSusChem* 6:1595–1614.

Antolini, E. 2007. Catalysts for direct ethanol fuel cells. *Journal of Power Sources* 170:1–12.

Antolini, E. 2009. Palladium in fuel cell catalysis. *Energy and Environmental Science* 2:915–931.

Beaudry, C. and V. Lariviere. 2016. Which gender gap? Factors affecting researchers' scientific impact in science and medicine. *Research Policy* 45:1790–1817.

Blankenship, K. M. 1993. Bringing gender and race in: US employment discrimination policy. *Gender & Society* 7:204–226.

Burnham, J. F. 2006. Scopus database: A review. *Biomedical Digital Libraries* 3:1–8.

Carlson, K. M., J. S. Gerber and D. Mueller, et al. 2017. Greenhouse gas emissions intensity of global croplands. *Nature Climate Change* 7:63–68.

Change, C. 2007. Climate change impacts, adaptation and vulnerability. *Science of the Total Environment* 326:95–112.

Dirth, T. P. and N. R. Branscombe. 2017. Disability models affect disability policy support through awareness of structural discrimination. *Journal of Social Issues* 73:413–442.

Ebadi, A. and A. Schiffauerova. 2016. How to boost scientific production? A statistical analysis of research funding and other influencing factors. *Scientometrics* 106:1093–1116.

Fang, X., Y. Shen, J. Zhao, X. Bao and Y. Qu. 2010. Status and prospect of lignocellulosic bioethanol production in China. *Bioresource Technology* 101:4814–4819.

Fauci, A. S., H. C. Lane and R. R. Redfield. 2020. Covid-19-navigating the uncharted. *New England Journal of Medicine* 382:1268–1269.

Fernando, S., S. Adhikari, C. Chandrapal and M. Murali. 2006. Biorefineries: Current status, challenges, and future direction. *Energy & Fuels* 20:1727–1737.

Formela, K., A. Hejna, L. Piszczyk, M. R. Saeb and X. Colom. 2016. Processing and structure-property relationships of natural rubber/wheat bran biocomposites. *Cellulose* 23:3157–3175.

Garfield, E. 1955. Citation indexes for science. *Science* 122:108–111.

Gillon, S. 2010. Fields of dreams: Negotiating an ethanol agenda in the Midwest United States. *Journal of Peasant Studies* 37:723–748.

Gnansounou, E. 2010. Production and use of lignocellulosic bioethanol in Europe: Current situation and perspectives. *Bioresource Technology* 101:4842–4850.

Hahn-Hagerdal, B., M. Galbe, M. F. Gorwa-Grauslund, G. Liden and G. Zacchi. 2006. Bio-ethanol - The fuel of tomorrow from the residues of today. *Trends in Biotechnology* 24:549–556.

Hamilton, J. D. 1983. Oil and the macroeconomy since World War II. *Journal of Political Economy* 91:228–248.

Hamilton, J. D. 2003. What is an oil shock? *Journal of Econometrics* 113:363–398.

Hamilton, J. D. 2009. Causes and consequences of the oil shock of 2007-08. *Brookings Papers on Economic Activity* 2009:215–261.

Hill, J., E. Nelson, D. Tilman, S. Polasky and D. Tiffany. 2006. Environmental, economic, and energetic costs and benefits of biodiesel and ethanol biofuels. *Proceedings of the National Academy of Sciences of the United States of America* 103:11206–11210.

Hill, J., S. Polasky and E. Nelson, et al. 2009. Climate change and health costs of air emissions from biofuels and gasoline. *Proceedings of the National Academy of Sciences of the United States of America* 106:2077–2082.

Hsieh, W. D., R. H. Chen, T. L. Wu and T. H. Lin. 2002. Engine performance and pollutant emission of an SI engine using ethanol-gasoline blended fuels. *Atmospheric Environment* 36:403–410.

Huang, H. J., S. Ramaswamy, U. W. Tschirner and B. V. Ramarao. 2008. A review of separation technologies in current and future biorefineries. *Separation and Purification Technology* 62:1–21.

Jones, T. C. 2012. America, oil, and war in the Middle East. *Journal of American History* 99:208–218.

Jonsson, L. J., E. Palmqvist, N. O. Nilvebrant and B. Hahn-Hagerdal. 1998. Detoxification of wood hydrolysates with laccase and peroxidase from the white-rot fungus *Trametes versicolor*. *Applied Microbiology and Biotechnology* 49:691–697.

Kerr, R. A. 2007. Global warming is changing the world. *Science* 316:188–190.

Kilian, L. 2008. Exogenous oil supply shocks: How big are they and how much do they matter for the US economy? *Review of Economics and Statistics* 90:216–240.

Kilian, L. 2009. Not all oil price shocks are alike: Disentangling demand and supply shocks in the crude oil market. *American Economic Review* 99:1053–1069.

Kilpelainen, I., H. Xie and A. King, et al. 2007. Dissolution of wood in ionic liquids. *Journal of Agricultural and Food Chemistry* 55:9142–9148.

Kim, J. 2018. Evaluating author name disambiguation for digital libraries: A case of DBLP. *Scientometrics* 116:1867–1886.

Konur, O. 2000. Creating enforceable civil rights for disabled students in higher education: An institutional theory perspective. *Disability & Society* 15:1041–1063.

Konur, O. 2002a. Access to nursing education by disabled students: Rights and duties of nursing programs. *Nurse Education Today* 22:364–374.

Konur, O. 2002b. Assessment of disabled students in higher education: Current public policy issues. *Assessment and Evaluation in Higher Education* 27:131–152.

Konur, O. 2002c. Access to employment by disabled people in the UK: Is the Disability Discrimination Act working? *International Journal of Discrimination and the Law* 5:247–279.

Konur, O. 2006a. Participation of children with dyslexia in compulsory education: Current public policy issues. *Dyslexia* 12:51–67.

Konur, O. 2006b. Teaching disabled students in higher education. *Teaching in Higher Education* 11:351–363.

Konur, O. 2007a. A judicial outcome analysis of the *Disability Discrimination Act*: A windfall for the employers? *Disability & Society* 22:187–204.

Konur, O. 2007b. Computer-assisted teaching and assessment of disabled students in higher education: The interface between academic standards and disability rights. *Journal of Computer Assisted Learning* 23:207–219.

Konur, O. 2011. The scientometric evaluation of the research on the algae and bio-energy. *Applied Energy* 88:3532–3540.

Konur, O. 2012a. The evaluation of the biogas research: A scientometric approach. *Energy Education Science and Technology Part A: Energy Science and Research* 29:1277–1292.

Konur, O. 2012b. The evaluation of the educational research: A scientometric approach. *Energy Education Science and Technology Part B: Social and Educational Studies* 4:1935–1948.

Konur, O. 2012c. The evaluation of the global energy and fuels research: A scientometric approach. *Energy Education Science and Technology Part A: Energy Science and Research* 30:613–628.

Konur, O. 2012d. The evaluation of the research on the biodiesel: A scientometric approach. *Energy Education Science and Technology Part A: Energy Science and Research* 28:1003–1014.

Konur, O. 2012e. The evaluation of the research on the bioethanol: A scientometric approach. *Energy Education Science and Technology Part A: Energy Science and Research* 28:1051–1064.

Konur, O. 2012f. The evaluation of the research on the biofuels: A scientometric approach. *Energy Education Science and Technology Part A: Energy Science and Research* 28:903–916.

Konur, O. 2012g. The evaluation of the research on the biohydrogen: A scientometric approach. *Energy Education Science and Technology Part A: Energy Science and Research* 29:323–338.

Konur, O. 2012h. The evaluation of the research on the microbial fuel cells: A scientometric approach. *Energy Education Science and Technology Part A: Energy Science and Research* 29:309–322.

Konur, O. 2012i. The scientometric evaluation of the research on the production of bioenergy from biomass. *Biomass and Bioenergy* 47:504–515.

Konur, O. 2015. Current state of research on algal bioethanol. In *Marine Bioenergy: Trends and Developments*, Ed. S. K. Kim and C. G. Lee, pp. 217–244. Boca Raton, FL: CRC Press.

Konur, O., Ed. 2018a. *Bioenergy and Biofuels*. Boca Raton, FL: CRC Press.

Konur, O. 2018b. Bioenergy and biofuels science and technology: Scientometric overview and citation classics. In *Bioenergy and Biofuels*, Ed. O. Konur, pp. 3–63. Boca Raton: CRC Press.

Konur, O. 2019. Cyanobacterial bioenergy and biofuels science and technology: A scientometric overview. In *Cyanobacteria: From Basic Science to Applications*, Ed. A. K. Mishra, D. N. Tiwari and A. N. Rai, pp. 419–442. Amsterdam: Elsevier.

Konur, O. 2020a. The scientometric analysis of the research on the bioethanol production from green macroalgae. In *Handbook of Algal Science, Technology and Medicine*, Ed. O. Konur, pp. 385–401. London: Academic Press.

Konur, O., Ed. 2020b. *Handbook of Algal Science, Technology and Medicine.* London: Academic Press.

Konur, O., Ed. 2021a. *Handbook of Biodiesel and Petrodiesel Fuels: Science, Technology, Health, and Environment.* Boca Raton, FL: CRC Press.

Konur, O., Ed. 2021b. *Handbook of Biodiesel and Petrodiesel Fuels: Science, Technology, Health, and Environment. Volume 1. Biodiesel Fuels: Science, Technology, Health, and Environment.* Boca Raton, FL: CRC Press.

Konur, O., Ed. 2021c. *Handbook of Biodiesel and Petrodiesel Fuels: Science, Technology, Health, and Environment. Volume 2. Biodiesel Fuels based on the Edible and Nonedible Feedstocks, Wastes, and Algae: Science, Technology, Health, and Environment.* Boca Raton, FL: CRC Press.

Konur, O., Ed. 2021d. *Handbook of Biodiesel and Petrodiesel Fuels: Science, Technology, Health, and Environment. Volume 3. Petrodiesel Fuels: Science, Technology, Health, and Environment.* Boca Raton, FL: CRC Press.

Konur, O. and F. L. Matthews. 1989. Effect of the properties of the constituents on the fatigue performance of composites: A review. *Composites* 20:317–328.

Kruyt, B., D. P. van Vuuren, H. J. de Vries and H. Groenenberg. 2009. Indicators for energy security. *Energy Policy* 37:2166–2181.

Kumar, R., G. Mago, V. Balan and C. E. Wyman. 2009. Physical and chemical characterizations of corn stover and poplar solids resulting from leading pretreatment technologies. *Bioresource Technology* 100:3948–3962.

Larsson, S., A. Reimann, N. O. Nilvebrant and L. J. Jonsson. 1999b. Comparison of different methods for the detoxification of lignocellulose hydrolyzates of spruce. *Applied Biochemistry and Biotechnology - Part A Enzyme Engineering and Biotechnology* 77–79:91–103.

Larsson, S., E. Palmqvist, E. and B. Hahn-Hagerdal, et al. 1999a. The generation of fermentation inhibitors during dilute acid hydrolysis of softwood. *Enzyme and Microbial Technology* 24:151–159.

Leydesdorff, L. 2000. Is the European Union becoming a single publication system? *Scientometrics* 47:265–280.

Leydesdorff, L. and C. Wagner. 2009. Is the United States losing ground in science? A global perspective on the world science system. *Scientometrics* 78:23–36.

Leydesdorff, L. and P. Zhou. 2005. Are the contributions of China and Korea upsetting the world system of science? *Scientometrics* 63:617–630.

Li, H., S. M. Liu, X. H. Yu, S. L. Tang and C. K. Tang. 2020. Coronavirus disease 2019 (COVID-19): Current status and future perspectives. *International Journal of Antimicrobial Agents* 55:105951.

Lin, Y. and S. Tanaka. 2006. Ethanol fermentation from biomass resources: Current state and prospects. *Applied Microbiology and Biotechnology* 69:627–642.

Ma, X., L. Sun and C. Song. 2002. A new approach to deep desulfurization of gasoline, diesel fuel and jet fuel by selective adsorption for ultra-clean fuels and for fuel cell applications. *Catalysis Today* 77:107–116.

Mooney, C. A., S. D. Mansfield, M. G. Touhy and J. N. Saddler. 1998. The effect of initial pore volume and lignin content on the enzymatic hydrolysis of softwoods. *Bioresource Technology* 64:113–119.

Morschbacker, A. 2009. Bio-ethanol based ethylene. *Polymer Reviews* 49:79–84.

Najafi, G., B. Ghobadian and T. Tavakoli, et al. 2009. Performance and exhaust emissions of a gasoline engine with ethanol blended gasoline fuels using artificial neural network. *Applied Energy* 86:630–639.

Newman, P. W. G. and J. R. Kenworthy. 1989. Gasoline consumption and cities: A comparison of U.S. cities with a global survey. *Journal of the American Planning Association* 55:24–37.

North, D. C. 1991. Institutions. *Journal of Economic Perspectives* 5:97–112.

Olsson, L. and B. Hahn-Hagerdal. 1996. Fermentation of lignocellulosic hydrolysates for ethanol production. *Enzyme and Microbial Technology* 18:312–331.

Palonen, H., F. Tjerneld, G. Zacchi and M. Tenkanen. 2004. Adsorption of *Trichoderma reesei* CBH I and EG II and their catalytic domains on steam pretreated softwood and isolated lignin. *Journal of Biotechnology* 107:65–72.

Pan, X., C. Arato and N. Gilkes, et al. 2005. Biorefining of softwoods using ethanol organosolv pulping: Preliminary evaluation of process streams for manufacture of fuel-grade ethanol and co-products. *Biotechnology and Bioengineering* 90:473–481.

Pan, X., N. Gilkes and J. Kadla, et al. 2006. Bioconversion of hybrid poplar to ethanol and co-products using an organosolv fractionation process: Optimization of process yields. *Biotechnology and Bioengineering* 94:851–861.

Pimentel, D. and T. W. Patzek. 2005. Ethanol production using corn, switchgrass, and wood; Biodiesel production using soybean and sunflower. *Natural Resources Research* 14:65–76.

Reeves, S. 2014. To Russia with love: How moral arguments for a humanitarian intervention in Syria opened the door for an invasion of the Ukraine. *Michigan State University International Law Review* 23:199.

Sanchez, O. J. and C. A. Cardona. 2008. Trends in biotechnological production of fuel ethanol from different feedstocks. *Bioresource Technology* 99:5270–5295.

Schwanninger, M., J. C. Rodrigues, H. Pereira and B. Hinterstoisser. 2004. Effects of short-time vibratory ball milling on the shape of FT-IR spectra of wood and cellulose. *Vibrational Spectroscopy* 36:23–40.

Studer, M. H., J. D. DeMartini and M. F. Davis, et al. 2011. Lignin content in natural populus variants affects sugar release. *Proceedings of the National Academy of Sciences of the United States of America* 108:6300–6305.

Sun, N., M. Rahman and Y. Qin, et al. 2009. Complete dissolution and partial delignification of wood in the ionic liquid 1-ethyl-3-methylimidazolium acetate. *Green Chemistry* 11:646–655.

Sun, Y. and J. Cheng. 2002. Hydrolysis of lignocellulosic materials for ethanol production: A review. *Bioresource Technology* 83:1–11.

Taherzadeh, M. J. and K. Karimi. 2007. Enzyme-based hydrolysis processes for ethanol from lignocellulosic materials: A review. *Bioresources* 2:707–738.

Taherzadeh, M. J. and K. Karimi. 2008. Pretreatment of lignocellulosic wastes to improve ethanol and biogas production: A review. *International Journal of Molecular Sciences* 9:1621–1651.

Wingren, A., M. Galbe and G. Zacchi. 2003. Techno-economic evaluation of producing ethanol from softwood: Comparison of SSF and SHF and identification of bottlenecks. *Biotechnology Progress* 19:1109–1117.

Winzer, C. 2012. Conceptualizing energy security. *Energy Policy* 46:36–48.

Yang, B. and C. E. Wyman. 2008. Pretreatment: The key to unlocking low-cost cellulosic ethanol. *Biofuels, Bioproducts and Biorefining* 2:26–40.

Zhu, J. Y., X. J. Pan, G. S. Wang and R. Gleisner, 2009. Sulfite pretreatment (SPORL) for robust enzymatic saccharification of spruce and red pine. *Bioresource Technology* 100:2411–2418.

52 Wood-based Bioethanol Fuels
Review

Ozcan Konur
(Formerly) Ankara Yildirim Beyazit University

52.1 INTRODUCTION

Crude oil-based gasoline fuels (Ma et al., 2002; Newman and Kenworthy, 1989) have been widely used in the transportation sector since the 1920s. However, there have been great public concerns over the adverse environmental and human impact of these fuels (Hill et al., 2006, 2009). Hence, biomass-based bioethanol fuels (Hill et al., 2006; Konur, 2012, 2015, 2019, 2020) have increasingly been used in blending gasoline fuels (Hsieh et al., 2002; Najafi et al., 2009), in fuel cells (Antolini, 2007, 2009), and in the biochemical production (Angelici et al., 2013; Morschbacker, 2009) in a biorefinery context (Fernando et al., 2006; Huang et al., 2008).

However, it is necessary to pretreat the biomass (Alvira et al., 2010; Taherzadeh and Karimi, 2008) to enhance the yield of bioethanol (Hahn-Hagerdal et al., 2006; Sanchez and Cardona, 2008) prior to bioethanol production through hydrolysis (Sun and Cheng, 2002; Taherzadeh and Karimi, 2007) and fermentation (Lin and Tanaka, 2006; Olsson and Hahn-Hagerdal, 1996) of biomass and hydrolysates, respectively.

One of the most studied biomass for bioethanol production has been wood. The research in the field of wood-based bioethanol fuels has intensified in this context in the key research fronts of the pretreatment of wood (Kilpelainen et al., 2007; Schwanninger et al., 2004; Sun et al., 2009), hydrolysis of wood (Mooney et al., 1998; Palonen et al., 2004; Studer et al., 2011), fermentation of wood hydrolysates (Jonsson et al., 1998; Larsson et al., 1999a,b), and bioethanol production from wood in general (Pan et al., 2005; Pimentel and Patzek, 2005; Wingren et al., 2003). Further, both softwood such as pine (Kilpelainen et al., 2007; Sun et al., 2009; Zhu et al., 2009) and hardwood such as poplar (Kumar et al., 2009; Pan et al., 2006; Studer et al., 2011) have been studied intensively.

However, it is essential to develop efficient incentive structures (North, 1991) for the primary stakeholders to enhance the research in this field (Konur, 2000, 2002a,b,c, 2006a,b, 2007a,b). Although there are a number of review papers on wood-based bioethanol fuels (Galbe and Zacchi, 2002; Zhu et al., 2010, Zhu and Pan, 2010), there has been no review of the 25 most-cited papers in this field.

Thus, this chapter presents a review of the 25 most-cited articles in the field of wood-based bioethanol fuels. Then, it discusses the key findings of these highly influential papers and comments on future research priorities in this field.

52.2 MATERIALS AND METHODS

The search for this study was carried out using the Scopus database (Burnham, 2006) in June 2022. As the first step for the search of the relevant literature, keywords were selected using the most-cited first 200 population papers. The selected keyword list was then optimized to obtain a representative sample of papers for the research field. This final keyword set was provided in the appendix of Konur (2023) for future studies.

As the second step, a sample dataset was used for this study. The first 25 articles with at least 268 citations each were selected for the review study. Key findings from each paper were taken from the abstracts of these papers and were discussed. Additionally, a number of brief conclusions were drawn and a number of relevant recommendations were made to enhance the future research landscape.

52.3 RESULTS

The brief information about the 25 most-cited papers with at least 268 citations each on wood-based bioethanol fuels is given below. The primary research fronts are the pretreatment and hydrolysis of wood biomass for bioethanol production, fermentation of wood hydrolysates, and bioethanol production from wood with 14, 5, 3, and 3 highly cited papers (HCPs), respectively.

52.3.1 Wood Biomass Pretreatments for Bioethanol Production

There are 14 HCPs for wood biomass pretreatments for bioethanol production (Table 52.1). The primary pretreatment is the chemical pretreatment of wood biomass with nine HCPs. Another prolific pretreatment is the hydrothermal pretreatment of wood biomass with 2 HCPs. Additionally, there is one HCP each for mechanical and enzymatic pretreatments, while there is also one HCP for a multiple set of pretreatments.

52.3.1.1 Biomass Pretreatments in General

Kumar et al. (2009) explored the changes in substrate chemical and physical features after pretreatment using untreated corn stover and poplar and their solids after pretreatments such as ammonia fiber expansion (AFEX), ammonia-recycled percolation (ARP), controlled pH, dilute acid, flowthrough, alkali, and SO_2 technologies in a paper with 711 citations. They observed that alkaline pretreatment removed the most acetyl groups from both corn stover and poplar, while AFEX removed the least. Low-pH pretreatments depolymerized cellulose and enhanced biomass crystallinity much more than higher-pH approaches. Alkaline-pretreated corn stover solids and flowthrough-pretreated poplar solids had the highest cellulase adsorption capacity, while dilute acid-pretreated corn stover solids and controlled-pH-pretreated poplar solids had the least. Furthermore, enzymatically extracted AFEX lignin preparations for both corn stover and poplar had the lowest cellulase adsorption capacity. SO_2-pretreated solids had the highest surface O/C ratio for poplar, but for corn stover, the highest value was observed for dilute acid pretreatment. Although dependent on pretreatment and substrate, along with changes in crosslinking and chemical changes, pretreatments might also decrystallize cellulose and change the ratio of crystalline cellulose polymorphs.

52.3.1.2 Mechanical Pretreatments

Schwanninger et al. (2004) studied the short-time vibratory ball milling of wood in a paper with 903 citations. They focused on the effects of temperature, particle size, oxygen, and mechanical treatment on chemical and structural changes. They observed that this process had a strong influence on the shape of FTIR spectra of wood and cellulose, while the mechanical treatment itself was the main influencing factor. They observed the most changes in the spectra of cellulose and wood at wave numbers 1,034, 1,059, 1,110, 1,162, 1,318, 1,335, 2,902 cm^{-1} and in the OH-stretching vibration region from 3,200 to 3,500 cm^{-1}. They attributed these changes to a decrease in the degree of crystallinity and/or a decrease in the degree of polymerization of the cellulose.

52.3.1.3 Chemical Pretreatments

The most prolific chemical pretreatment is the pretreatment of wood biomass with ionic liquids (ILs) with five HCPs. Other chemical pretreatments include acid, alkaline, sulfite, and ethanol solvent pretreatments of wood biomass with one HCP each.

TABLE 52.1
Wood Biomass Pretreatments for Bioethanol Production

No.	Papers	Biomass/ Hydrolysate	Prts.	Parameters	Keywords	Lead Author	Affil.	Cits
1	Schwanninger et al. (2004)	Wood, cellulose	Ball milling	Ball milling, chemical and structural changes	Wood, milling	Schwanninger, Manfred 6602877236	Univ. Natr. Res. Life Sci. Austria	903
2	Kilpelainen et al. (2007)	Pine, spruce	[C_4MIM][Cl], [EMIM][BF_4], [BzMIM][Cl], cellulase	Wood dissolution in ILs, IL and wood type, wood acetylation, hydrolysis	Wood, ionic liquids	Kilpelainen, Ilkka 7006830888	Univ. Helsinki Finland	821
3	Sun et al. (2009)	Pine, oak	[C_2MIM][OAc], [C_4MIM][Cl]	Wood dissolution in ILs, IL and wood type	Wood, ionic liquids	Rogers, Robin D. 35474829200	Univ. Alabama USA	817
4	Lee et al. (2009)	Wood	[EMIM][OAc], cellulase	Lignin extraction by IL, hydrolysis, chemical and structural changes	Wood, ionic liquids, hydrolysis	Dordick, Jonathan S. 7102545507	Rensselaer Polytech. Inst. USA	800
5	Fort et al. (2007)	Wood	[C_4MIM][Cl]	Wood dissolution in ILs, wood type, dissolution profiles, cellulose reconstitution	Wood, ionic liquids	Rogers, Robin D. 35474829200	Univ. Alabama USA	717
6	Kumar et al. (2009)	Poplar, corn stover	Ammonia, acid, alkali, SO_2,	Structural effects of pretreatments	Poplar, pretreatment	Wyman, Charles E. 7004396809	Univ. Calif. Riverside USA	711
7	Li et al. (2007)	Aspen	Steam explosion	Lignin depolymerization and repolymerization, wood delignification	Aspen, wood, steam	Li, Jiebing 7410069979	Res. Inst. Sweden Sweden	645

(Continued)

TABLE 52.1 (*Continued*)
Wood Biomass Pretreatments for Bioethanol Production

No.	Papers	Biomass/ Hydrolysate	Prts.	Parameters	Keywords	Lead Author	Affil.	Cits
8	Berlin et al. (2006)	Softwood	Solvent, cellulase, xylanase, β-glucosidase	Lignin-enzyme interactions, lignin and enzyme type	Softwood, cellulase	Berlin, Alex 8639650700	Novozymes Biotech. Inc. USA	535
9	Esteghlalian et al. (1997	Poplar, switchgrass, corn stover	H_2SO_4	Modeling and optimization of pretreatment, parameter determination, xylose yield	Poplar, acid	Hashimoto, Andrew G. 7202605473	Univ. Hawai'i USA	450
10	Zhu et al. (2009)	Spruce, pine	Sulfite, milling, H_2SO_4, cellulase, β-glucosidase	Pretreatments, hydrolysis, sugar yield, wood type, energy consumption	Spruce, pine, saccharification, SPORL	Zhu, Junyong 7405692678	USDA Forest Serv. USA	440
11	Garrote et al. (1999)	Eucalyptus	LHW	LHW pretreatment effect on cellulose and lignin, modeling	Wood, autohydrolysis	Parajo, Juan C. 7005594129	Univ. Vigo Spain	376
12	Pan et al. (2006)	Poplar	Ethanol solvent, cellulase, H_2SO_4	ethanol solvent pretreatment, composite design, sugar recovery	Poplar, ethanol, organosolv	Pan, Xuejun 57203296000	Univ. Wisconsin Madison USA	373
13	Brandt et al. (2010)	Pine	Ionic liquids	IL anion effect on biomass swelling and dissolution, anion type and basicity	Wood, pine, ionic liquids	Welton, Tom 7003503272	Imperial Coll. UK	268
14	Zhao et al. (2008)	Spruce	NaOH, urea	Alkaline pretreatment, alkali content, temperature, hydrolysis efficiency, sugar yield	Spruce, hydrolysis, alkaline	Deng, Yulin 55261848900	Georgia Inst. Technol. USA	268

Cits., Number of citations received for each paper; Prt, Biomass pretreatments.

52.3.1.3.1   IL Pretreatments

Kilpelainen et al. (2007) studied the dissolution of wood in ILs in a paper with 821 citations. They showed that both hardwood and softwood were readily soluble in various imidazolium-based ILs under gentle conditions as 1-butyl-3-methylimidazolium chloride ([C$_4$MIM]Cl) and 1-allyl-3-me-thylimidazolium chloride ([EMIM]BF$_4$) had good solvating power for Norway spruce sawdust and Norway spruce and Southern pine thermomechanical pulp fibers. These ILs provided solutions that permitted the complete acetylation of the wood. Alternatively, transparent amber solutions of wood were obtained when dissolution of the same lignocellulosic samples was attempted in 1-benzyl-3-methylimidazolium chloride ([BzMIM]Cl). Cellulose of the regenerated wood was efficiently digested to glucose by a cellulase enzymatic hydrolysis pretreatment. Furthermore, completely acetylated wood was readily soluble in chloroform.

Sun et al. (2009) studied the complete dissolution and partial delignification of wood in 1-ethyl-3-methylimidazolium acetate ([C$_2$MIM]OAc) in a paper with 817 citations. They showed that both southern yellow pine and red oak could be completely dissolved in this IL after mild grinding. Further, they obtained complete dissolution of wood by heating the sample using oil irradiation. They then showed that [C$_2$MIM]OAc was a better solvent for wood than 1-butyl-3-methylimidazolium chloride ([C$_4$MIM]Cl) and that variables such as type of wood, initial wood load, and particle size affected dissolution and dissolution rates. For example, red oak dissolved better and faster than southern yellow pine. Carbohydrate-free lignin and cellulose-rich materials could be obtained by using proper reconstitution solvents (e.g., acetone/water 1: 1 v/v), and approximately 26.1% and 34.9% reductions of lignin content in the reconstituted cellulose-rich materials were achieved from pine and oak, respectively, in one dissolution/reconstitution cycle. Further, for pine, 59% of holocellulose in the original wood could be recovered in the cellulose-rich reconstituted material, whereas 31% and 38% of original lignin were recovered, respectively, as carbohydrate-free lignin and as carbohydrate-bonded lignin in the cellulose-rich material.

Lee et al. (2009) studied IL-mediated selective extraction of lignin from wood for enhanced enzymatic cellulose hydrolysis in a paper with 800 citations. They used 1-ethyl-3-methylimid-azolium acetate ([EMIM]OAc) to extract lignin from wood flour. They observed that cellulose in the pretreated wood flour became far less crystalline without undergoing solubilization. When 40% of lignin was removed, the cellulose crystallinity index dropped below 45 wherein more than 90% of the cellulose in wood flour was hydrolyzed by *Trichoderma viride* cellulase. This IL was easily reused, thereby resulting in a highly concentrated solution of chemically unmodified lignin.

Fort et al. (2007) dissolved wood in 1-*n*-butyl-3-methylimidazolium chloride ([C$_4$MIM]Cl) in a paper with 717 citations. They presented the dissolution profiles for wood with different hardness values focusing on the direct analysis of the cellulosic material and lignin content in the resulting liquors by means of conventional[13]C NMR techniques. They then showed that cellulose could be readily reconstituted from IL-based wood liquors in fair yields by the addition of a variety of pre-cipitating solvents. They found that the polysaccharide obtained in this manner was virtually free of lignin and hemicellulose and had characteristics that were comparable to those of pure cellulose samples subjected to similar processing conditions.

Brandt et al. (2010) studied the effect of the IL anion in the pretreatment of pine wood chips in a paper with 268 citations. All ILs contained the 1-butyl-3-methylimidazolium [C$_4$MIM] cation; the anions were trifluoromethanesulfonate [triflate], methylsulfate [CH$_3$SO$_4$], dimethylphosphate [DMP], dicyanamide [N(CN)$_2$], chloride [Cl], and acetate [OAc]. They showed that the anion had a profound impact on the ability to promote both swelling and dissolution of biomass, while viscos-ity, temperature, and water content were also important parameters influencing the swelling pro-cess. Anion basicity correlated with the ability to expand and dissolve pine lignocellulose. Finally, 1-butyl-3-methylimidazolium dicyanamide ([C$_4$MIM]N(CN)$_2$) dissolved neither cellulose nor lig-nocellulosic material.

52.3.1.3.2 Acid Pretreatments

Esteghlalian et al. (1997) modeled and optimized the dilute sulfuric acid pretreatment of corn stover, poplar, and switchgrass in a paper with 450 citations. They pretreated these feedstocks with dilute sulfuric acid (0.6%, 0.9%, and 1.2% w/w) at relatively high temperatures (140°C, 160°C, and 180°C) in a Parr batch reactor. They then modeled the hydrolysis of hemicellulose to its monomeric constituents and possible degradation of these monomers by a series of first-order reactions. Further, they determined the kinetic parameters of two mathematical models for predicting the percentage of xylan remaining in the substrate after pretreatment and the net xylose yield in the liquid stream using the actual acid concentration in the reactor after accounting for the neutralization effect of the substrates.

52.3.1.3.3 Alkaline Pretreatments

Zhao et al. (2008) studied the enzymatic hydrolysis of spruce by alkaline pretreatment at low temperatures in both the presence and absence of urea in a paper with 268 citations. They observed that the enzymatic hydrolysis rate and efficiency could be significantly improved by this pretreatment. At low temperatures, the pretreatment chemical, either NaOH alone or NaOH/urea mixture solution, could slightly remove lignin, hemicelluloses, and cellulose in lignocellulosic materials; disrupt the connections between hemicelluloses, cellulose, and lignin; and alter the structure of treated biomass to make cellulose more accessible to hydrolysis enzymes. Therefore, the enzymatic hydrolysis efficiency of untreated mechanical fibers could also be remarkably enhanced by NaOH or NaOH/urea solution pretreatment. For spruce, up to 70% glucose yield could be obtained for the cold temperature pretreatment (–15°C) using 7% NaOH/12% urea solution, but only 20% and 24% glucose yields were obtained at temperatures of 23°C and 60°C, respectively, when other conditions remained the same. The best condition for this pretreatment was 3% NaOH/12% urea and –15°C where over 60% glucose conversion was achieved.

52.3.1.3.4 Sulfite Pretreatments

Zhu et al. (2009) studied the sulfite pretreatment (SPORL) for the enzymatic hydrolysis of spruce and red pine in a paper with 440 citations. This process consisted of SPORL of wood chips under acidic conditions followed by mechanical size reduction using disk refining. After the SPORL of spruce chips with 8%–10% bisulfite and 1.8%–3.7% sulfuric acid on oven-dried (OD) wood at 180°C for 30 min, they obtained more than 90% cellulose conversion of the substrate with an enzyme loading of about 14.6 FPU cellulase plus 22.5 CBU β-glucosidase per gram of OD substrate after 48 h hydrolysis. Glucose yield from enzymatic hydrolysis of the substrate per 100 g of untreated OD spruce wood (glucan content 43%) was about 37 g (excluding the dissolved glucose during pretreatment). Hemicellulose removal was as critical as lignin sulfonation for cellulose conversion in the SPORL process. Pretreatment altered the wood chips, which reduced electric energy consumption for size reduction to about 19 Wh/kg OD untreated wood or about 19 g glucose/Wh electricity. Furthermore, the SPORL produced low amounts of fermentation inhibitors, hydroxymethylfurfural (HMF) and furfural, of about 5 and 1 mg/g of untreated OD wood, respectively. In addition, they obtained similar results when the SPORL was applied to red pine.

52.3.1.3.5 Ethanol Solvent Pretreatments

Pan et al. (2006) studied the ethanol solvent pretreatment of hybrid poplar using the organosolv fractionation process in a paper with 373 citations. This organosolv process involved extraction with hot aqueous ethanol. The process resulted in the fractionation of poplar chips into a cellulose-rich solids fraction; an ethanol organosolv lignin (EOL) fraction; and a water-soluble fraction containing hemicellulosic sugars, sugar breakdown products, degraded lignin, and other components. They found that center point conditions for the composite design (180°C, 60 min, 1.25% H_2SO_4, and 60% ethanol) yielded a solids fraction containing ~88% of the cellulose present in the untreated poplar. Approximately 82% of the total cellulose in the untreated poplar was recovered as monomeric glucose after hydrolysis of the solids fraction for 24 h using a low enzyme loading (20 filter paper units (FPUs) of cellulase/g cellulose), while ~85% was recovered after 48-h hydrolysis. The total recovery of xylose (soluble and insoluble) was equivalent to ~72% of the xylose present in untreated wood. Approximately 74% of lignin in untreated wood was recovered as EOL.

52.3.1.4 Hydrothermal Pretreatments

There is one HCP each for the pretreatment of the wood biomass with steam explosion and liquid hot water (LHW).

52.3.1.4.1 Steam Pretreatments

Li et al. (2007) studied the role of lignin depolymerization and repolymerization in delignification of aspen wood by steam explosion in a paper with 645 citations. They subjected aspen wood and its isolated lignin to steam pretreatment under various conditions. They observed the competition between lignin depolymerization and repolymerization and identified the conditions required for these two types of reaction. Further, the addition of a reactive phenol, 2-naphthol, inhibited the repolymerization reaction strongly, resulting in a highly improved delignification by subsequent solvent extraction and an extracted lignin of uniform structure.

52.3.1.4.2 LHW Pretreatments

Garrote et al. (1999) studied the mild LHW pretreatment of eucalyptus in a paper with 377 citations. They subjected *Eucalyptus globulus* wood samples to LHW pretreatment under mild operational conditions (145°C–190°C, liquor to solid ratio 6–10 g/g, reaction times up to 7.5 h). They then determined residual xylan, xylooligosaccharides, other sugars, furfural, glucan, and lignin contents. They observed that negligible effects were caused by LHW on both cellulose and lignin. They next developed kinetic models which described the hydrolysis of hemicelluloses. They found that xylan degradation, xylooligosaccharide and xylose generation, and xylose dehydration to furfural were accurately described by models based on pseudo-homogeneous first-order kinetics with Arrhenius-type temperature dependence.

52.3.1.5 Enzymatic Pretreatments

There is one HCP for the enzymatic pretreatment of the biomass. Berlin et al. (2006) studied the inhibition of cellulase, xylanase, and β-glucosidase activities by softwood lignin preparations in a paper with 535 citations. They examined the inhibition of seven cellulase preparations, three xylanase preparations, and a β-glucosidase preparation by two purified, particulate lignin preparations derived from softwood using the organosolv pretreatment process followed by enzymatic hydrolysis. They observed that the two lignin preparations had similar particle sizes and surface areas but differed significantly in other physical properties and in their chemical compositions determined. Further, various cellulases differed by up to 3.5-fold in their inhibition by lignin, while xylanases showed less variability (less than 1.7-fold). Of all the enzymes tested, β-glucosidase was least affected by lignin.

52.3.2 Wood Biomass Hydrolysis for Bioethanol Production

There are five HCPs for the hydrolysis of wood biomass (Table 52.2). Studer et al. (2011) studied the effect of the lignin content in natural poplar variants on sugar release in a paper with 462 citations. They selected 47 extreme phenotypes across the measured lignin content and the ratio of syringyl and guaiacyl units (S/G ratio) from a sample set of wood cores representing 1,100 individual undomesticated *Populus trichocarpa* trees. They tested this sample for total sugar release through enzymatic hydrolysis alone as well as through combined LHW pretreatment and enzymatic hydrolysis. They found that the total amount of glucan and xylan released varied widely among samples, with total sugar yields of up to 92% of the theoretical maximum. There was a strong negative correlation between sugar release and the lignin content for pretreated samples with an S/G ratio less than 2.0.

TABLE 52.2

Wood Biomass Hydrolysis for Bioethanol Production

No.	Papers	Biomass/ Hydrolysate	Prts.	Parameters	Keywords	Lead Author	Affil.	Cits
1	Studer et al. (2011)	Poplar	Enzymes, LHW	Lignin content effect on sugar release, lignin content, S/G ratio, hydrolysis, LHW	Populus, sugar	Wyman, Charles E. 7004396809	Univ. Calif. Riverside USA	462
2	Palonen et al. (2004)	Softwood	Steam, *T. reesei* cellulases	Lignin-enzyme interactions	Softwood, steam	Palonen, Hetti* 56575011300	VTT Tech. Res. Ctr. Finland	401
3	Mooney et al. (1998)	Douglas fir, pulps	Cellulases	Lignin-enzyme interactions, lignin content, pore volume, substrate type, cellulase adsorption	Softwood, hydrolysis	Saddler, Jack N. 7005297559	Univ. British Columbia Canada	363
4	Xiao et al. (2004)	Softwood, cellulose	Cellulases, β-glucosidase, acetic acid	Sugar inhibition on enzymes, hydrolysis, sugar content	Softwood, hydrolysis, cellulase	Saddler, Jack N. 7005297559	Univ. British Columbia Canada	320
5	Kumar et al. (2012)	Douglas fir	Steam, enzymes	Lignin effect on cellulose accessibility, lignin content, enzyme loadings, hydrolysis	Softwood, steam	Saddler, Jack N. 7005297559	Univ. British Columbia Canada	300

*, Female; Cits., Number of citations received for each paper; Prt, Biomass pretreatments.

For higher S/G ratios, sugar release was generally higher, and the negative influence of lignin was less pronounced. When examined separately, only glucose release was correlated with the lignin content and the S/G ratio in this manner, whereas xylose release depended on the S/G ratio alone. For enzymatic hydrolysis without the LHW pretreatment, sugar release increased significantly with a decreasing lignin content below 20%, irrespective of the S/G ratio. Furthermore, certain samples featuring average lignin content and S/G ratios exhibited exceptional sugar release. They suggested that factors beyond the lignin content and S/G ratio influenced recalcitrance to sugar release.

Palonen et al. (2004) studied the adsorption of *Trichoderma reesei* cellulases and their catalytic domains on steam-pretreated softwood and isolated lignin in a paper with 401 citations. They used two purified cellulases from *T. reesei*, CBH I (Cel7A) and EG II (Cel5A). They found that both CBH I and its catalytic domain exhibited a higher affinity to SPS than EG II or its catalytic domain. Removal of the cellulose-binding domain markedly decreased the binding efficiency. Significant amounts of CBH I and EG II also bound to isolated lignin. Surprisingly, the catalytic domains of the two enzymes of *T. reesei* differed essentially in the adsorption to isolated lignin. The catalytic domain of EG II was able to adsorb to alkaline-isolated lignin with a high affinity, whereas the catalytic domain of CBH I did not adsorb to any of the lignins tested. They concluded that the cellulose-binding domain had a significant role in the unspecific binding of cellulases to lignin.

Mooney et al. (1998) studied the effect of the initial pore volume and lignin content on the enzymatic hydrolysis of softwood in a paper with 363 citations. They used four Douglas fir pulps, a refiner mechanical pulp (RMP), sulfonated RMP, delignified RMP, and a kraft pulp to determine whether the lignin content and initial pore volume affected cellulase adsorption and substrate hydrolysis. When compared on the basis of lignin content, they found from the cellulase treatment of the sulfonated RMP that the proportion of lignin did not affect enzyme adsorption when the fibers were sufficiently swollen. However, they observed that the initial adsorption of cellulase did not always translate to fast and complete hydrolysis. Further, although modification of lignin resulted in a dramatically increased fiber saturation point, the median pore width was not increased accordingly. In contrast, the delignified RMP had a higher median pore width and was hydrolyzed more completely, suggesting that steric hindrance from the residual lignin might be the rate-limiting characteristic in this situation. Hydrolysis of the kraft pulp indicated that the larger average particle size of this substrate might have been an inhibiting factor, since it was hydrolyzed more slowly than the delignified RMP despite having a higher median pore width and lower lignin content.

Xiao et al. (2004) determined the effects of sugar inhibition on cellulases and β-glucosidase during enzymatic hydrolysis of acetic acid-pretreated softwood in a paper with 320 citations. They observed that the increased glucose content in the hydrolysate resulted in a dramatic increase in the degrees of inhibition on both β-glucosidase and cellulase activities. Supplementation of mannose, xylose, and galactose during cellobiose hydrolysis did not show any inhibitory effects on β-glucosidase activity, but had significant inhibitory effects on cellulase activity during cellulose hydrolysis. High-substrate consistency hydrolysis with the supplementation of hemicellulose would be a practical solution to minimize end-product inhibition effects while producing hydrolysate with high glucose concentration.

Kumar et al. (2012) studied the effect of lignin present in steam-pretreated Douglas fir wood on enzymes and cellulose accessibility in a paper with 300 citations. They assessed the influence of cellulose accessibility and protein loading on the efficiency of enzymatic hydrolysis of steam-pretreated Douglas fir wood. They observed that the lignin component significantly influenced the swelling and accessibility of cellulose as at low protein loadings (5 FPU/g cellulose), only 16% of the cellulose present in the feedstock was hydrolyzed, while almost complete hydrolysis was achieved with the delignified substrate. When lignin was added back in the same proportions to the highly accessible, swollen, delignified steam-pretreated softwood and to the Avicel cellulose, the hydrolysis yields decreased by 9% and 46%, respectively. However, when higher enzyme loadings were employed, the greater availability of the enzyme could overcome the limitations imposed by both lignin's restrictions on cellulose accessibility and direct binding of the enzymes, resulting in a near-complete hydrolysis of the cellulose.

52.3.3 WOOD HYDROLYSATE FERMENTATION FOR BIOETHANOL PRODUCTION

There are two research fronts for the fermentation of wood hydrolysates in bioethanol production: fermentation inhibitors and detoxification of the fermentation with one and two HCPs, respectively.

52.3.3.1 Fermentation Inhibitors

Larsson et al. (1999a) studied the generation of fermentation inhibitors during dilute sulfuric acid (H_2SO_4) hydrolysis of spruce in a paper with 872 citations. They focused on the influence of the severity of the hydrolysis on sugar yield and on fermentability of the hydrolysate by *Saccharomyces cerevisiae*. They assessed fermentability as the ethanol yield on fermentable sugars (mannose and glucose) and the mean volumetric productivity (4 h) as a function of the combined severity (CS) of the hydrolysis. When the CS of hydrolysis conditions increased, they observed that the yield of fermentable sugars increased to a maximum between CS 2.0–2.7 for mannose and 3.0–3.4 for glucose above which it decreased. The decrease in the yield of monosaccharides coincided with the maximum concentrations of furfural and 5-HMF. With the further increase in CS, the concentrations of furfural and 5-HMF decreased, while the formation of formic acid and levulinic acid increased. The yield of ethanol decreased at approximately CS 3; however, the volumetric productivity decreased at lower CS. Further, furfural and 5-HMF decreased the volumetric productivity but did not influence the final yield of ethanol. The decrease in volumetric productivity was more pronounced when 5-HMF was added to fermentation, and this compound was depleted at a lower rate than furfural. Finally, they found that the inhibition observed in hydrolysates produced in higher CS could not be fully explained by the effect of the by-products furfural, 5-HMF, acetic acid, formic acid, and levulinic acid.

52.3.3.2 Hydrolysate Detoxification

Larsson et al. (1999b) compared different methods for the detoxification of lignocellulose hydro-lysates of spruce to improve both cell growth and ethanol production by *S. cerevisiae* in a paper with 425 citations. They used a dilute acid hydrolysate of spruce for all the detoxification methods tested; determined the changes in the concentrations of fermentable sugars and three groups of inhibitory compounds of aliphatic acids, furan derivatives, and phenolic compounds; and assayed the fermentability of the detoxified hydrolysate. The applied detoxification methods included: treatment with alkali (sodium hydroxide or calcium hydroxide); treatment with sulfite (0.1% [w/v] or 1% [w/v] at pH 5.5 or 10); evaporation of 10% or 90% of the initial volume; anion exchange (at pH 5.5 or 10); enzymatic detoxification with the phenoloxidase laccase; and detoxification with the *T. reesei*. They found that anion exchange at pH 5.5 or 10 and pretreatment with laccase, cal-cium hydroxide, and *T. reesei* were the most efficient detoxification methods, while evaporation of 10% of the initial volume and pretreatment with 0.1% sulfite were the least efficient detoxification methods. Further, pretreatment with laccase was the only detoxification method that specifically removed only phenolic compounds, while anion exchange at pH 10 was the most efficient method for removing all three major groups of inhibitory compounds although it also resulted in the loss of fermentable sugars.

Jonsson et al. (1998) detoxified wood hydrolysates with laccase and lignin peroxidase from the *Trametes versicolor* to increase the fermentability of these hydrolysates in a paper with 342 cita-tions. They used a lignocellulosic hydrolysate from willow pretreated with steam and SO_2 and *S. cerevisiae*. They observed more rapid consumption of glucose and increased ethanol productivity for samples treated with laccase. Treatment of the hydrolysate with lignin peroxidase also resulted in improved fermentability. The mechanism of laccase detoxification involved the removal of mono-aromatic phenolic compounds present in the hydrolysate as these phenolic compounds were impor-tant inhibitors of the fermentation process (Table 52.3).

52.3.4 Bioethanol Fuels in General

There are three HCPs for bioethanol fuels in general.

Pimentel and Patzek (2005) evaluated the energy output from ethanol produced using corn grains, switchgrass, and wood in a paper with 1,010 citations. They found that the energy output from ethanol produced using these feedstocks was each less than the respective fossil energy inputs. Ethanol production using corn grain, switchgrass, and wood required 29%, 50%, and 57% more fossil energy than the ethanol fuel produced, respectively.

Wingren et al. (2003) carried out the techno-economic evaluation of ethanol production from softwood in a paper with 540 citations. They focused on the enzymatic processes as well as simul-taneous saccharification and fermentation (SSF) and separate hydrolysis and fermentation (SHF) processes. They found that the ethanol production costs for the SSF and SHF base cases were 0.57 and 0.63 USD/L, respectively. The main reason for SSF being lower was that the capital cost was lower and the overall ethanol yield was higher. However, a major drawback of the SSF process was the problem with the recirculation of yeast following the SSF step. Further, major economic improvements in both SSF and SHF could be achieved by increasing the income from the solid fuel coproduct. This was done by lowering the energy consumption in the process by running the enzy-matic hydrolysis or the SSF step at a higher substrate concentration and by recycling the process streams. For example, running SSF using 8% rather than 5% non-soluble solid material would result in a 19% decrease in production cost. If after distillation, 60% of the stillage stream was recycled back to the SSF step, the production cost would be reduced by 14%. The cumulative effect of these various improvements resulted in a production cost of 0.42 USD/L for the SSF process.

Pan et al. (2005) biorefined softwoods using ethanol organosolv pulping for bioethanol produc-tion in a paper with 488 citations. They prepared pulps with residual lignin ranging from 6.4% to 27.4% (w/w) from mixed softwood using the lignol process based on aqueous ethanol organosolv extraction. They found that all pulps were readily hydrolyzed without further delignification. More

TABLE 52.3

Wood Hydrolysate Fermentation for Bioethanol Production

No.	Papers	Biomass/ Hydrolysate	Prts.	Yeasts	Parameters	Keywords	Lead Author	Affil.	Cits
1	Larsson et al. (1999a)	Spruce	H_2SO_4	S. cerevisiae	Fermentation inhibitors, sugar and ethanol yield, volumetric productivity, hydrolysis severity	Softwood, fermentation, hydrolysis, acid, inhibitors	Hahn-Hagerdal, Barbel* 7005389381	Lund Univ. Sweden	872
2	Larsson et al. (1999b)	Spruce	Acid	S. cerevisiae	Hydrolysate detoxification, detoxification methods	Spruce, hydrolysates, detoxification	Jonsson, Leif J. 7102349315	Umea Univ. Sweden	425
3	Jonsson et al. (1998)	Willow	Laccase, peroxidase, steam, SO_2	S. cerevisiae	Hydrolysate detoxification, enzyme type, hydrolysate fermentability, fermentation inhibitors	Wood, hydrolysate, detoxification	Jonsson, Leif J. 7102349315	Umea Univ. Sweden	342

*, Female; Cits., Number of citations received for each paper; Prt, Biomass pretreatments.

than 90% of the cellulose in low lignin pulps (less than 18.4% residual lignin) was hydrolyzed to glucose in 48 h using an enzyme loading of 20 filter paper units (FPUs)/g cellulose. Similarly, cellulose in a high-lignin pulp (27.4% residual lignin) was hydrolyzed to more than 90% conversion within 48 h using 40 FPUs/g. Further, these pulps performed well in both sequential and SSF processes, indicating an absence of fermentation inhibitors. Finally, lignin extracted during organosolv pulping of softwood was a suitable feedstock for the production of lignin-based products due to its high purity, low molecular weight, and abundance of reactive groups (Table 52.4).

52.4 DISCUSSION

52.4.1 INTRODUCTION

Crude oil-based gasoline fuels have been widely used in the transportation sector since the 1920s. However, there have been great public concerns over the adverse environmental and human impact of these fuels. Hence, biomass-based bioethanol fuels have increasingly been used in blending gasoline and petrodiesel fuels, in fuel cells, and in biochemical production in a biorefinery context.

However, it is necessary to pretreat the biomass to enhance the yield of the bioethanol prior to bioethanol production through the hydrolysis and fermentation of biomass. One of the most-studied biomasses for bioethanol production has been wood. The research in the field of wood-based bioethanol fuels has intensified in this context in the key research fronts of the pretreatment and hydrolysis of wood, fermentation of wood hydrolysates, and bioethanol production from wood in general, including the techno-economics studies of bioethanol fuels. Further, both softwood such as pine and hardwood such as poplar have been studied intensively.

However, it is essential to develop efficient incentive structures for the primary stakeholders to enhance the research in this field. Although there are a number of review papers in this field, there has been no review of the 25 most-cited articles in this field.

TABLE 52.4

Bioethanol Fuels in General

No.	Papers	Biomass/ Hydrolysate	Prts.	Yeasts	Parameters	Keywords	Lead Author	Affil.	Cits
1	Pimentel and Patzek (2005)	Wood, switchgrass, corn grains	Na	Na	Energy output	Wood, ethanol	Pimentel, David 7005471319	Cornell Univ. USA	1010
2	Wingren et al. (2003)	Softwood	Enzymes	Na	Techno-economics of ethanol production, SSF/ SHF processes, substrate concentration, stillage stream recycling	Softwood, ethanol, SSF, SHF	Zacchi, Guido 7006727748	Lund Univ. Sweden	540
3	Pan et al. (2005)	Softwood	Enzymes, ethanol solvent	Yeasts	Ethanol production, pretreatments, hydrolysis, SSF, lignin concentration	Softwood, biorefining, bioethanol, organosolv	Pan, Xuejun 57203296000	Univ. Wisconsin Madison USA	488

Cits., Number of citations received for each paper; Na, non-available; Prt, Biomass pretreatments.

Thus, this chapter presents a review of the 25 most-cited articles on wood-based bioethanol fuels. Then, it discusses the key findings of these highly influential papers and comments on future research priorities in this field.

As the first step for the search of the relevant literature, the keywords were selected using the first 200 most-cited population papers. The selected keyword list was then optimized to obtain a representative sample of papers in the research field. This keyword list was provided in the appendix of Konur (2023) for future replicative studies.

As the second step, a sample dataset was used for this study. The first 25 articles with at least 268 citations each were selected for the review study. Key findings from each paper were taken from the abstracts of these papers and were discussed. Additionally, a number of brief conclusions were drawn, and a number of relevant recommendations were made to enhance the future research landscape.

Information about the research fronts of sample papers in wood-based bioethanol fuels with regard to the wood biomass used in these processes is given in Table 52.5. As this table shows, there are three primary research fronts for this field: softwood, hardwood, and wood-based bioethanol fuels with 64%, 32%, and 16% of these HCPs, respectively.

Further, the most prolific research fronts are softwood and spruce-based bioethanol fuels with 20% of these HCPs each, followed by pine- and poplar-based bioethanol fuels with 16% of these HCPs each.

Further, research on softwood-based bioethanol fuels is the most influential research front with a 29% surplus, while that on hardwood-based bioethanol fuels is the least influential research front with a 17% deficit.

Further, research on softwood and spruce-based bioethanol fuels is the most influential research fronts with 12% and 11% surplus, respectively. Similarly, eucalyptus, hardwood, birch, and bamboo-based biohydrogen production is the least influential research front with a 4%–6% deficit each.

Information about the thematic research fronts for sample papers on wood-based bioethanol fuels is given in Table 52.6. As this table shows, there are four primary research fronts for this field:

TABLE 52.5
Most Prolific Research Fronts for Wood-based Bioethanol Fuels

No.	Research Fronts	N Paper (%) Review	N Paper (%) Sample	Surplus (%)
1	Hardwood	32.0	49.0	−17.0
	Poplar	16.0	18.0	−2.0
	Eucalyptus	4.0	9.5	−5.5
	Aspen	4.0	5.0	−1.0
	Hardwood in general	0.0	4.5	−4.5
	Birch	0.0	4.0	−4.0
	Bamboo	0.0	3.5	−3.5
	Willow	4.0	3.5	0.5
	Other hardwoods	4.0	1.0	3.0
2	Softwood	64.0	35.5	28.5
	Pine	16.0	13.0	3.0
	Spruce	20.0	9.5	10.5
	Softwood in general	20.0	8.5	11.5
	Douglas fir	8.0	4.5	3.5
3	Wood in general	16.0	20.0	−4.0

N Paper (%) review, The number of papers in the sample of 25 reviewed papers; N paper (%) sample, The number of papers in the population sample of 200 papers.

pretreatment and hydrolysis of wood biomass, fermentation of wood hydrolysates, and bioethanol production in general with 96%, 20%, 12%, and 12% of these HCPs, respectively.

Further, the most prolific pretreatments are chemical, enzymatic, and hydrothermal pretreatments with 60%, 52%, and 20% of these HCPs, respectively. Another research front is the mechanical pretreatments of the wood biomass with 8% of the HCPs. Similarly, the most prolific hydrolysis is the enzymatic hydrolysis of the wood biomass with 20% of these HCPs, while the prolific research fronts in the fermentation of wood hydrolysates are detoxification of wood hydrolysates and fermentation inhibitors with 8% and 4% of these HCPs, respectively.

Further, the enzymatic pretreatment is the most influential front with a 45% surplus, followed by IL pretreatment, wood biomass pretreatment, and hydrolysate detoxification with 14%, 9%, and 6% surplus, respectively. Similarly, wood biomass hydrolysis is the least influential research front with a 43% deficit, followed by enzymatic hydrolysis, hydrothermal pretreatments, and steam explosion pretreatment with 33%, 18%, and 11% deficit, respectively.

52.4.2 Wood Biomass Pretreatments for Bioethanol Production

There are 14 HCPs for wood biomass pretreatments in bioethanol production (Table 52.1). The primary pretreatment is the chemical pretreatment of wood biomass with nine HCPs. Another prolific pretreatment is the hydrothermal pretreatment of wood biomass with two HCPs. Additionally, there is one HCP each for mechanical and enzymatic pretreatments, while there is also one HCP for a multiple set of pretreatments.

These HCPs show a sample of research on the pretreatment of wood biomass in bioethanol production. These studies hint that mechanical, chemical, hydrothermal, and enzymatic pretreatments enhance both sugar and ethanol yields during bioethanol production from wood. In other words, the pretreatment stage is one of the most important phases of wood-based bioethanol production (Alvira et al., 2010; Taherzadeh and Karimi, 2008).

TABLE 52.6
The Most Prolific Thematic Research Fronts for Wood-based Bioethanol Fuels

No.	Research Fronts	N Paper (%) Review	N Paper (%) Sample	Surplus (%)
1	Wood Biomass Pretreatments	96.0	87.5	8.5
	Mechanical pretreatments	8.0	8.0	0.0
	Milling pretreatments	8.0	6.0	2.0
	Other pretreatments	0.0	2.0	−2.0
	Hydrothermal pretreatments	24.0	42.0	−18.0
	Steam explosion pretreatment	16.0	27.0	−11.0
	Liquid hot water pretreatment	8.0	7.0	1.0
	Autohydrolysis pretreatment	0.0	5.0	−5.0
	Hot compressed water pretreatment	0.0	2.5	−2.5
	Wet oxidation pretreatment	0.0	0.5	−0.5
	Chemical pretreatments	60.0	60.0	0.0
	Acid pretreatment	24.0	24.5	−0.5
	Solvent pretreatment	12.0	12.0	0.0
	Alkaline pretreatment	8.0	9.5	−1.5
	Sulfur dioxide pretreatment	8.0	7.5	0.5
	Ionic liquid pretreatment	20.0	6.5	13.5
	Sulfite pretreatment	4.0	4.0	0.0
	Ammonia pretreatment	4.0	2.5	1.5
	Other chemical pretreatments	0.0	2.5	−2.5
	Enzymatic pretreatments	52.0	7.0	45.0
2	Wood Biomass Hydrolysis	20.0	62.5	−42.5
	Enzymatic hydrolysis	20.0	53.0	−33.0
	Autohydrolysis	0.0	5.0	−5.0
	Acid hydrolysis	0.0	4.5	−4.5
3	Wood Hydrolystae Fermentation	12.0	18.0	−6.0
	Fermentation inhibitors	4.0	7.0	−3.0
	SSF	0.0	6.5	−6.5
	Hydrolysate detoxification	8.0	2.5	5.5
	Fermentation in general	0.0	2.0	−2.0
4	Ethanol Fuels in General	12.0	18.0	−6.0

N Paper (%) review, The number of papers in the sample of 25 reviewed papers; N paper (%) sample, The number of papers in the population sample of 200 papers

Further, the primary research front, as expected, is the chemical pretreatment of wood biomass with nine HCPs, and the most prolific chemical pretreatment is the pretreatment of wood biomass with ILs (Plechkova and Seddon, 2008) with five HCPs. Other chemical pretreatments include acid, alkaline, sulfite, and ethanol solvent pretreatments of wood biomass with one HCP each.

52.4.2.1 Biomass Pretreatments in General
Kumar et al. (2009) explored the changes in substrate chemical and physical features after pretreatment using untreated corn stover and poplar and their solids using pretreatments by AFEX, ARP, controlled pH, dilute acid, flowthrough, alkali, and SO_2 technologies. They observed that alkaline pretreatment removed the most acetyl groups from both corn stover and poplar, while AFEX removed the least.

52.4.2.2   Mechanical Pretreatments

Schwanninger et al. (2004) studied the short-term vibratory ball milling of wood and observed that this process had a strong influence on the shape of FTIR spectra of wood and cellulose, while the mechanical treatment itself was the main influencing factor.

52.4.2.3   Chemical Pretreatments

The most prolific chemical pretreatment is the pretreatment of wood biomass with ILs with five HCPs. Other chemical pretreatments include acid, alkaline, sulfite, and ethanol solvent pretreatments of wood biomass with one HCP each.

52.4.2.3.1   IL Pretreatments

Kilpelainen et al. (2007) studied the dissolution of wood in ILs and showed that both hardwood and softwood were readily soluble in various imidazolium-based ILs under gentle conditions as [C$_4$MIM]Cl and [EMIM]BF$_4$ had good solvating power for Norway spruce sawdust and Norway spruce and Southern pine thermomechanical pulp fibers. Further, Sun et al. (2009) studied the complete dissolution and partial delignification of wood in [C$_2$MIM]OAc and showed that both southern yellow pine and red oak could be completely dissolved in this IL after mild grinding.

Lee et al. (2009) studied the IL-mediated selective extraction of lignin from wood for enhanced enzymatic cellulose hydrolysis and observed that the cellulose in the pretreated wood flour became far less crystalline without undergoing solubilization in [EMIM]OAc. Further, Fort et al. (2007) dissolved wood in [C$_4$MIM]Cl and found that the polysaccharide obtained in this manner was virtually free of lignin and hemicellulose and had characteristics that were comparable to those of pure cellulose samples subjected to similar processing conditions.

Brandt et al. (2010) studied the effect of the IL anion in the pretreatment of pine wood chips and showed that the anion had a profound impact on the ability to promote both swelling and dissolution of the biomass, while viscosity, temperature, and water content were also important parameters influencing the swelling process.

52.4.2.3.2   Acid Pretreatments

Esteghlalian et al. (1997) modeled and optimized the dilute sulfuric acid pretreatment of corn stover, poplar, and switchgrass and modeled the hydrolysis of hemicellulose to its monomeric constituents and possible degradation of these monomers by a series of first-order reactions.

52.4.2.3.3   Alkali Pretreatments

Zhao et al. (2008) studied the enzymatic hydrolysis of spruce by alkaline pretreatment at low temperatures in both the presence and absence of urea and observed that the enzymatic hydrolysis rate and efficiency could be significantly improved by this pretreatment.

52.4.2.3.4   Sulfite Pretreatments

Zhu et al. (2009) studied the SPORL for the enzymatic hydrolysis of spruce and red pine and obtained more than 90% cellulose conversion of the substrate with an enzyme loading of about 14.6 FPU cellulase plus 22.5 CBU β-glucosidase per gram of OD substrate after 48-h hydrolysis.

52.4.2.3.5   Ethanol Solvent Pretreatments

Pan et al. (2006) studied the ethanol solvent pretreatment of hybrid poplar using an organosolv fractionation process and observed that this organosolv process involved extraction with hot aqueous ethanol.

52.4.2.4   Hydrothermal Pretreatments

There is one HCP each for the pretreatment of wood biomass with steam explosion and LHW.

52.4.2.4.1 Steam Pretreatments

Li et al. (2007) studied the role of lignin depolymerization and repolymerization in delignification of aspen wood by steam explosion, observed the competition between lignin depolymerization and repolymerization, and identified the conditions required for these two types of reaction.

52.4.2.4.2 LHW Pretreatments

Garrote et al. (1999) studied the mild LHW pretreatment of eucalyptus and observed that negligible effects were caused by LHW on both cellulose and lignin.

52.4.2.5 Enzymatic Pretreatments

There is one HCP for the enzymatic pretreatment of biomass. Berlin et al. (2006) studied the inhibition of cellulase, xylanase, and β-glucosidase activities by softwood lignin preparations and observed that the two lignin preparations had similar particle sizes and surface areas but differed significantly in other physical properties and chemical compositions.

52.4.3 Wood Biomass Hydrolysis for Bioethanol Production

There are five HCPs for the enzymatic hydrolysis of wood biomass (Table 52.2). These HCPs show a sample of research on the hydrolysis of wood for bioethanol production. These studies hint that the enzymatic hydrolysis of biomass enhances the sugar and ethanol yield during the bioethanol production process. In other words, the hydrolysis stage is one of the most critical phases of bioethanol production from wood (Sun and Cheng, 2002; Taherzadeh and Karimi, 2007).

Studer et al. (2011) studied the effect of the lignin content in natural poplar variants on the sugar release and found that the total amount of glucan and xylan released varied widely among samples, with total sugar yields of up to 92% of the theoretical maximum. Further, Palonen et al. (2004) studied the adsorption of *T. reesei* cellulases and their catalytic domains on steam-pretreated softwood and isolated lignin and found that both CBH I and its catalytic domain exhibited a higher affinity to SPS than EG II or its catalytic domain.

Mooney et al. (1998) studied the effect of the initial pore volume and lignin content on the enzymatic hydrolysis of softwood and found from the cellulase treatment of the sulfonated RMP that the proportion of lignin did not affect enzyme adsorption when the fibers were sufficiently swollen. Further, Xiao et al. (2004) determined the effects of sugar inhibition on cellulases and β-glucosidase during enzymatic hydrolysis of acetic acid-pretreated softwood and observed that the increased glucose content in the hydrolysate resulted in a dramatic increase in the degrees of inhibition on both β-glucosidase and cellulase activities.

Kumar et al. (2012) studied the effect of the lignin present in steam-pretreated Douglas fir wood on enzymes and cellulose accessibility and observed that the lignin component significantly influenced the swelling and accessibility of cellulose.

52.4.4 Wood Hydrolysate Fermentation for Bioethanol Production

There are two research fronts for the fermentation of wood hydrolysates in bioethanol production: fermentation inhibitors and detoxification of the fermentation with one and two HCPs, respectively.

These HCPs show a sample of research on the fermentation of wood hydrolysates in bioethanol production. These studies hint that fermentation of these hydrolysates is one of the most critical phases in bioethanol production from wood (Lin and Tanaka, 2006; Olsson and Hahn-Hagerdal, 1996). Further, the most studied research issues in the fermentation of wood hydrolysates are the emergence of fermentation inhibitors and the detoxification of these inhibitors to improve the ethanol yield.

52.4.4.1 Fermentation Inhibitors

Larsson et al. (1999a) studied the generation of fermentation inhibitors during dilute H_2SO_4 hydrolysis of spruce and observed that the yield of fermentable sugars increased to a maximum between CS 2.0–2.7 for mannose and 3.0–3.4 for glucose, above which it decreased.

52.4.4.2   Hydrolysate Detoxification

Larsson et al. (1999b) compared different methods for the detoxification of lignocellulose hydroly-sates of spruce to improve both cell growth and ethanol production by *S. cerevisiae* and found that anion exchange at pH 5.5 or 10, and pretreatment with laccase, lime, and *T. reesei* were the most efficient detoxification methods, while the evaporation of 10% of the initial volume and pretreat-ment with 0.1% sulfite were the least efficient detoxification methods. Further, Jonsson et al. (1998) detoxified wood hydrolysates with laccase and lignin peroxidase from *T. versicolor* to increase the fermentability of these hydrolysates and observed a more rapid consumption of glucose and increased ethanol productivity in samples treated with laccase.

52.4.5   Bioethanol Fuels in General

There are three HCPs for bioethanol production in general. These HCPs show a sample of the research on bioethanol production from wood. These studies hint that bioethanol production involves the phases of pretreatment and hydrolysis of wood biomass and fermentation of wood hydrolysates (Hahn-Hagerdal et al., 2006; Sanchez and Cardona, 2008). Further, the most studied research issues in wood bioethanol production are techno-economics of bioethanol fuels as the cost of bioethanol fuels is a critical issue for the investment and consumption decisions by companies and consumers, respectively.

Pimentel and Patzek (2005) evaluated the energy output from ethanol produced using corn grains, switchgrass, and wood and found that the energy outputs from ethanol produced using these feedstocks were each less than the respective fossil energy inputs. Further, Wingren et al. (2003) carried out the techno-economic evaluation of ethanol production from softwood and found that ethanol production costs for SSF and SHF base cases were 0.57 and 0.63 USD/L, respectively. Finally, Pan et al. (2005) biorefined softwood using ethanol organosolv pulping for bioethanol pro-duction and found that all pulps were readily hydrolyzed without further delignification.

52.5   CONCLUSION AND FUTURE RESEARCH

The brief information about the key research fronts covered by the 25 most-cited papers with at least 268 citations each is given under four primary headings: pretreatments of wood biomass and to a lesser extent hydrolysis of wood biomass, fermentation of wood hydrolysates, and bioethanol production in general.

The usual characteristics of these HCPs are that the pretreatments and hydrolysis of wood bio-mass and fermentation of wood hydrolysates are the primary processes for wood-based bioethanol production to improve the ethanol yield as wood biomass is one of the most important feedstocks in bioethanol production, especially in countries with the large forests, such as Sweden, Canada, and the USA.

The key findings of these research fronts should be read in light of the increasing public concerns about climate change, GHG emissions, and global warming as these concerns have been certainly behind the boom in the research on bioethanol fuels as alternatives to crude oil-based gasoline and petrodiesel fuels in the past decades. The recent supply shocks caused by the COVID-19 pandemic and the Russian invasion of Ukraine also highlight the importance of the production and utilization of bioethanol fuels as alternatives to crude oil-based gasoline and petrodiesel fuels.

There are three primary research fronts in this field: softwood-, hardwood-, and wood-based bioethanol fuels. The most prolific research fronts are softwood- and spruce-based bioethanol fuels, followed by pine- and poplar-based bioethanol fuels. Further, softwood-based bioethanol fuels are the most influential research front, while hardwood-based bioethanol fuels are the least influential research front. On the other hand, softwood- and spruce-based bioethanol fuels are the most influ-ential research fronts. Similarly, eucalyptus, hardwood, birch, and bamboo-based bioethanol fuels are the least influential research fronts.

Similarly, there are four primary research fronts in this field: pretreatments and hydrolysis of wood biomass, fermentation of wood hydrolysates, and bioethanol fuels in general. Further, the most prolific pretreatments are chemical, enzymatic, and hydrothermal pretreatments, while another research front is the mechanical pretreatment of the wood biomass. Similarly, the most prolific hydrolysis is the enzymatic hydrolysis of wood biomass, while the prolific research fronts for the fermentation of wood hydrolysates are detoxification of wood hydrolysates and fermentation inhibitors.

These studies emphasize the importance of proper incentive structures for efficient wood-based bioethanol fuels in light of North's institutional framework (North, 1991). In this context, the major producers and users of bioethanol fuels such as Europe, the USA, and to a lesser extent Canada and Japan have developed strong incentive structures for efficient bioethanol production.

In light of the supply shocks caused primarily by the COVID-19 pandemic and the Russian invasion of Ukraine, it is expected that the incentive structures such as public funding would be enhanced to increase the share of bioethanol fuels in the global fuel portfolio as a strong alternative to crude oil-based gasoline and petrodiesel fuels. In this context, it is expected that the most prolific researchers, institutions, countries, funding bodies, and journals in this field would have a first-mover advantage to benefit from such potential incentives.

It is recommended that such review studies are performed for the primary research fronts of wood-based bioethanol fuels.

ACKNOWLEDGMENTS

The contribution of the highly cited researchers in the field of wood-based bioethanol fuels has been gratefully acknowledged.

REFERENCES

Alvira, P., E. Tomas-Pejo, M. Ballesteros and M. J. Negro. 2010. Pretreatment technologies for an efficient bioethanol production process based on enzymatic hydrolysis: A review. *Bioresource Technology* 101:4851–4861.

Angelici, C., B. M. Weckhuysen and P. C. A. Bruijnincx. 2013. Chemocatalytic conversion of ethanol into butadiene and other bulk chemicals. *ChemSusChem* 6:1595–1614.

Antolini, E. 2007. Catalysts for direct ethanol fuel cells. *Journal of Power Sources* 170:1–12.

Antolini, E. 2009. Palladium in fuel cell catalysis. *Energy and Environmental Science* 2:915–931.

Berlin, A., M. Balakshin and N. Gilkes, N., et al. 2006. Inhibition of cellulase, xylanase and β-glucosidase activities by softwood lignin preparations. *Journal of Biotechnology* 125:198–209.

Brandt, A., J. P. Hallett, D. J. Leak, R. J. Murphy and T. Welton. 2010. The effect of the ionic liquid anion in the pretreatment of pine wood chips. *Green Chemistry* 12:672–679.

Burnham, J. F. 2006. Scopus database: A review. *Biomedical Digital Libraries* 3:1–8.

Esteghlalian, A., A. G. Hashimoto, J. J. Fenske and M. H. Penner. 1997. Modeling and optimization of the dilute-sulfuric-acid pretreatment of corn stover, poplar and switchgrass. *Bioresource Technology* 59:129–136.

Fernando, S., S. Adhikari, C. Chandrapal and M. Murali. 2006. Biorefineries: Current status, challenges, and future direction. *Energy & Fuels* 20:1727–1737.

Fort, D. A., R. C. Remsing and R. P. Swatloski, et al. 2007. Can ionic liquids dissolve wood? Processing and analysis of lignocellulosic materials with 1-*n*-butyl-3-methylimidazolium chloride. *Green Chemistry* 9:63–69.

Galbe, M. and G. Zacchi. 2002. A review of the production of ethanol from softwood. *Applied Microbiology and Biotechnology* 59:618–628.

Garrote, G., H. Domínguez and J. C. Parajo. 1999. Mild autohydrolysis: An environmentally friendly technology for xylooligosaccharide production from wood. *Journal of Chemical Technology and Biotechnology* 74:1101–1109.

Hahn-Hagerdal, B., M. Galbe, M. F. Gorwa-Grauslund, G. Liden and G. Zacchi. 2006. Bio-ethanol - The fuel of tomorrow from the residues of today. *Trends in Biotechnology* 24:549–556.

Hill, J., E. Nelson, D. Tilman, S. Polasky and D. Tiffany. 2006. Environmental, economic, and energetic costs and benefits of biodiesel and ethanol biofuels. *Proceedings of the National Academy of Sciences of the United States of America* 103:11206–11210.

Hill, J., S. Polasky and E. Nelson, et al. 2009. Climate change and health costs of air emissions from biofuels and gasoline. *Proceedings of the National Academy of Sciences of the United States of America* 106:2077–2082.

Hsieh, W. D., R. H. Chen, T. L. Wu and T. H. Lin. 2002. Engine performance and pollutant emission of an SI engine using ethanol-gasoline blended fuels. *Atmospheric Environment* 36:403–410.

Huang, H. J., S. Ramaswamy, U. W. Tschirner and B. V. Ramarao. 2008. A review of separation technologies in current and future biorefineries. *Separation and Purification Technology* 62:1–21.

Jonsson, L. J., E. Palmqvist, N. O. Nilvebrant and B. Hahn-Hagerdal. 1998. Detoxification of wood hydroly-sates with laccase and peroxidase from the white-rot fungus *Trametes versicolor. Applied Microbiology and Biotechnology* 49:691–697.

Kilpelainen, I., H. Xie and A. King, et al. 2007. Dissolution of wood in ionic liquids. *Journal of Agricultural and Food Chemistry* 55:9142–9148.

Konur, O. 2000. Creating enforceable civil rights for disabled students in higher education: An institutional theory perspective. *Disability & Society* 15:1041–1063.

Konur, O. 2002a. Access to nursing education by disabled students: Rights and duties of nursing programs. *Nurse Education Today* 22:364–374.

Konur, O. 2002b. Assessment of disabled students in higher education: Current public policy issues. *Assessment and Evaluation in Higher Education* 27:131–52.

Konur, O. 2002c. Access to employment by disabled people in the UK: Is the Disability Discrimination Act working? *International Journal of Discrimination and the Law* 5:247–279.

Konur, O. 2006a. Participation of children with dyslexia in compulsory education: Current public policy issues. *Dyslexia* 12:51–67.

Konur, O. 2006b. Teaching disabled students in higher education. *Teaching in Higher Education* 11:351–363.

Konur, O. 2007a. A judicial outcome analysis of the *Disability Discrimination Act*: A windfall for the employ-ers? *Disability & Society* 22:187–204.

Konur, O. 2007b. Computer-assisted teaching and assessment of disabled students in higher education: The interface between academic standards and disability rights. *Journal of Computer Assisted Learning* 23:207–219.

Konur, O. 2012. The evaluation of the research on the bioethanol: A scientometric approach. *Energy Education Science and Technology Part A: Energy Science and Research* 28:1051–1064.

Konur, O. 2015. Current state of research on algal bioethanol. In *Marine Bioenergy: Trends and Developments*, Ed. S. K. Kim and C. G. Lee, pp. 217–244. Boca Raton, FL: CRC Press.

Konur, O. 2019. Cyanobacterial bioenergy and biofuels science and technology: A scientometric overview. In *Cyanobacteria: From Basic Science to Applications*, Ed. A. K. Mishra, D. N. Tiwari and A. N. Rai, pp. 419–442. Amsterdam: Elsevier.

Konur, O. 2020. The scientometric analysis of the research on the bioethanol production from green mac-roalgae. In *Handbook of Algal Science, Technology and Medicine*, Ed. O. Konur, pp. 385–401. London: Academic Press.

Konur, O. 2023. Wood-based bioethanol fuels: Scientometric study. In *Feedstock-based Bioethanol Fuels. I. Non-Waste Feedstocks: Starch, Sugar, Grass, Wood, Cellulose, Algae, and Biosyngas-based Bioethanol Fuels. Handbook of Bioethanol Fuels Volume 3*, Ed. O. Konur. Boca Raton, FL: CRC Press.

Kumar, L., V. Arantes, R. Chandra and J. Saddler. 2012. The lignin present in steam pretreated softwood binds enzymes and limits cellulose accessibility. *Bioresource Technology* 103:201–208.

Kumar, R., G. Mago, V. Balan and C. E. Wyman. 2009. Physical and chemical characterizations of corn stover and poplar solids resulting from leading pretreatment technologies. *Bioresource Technology* 100:3948–3962.

Larsson, S., A. Reimann, N. O. Nilvebrant and L. J. Jonsson. 1999b. Comparison of different methods for the detoxification of lignocellulose hydrolyzates of spruce. *Applied Biochemistry and Biotechnology - Part A Enzyme Engineering and Biotechnology* 77–79:91–103.

Larsson, S., E. Palmqvist, E. and B. Hahn-Hagerdal, et al. 1999a. The generation of fermentation inhibitors during dilute acid hydrolysis of softwood. *Enzyme and Microbial Technology* 24:151–159.

Lee, S. H., T.V. Doherty, R. J. Linhardt and J. S. Dordick. 2009. Ionic liquid-mediated selective extraction of lignin from wood leading to enhanced enzymatic cellulose hydrolysis. *Biotechnology and Bioengineering* 102:1368–1376.

Li, J., G. Henriksson and G. Gellerstedt. 2007. Lignin depolymerization/repolymerization and its critical role for delignification of aspen wood by steam explosion. *Bioresource Technology* 98:3061–3068.

Lin, Y. and S. Tanaka. 2006. Ethanol fermentation from biomass resources: Current state and prospects. *Applied Microbiology and Biotechnology* 69:627–642.

Ma, X., L. Sun and C. Song. 2002. A new approach to deep desulfurization of gasoline, diesel fuel and jet fuel by selective adsorption for ultra-clean fuels and for fuel cell applications. *Catalysis Today* 77:107–116.

Mooney, C. A., S. D. Mansfield, M. G. Touhy and J. N. Saddler. 1998. The effect of initial pore volume and lignin content on the enzymatic hydrolysis of softwoods. *Bioresource Technology* 64:113–119.

Morschbacker, A. 2009. Bio-ethanol based ethylene. *Polymer Reviews* 49:79–84.

Najafi, G., B. Ghobadian and T. Tavakoli, et al. 2009. Performance and exhaust emissions of a gasoline engine with ethanol blended gasoline fuels using artificial neural network. *Applied Energy* 86:630–639.

Newman, P. W. G. and J. R. Kenworthy. 1989. Gasoline consumption and cities: A comparison of U.S. cities with a global survey. *Journal of the American Planning Association* 55:24–37.

North, D. C. 1991. Institutions. *Journal of Economic Perspectives* 5:97–112.

Olsson, L. and B. Hahn-Hagerdal. 1996. Fermentation of lignocellulosic hydrolysates for ethanol production. *Enzyme and Microbial Technology* 18:312–331.

Palonen, H., F. Tjerneld, G. Zacchi and M. Tenkanen. 2004. Adsorption of *Trichoderma reesei* CBH I and EG II and their catalytic domains on steam pretreated softwood and isolated lignin. *Journal of Biotechnology* 107:65–72.

Pan, X., C. Arato and N. Gilkes, et al. 2005. Biorefining of softwoods using ethanol organosolv pulping: Preliminary evaluation of process streams for manufacture of fuel-grade ethanol and co-products. *Biotechnology and Bioengineering* 90:473–481.

Pan, X., N. Gilkes and J. Kadla et al. 2006. Bioconversion of hybrid poplar to ethanol and co-products using an organosolv fractionation process: Optimization of process yields. *Biotechnology and Bioengineering* 94:851–861.

Pimentel, D. and T. W. Patzek. 2005. Ethanol production using corn, switchgrass, and wood; biodiesel production using soybean and sunflower. *Natural Resources Research* 14:65–76.

Plechkova, N. V. and K. R. Seddon. 2008. Applications of ionic liquids in the chemical industry. *Chemical Society Reviews* 37:123–150.

Sanchez, O. J. and C. A. Cardona. 2008. Trends in biotechnological production of fuel ethanol from different feedstocks. *Bioresource Technology* 99:5270–5295.

Schwanninger, M., J. C. Rodrigues, H. Pereira and B. Hinterstoisser. 2004. Effects of short-time vibratory ball milling on the shape of FT-IR spectra of wood and cellulose. *Vibrational Spectroscopy* 36:23–40.

Studer, M. H., J. D. DeMartini and M. F. Davis, et al. 2011. Lignin content in natural populus variants affects sugar release. *Proceedings of the National Academy of Sciences of the United States of America* 108:6300–6305.

Sun, N., M. Rahman and Y. Qin, et al. 2009. Complete dissolution and partial delignification of wood in the ionic liquid 1-ethyl-3-methylimidazolium acetate. *Green Chemistry* 11:646–655.

Sun, Y. and J. Cheng. 2002. Hydrolysis of lignocellulosic materials for ethanol production: A review. *Bioresource Technology* 83:1–11.

Taherzadeh, M. J. and K. Karimi. 2007. Enzyme-based hydrolysis processes for ethanol from lignocellulosic materials: A review. *Bioresources* 2:707–738.

Taherzadeh, M. J. and K. Karimi. 2008. Pretreatment of lignocellulosic wastes to improve ethanol and biogas production: A review. *International Journal of Molecular Sciences* 9:1621–1651.

Wingren, A., M. Galbe and G. Zacchi. 2003. Techno-economic evaluation of producing ethanol from softwood: Comparison of SSF and SHF and identification of bottlenecks. *Biotechnology Progress* 19:1109–1117.

Xiao, Z., X. Zhang, D. J. Gregg and J. N. Saddler. 2004. Effects of sugar inhibition on cellulases and β-glucosidase during enzymatic hydrolysis of softwood substrates. *Applied Biochemistry and Biotechnology - Part A Enzyme Engineering and Biotechnology* 115:1115–1126.

Zhao, Y., Y. Wang, J. Y. Zhu, A. Ragauskas and Y. Deng. 2008. Enhanced enzymatic hydrolysis of spruce by alkaline pretreatment at low temperature. *Biotechnology and Bioengineering* 99:1320–1328.

Zhu, J. Y. and X. J. Pan. 2010. Woody biomass pretreatment for cellulosic ethanol production: Technology and energy consumption evaluation. *Bioresource Technology* 101:4992–5002.

Zhu, J. Y., X. J. Pan, G. S. Wang and R. Gleisner, 2009. Sulfite pretreatment (SPORL) for robust enzymatic saccharification of spruce and red pine. *Bioresource Technology* 100:2411–2418.

Zhu, J. Y., X. J. Pan and R. S. Zalesny. 2010. Pretreatment of woody biomass for biofuel production: Energy efficiency, technologies, and recalcitrance. *Applied Microbiology and Biotechnology* 87:847–857.

Part 13

Cellulose-based Bioethanol Fuels

53 Cellulose-based Bioethanol Fuels
Scientometric Study

Ozcan Konur
(Formerly) Ankara Yildirim Beyazit University

53.1 INTRODUCTION

Crude oil-based gasoline fuels (Ma et al., 2002; Newman and Kenworthy, 1989) have been widely used in the transportation sector since the 1920s. However, there have been great public concerns over the adverse environmental and human impact of these fuels (Hill et al., 2006, 2009). Hence, biomass-based bioethanol fuels (Hill et al., 2006; Konur, 2012e, 2015, 2019, 2020a) have increasingly been used in blending gasoline fuels (Hsieh et al., 2002; Najafi et al., 2009), in fuel cells (Antolini, 2007, 2009), and in biochemical production (Angelici et al., 2013; Morschbacker, 2009) in a biorefinery context (Fernando et al., 2006; Huang et al., 2008).

Bioethanol fuels also play a critical role in maintaining energy security (Kruyt et al., 2009; Winzer, 2012) in supply shocks (Kilian, 2008, 2009) related to oil price shocks (Hamilton, 1983, 2003), COVID-19 pandemic (Fauci et al., 2020; Li et al., 2020), or wars (Hamilton, 1983; Jones, 2012) in the aftermath of the Russian invasion of Ukraine (Reeves, 2014).

However, it is necessary to pretreat the biomass (Taherzadeh and Karimi, 2008; Yang and Wyman, 2008) to enhance the yield of bioethanol (Hahn-Hagerdal et al., 2006; Sanchez and Cardona, 2008) prior to bioethanol production through hydrolysis (Sun and Cheng, 2002; Taherzadeh and Karimi, 2007) and fermentation (Lin and Tanaka, 2006; Olsson and Hahn-Hagerdal, 1996) of biomass.

One of the most studied feedstocks for bioethanol fuels has been cellulose, a key constituent of lignocellulosic feedstocks. Research in the field of cellulose-based bioethanol fuels has intensified in this context in the key research fronts of pretreatment (Cai and Zhang, 2005; Fukaya et al., 2008) and hydrolysis (Park et al., 2010; Suganuma et al., 2008) of cellulose. Thus, it emerges as a distinctive research field, complementing primarily the research on lignocellulosic biomass-based bioethanol fuels from agricultural residues, other wastes, wood, grass, and sugar and starch feedstocks.

However, it is essential to develop efficient incentive structures (North, 1991) for the primary stakeholders to enhance the research in this field (Konur, 2000, 2002a,b,c, 2006a,b, 2007a,b). Scientometric analysis has been used in this context to inform the primary stakeholders about the current state of research in this research field (Garfield, 1955; Konur, 2011, 2012a,b,c,d,e,f,g,h,i, 2015, 2018b, 2019, 2020a).

As there have been no published scientometric studies in this field, this chapter presents a scientometric study of research on cellulose-based bioethanol fuels. It examines the scientometric characteristics of both the sample and population data and presents them in the order of documents, authors, publication years, institutions, funding bodies, source titles, countries, Scopus subject categories, Scopus keywords, and research fronts.

53.2 MATERIALS AND METHODS

The search for this study was carried out using the Scopus database (Burnham, 2006) in October 2022.

As the first step for the search of the relevant literature, keywords were selected using the 200 most-cited sample papers. The selected keyword list was then optimized to obtain a representative sample of papers for this research field. This keyword list was provided in the Appendix for future replicative studies.

As the second step, two sets of data were used in this study. First, a population sample of 3,941 papers was used to examine the scientometric characteristics of the population data. Second, a sample of 197 most-cited papers, corresponding to 5% of the population papers, was used to examine the scientometric characteristics of these citation classics.

The scientometric characteristics of both these sample and population datasets were presented in the order of documents, authors, publication years, institutions, funding bodies, source titles, countries, Scopus subject categories, Scopus keywords, and research fronts.

Lastly, the key scientometric findings for both datasets were discussed to highlight the research landscape for cellulose-based bioethanol fuels. Additionally, a number of brief conclusions were drawn, and a number of relevant recommendations were made to enhance the future research landscape.

53.3 RESULTS

53.3.1 THE MOST PROLIFIC DOCUMENTS ON CELLULOSE-BASED BIOETHANOL FUELS

Information on the types of documents for both datasets is given in Table 53.1. Articles and conference papers, published in journals, dominate both the sample (86%) and population (94%) papers with an 8% deficit. Further, review papers and short surveys have an 11% surplus as they are over-represented in the sample papers as they constitute 14% and 3% of the sample and population papers, respectively.

It is further notable that 98%, 1%, and 1% of the population papers were published in journals, books, and book series, respectively. Similarly, 98% and 3% of the sample papers were published in journals and book series, respectively.

TABLE 53.1
Documents in Cellulose-based Bioethanol Fuels

Documents	Sample Dataset (%)	Population Dataset (%)	Surplus (%)
Article	81.7	92.2	−10.5
Review	13.7	2.7	11.0
Conference paper	3.6	2.1	1.5
Letter	0.5	1.1	−0.6
Short Survey	0.5	0.2	0.3
Book chapter	0.0	0.8	−0.8
Note	0.0	0.7	−0.7
Book	0.0	0.6	−0.6
Editorial	0.0	0.1	−0.1
Sample size	197	3,941	

Population dataset, The number of papers (%) in the set of the 3,941 population papers; Sample dataset, The number of papers (%) in the set of 197 highly cited papers.

53.3.2 The Most Prolific Authors on Cellulose-based Bioethanol Fuels

Information about the 26 most prolific authors with at least 1.5% of sample papers each is given in Table 53.2. The most prolific authors are Lee R. Lynd and Charles E. Wyman, with 3% of the sample papers each, followed by Roberto Rinaldi with 2.5% of the sample papers. Other prolific authors are Thomas Heinze, Michael E. Himmel, Arthur J. Ragauskas, Jianji Wang, Liangtseng Fan, Ferdi Schuth, and Yi-Heng P. Zhang, with 2% of the sample papers each.

On the other hand, the most influential author is Lee R. Lynd with 2.7% of the sample papers, followed by Charles E. Wyman with 2.5% of the sample papers. Other influential authors are Roberto Rinaldi, Ferdi Schuth, Yi-Heng P. Zhang, Liangtseng Fan, and Jianji Wang, with a 1.8%–2.2% surplus each.

The most prolific institution for the sample dataset is the University of British Columbia with three authors, followed by the National Renewable Energy Laboratory (NREL) and the University of Natural Science and Life Sciences with two authors each. On the other hand, the most prolific country for the sample dataset is the USA with eight authors, followed by China and Canada with three authors each. Other prolific countries are Austria, Germany, Israel, and the UK, with two authors each. In total, only 11 countries house these top authors.

The most prolific research front for these top authors is the pretreatment of cellulose with 26 authors followed by hydrolysis of cellulose with 23 authors. Other research fronts are fermentation of cellulose-based hydrolysates and bioethanol production with one author each.

On the other hand, there is a significant gender deficit (Beaudry and Lariviere, 2016) for the sample dataset as surprisingly only one of these top researchers is female with a representation rate of 4%.

Additionally, there are other authors with a relatively low citation impact and with 0.3%–0.5% of the population papers each: Arkady P. Sinitsyn, David B. Wilson, Antje Potthast, Haibo Xie, Tim Liebert, A. Whaaed Khan, Thomas Rosenau, George T. Tsao, Jun Zhang, Omar A. el Seoud, Alexander V. Gusakov, Herbert Sixta, Atsushi Fukuoa, Takahisa Kanda, Frank Meister, Bern Nidetzky, Burkart Philipp, Sheng Zhang, Hirokazu Kobayashi, Akihiko Kondo, Goran Pettersson, Kenji Takahashi, Wilfred Chen, Maria J. Cocero, Bruce E. Dale, Gunnar Johansson, Ilkka Kilpelainen, Alistair W T King, Kosuke Kuroda, Y. Ogiwara, Masahisa Wada, Larry P. Walker, and Vladimir V. Zverlov.

53.3.3 The Most Prolific Research Output by Years on Cellulose-based Bioethanol Fuels

Information about papers published between 1970 and 2022 is given in Figure 53.1. This figure clearly shows that the bulk of the research papers in the population dataset was published primarily in the 2010s and early 2020s, with 42% and 13% of the population dataset, respectively. Similarly, publication rates for the 2000s, 1990s, 1980s, and 1970s were 13%, 10%, 10%, and 7%, respectively. Further, the publication rate for the pre-1970s was 5%.

Similarly, the bulk of the research papers in the sample dataset was published in the 2010s and 2000s, with 36% and 33% of the sample dataset, respectively. Similarly, publication rates for the early 1990s, 1980s, and 1970s were 18%, 7%, and 6% of the sample papers, respectively.

The most prolific publication years for the population dataset were 2021, 2018, and 2016, with 4.7%, 4.6%, and 4.7% of the dataset, respectively, while 59% of the population papers were published between 2008 and 2022. Similarly, 61% of the sample papers were published between 2005 and 2016, while the most prolific publication years were 2010, 2011, and 2008, with 11.7%, 7.1%, and 7.1% of the sample papers, respectively. The number of population papers rose between 2008 and 2012 and it lost its momentum after 2012.

TABLE 53.2
The Most Prolific Authors on Cellulose-based Bioethanol Fuels

No.	Author Name	Author Code	Sample Papers (%)	Population Papers (%)	Surplus	Institution	Country	HI	N	Res. Front
1	Wyman, Charles E.	7004396809	3.0	0.5	2.5	Univ. Calf. Riverside	USA	80	287	P, H
2	Lynd, Lee R.	35586183800	3.0	0.3	2.7	Dartmouth Coll.	USA	70	291	P, H, F, R
3	Rinaldi, Roberto	7101680610	2.5	0.3	2.2	Imperial Coll.	UK	39	86	P, H
4	Heinze, Thomas	7006547465	2.0	0.7	1.3	Friedrich Schiller Univ.	Germany	66	486	P
5	Himmel, Michael E.	7007125552	2.0	0.5	1.5	Natl. Renew. Ener. Lab.	USA	73	423	P, H
6	Ragauskas, Arthur J.	7006265204	2.0	0.3	1.7	Univ. Tennessee	USA	93	762	P, H
7	Wang, Jianji	55904673200	2.0	0.2	1.8	Henan Normal Univ.	China	69	619	P, H
8	Fan, Liangtseng	55743058500	2.0	0.2	1.8	Kansas State Univ.	USA	70	670	P, H
9	Schuth, Ferdi	7006419362	2.0	0.1	1.9	Max Planck Inst.	Germany	111	490	P, H
10	Zhang, Yi-Heng P.	34876090400	2.0	0.1	1.9	Tianjin Inst. Ind. Biotech.	China	58	178	P, H
11	Bayer, Edward A.	7201553352	1.5	0.5	1.0	Weizmann Inst. Sci.	Israel	80	417	P, H
12	Saddler, John N.	7005297559	1.5	0.4	1.1	Univ. British Columbia	Canada	99	420	P, H
13	Zhang, Lina*	55917992100	1.5	0.4	1.1	Wuhan Univ.	China	99	712	P
14	Lamed, Raphael J.	16215972100	1.5	0.4	1.1	Tel Aviv Univ.	Israel	65	228	P, H
15	Gilbert, Harry J	7202943610	1.5	0.4	1.1	Univ. Newcastle	UK	78	305	P, H
16	Kilburn, Douglas G.	35500772400	1.5	0.3	1.2	Univ. British Columbia	Canada	60	195	P, H
17	Lindman, Bjorn	35478309200	1.5	0.3	1.2	Lund Univ.	Sweden	87	525	P
18	Gilkes, Neil R.	35493889000	1.5	0.2	1.3	Univ. British Columbia	Canada	51	87	P, H
19	Henrissat, Bernard	7005911606	1.5	0.2	1.3	King Abdulaziz Univ.	S. Arabia	136	598	P, H
20	Ludwig, Roland	7201360898	1.5	0.2	1.3	Univ. Natr. Res. Life Sci.	Austria	52	229	P, H
21	Beguin, Pierre	7004815865	1.5	0.2	1.3	Inst. Pasteur	France	44	97	P, H
22	Bommarius, Andreas S.	6701329771	1.5	0.1	1.4	Georgia Inst. Technol.	USA	49	206	P, H
23	Eijsink, Vincent G. H.	7006656581	1.5	0.1	1.4	Norwegian Univ. Life Sci.	Norway	78	390	P, H
24	Haltrich, Dietmar	7007184861	1.5	0.1	1.4	Univ. Natr. Res. Life Sci.	Austria	58	266	P, H
25	Johnson, David K.	24550868900	1.5	0.1	1.4	Natl. Renew. Ener. Lab.	USA	38	100	P, H
26	Yang, Bin	7404473046	1.5	0.1	1.4	Washington State Univ.	USA	42	97	P, H

*, Female; Author code, the unique code given by Scopus to the authors; F, Fermentation of the cellulose-based hydrolysates; H, Hydrolysis of the cellulose; P, Pretreatment of the cellulose; Population papers, the number of papers authored in the population dataset; R, Bioethanol fuel production; Sample papers, the number of papers authored in the sample dataset.

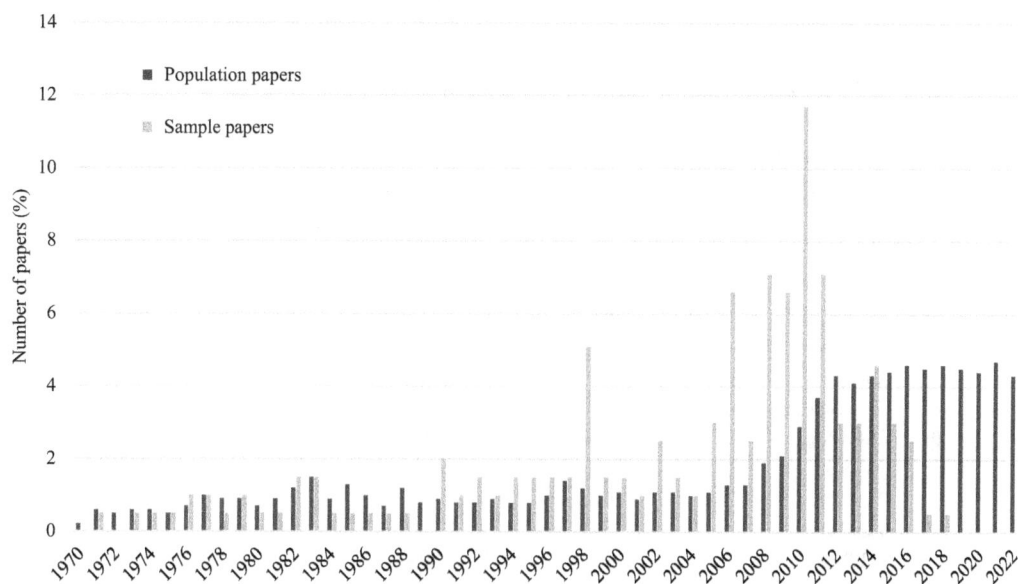

FIGURE 53.1 Research output by years regarding cellulose-based bioethanol fuels.

53.3.4 THE MOST PROLIFIC INSTITUTIONS ON CELLULOSE-BASED BIOETHANOL FUELS

Information about the 20 most prolific institutions publishing papers on cellulose-based bioethanol fuels with at least 2% of the sample papers each is given in Table 53.3.

The most prolific institutions are the Chinese Academy of Sciences and the University of British Columbia, with 5.1% of the sample papers each, followed by Dartmouth College, NREL, the National Scientific Research Centre (CNRS), and the Georgia Institute of Technology, with 3.6%–4.6% of the sample papers each. Other prolific institutions are the University of Tokyo, Friedrich Schiller University, the University of Natural Resources and Life Sciences, NC State University, and Max Planck Institute, with 2.5%–3% of the sample papers each. Similarly, the top country for these most prolific institutions is the USA with seven institutions. Other prolific institutions are China and Germany with four and two institutions, respectively. In total, nine countries house these top institutions.

On the other hand, institutions with the highest citation impact are Dartmouth College and the University of British Columbia, with a 4.1% and 4.0% surplus, respectively. Other influential institutions are the Georgia Institute of Technology, NREL, and Max Planck Institute, with a 3%, 2.9%, and 2.1% surplus each, respectively.

Additionally, there are other institutions with a relatively low citation impact and with 0.5%–1.6% of the population papers each: Russian Academy of Sciences, South China University of Technology, University of Sao Paulo, Oak Ridge National Laboratory, Kyoto University, Aalto University, National Research Council Canada, State Key Laboratory of Pulp and Paper Engineering, Cornell University, Shandong University, Purdue University, University of California Berkeley, Shinshu University, Beijing Forestry University, Lomonosov Moscow State University, Donghua University, Tohoku University, University of Science and Technology of China, University of Helsinki, RWTH Aachen University, Technical University of Munich, State University of Campinas, Uppsala University, University of Tennessee, Tianjin University, University of Illinois Urbana-Champaign, Royal Institute of Technology, Hokkaido University, Beijing University of Chemical Technology, Nanjing Forestry University, Tianjin University of Science and Technology, Weizmann Institute of Science Israel, and Tokyo Institute of Technology.

TABLE 53.3

The Most Prolific Institutions on Cellulose-based Bioethanol Fuels

No.	Institutions	Country	Sample Papers (%)	Population Papers (%)	Surplus (%)
1	Chinese Acad. Sci.	China	5.1	3.8	1.3
2	Univ. British Columbia	Canada	5.1	1.1	4.0
3	Dartmouth Coll.	USA	4.6	0.5	4.1
4	Natl. Renew. Ener. Lab.	USA	4.1	1.2	2.9
5	CNRS	France	3.6	1.7	1.9
6	Georgia Inst. Technol.	USA	3.6	0.6	3.0
7	Univ. Tokyo	Japan	3.0	1.1	1.9
8	Friedrich Schiller Univ.	Germany	2.5	0.9	1.6
9	Univ. Natr. Res. Life Sci.	Austria	2.5	0.8	1.7
10	NC State Univ.	USA	2.5	0.7	1.8
11	Max Planck Inst.	Germany	2.5	0.4	2.1
12	Lund Univ.	Sweden	2.0	1.0	1.0
14	VTT Tech. Res. Ctr.	Finland	2.0	1.0	1.0
15	Univ. Wisconsin Madison	USA	2.0	0.8	1.2
16	Univ. Calif. Riverside	USA	2.0	0.5	1.5
17	Wuhan Univ.	China	2.0	0.5	1.5
18	Beijing Natl. lab. Mol. Sci.	China	2.0	0.4	1.6
19	Kansas State Univ.	USA	2.0	0.4	1.6
20	Henan Normal Univ.	China	2.0	0.2	1.8

53.3.5 THE MOST PROLIFIC FUNDING BODIES ON CELLULOSE-BASED BIOETHANOL FUELS

Information about the 11 most prolific funding bodies funding at least 1% of the sample papers each is given in Table 53.4. Further, only 22% and 37% of the sample and population papers each were funded, respectively.

The most prolific funding body is the U.S. Department of Energy (DOE) with 2.5% of the sample papers, followed by the National Natural Science Foundation of China, National Science Foundation, European Commission, and the Seventh Framework Program, with 2% of the sample papers each.

On the other hand, the most prolific countries for these top funding bodies are the USA, China, and Europe, with two to three funding bodies each. In total, only six countries and the EU house these top funding bodies.

The funding bodies with the highest citation impact are the Seventh Framework Program and the European Commission, with a 1.6% and 1.4% surplus, respectively. Further, the funding body with the lowest citation impact is the National Natural Science Foundation of China with a 7.2% deficit. This funding body is the largest funder of the population papers with a 9.2% funding rate.

Other funding bodies with a relatively low citation impact and with 0.5%–2.7% of the population papers each are the Japan Society for the Promotion of Science, National Council for Scientific and Technological Development, Research Support Foundation of the State of Sao Paulo, Office of Science, Ministry of Education, Culture, Sports, Science and Technology, Fundamental Research Funds for the Central Universities, Chinese Academy of Sciences, Biological and Environmental Research, Higher Education Personnel Improvement Coordination, National Research Foundation of Korea, China Postdoctoral Science Foundation, National Basic Research Program of China (973 Program), Engineering and Physical Sciences Research Council, German Research Foundation, Japan Science and Technology Agency, China Scholarship Council, National High-tech Research and Development Program, and the U.S. Department of Agriculture.

TABLE 53.4
The Most Prolific Funding Bodies on Cellulose-based Bioethanol Fuels

No.	Funding Bodies	Country	Sample Paper No. (%)	Population Paper No. (%)	Surplus (%)
1	U.S. Department of Energy	USA	2.5	1.7	0.8
2	National Natural Science Foundation of China	China	2.0	9.2	−7.2
3	National Science Foundation	USA	2.0	2.1	−0.1
4	European Commission	EU	2.0	0.6	1.4
5	Seventh Framework Program	EU	2.0	0.4	1.6
6	National Key Research and Development Program of China	China	1.0	1.6	−0.6
7	Natural Sciences and Engineering Research Council of Canada	Canada	1.0	0.7	0.3
8	Foundation for Science and Technology	Portugal	1.0	0.3	0.7
9	Austrian Science Fund	Austria	1.0	0.2	0.8
10	American Chemical Society Petroleum Research Fund	USA	1.0	0.1	0.9
11	Nils and Dorthi Troedsson Research Foundation	Sweden	1.0	0.1	0.9

53.3.6 THE MOST PROLIFIC SOURCE TITLES ON CELLULOSE-BASED BIOETHANOL FUELS

Information about the 16 most prolific source titles publishing at least 1.5% of the sample papers each on cellulose-based bioethanol fuels is given in Table 53.5.

The most prolific source title is Biotechnology and Bioengineering with 12% of the sample papers, followed by Green Chemistry, Cellulose, Applied and Environmental Microbiology, and Bioresource Technology, with 5%–8% of the sample papers each. Other prolific titles are the Journal of Biological Chemistry, ChemSusChem, Biotechnology for Biofuels, and Journal of Physical Chemistry B, with 3%–4% of the sample papers each.

On the other hand, the source title with the highest citation impact is Biotechnology and Bioengineering with an 8% surplus, followed by Green Chemistry, Applied and Environmental Microbiology, and the Journal of Biological Chemistry, with a 5%, 4%, and 3% surplus, respectively. Other influential titles are ChemSusChem, Bioresource Technology, Chemical Communications, and Proceedings of the National Academy of Sciences of the United States of America, with a 2% surplus each.

Other source titles with a relatively low citation impact with 0.5%–1.7% of the population papers each are Bioresources, Biotechnology Letters, RSC Advances, ACS Sustainable Chemistry and Engineering, Enzyme and Microbial Technology, Applied Microbiology and Biotechnology, Applied Biochemistry and Biotechnology, Cellulose Chemistry and Technology, Sen I Gakkaishi, Journal of Molecular Liquids, Industrial Crops and Products, Process Biochemistry, Journal of Membrane Science, FEMS Microbiology Letters, Biomass and Bioenergy, Journal of Biotechnology, Scientific Reports, ACS Symposium Series, Applied Biochemistry and Biotechnology Part A Enzyme Engineering and Biotechnology, Chemical Engineering Journal, Biomass Conversion and Biorefinery, Archives of Biochemistry and Biophysics, Biotechnology Progress, Journal of Chemical Technology and Biotechnology, Physical Chemistry Chemical Physics, Plos One, and Macromolecular Symposia.

53.3.7 THE MOST PROLIFIC COUNTRIES ON CELLULOSE-BASED BIOETHANOL FUELS

Information about the 17 most prolific countries publishing at least 1.5% of sample papers each on cellulose-based bioethanol fuels is given in Table 53.6.

TABLE 53.5
The Most Prolific Source Titles on Cellulose-based Bioethanol Fuels

No.	Source Titles	Sample Papers (%)	Population Papers (%)	Surplus (%)
1	Biotechnology and Bioengineering	12.2	3.8	8.4
2	Green Chemistry	7.6	2.3	5.3
3	Cellulose	6.1	7.0	-0.9
4	Applied and Environmental Microbiology	5.6	1.6	4.0
5	Bioresource Technology	5.1	3.5	1.6
6	Journal of Biological Chemistry	3.6	0.8	2.8
7	ChemSusChem	3.0	0.9	2.1
8	Biotechnology for Biofuels	2.5	1.4	1.1
9	Journal of Physical Chemistry B	2.5	1.0	1.5
10	Industrial and Engineering Chemistry Research	2.0	0.9	1.1
11	Biochemical Journal	2.0	0.5	1.5
12	Proceedings of the National Academy of Sciences of the United States of America	2.0	0.4	1.6
13	Chemical Communications	2.0	0.4	1.6
14	Journal of the American Chemical Society	1.5	0.9	0.6
15	Journal of Bacteriology	1.5	0.7	0.8
16	European Journal of Biochemistry	1.5	0.3	1.2

TABLE 53.6
The Most Prolific Countries on Cellulose-based Bioethanol Fuels

No.	Countries	Sample Papers (%)	Population Papers (%)	Surplus (%)
1	USA	36.5	17.8	18.7
2	China	13.7	19.3	−5.6
3	Japan	11.7	11.0	0.7
4	Germany	10.2	5.5	4.7
5	UK	6.6	3.9	2.7
6	France	6.6	3.8	2.8
7	Sweden	6.6	3.2	3.4
8	Canada	6.1	3.7	2.4
9	Finland	3.0	2.6	0.4
10	Austria	2.5	1.5	1.0
11	India	2.0	4.4	−2.4
12	South Korea	2.0	3.0	−1.0
13	Russia	1.5	2.5	−1.0
14	Spain	1.5	2.0	−0.5
15	Israel	1.5	1.0	0.5
16	Portugal	1.5	1.0	0.5
17	Norway	1.5	0.4	1.1

The most prolific country is the USA with 37% of the sample papers, followed by China, Japan, and Germany, with 14%, 12%, and 10% of the sample papers, respectively. Other prolific countries are India, Denmark, Canada, and the UK, with 5%–9% of the sample papers each. China is also the largest producer of population papers with a 19.3% publication rate. Further, ten European countries listed in Table 53.6 produce 42% and 26% of the sample and population papers, respectively, as the largest producer of the sample papers.

On the other hand, the country with the highest citation impact is the USA with a 14% surplus, followed by Sweden and Denmark with a 10% and 5% surplus, respectively. Other influential countries are the UK, France, Sweden, and Canada with a 6%–7% surplus each. Similarly, the country with the lowest citation impact is China with a 6% deficit, followed by India, South Korea, and Russia, with a 2%, 1%, and 1% deficit, respectively.

Additionally, there are other countries with a relatively low citation impact and with 0.5%–2.8% of the sample papers each: Brazil, Australia, Malaysia, Italy, Denmark, Poland, Thailand, Iran, Egypt, Taiwan, the Netherlands, South Africa, Turkey, Singapore, the Czech Republic, Indonesia, Hungary, Romania, and Saudi Arabia.

53.3.8 THE MOST PROLIFIC SCOPUS SUBJECT CATEGORIES ON CELLULOSE-BASED BIOETHANOL FUELS

Information about the eight most prolific Scopus subject categories indexing at least 6.6% of the sample papers each is given in Table 53.7.

The most prolific Scopus subject category on cellulose-based bioethanol fuels is Biochemistry, Genetics, and Molecular Biology, with 50% of the sample papers, followed by Chemical Engineering with 44% of the sample papers. Other prolific subjects are Immunology and Microbiology, Environmental Science, and Chemistry, with 33%, 27%, and 22% of the sample papers, respectively. It is notable that Social Sciences including Economics and Business accounts for 0% and 1% of the sample and population studies, respectively.

On the other hand, the Scopus subject category with the highest citation impact is Biochemistry, Genetics, and Molecular Biology. Other influential subject areas are Immunology and Microbiology and Environmental Science, with 8% and 7% of the sample papers, respectively. Similarly, the least influential subject categories are Materials Science, Chemistry, and Agricultural and Biological Sciences, with a 6%, 5%, and 2% deficit, respectively.

53.3.9 THE MOST PROLIFIC KEYWORDS ON CELLULOSE-BASED BIOETHANOL FUELS

Information about the Scopus keywords used with at least 5.6% or 4.0% of the sample or population papers, respectively, is given in Table 53.8. For this purpose, keywords related to the keyword sets given in the Appendix are selected from a list of the most prolific keyword set provided by the Scopus database.

These keywords are grouped under five headings: cellulose, pretreatments, fermentation, hydrolysis and hydrolysates, and products.

TABLE 53.7

The Most Prolific Scopus Subject Categories on Cellulose-based Bioethanol Fuels

No.	Scopus Subject Categories	Sample Papers (%)	Population Papers (%)	Surplus (%)
1	Biochemistry, Genetics and Molecular Biology	49.7	35.1	14.6
2	Chemical Engineering	43.7	38.8	4.9
3	Immunology and Microbiology	33.0	24.4	8.6
4	Environmental Science	27.4	19.7	7.7
5	Chemistry	22.3	27.1	−4.8
6	Materials Science	18.3	24.0	−5.7
7	Energy	14.2	12.4	1.8
8	Agricultural and Biological Sciences	6.6	8.7	−2.1

TABLE 53.8

The Most Prolific Keywords on Cellulose-based Bioethanol Fuels

No.	Keywords	Sample Papers (%)	Population Papers (%)	Surplus (%)
1	Cellulose			
	Cellulose	78.7	74.1	4.6
	Biomass	17.8	8.3	9.5
	Microcrystalline cellulose	12.2	6.3	5.9
	Cellobiose	9.1	4.6	4.5
	Lignin	7.1	4.6	2.5
2	Pretreatments			
	Cellulases	34.0	21.8	12.2
	Ionic liquids	23.4	17.8	5.6
	Bacteria	17.8	11.8	6.0
	Solvents	17.8	10.3	7.5
	Enzymes	17.3	11.0	6.3
	Dissolution	16.2	8.1	8.1
	Trichoderma	14.7	7.7	7.0
	Catalysts	12.7	6.6	6.1
	Fungi	11.2	8.5	2.7
	Clostridium	9.6	4.6	5.0
	Degradation	8.1	7.1	1.0
	β-glucosidase	8.1	3.7	4.4
	Hypocrea	7.6	3.6	4.0
	Temperature	7.1	5.6	1.5
	Biodegradation	5.1	5.3	-0.2
3	Fermentation			
	Fermentation	10.2	6.9	3.3
4	Hydrolysis and hydrolysates			
	Hydrolysis	41.6	24.3	17.3
	Glucose	20.8	11.9	8.9
	Enzyme activity	17.3	11.3	6.0
	Saccharification	9.1	4.9	4.2
	Enzymatic hydrolysis	8.6	8.0	0.6
	Cellulose hydrolysis	8.6	6.4	2.2
	Enzymology	7.6	4.4	3.2
5	Products			
	Biofuels	9.1	5.0	4.1
	Ethanol	9.6	6.4	3.2

The most prolific keyword related to cellulose is cellulose, with 79% of the sample papers. Other prolific keywords are biomass and microcrystalline cellulose, with 18% and 12% of the sample papers, respectively.

Further, the most prolific keyword related to pretreatments is cellulases, with 34% of the sample papers. Other prolific keywords are ionic liquids, bacteria, solvents, enzymes, dissolution, and trichoderma, with 15%–23% of the sample papers each. Further, the most prolific keyword related to fermentation is fermentation, with 10% of the sample papers.

TABLE 53.9
The Most Prolific Thematic Research Fronts on Cellulose-based Bioethanol Fuels

No.	Research Fronts	N Paper (%) Sample
1	Biomass pretreatments	95.4
2	Biomass hydrolysis	61.4
3	Hydrolysate fermentation	4.6
4	Bioethanol production	4.6
5	Bioethanol fuel evaluation	0.0

N paper (%) sample, The number of papers in the population sample of 197 papers.

Further, the most prolific keyword related to hydrolysis and hydrolysates is hydrolysis, with 42% of the sample papers. Other prolific keywords are glucose and enzyme activity, with 21% and 17% of the sample papers, respectively. Finally, the most prolific keywords related to products are ethanol and biofuels, with 10% and 9% of the sample papers, respectively.

On the other hand, the most prolific keywords across all research fronts are cellulose, hydrolysis, cellulases, ionic liquids, glucose, bacteria, solvents, and biomass, with 18%–79% of the sample papers each. Similarly, the most influential keywords are hydrolysis, cellulases, biomass, glucose, and dissolution, with an 8%–17% surplus each.

53.3.10 THE MOST PROLIFIC RESEARCH FRONTS ON CELLULOSE-BASED BIOETHANOL FUELS

Information about the thematic research fronts for the sample papers on cellulose-based bioethanol fuels is given in Table 53.9. As this table shows, the most prolific research front is the pretreatment of lignocellulosic feedstocks, with 95% of the sample papers, followed by the hydrolysis of lignocellulosic feedstocks, with 61% of the sample papers. Other research fronts are hydrolysate fermentation and production of bioethanol fuels, with 5% of the sample papers each.

53.4 DISCUSSION

53.4.1 INTRODUCTION

Crude oil-based gasoline fuels have been widely used in the transportation sector since the 1920s. However, there have been great public concerns over the adverse environmental and human impacts of these fuels. Hence, biomass-based bioethanol fuels have increasingly been used in blending gasoline fuels, in fuel cells, and in biochemical production in a biorefinery context.

However, it is necessary to pretreat the biomass to enhance the yield of bioethanol prior to its bioethanol production through hydrolysis and fermentation. One of the most studied feedstocks for bioethanol fuels has been cellulose, a key constituent of lignocellulosic feedstocks. The research in the field of cellulose-based bioethanol fuels has intensified in this context on the key research fronts of the pretreatment and hydrolysis of cellulose. Thus, it emerges as a distinctive research field, complementing primarily the research on lignocellulosic biomass-based bioethanol fuels from agricultural residues, other wastes, wood, grass, and sugar and starch feedstocks.

However, it is essential to develop efficient incentive structures for the primary stakeholders to enhance research in this field. This is especially important to maintain energy security in the case of supply shocks such as oil price shocks, war-related shocks as in the case of the Russian invasion of Ukraine, or COVID-19 shocks.

Scientometric analysis has been used in this context to inform the primary stakeholders about the current state of research in this research field. As there has been no scientometric study in this field, this chapter presents a scientometric study of research on cellulose-based bioethanol fuels. It examines the scientometric characteristics of both the sample and population data and presents them in the order of documents, authors, publication years, institutions, funding bodies, source titles, countries, Scopus subject categories, Scopus keywords, and research fronts.

As the first step for the search of the relevant literature, keywords were selected using the 200 most-cited papers. The selected keyword list was then optimized to obtain a representative sample of papers in this research field. A copy of this extended keyword list is provided in the appendix for future replicative studies. Further, a selected list of the keywords is presented in Table 53.8.

As the second step, two sets of data were used in this study. First, a population sample of 3,941 papers was used to examine the scientometric characteristics of the population data. Second, a sample of 197 most-cited papers, corresponding to 5% of the population dataset, was used to examine the scientometric characteristics of these citation classics.

The scientometric characteristics of these sample and population datasets were presented in the order of documents, authors, publication years, institutions, funding bodies, source titles, countries, Scopus subject categories, Scopus keywords, and research fronts.

Lastly, the key scientometric findings for both datasets were discussed to highlight the research landscape for cellulose-based bioethanol fuels. Additionally, a number of brief conclusions were drawn, and a number of relevant recommendations were made to enhance the future research landscape.

53.4.2 The Most Prolific Documents on Cellulose-based Bioethanol Fuels

Articles (together with conference papers) dominate both the sample (86%) and population (94%) papers, with an 8% deficit (Table 53.1). Further, review papers have a surplus (11%), and the representation of the reviews in the sample papers is quite extraordinary (14%).

Scopus differs from the Web of Science database in differentiating and showing articles (82%) and conference papers (4%) published in journals separately. However, it should be noted that these conference papers are also published in journals as articles, compared to those published only in conference proceedings. Hence, the total number of articles and review papers in the sample dataset is 86% and 14%, respectively.

It is observed during the search process that there has been inconsistency in the classification of documents in Scopus as well as in other databases such as Web of Science. This is especially relevant for the classification of papers as reviews or articles, as papers not involving a literature review may be erroneously classified as review papers. There is also a case of review papers being classified as articles. For example, the total number of reviews in the sample data set was manually found as nearly 23% compared to 14% as indexed by Scopus, decreasing the number of articles and conference papers to 77% for the sample dataset.

In this context, it would be helpful to provide a classification note for published papers in books and journals at the first instance. It would also be helpful to use document types listed in Table 53.1 for this purpose. Book chapters may also be classified as articles or reviews as an additional classification to differentiate review chapters from experimental chapters as it is done by the Web of Science. It would be further helpful to additionally classify the conference papers as articles or review papers as well as it is done in the Web of Science database.

53.4.3 The Most Prolific Authors on Cellulose-based Bioethanol Fuels

There have been 26 most prolific authors with at least 1.5% of the sample papers each, as given in Table 53.2. These authors have shaped the development of research in this field.

The most prolific authors are Lee R. Lynd, Charles E. Wyman, and, to a lesser extent, Roberto Rinaldi, Thomas Heinze, Michael E. Himmel, Arthur J. Ragauskas, Jianji Wang, Liangtseng Fan, Ferdi Schuth, and Yi-Heng P. Zhang. Further, the most influential authors are Lee R. Lynd, Charles E. Wyman, and, to a lesser extent, Roberto Rinaldi, Ferdi Schuth, Yi-Heng P. Zhang, Liangtseng Fan, and Jianji Wang.

It is important to note the inconsistencies in the indexing of author names in Scopus and other databases. It is especially an issue for names with more than two components, such as 'Blake Sam de Hyun Lynd'. The probable outcomes are 'Lynd, B.S.D.H.', 'de Hyun Lynd, B.S.', or 'Hyun Lynd, B.S.D.'. The first choice is the gold standard of the publishing sector as the last word in the name is taken as the last name. In most of the academic databases, such as PUBMED and EBSCO databases, this version is used predominantly. The second choice is a strong alternative, while the last choice is an undesired outcome as two last words are taken as the last name. It is good practice to combine the words of the last name with a hyphen: 'Hyun- Lynd, B.S.D.'. It is notable that inconsistent indexing of author names may cause substantial inefficiencies in the search process for papers as well as allocating credit to authors as there are different author entries for each outcome in the databases.

There are also inconsistencies in the shortening of Chinese names. For example, 'YangYing Zhang' is often shortened as 'Zhang, Y.', 'Zhang, Y.-Y.', and 'Zhang, Y.Y.', as it is done in the Web of Science database as well. However, the gold standard in this case is 'Zhang, Y', where the last word is taken as the last name and the first word is taken as a single forename. In most of the academic databases, such as PUBMED and EBSCO, this first version is used predominantly. However, it makes sense to use the third option to differentiate Chinese names efficiently: 'Zhang, Y.Y.'. Therefore, there have been difficulties in locating papers for Chinese authors. In such cases, the use of the unique author codes provided for each author by the Scopus database has been helpful.

There is also a difficulty in allocating credit for authors, especially for those with common names such as 'Zhang, X.', in conducting scientometric studies. These difficulties strongly influence the efficiency of scientometric studies as well as allocating credit to authors as there are the same author entries for different authors with the same name, e.g., 'Zhang, X.' in the databases.

In this context, the coding of authors in the Scopus database is a welcome innovation compared to other databases such as Web of Science. In this process, Scopus allocates a unique number to each author in the database (Aman, 2018). However, there might still be substantial inefficiencies in this coding system, especially for common names. For example, some of the papers by a certain author may be allocated to another researcher with a different author code. It is possible that Scopus uses a number of software programs to differentiate author names, and the program may not be false-proof (Kim, 2018).

In this context, it does not help that author names are not given in full in some journals and books. This makes it difficult to differentiate authors with common names and makes scientometric studies further difficult in the author domain. Therefore, author names should be given in all books and journals at the first instance. There is also a cultural issue where some authors do not use their full names in their papers. Instead, they use initials for their forenames: 'Lynd, H.J.', 'Lynd, H.', or 'Lynd, J.', instead of 'Lynd, Hyun Jae'.

There are also inconsistencies in the naming of authors with more than two components by authors themselves in journal papers and book chapters. For example. 'Lynd, A.P.C.' might be given as 'Lynd, A.' or 'Lynd, A.C.' or 'Lynd, A.P.' or 'Lynd, C.' in journals and books. This also makes scientometric studies difficult in the author domain. Hence, contributing authors should use their name consistently in their publications.

Another critical issue regarding author names is the inconsistencies in the spelling of author names in the national spellings (e.g., Özğümüş, Şenkökçe) rather than in the English spellings (e.g., Ozgumus, Senkokce) in the Scopus database. Scopus differs from the Web of Science database and many other databases in this respect, where author names are given only in their English spelling. It is observed that the national spelling of author names do not help much in conducting scientometric studies as well as in allocating credit to authors as sometimes there are different author entries for English and National spellings in the Scopus database.

The most prolific institutions for the sample dataset are the University of British Columbia and, to a lesser extent, NREL and the University of Natural Science and Life Sciences. Further, the most prolific countries for the sample dataset are the USA and, to a lesser extent, China, Canada, Austria, Germany, Israel, and the UK. These findings confirm the dominance of the USA, Europe, and, to a lesser extent, China, Canada, and Israel in this field. On the other hand, the primary research fronts are the pretreatment and hydrolysis of cellulose and, to a lesser extent, fermentation of cellulose-based hydrolysates and production of bioethanol fuels from cellulose.

It is also notable that there is a significant gender deficit in the sample dataset surprisingly with a representation rate of 14%. This finding is the most thought-provoking with strong public policy implications. Hence, institutions, funding bodies, and policymakers should take efficient measures to reduce the gender deficit in this field as well as in other scientific fields with a strong gender deficit. In this context, it is worth noting that the level of representation of researchers from minority groups in science on the basis of race, sexuality, age, and disability, besides gender (Blankenship, 1993; Dirth and Branscombe, 2017; Konur, 2000, 2002a,b,c, 2006a,b, 2007a,b).

53.4.4   THE MOST PROLIFIC RESEARCH OUTPUT BY YEARS ON CELLULOSE-BASED BIOETHANOL FUELS

The research output observed between 1970 and 2022 is illustrated in Figure 53.1. This figure clearly shows that the bulk of the research papers in the population dataset was published primarily in the 2010s and early 2020s. Similarly, the bulk of the research papers in the sample dataset was published in the 2010s and 2000s.

These findings suggest that the most prolific sample and population papers were primarily published in the 2010s. Further, a significant portion of the sample and population papers were published in the early 2020s and 2000s, respectively.

These are the thought-provoking findings as there has been a significant research boom since 2009 and 2005 in the population and sample papers, respectively. In this context, the increasing public concerns about climate change (Change, 2007), greenhouse gas emissions (Carlson et al., 2017), and global warming (Kerr, 2007) have been certainly behind the research boom in this field since 2007. Furthermore, the recent supply shocks due to the COVID-19 pandemic and the Ukrainian war might also be behind the research boom in this field since 2019. However, it is notable that the research output lost its momentum after 2012.

Based on these findings, the size of the population papers is likely to more than double in the current decade, provided that the public concerns about climate change, greenhouse gas emissions, and global warming, as well as supply shocks, are translated efficiently to the research funding in this field.

53.4.5   THE MOST PROLIFIC INSTITUTIONS ON CELLULOSE-BASED BIOETHANOL FUELS

The 20 most prolific institutions publishing papers on cellulose-based bioethanol fuels, with at least 2% of the sample papers each given in Table 53.3, have shaped the development of research in this field.

The most prolific institutions are the Chinese Academy of Sciences, the University of British Columbia, and, to a lesser extent, Dartmouth College, NREL, CNRS, the Georgia Institute of Technology, the University of Tokyo, Friedrich Schiller University, the University of Natural Resources and Life Sciences, NC State University, and the Max Planck Institute. Similarly, top countries for these most prolific institutions are the USA and, to a lesser extent, China and Germany. In total, nine countries house these top institutions.

On the other hand, institutions with the highest citation impact are Dartmouth College, the University of British Columbia, and, to a lesser extent, the Georgia Institute of Technology, NREL, and Max Planck Institute. These findings confirm the dominance of institutions from the USA, Europe, and, to a lesser extent, China, Japan, and Canada.

53.4.6   THE MOST PROLIFIC FUNDING BODIES ON CELLULOSE-BASED BIOETHANOL FUELS

The 11 most prolific funding bodies funding at least 1% of the sample papers each are given in Table 53.4. It is notable that only 22% and 37% of the sample and population papers were funded, respectively.

The most prolific funding bodies are the US DOE and, to a lesser extent, the National Natural Science Foundation of China, the National Science Foundation, the European Commission, and the Seventh Framework Program. On the other hand, the most prolific countries for these top funding bodies are the USA, China, and Europe. In total, only six countries and the EU house these top funding bodies. The National Natural Science Foundation of China is the largest funder of the population papers.

Funding bodies with the highest citation impact are the Seventh Framework Program and the European Commission, while the funding body with the lowest citation impact is the National Natural Science Foundation of China.

These findings on the funding of the research in this field suggest that the level of funding, mostly since 2010, is relatively small, but nevertheless it has been largely instrumental in enhancing research in this field (Ebadi and Schiffauerova, 2016) in light of North's institutional framework (North, 1991). It is also notable that the funding rate in this field is relatively small compared to those in other research fronts of bioethanol fuels, such as algal bioethanol fuels. Further, it is expected that this funding rate would improve in light of the recent supply shocks and the stagnation of the research output for population papers after 2012. Further, it emerges that the USA, China, and Europe have heavily funded the research on cellulose-based bioethanol fuels.

53.4.7 The Most Prolific Source Titles on Cellulose-based Bioethanol Fuels

The 16 most prolific source titles publishing at least 1.5% of the sample papers each on cellulose-based bioethanol fuels have shaped the development of the research in this field (Table 53.5).

The most prolific source titles are Biotechnology and Bioengineering and, to a lesser extent, Green Chemistry, Cellulose, Applied and Environmental Microbiology, Bioresource Technology, Journal of Biological Chemistry, ChemSusChem, Biotechnology for Biofuels, and Journal of Physical Chemistry B. On the other hand, the source titles with the highest citation impact are Biotechnology and Bioengineering and, to a lesser extent, Green Chemistry, Applied and Environmental Microbiology, Journal of Biological Chemistry, ChemSusChem, Bioresource Technology, Chemical Communications, and Proceedings of the National Academy of Sciences of the United States of America.

It is notable that these top source titles are primarily related to biotechnology and, to a lesser extent, microbial technology, bioresources, and cellulose. This finding suggests that Biotechnology and Bioengineering and other prolific journals in these fields have significantly shaped the development of research in this field as they focus primarily on cellulose-based bioethanol fuels with a high yield. In this context, the influence of top journals is quite extraordinary.

53.4.8 The Most Prolific Countries on Cellulose-based Bioethanol Fuels

The 17 most prolific countries publishing at least 1.5% of the sample papers each have significantly shaped the development of research in this field (Table 53.6).

The most prolific countries are the USA, China, Japan, Germany, and, to a lesser extent, India, Denmark, Canada, and the UK. China is also the largest producer of population papers. Further, ten European countries listed in Table 53.6 produce 42% and 26% of the sample and population papers, respectively, as the largest producer of the sample papers.

On the other hand, countries with the highest citation impact are the USA, Sweden, and, to a lesser extent, Denmark, the UK, France, Sweden, and Canada. Similarly, countries with the lowest impact are China and, to a lesser extent, India, South Korea, and Russia.

A close examination of these findings suggests that the USA, Europe, China, Japan, and, to a lesser extent, Canada and India are the major producers of research in this field. It is a fact that the USA has been a major player in science (Leydesdorff and Wagner, 2009). The USA has further developed a strong research infrastructure to support its corn- and grass-based bioethanol industries (Gillon, 2010).

However, China has been a rising megastar in scientific research in competition with the USA and Europe (Leydesdorff and Zhou, 2005). China is also a major player in this field as a major producer of bioethanol (Fang et al., 2010).

Next, Europe has been a persistent player in scientific research in competition with both the USA and China (Leydesdorff, 2000). Europe has also been a persistent producer of bioethanol along with the USA and Brazil (Gnansounou, 2010).

Further, Japan (Negishi et al., 2004), Canada (Tahmooresnejad et al., 2015), and India (Basu and Kumar, 2000) are the other countries with substantial research activities in bioethanol fuels.

53.4.9 THE MOST PROLIFIC SCOPUS SUBJECT CATEGORIES ON CELLULOSE-BASED BIOETHANOL FUELS

The eight most prolific Scopus subject categories indexing at least 6.6% of the sample papers each, given in Table 53.7, have shaped the development of research in this field.

The most prolific Scopus subject categories on cellulose-based bioethanol fuels are Biochemistry, Genetics and Molecular Biology, Chemical Engineering, and, to a lesser extent, Immunology and Microbiology, Environmental Science, and Chemistry. It is also notable that Social Sciences including Economics and Business have a minimal presence in both sample and population studies.

On the other hand, Scopus subject categories with the highest citation impact are Biochemistry, Genetics, and Molecular Biology, and, to a lesser extent, Immunology and Microbiology, and Environmental Science. Similarly, the least influential subject categories are Materials Science, Chemistry, and Agricultural and Biological Sciences.

These findings are thought-provoking, suggesting that the primary subject categories are related to molecular biology, chemical engineering, and, to a lesser extent, microbiology and environmental sciences as the core of the research in this field concerns the production and utilization of cellulose-based bioethanol fuels. Another finding is that social sciences are not well represented in both sample and population papers in line with most of the fields in bioethanol fuels. Social, environmental, and economics studies account for the field of social sciences.

53.4.10 THE MOST PROLIFIC KEYWORDS ON CELLULOSE-BASED BIOETHANOL FUELS

A limited number of keywords have shaped the development of research in this field, as shown in Table 53.8 and the Appendix. These keywords are grouped under five headings: cellulose, pretreatments, fermentation, hydrolysis and hydrolysates, and products.

The most prolific keywords across all research fronts are cellulose, hydrolysis, cellulases, ionic liquids, glucose, bacteria, solvents, and biomass. Similarly, the most influential keywords are hydrolysis, cellulases, biomass, glucose, and dissolution.

These findings suggest that it is necessary to determine the keyword set carefully to locate the relevant research in each of these research fronts. Additionally, the size of the samples for each keyword highlights the intensity of research in the relevant research areas for both sample and population datasets. For example, Table 53.8 shows that the key research fronts for both sample and population papers are pretreatment and hydrolysis of cellulose. On the contrary, research fronts of fermentation and bioethanol production are not extensive.

These findings also highlight different spellings of some strategic keywords such as pretreatment vs. pre-treatment and ethanol vs. bio-ethanol, etc. However, there is a tendency toward the use of the connected keywords without using a hyphen.

53.4.11 THE MOST PROLIFIC RESEARCH FRONTS ON CELLULOSE-BASED BIOETHANOL FUELS

Information about thematic research fronts for sample papers on cellulose-based bioethanol fuels is given in Table 53.9. As this table shows, the most prolific research fronts are pretreatment and hydrolysis of lignocellulosic feedstocks and, to a lesser extent, hydrolysate fermentation and production of bioethanol fuels. There is no highly cited paper (HCP) for the evaluation and utilization of cellulose-bioethanol fuels.

These findings are thought-provoking in seeking ways to increase cellulose feedstock-based bio-ethanol yield at the global scale. It is clear that all of these research fronts have public importance and merit substantial funding and other incentives. Further, it is notable that cellulose-based bio-ethanol fuels have become a core unit of bioethanol research to make it more competitive with crude oil-based gasoline and petrodiesel fuels, especially for the USA, Europe, and China.

In comparison to the other feedstock-based research fronts, it is notable that the pretreatment and hydrolysis of cellulose emerge as primary research fronts for this field. However, the research fronts of fermentation of cellulose-based hydrolysates and bioethanol production from cellulose-based hydrolysates are also important.

Further, the field of evaluation and utilization of bioethanol fuels is a neglected area. This suggests that the primary stakeholders have been primarily interested in these key processes of bioethanol production. It is also notable that evaluation of cellulose-based bioethanol fuels, such as technoeconomics, life cycle, economics, social science, land use, labor, and environment-related studies, emerges as a case study for bioethanol fuels in general. Similarly, utilization of these biofuels in gasoline or diesel engines is also an important research field from a societal perspective. In this context, the USA and Brazil have been the global leaders in the production and use of corn- and sugarcane-based bioethanol fuels since the 1970s in the aftermath of the global crude oil crisis in the early 1970s.

It is further notable that research on cellulose-based bioethanol fuels complements research on lignocellulosic biomass-based bioethanol fuels from sugar and starch feedstocks, wood, grass, and wastes as cellulose is the main constituent of lignocellulosic biomass.

In the end, these most-cited papers in this field hint that the production of cellulose-based bio-ethanol fuels could be optimized using the structure, processing, and property relationships of cellulose in the fronts of feedstock pretreatment and hydrolysis, and hydrolysate fermentation (Formela et al., 2016; Konur, 2018a, 2020b, 2021a,b,c,d; Konur and Matthews, 1989).

53.5 CONCLUSION AND FUTURE RESEARCH

The research on cellulose-based bioethanol fuels has been mapped through a scientometric study of both sample (197 papers) and population (3,941 papers) datasets.

The critical issue in this study has been obtaining a representative sample of research as in any other scientometric study. Therefore, the keyword set has been carefully devised and optimized after a number of runs in the Scopus database. It is a representative sample of wider population studies. This keyword set was provided in the Appendix, and the relevant keywords are presented in Table 53.8. However, it should be noted that it has been very difficult to compile a representative keyword set since this research field has been closely connected with many other fields. Therefore, it has been necessary to compile a keyword list to exclude papers concerned with other research fields.

Another issue has been the selection of a multidisciplinary database to carry out the sciento-metric study of research in this field. For this purpose, the Scopus database has been selected. The journal coverage of this database has been notably wider than that of Web of Science and other multisubject databases.

The key scientometric properties of research in this field have been determined and discussed in this chapter. It is evident that a limited number of documents, authors, institutions, publication years, institutions, funding bodies, source titles, countries, Scopus subject categories, Scopus keywords, and research fronts have shaped the development of research in this field.

There is ample scope to increase the efficiency of scientometric studies in this field in the author and document domains by developing consistent policies and practices in both domains across all academic databases. In this respect, it seems that authors, journals, and academic databases have a lot to do. Furthermore, the significant gender deficit, as in most scientific fields, emerges as a public policy issue. Potential deficits on the basis of age, race, disability, and sexuality also need to be explored in this field as in other scientific fields.

Research in this field has boomed since 2009 and 2005 for population and sample papers, respectively, possibly promoted by public concerns about global warming, greenhouse gas emissions, and climate change. Furthermore, the recent COVID-19 pandemic and the Russian invasion of Ukraine have resulted in global supply shocks shifting the recent focus of stakeholders from crude oil-based fuels to biomass-based fuels such as bioethanol fuels. It is expected that there would be further incentives for the key stakeholders to carry out research on cellulose-based bioethanol fuels to increase the ethanol yield and to make it more competitive with crude oil-based gasoline and petrodiesel fuels. This might be true for crude oil- and foreign exchange-deficient countries to maintain energy and food security in the face of global supply shocks. However, it is notable that the research output of population papers stagnated after 2012 raising questions about its source.

The relatively modest funding rate of 22% and 37% for sample and population papers, respectively, suggests that funding in this field has significantly enhanced research in this field primarily since 2009, possibly more than doubling in the current decade. However, it is evident that there is ample room for more funding and other incentives to enhance the research in this field further in light of the recent supply shocks and the stagnation of the research output after 2012.

Institutions from the USA and, to a lesser extent, China and Germany have mostly shaped the research in this field. Further, the USA, Europe, China, Japan, and, to a lesser extent, Canada and India have been the major producers of the research in this field as the major producers and users of bioethanol fuels. It is evident that these countries have well-developed research infrastructure in bioethanol fuels and their derivatives.

It emerges that ethanol is more popular than bioethanol as a keyword, with strong implications for the search strategy. In other words, the search strategy using only bioethanol as the keyword would not be much helpful. On the other hand, the Scopus keywords are grouped under five headings: biomass, pretreatments, fermentation, hydrolysis and hydrolysates, and products.

Further, the most prolific research fronts are pretreatment and hydrolysis of cellulose and, to a lesser extent, fermentation of cellulose-based hydrolysates and bioethanol production. The first two research fronts dominate the research in this field, while the fields of evaluation and utilization of cellulose-based bioethanol fuels are a relatively neglected research field. In this context, it is notable that there is ample room for improvement in social and humanitarian aspects of research on bioethanol fuels from cellulose such as scientometric and user studies.

These findings are thought-provoking in seeking ways to increase cellulose-based bioethanol yield on a global scale. It is clear that all these research fronts have public importance and merit substantial funding and other incentives. Further, it is notable that cellulose-based bioethanol fuels have become a core unit of bioethanol research to make it more competitive with crude oil-based gasoline and petrodiesel fuels, especially in the USA, Europe, Brazil, and China. It is further notable that research on cellulose-based bioethanol emerges as a distinctive research field, complementing the research on lignocellulosic biomass-based bioethanol fuels from agricultural residues, other wastes, grass, wood, and sugar and starch feedstocks.

Thus, scientometric analysis has a great potential to gain valuable insights into the evolution of the research in this field as in other scientific fields, especially in the aftermath of significant global supply shocks such as the COVID-19 pandemic and the Russian invasion of Ukraine.

It is recommended that further scientometric studies are carried out on the primary research fronts. It is further recommended that reviews of the most-cited papers are carried out for each primary research front to complement these scientometric studies. Next, the scientometric studies of the hot papers in these primary fields need to be carried out.

ACKNOWLEDGMENTS

The contribution of the highly cited researchers in the field of cellulose-based bioethanol fuels has been gratefully acknowledged.

APPENDIX: THE KEYWORD SET FOR CELLULOSE-BASED BIOETHANOL FUELS

((TITLE (cellulose OR avicel) AND TITLE (ethanol* OR bioethanol OR hydroly* OR pretreat* OR "pre treat*" OR saccharif* OR ferment* OR ssf OR fractionat* OR detoxif* OR *cellulase OR delignif* OR "consolidated bioprocessing" OR recalcitrance OR "ionic liquid*" OR decompos* OR degrad* OR dissol* OR *degradation OR solvents OR xylanase* OR deconstruct* OR enzyme* OR xylose OR sugar OR glucosidase OR "steam treat*" OR "cellulose accessibility" OR glucanase* OR "hydrothermal treat*" OR digestibility OR breakdown)) AND NOT (TITLE (nanoscale OR nanopart* OR switchgrass OR *wood OR woody OR ligno* OR corn OR straw OR *furfural OR waste* OR stillage OR *crystals OR maize OR acetic OR paper OR bagasse OR *grass OR miscanthus OR *gels OR hamster OR husks OR forest* OR *whiskers OR derivatives OR transformers OR hydrogenation OR removal OR effluent* OR films OR "thermal decompos*" OR levulinic OR *hydrogen OR *fibres OR nanocrystal* OR alcohols OR electricity OR phenol OR *fibers OR rumen OR *sorbents OR fir OR toluene* OR *composites OR poplar OR protein OR eucalyptus OR cotton OR arabidopsis OR pine OR pinus OR *algae OR nano OR polynucleotide OR dye OR bamboo OR batteries OR pyrolysis OR spruce OR stones OR populous OR hmf OR *butanol OR *beet OR caproic OR lactose OR nanofibrils) OR SUBJAREA (medi OR phar OR vete OR heal OR neur) OR SRCTITLE (polymer* OR macromolecules OR biomacromolecules OR hydrogen OR materials OR food* OR pyrolysis OR organic OR plant* OR composites OR soil OR insect* OR biomaterials OR poultry OR ecology OR animal OR chromat* OR carbohydrate* OR proteins OR botan* OR drug* OR textile* OR fibre* OR holz* OR thermal* OR dairy OR wood OR nano* OR paper OR waste OR makromol* OR dye* OR anal* OR lwt))) AND (LIMIT-TO (SRCTYPE, "j") OR LIMIT-TO (SRCTYPE, "b") OR LIMIT-TO (SRCTYPE, "k")) AND (LIMIT-TO (DOCTYPE, "ar") OR LIMIT-TO (DOCTYPE, "re") OR LIMIT-TO (DOCTYPE, "cp") OR LIMIT-TO (DOCTYPE, "ch") OR LIMIT-TO (DOCTYPE, "ed") OR LIMIT-TO (DOCTYPE, "sh") OR LIMIT-TO (DOCTYPE, "le") OR LIMIT-TO (DOCTYPE, "no") OR LIMIT-TO (DOCTYPE, "bk")) AND (LIMIT-TO (LANGUAGE, "English")))

REFERENCES

Aman, V. 2018. Does the Scopus author ID suffice to track scientific international mobility? A case study based on Leibniz laureates. *Scientometrics* 117:705–720.

Angelici, C., B. M. Weckhuysen and P. C. A. Bruijnincx. 2013. Chemocatalytic conversion of ethanol into butadiene and other bulk chemicals. *ChemSusChem* 6:1595–1614.

Antolini, E. 2007. Catalysts for direct ethanol fuel cells. *Journal of Power Sources* 170:1–12.

Antolini, E. 2009. Palladium in fuel cell catalysis. *Energy and Environmental Science* 2:915–931.

Basu, A. and B. V. Kumar. 2000. International collaboration in Indian scientific papers. *Scientometrics* 48:381–402.

Beaudry, C. and V. Lariviere. 2016. Which gender gap? Factors affecting researchers' scientific impact in science and medicine. *Research Policy* 45:1790–1817.

Blankenship, K. M. 1993. Bringing gender and race in: US employment discrimination policy. *Gender & Society* 7:204–226.

Burnham, J. F. 2006. Scopus database: A review. *Biomedical Digital Libraries* 3:1–8.

Cai, J. and L. Zhang. 2005. Rapid dissolution of cellulose in LiOH/urea and NaOH/urea aqueous solutions. *Macromolecular Bioscience* 5:539–548.

Carlson, K. M., J. S. Gerber and D. Mueller, et al. 2017. Greenhouse gas emissions intensity of global croplands. *Nature Climate Change* 7:63–68.

Change, C. 2007. Climate change impacts, adaptation and vulnerability. *Science of the Total Environment* 326:95–112.

Dirth, T. P. and N. R. Branscombe. 2017. Disability models affect disability policy support through awareness of structural discrimination. *Journal of Social Issues* 73:413–442.

Ebadi, A. and A. Schiffauerova. 2016. How to boost scientific production? A statistical analysis of research funding and other influencing factors. *Scientometrics* 106:1093–1116.

Fang, X., Y. Shen, J. Zhao, X. Bao and Y. Qu. 2010. Status and prospect of lignocellulosic bioethanol production in China. *Bioresource Technology* 101:4814–4819.

Fauci, A. S., H. C. Lane and R. R. Redfield. 2020. Covid-19-navigating the uncharted. *New England Journal of Medicine* 382:1268–1269.

Fernando, S., S. Adhikari, C. Chandrapal and M. Murali. 2006. Biorefineries: Current status, challenges, and future direction. *Energy & Fuels* 20:1727–1737.

Formela, K., A. Hejna, L. Piszczyk, M. R. Saeb and X. Colom. 2016. Processing and structure-property relationships of natural rubber/wheat bran biocomposites. *Cellulose* 23:3157–3175.

Fukaya, Y., K. Hayashi, M. Wada and H. Ohno. 2008. Cellulose dissolution with polar ionic liquids under mild conditions: Required factors for anions. *Green Chemistry* 10:44–46.

Garfield, E. 1955. Citation indexes for science. *Science* 122:108–111.

Gillon, S. 2010. Fields of dreams: Negotiating an ethanol agenda in the Midwest United States. *Journal of Peasant Studies* 37:723–748.

Gnansounou, E. 2010. Production and use of lignocellulosic bioethanol in Europe: Current situation and perspectives. *Bioresource Technology* 101:4842–4850.

Hahn-Hagerdal, B., M. Galbe, M. F. Gorwa-Grauslund, G. Liden and G. Zacchi. 2006. Bio-ethanol - The fuel of tomorrow from the residues of today. *Trends in Biotechnology* 24:549–556.

Hamilton, J. D. 1983. Oil and the macroeconomy since World War II. *Journal of Political Economy* 91:228–248.

Hamilton, J. D. 2003. What is an oil shock? *Journal of Econometrics* 113:363–398.

Hill, J., E. Nelson, D. Tilman, S. Polasky and D. Tiffany. 2006. Environmental, economic, and energetic costs and benefits of biodiesel and ethanol biofuels. *Proceedings of the National Academy of Sciences of the United States of America* 103:11206–11210.

Hill, J., S. Polasky and E. Nelson, et al. 2009. Climate change and health costs of air emissions from biofuels and gasoline. *Proceedings of the National Academy of Sciences of the United States of America* 106:2077–2082.

Hsieh, W. D., R. H. Chen, T. L. Wu and T. H. Lin. 2002. Engine performance and pollutant emission of an SI engine using ethanol-gasoline blended fuels. *Atmospheric Environment* 36:403–410.

Huang, H. J., S. Ramaswamy, U. W. Tschirner and B. V. Ramarao. 2008. A review of separation technologies in current and future biorefineries. *Separation and Purification Technology* 62:1–21.

Jones, T. C. 2012. America, oil, and war in the Middle East. *Journal of American History* 99:208–218.

Kerr, R. A. 2007. Global warming is changing the world. *Science* 316:188–190.

Kilian, L. 2008. Exogenous oil supply shocks: How big are they and how much do they matter for the US economy? *Review of Economics and Statistics* 90:216–240.

Kilian, L. 2009. Not all oil price shocks are alike: Disentangling demand and supply shocks in the crude oil market. *American Economic Review*, 99:1053–1069.

Kim, J. 2018. Evaluating author name disambiguation for digital libraries: A case of DBLP. *Scientometrics* 116:1867–1886.

Konur, O. 2000. Creating enforceable civil rights for disabled students in higher education: An institutional theory perspective. *Disability & Society* 15:1041–1063.

Konur, O. 2002a. Access to nursing education by disabled students: Rights and duties of nursing programs. *Nurse Education Today* 22:364–374.

Konur, O. 2002b. Assessment of disabled students in higher education: Current public policy issues. *Assessment and Evaluation in Higher Education* 27:131–152.

Konur, O. 2002c. Access to employment by disabled people in the UK: Is the Disability Discrimination Act working? *International Journal of Discrimination and the Law* 5:247–279.

Konur, O. 2006a. Participation of children with dyslexia in compulsory education: Current public policy issues. *Dyslexia* 12:51–67.

Konur, O. 2006b. Teaching disabled students in higher education. *Teaching in Higher Education* 11:351–363.

Konur, O. 2007a. A judicial outcome analysis of the *Disability Discrimination Act*: A windfall for the employers? *Disability & Society* 22:187–204.

Konur, O. 2007b. Computer-assisted teaching and assessment of disabled students in higher education: The interface between academic standards and disability rights. *Journal of Computer Assisted Learning* 23:207–219.

Konur, O. 2011. The scientometric evaluation of the research on the algae and bio-energy. *Applied Energy* 88:3532–3540.

Konur, O. 2012a. The evaluation of the biogas research: A scientometric approach. *Energy Education Science and Technology Part A: Energy Science and Research* 29:1277–1292.

Konur, O. 2012b. The evaluation of the educational research: A scientometric approach. *Energy Education Science and Technology Part B: Social and Educational Studies* 4:1935–1948.

Konur, O. 2012c. The evaluation of the global energy and fuels research: A scientometric approach. *Energy Education Science and Technology Part A: Energy Science and Research* 30:613–628.

Konur, O. 2012d. The evaluation of the research on the biodiesel: A scientometric approach. *Energy Education Science and Technology Part A: Energy Science and Research* 28:1003–1014.

Konur, O. 2012e. The evaluation of the research on the bioethanol: A scientometric approach. *Energy Education Science and Technology Part A: Energy Science and Research* 28:1051–1064.

Konur, O. 2012f. The evaluation of the research on the biofuels: A scientometric approach. *Energy Education Science and Technology Part A: Energy Science and Research* 28:903–916.

Konur, O. 2012g. The evaluation of the research on the biohydrogen: A scientometric approach. *Energy Education Science and Technology Part A: Energy Science and Research* 29:323–338.

Konur, O. 2012h. The evaluation of the research on the microbial fuel cells: A scientometric approach. *Energy Education Science and Technology Part A: Energy Science and Research* 29:309–322.

Konur, O. 2012i. The scientometric evaluation of the research on the production of bioenergy from biomass. *Biomass and Bioenergy* 47:504–515.

Konur, O. 2015. Current state of research on algal bioethanol. In *Marine Bioenergy: Trends and Developments*, Ed. S. K. Kim and C. G. Lee, pp. 217–244. Boca Raton, FL: CRC Press.

Konur, O., Ed. 2018a. *Bioenergy and Biofuels*. Boca Raton, FL: CRC Press.

Konur, O. 2018b. Bioenergy and biofuels science and technology: Scientometric overview and citation classics. In *Bioenergy and Biofuels*, Ed. O. Konur, pp. 3–63. Boca Raton: CRC Press.

Konur, O. 2019. Cyanobacterial bioenergy and biofuels science and technology: A scientometric overview. In *Cyanobacteria: From Basic Science to Applications*, Ed. A. K. Mishra, D. N. Tiwari and A. N. Rai, pp. 419–442. Amsterdam: Elsevier.

Konur, O. 2020a. The scientometric analysis of the research on the bioethanol production from green macroalgae. In *Handbook of Algal Science, Technology and Medicine*, Ed. O. Konur, pp. 385–401. London: Academic Press.

Konur, O., Ed. 2020b. *Handbook of Algal Science, Technology and Medicine*. London: Academic Press.

Konur, O., Ed. 2021a. *Handbook of Biodiesel and Petrodiesel Fuels: Science, Technology, Health, and Environment*. Boca Raton, FL: CRC Press.

Konur, O., Ed. 2021b. *Handbook of Biodiesel and Petrodiesel Fuels: Science, Technology, Health, and Environment. Volume 1. Biodiesel Fuels: Science, Technology, Health, and Environment*. Boca Raton, FL: CRC Press.

Konur, O., Ed. 2021c. *Handbook of Biodiesel and Petrodiesel Fuels: Science, Technology, Health, and Environment. Volume 2. Biodiesel Fuels based on the Edible and Nonedible Feedstocks, Wastes, and Algae: Science, Technology, Health, and Environment*. Boca Raton, FL: CRC Press.

Konur, O., Ed. 2021d. *Handbook of Biodiesel and Petrodiesel Fuels: Science, Technology, Health, and Environment. Volume 3. Petrodiesel Fuels: Science, Technology, Health, and Environment*. Boca Raton, FL: CRC Press.

Konur, O. and F. L. Matthews. 1989. Effect of the properties of the constituents on the fatigue performance of composites: A review. *Composites* 20:317–328.

Kruyt, B., D. P. van Vuuren, H. J. de Vries and H. Groenenberg. 2009. Indicators for energy security. *Energy Policy* 37:2166–2181.

Leydesdorff, L. 2000. Is the European Union becoming a single publication system? *Scientometrics* 47:265–280.

Leydesdorff, L. and C. Wagner. 2009. Is the United States losing ground in science? A global perspective on the world science system. *Scientometrics* 78:23–36.

Leydesdorff, L. and P. Zhou. 2005. Are the contributions of China and Korea upsetting the world system of science? *Scientometrics* 63:617–630.

Li, H., S. M. Liu, X. H. Yu, S. L. Tang and C. K. Tang. 2020. Coronavirus disease 2019 (COVID-19): Current status and future perspectives. *International Journal of Antimicrobial Agents* 55:105951.

Lin, Y. and S. Tanaka. 2006. Ethanol fermentation from biomass resources: Current state and prospects. *Applied Microbiology and Biotechnology* 69:627–642.

Ma, X., L. Sun and C. Song. 2002. A new approach to deep desulfurization of gasoline, diesel fuel and jet fuel by selective adsorption for ultra-clean fuels and for fuel cell applications. *Catalysis Today* 77:107–116.

Morschbacker, A. 2009. Bio-ethanol based ethylene. *Polymer Reviews* 49:79–84.

Najafi, G., B. Ghobadian and T. Tavakoli, et al. 2009. Performance and exhaust emissions of a gasoline engine with ethanol blended gasoline fuels using artificial neural network. *Applied Energy* 86:630–639.

Negishi, M., Y. Sun and K. Shigi. 2004. Citation database for Japanese papers: A new bibliometric tool for Japanese academic society. *Scientometrics* 60:333–351.

Newman, P. W. G. and J. R. Kenworthy. 1989. Gasoline consumption and cities: A comparison of U.S. cities with a global survey. *Journal of the American Planning Association* 55:24–37.

North, D. C. 1991. Institutions. *Journal of Economic Perspectives* 5:97–112.

Olsson, L. and B. Hahn-Hagerdal. 1996. Fermentation of lignocellulosic hydrolysates for ethanol production. *Enzyme and Microbial Technology* 18:312–331.

Park, S., J. O. Baker, M. E. Himmel, P. A. Parilla and D. K. Johnson. 2010. Cellulose crystallinity index: Measurement techniques and their impact on interpreting cellulase performance. *Biotechnology for Biofuels* 3:10.

Reeves, S. 2014. To Russia with love: How moral arguments for a humanitarian intervention in Syria opened the door for an invasion of the Ukraine. *Michigan State University International Law Review* 23:199.

Sanchez, O. J. and C. A. Cardona. 2008. Trends in biotechnological production of fuel ethanol from different feedstocks. *Bioresource Technology* 99:5270–5295.

Suganuma, S., K. Nakajima and M. Kitano, et al. 2008. Hydrolysis of cellulose by amorphous carbon bearing SO_3H, COOH, and OH groups. *Journal of the American Chemical Society* 130:12787–12793.

Sun, Y. and J. Cheng. 2002. Hydrolysis of lignocellulosic materials for ethanol production: A review. *Bioresource Technology* 83:1–11.

Taherzadeh, M. J. and K. Karimi. 2007. Enzyme-based hydrolysis processes for ethanol from lignocellulosic materials: A review. *Bioresources* 2:707–738.

Taherzadeh, M. J. and K. Karimi. 2008. Pretreatment of lignocellulosic wastes to improve ethanol and biogas production: A review. *International Journal of Molecular Sciences* 9:1621–1651.

Tahmooresnejad, L., C. Beaudry, C and A. Schiffauerova. 2015. The role of public funding in nanotechnology scientific production: Where Canada stands in comparison to the United States. *Scientometrics* 102:753–787.

Winzer, C. 2012. Conceptualizing energy security. *Energy Policy* 46:36–48.

Yang, B. and C. E. Wyman. 2008. Pretreatment: The key to unlocking low-cost cellulosic ethanol. *Biofuels, Bioproducts and Biorefining* 2:26–40.

54 Cellulose-based Bioethanol Fuels

Review

Ozcan Konur

(Formerly) Ankara Yildirim Beyazit University

54.1 INTRODUCTION

The crude oil-based gasoline fuels (Ma et al., 2002; Newman and Kenworthy, 1989) have been widely used in the transportation sector since the 1920s. However, there have been great public concerns over the adverse environmental and human impact of these fuels (Hill et al., 2006, 2009). Hence, biomass-based bioethanol fuels (Hill et al., 2006; Konur, 2012, 2015, 2020) have increasingly been used in blending gasoline fuels (Hsieh et al., 2002; Najafi et al., 2009), in the fuel cells (Antolini, 2007, 2009), and in the biochemical production (Angelici et al., 2013; Morschbacker, 2009) in a biorefinery context (Fernando et al., 2006; Huang et al., 2008).

However, it is necessary to pretreat the biomass (Alvira et al., 2010; Taherzadeh and Karimi, 2008) to enhance the yield of the bioethanol (Hahn-Hagerdal et al., 2006; Sanchez and Cardona, 2008) prior to the bioethanol fuel production from the feedstocks through the hydrolysis (Sun and Cheng, 2002; Taherzadeh and Karimi, 2007) and fermentation (Lin and Tanaka, 2006; Olsson and Hahn-Hagerdal, 1996) of the biomass and hydrolysates, respectively.

One of the most-studied feedstocks for the bioethanol fuels has been the cellulose, a key constituent of the lignocellulosic feedstocks. The research in the field of the cellulose-based bioethanol fuels has intensified in this context in the key research fronts of the pretreatment (Cai and Zhang, 2005; Fukaya et al., 2008) and hydrolysis (Park et al., 2010; Suganuma et al., 2008) of the cellulose. Thus, it emerges as a distinctive research field, complementing primarily the research on the lignocellulosic biomass-based bioethanol fuels from the agricultural residues, other wastes, wood, grass, and sugar and starch feedstocks.

However, it is essential to develop efficient incentive structures (North, 1991) for the primary stakeholders to enhance the research in this field (Konur, 2000, 2002a,b,c, 2006a,b, 2007a,b). Although there have been a number of review papers on the cellulose-based bioethanol fuels (Pinkert et al., 2009; Zhang and Lynd, 2004; Zhu et al., 2006), there has been no review of the most-cited 25 papers in this field.

Thus, this book chapter presents a review of the most-cited 25 articles in the field of the cellulose-based bioethanol fuels. Then, it discusses the key findings of these highly influential papers and comments on the future research priorities in this field.

54.2 MATERIALS AND METHODS

The search for this study was carried out using Scopus database (Burnham, 2006) in October 2022.

As a first step for the search of the relevant literature, the keywords were selected using the most-cited first 200 population papers. The selected keyword list was then optimized to obtain a representative sample of papers for the searched research field. This final keyword set was provided in Appendix of Konur (2023) for future replication studies.

DOI: 10.1201/9781003226451-71

As a second step, a sample dataset was used for this study. The first 25 articles with at least 363 citations each were selected for the review study. Key findings from each paper were taken from the abstracts of these papers and were discussed. Additionally, a number of brief conclusions were drawn and a number of relevant recommendations were made to enhance the future research landscape.

54.3   RESULTS

The brief information about 25 most-cited papers with at least 363 citations each on the cellulose-based bioethanol fuels is given later. The primary research fronts are the pretreatment and hydrolysis of the cellulose with nine and 16 highly cited papers (HCPs), respectively.

54.3.1   Pretreatments of Cellulose

The brief information about nine most-cited papers on the pretreatments of cellulose with at least 363 citations each is given in Table 54.1.

Cai and Zhang (2005) dissolved cellulose in LiOH/urea and NaOH/urea aqueous solutions and evaluated the dissolution behavior and solubility of cellulose in a paper with 751 citations. They observed that cellulose having viscosity average molecular weight of 11.4×10^4 and 37.2×10^4 could be dissolved in 7% NaOH/12% urea and 4.2% LiOH/12% urea aqueous solutions, respectively, pre-cooled to $-10°C$ within 2 min, whereas all of them could not be dissolved in KOH/urea aqueous solution. The dissolution power of the solvent systems was in the order of LiOH/urea > NaOH/urea ≫ KOH/urea aqueous solution. Further, LiOH/urea and NaOH/urea aqueous solutions as non-derivatizing solvents broke the intra- and inter-molecular hydrogen bonding of cellulose and prevented the approach toward each other of the cellulose molecules, leading to the good dispersion of cellulose to form an actual solution.

Fukaya et al. (2008) dissolved cellulose with polar ionic liquids (ILs) with different cations under mild conditions in a paper with 672 citations. They prepared a series of alkylimidazolium salts containing dimethyl phosphate, methyl methylphosphonate, or methyl phosphonate by a facile, one-pot procedure as room-temperature ILs, which had the potential to solubilize cellulose under mild conditions. They found that especially, N-ethyl-N'-methylimidazolium methylphosphonate ([C_2mim]MeO) enabled the preparation of 10 wt% cellulose solution by keeping it at 45°C for 30 min with stirring and rendered soluble 2–4 wt% cellulose without pretreatments and heating.

Quinlan et al. (2011) performed the oxidative degradation of cellulose by a copper metalloenzyme in a paper with 669 citations. They showed that glycoside hydrolase (CAZy) GH61 enzymes were a unique family of copper-dependent oxidases. They further showed that copper was needed for GH61 maximal activity and that the formation of cellodextrin and oxidized cellodextrin products by GH61 was enhanced in the presence of small-molecule redox-active cofactors such as ascorbate and gallate. Further, the active site of GH61 contained a type II copper and, uniquely, a methylated histidine in the copper's coordination sphere, thus providing an innovative paradigm in bioinorganic enzymatic catalysis.

Remsing et al. (2006) explored the mechanism of cellulose dissolution in the 1-n-butyl-3- methylimidazolium chloride ([C_4mim]Cl) through A^{13}C and$^{35/37}$Cl NMR relaxation study on model systems in a paper with 656 citations. They showed that the solvation of cellulose by this IL involved hydrogen bonding between the carbohydrate hydroxyl protons and the IL chloride ions in a 1 : 1 stoichiometry.

Heinze et al. (2005) applied 1-N-butyl-3-methylimidazolium chloride ([C_4mim]$^+$Cl$^-$), 3-methyl-N-butyl-pyridinium chloride, and benzyldimethyl(tetradecyl)ammonium chloride as solvents for cellulose in a paper with 538 citations. They observed that these ILs had the ability to dissolve cellulose with a degree of polymerization in the range from 290 to 1,200 to a very high concentration. However, using [C_4mim]$^+$Cl$^-$, no degradation of the polymer appeared. This IL was a non-derivatizing solvent. This IL could be applied as a reaction medium for the synthesis of

TABLE 54.1
The Pretreatments of Cellulose

No.	Papers	Prts.	Parameters	Keywords	Lead Authors	Affil.	Cits
3	Cai and Zhang (2005)	Alkali, urea	Cellulose dissolution, alkaline solutions, solvent dissolution power, cellulose hydrogen bonding, LiOH, NaOH	Cellulose, dissolution	Zhang, Lina* 55917992100	Wuhan Univ. China	751
4	Fukaya et al. (2008)	ILs	Cellulose dissolution, ILs, cations, methylphosphonate	Cellulose, dissolution	Ohno, Hiroyuki 7403244652	Tokyo Univ. Agr. Technol. Japan	672
5	Quinlan et al. (2011)	Enzymes	Cellulose oxidative degradation, glycoside hydrolase (CAZy) GH61 enzymes, copper metalloenzyme, cellodextrin, redox-active cofactors	Cellulose, degradation	Johansen, Katja Salomon* 36473579400	Univ. Copenhagen Denmark	
6	Remsing et al. (2006)	ILs	Cellulose dissolution, ILs, mechanisms, hydrogen bonding, carbohydrate hydroxyl protons	Cellulose, dissolution, IL	Rogers, Robin D. 35474829200	Univ. Alabama USA	656
9	Heinze et al. (2005)	ILs	Cellulose dissolution, ILs, IL types	Cellulose, ILs	Heinze, Thomas 7006547465	Friedrich-Schiller-Univ. Germany	538
10	Kosan et al. (2008)	ILs	Cellulose dissolution, ILs, IL types	Cellulose, dissolution, IL	Kosan, Birgit* 6602356689	Thuringian Inst. Text. Plast. Res. Germany	522
23	Vitz et al. (2009)	ILs	Cellulose dissolution, IL screening	Cellulose, dissolution	Schubert, Ulrich S. 55154772500	Friedrich-Schiller Univ. Germany	386
24	Isogai and Atalla (1998)	Alkali	Cellulose dissolution, molecular weight, crystalline form, degree of crystallinity, cellulose structure	Cellulose, dissolution	Isogai, Akira 56492403300	Shinshu Univ. Japan	378
25	Barthel and Heinze (2006)	ILs	Cellulose dissolution, ILs, acylation carbanilation, cellulose accessibility	Cellulose, ILs	Heinze, Thomas 7006547465	Friedrich-Schiller-Univ. Germany	363

*: Female; Cits.: number of citations received for each paper; Prt: biomass pretreatments.

carboxymethyl cellulose and cellulose acetate. Without using any catalyst, cellulose derivatives with high degree of substitution (DS) could be prepared.

Kosan et al. (2008) dissolved cellulose with the ILs in a paper with 522 citations. They prepared the dopes starting from slurry of cellulose in the aqueous IL by removing the water at elevated temperature, vacuum, and high shearing rates. They used the 1-N-butyl-3-methylimidazolium chloride ([C$_4$mim]Cl), the 1-ethyl-3-methylimidazolium chloride ([Emim]Cl), the 1-N-butyl-2,3-dimethylimidazolium chloride ([Bmim]Cl), the 1-N-butyl-3-methylimidazolium acetate ([Bmim] Ac), and the 1-ethyl-3-methylimidazolium acetate ([Emim]Ac). They compared these results with cellulose solutions in N-methyl-morpholine-N-oxide monohydrate. Finally, they shaped the cellulose dopes by a dry–wet spinning process to manufacture cellulose fibers.

Vitz et al. (2009) screened a wide range of potentially suitable imidazolium-based ILs for the cellulose dissolution in a paper with 386 citations. They found an odd–even effect for different alkyl side-chain lengths of the imidazolium chlorides which could not be observed for the bromides. Furthermore, 1-ethyl-3-methylimidazolium diethyl phosphate ([C$_2$mim](C$_2$)$_2$OPO$_3$) was the best suitable for the dissolution of cellulose. Further, the dissolution under microwave irradiation resulted in almost no color change. No degradation of cellulose could be observed. In addition, this IL had a low melting point which made the viscosity of the cellulose solution lower and, thus, easier to handle.

Isogai and Atalla (1998) dissolved cellulose in aqueous NaOH solutions to study the effect of molecular weight, crystalline form, and the degree of crystallinity of the source samples in a paper with 378 citations. They developed a procedure for dissolving microcrystalline cellulose and then applied to other cellulose samples of different crystalline forms, crystallinity indices, and molecular weights. They found that the optimum conditions involved swelling cellulose in 8–9 wt % NaOH and then freezing it into a solid mass by holding it at −20°C. This was followed by thawing the frozen mass at room temperature and diluting with water to 5% NaOH. They found that all samples prepared from microcrystalline cellulose were completely dissolved in the NaOH solution by this procedure. Further, all regenerated celluloses having either cellulose II or an amorphous structure prepared from linter cellulose and kraft pulps were also essentially dissolved in the aqueous NaOH by this process. However, the original linter cellulose, its mercerized form, and cellulose III samples prepared from it had limited solubility values of only 26%–37%, when the same procedure was applied.

Barthel and Heinze (2006) studied the acylation and carbanilation of cellulose in ILs in a paper with 363 citations. They used 1-N-butyl-3-methylimidazolium chloride ([C$_4$mim]$^+$Cl$^-$), 1-N-ethyl-3-methylimidazolium chloride ([C$_2$mim]$^+$Cl$^-$), 1-N-butyldimethylimidazolium chloride ([C$_4$dmim]$^+$Cl$^-$), and 1-N-allyl-2,3-dimethylimidazolium bromide ([Admim]$^+$Br$^-$). They observed that cellulose acetates with a DS in the range from 2.5 to 3.0 were accessible within 2 h at 80°C in a complete homogeneous procedure. The acylation of cellulose with the fatty acid chloride (lauroyl chloride) led to cellulose laurate with DS from 0.34 to 1.54. The reaction started homogeneously and continues heterogeneously. The synthesis of cellulose carbanilates succeeds in the [C4mim]+Cl– without any catalyst. The new homogeneous path gave pure cellulose carbanilates. All reactions were carried out under mild conditions, low excess of reagent, and a short reaction time.

54.3.2 Hydrolysis of Cellulose

The brief information about 16 most-cited papers on the hydrolysis of cellulose with at least 403 citations each is given below (Table 54.2). These papers also cover the research on the pretreatment of the cellulose.

Park et al. (2010) compared four different techniques for measuring the crystallinity index (CI) using eight different cellulose preparations and explored their impact on interpreting cellulase performance in a paper with 2,025 citations. They found that the simplest and most widely used method which involved measurement of just two heights in the X-ray diffractogram produced significantly higher crystallinity values than did the other methods for Avicel PH-101. The alternative X-ray diffraction (XRD) and nuclear magnetic resonance (NMR) methods, which considered the contributions from amorphous and crystalline cellulose to the entire XRD and NMR spectra, provided a more accurate measure of the crystallinity of cellulose. Cellulose accessibility should be affected by crystallinity, but was also likely to be affected by several other parameters, such as lignin/hemicellulose contents and distribution, porosity, and particle size. Given the methodological dependency

TABLE 54.2
The Hydrolysis of Cellulose

No.	Papers	Prts.	Parameters	Keywords	Lead Authors	Affil.	Cits
1	Park et al. (2010)	Enzymes	CI, measurement methods, XRD, NMR, amorphous and crystalline cellulose, cellulose accessibility, cellulase–cellulose interactions	Cellulose, cellulases	Johnson, David K. 24550868900	Natl. Renew. Ener. Lab. USA	2,025
2	Suganuma et al. (2008)	C catalysts	Crystalline pure cellulose, hydrolysis, solid amorphous carbon catalysts, activation energy	Cellulose, hydrolysis	Hara, Michikazu 7403345875	Tokyo Inst. Technol. Japan	885
7	Sasaki et al. (2000)	Water	Microcrystalline cellulose dissolution and hydrolysis, cellulose decomposition and hydrolysis rate, hydrogen linkage cleavage	Cellulose, dissolution, hydrolysis	Arai, Kunio 7403965625	Tohoku Univ. Japan	608
8	Sasaki et al. (1998)	Water	Cellulose hydrolysis, hydrolysis rate, hydrolysis products, glucose decomposition rate	Cellulose, hydrolysis	Arai, Kunio 7403965625	Tohoku Univ. Japan	579
11	Onda et al. (2008)	C catalysts	Cellulose hydrolysis, glucose yield, SO_3H functional groups, hydrothermal stability	Cellulose, hydrolysis	Onda, Ayumu 56689677300	Kochi Univ. Japan	519
12	Dadi et al. (2006)	ILs	Cellulose hydrolysis, hydrolysis rates	Cellulose, saccharification, IL	Schall, Constance A.* 6603671396	Univ. Toledo USA	502
13	Rinaldi et al. (2008)	C catalysts, ILs	Cellulose hydrolysis, solid catalysts	Cellulose, ILs	Schuth, Ferdi 7006419362	Max Planck Inst. Germany	498
14	Phillips et al. (2011)	Enzymes	Cellulose degradation and hydrolysis, cellobiose dehydrogenase, Cu-dependent PMO, enzyme-catalyzed oxidation of cellulose	Cellulose, degradation	Marletta, Michael A. 7101716736	Univ. Calif. Berkeley USA	472
15	Yang and Wyman (2006)	Enzymes, BSA, acids	Cellulose hydrolysis, BSA adsorption, cellulases, cellulase adsorption, glucose yields	Cellulose, hydrolysis	Wyman, Charles E. 7004396809	Univ. Calif. Riverside USA	438

(Continued)

TABLE 54.2 (*Continued*)
The Hydrolysis of Cellulose

No.	Papers	Prts.	Parameters	Keywords	Lead Authors	Affil.	Cits
16	Hall et al. (2010)	Enzymes	Cellulose hydrolysis, cellulose crystallinity, hydrolysis rate, CI, cellulase adsorption	Cellulose, hydrolysis	Hall, Melanie* 55900277400	Univ. Graz Austria	430
17	Jeoh et al. (2007)	Enzymes	Cellulase hydrolysis, cellulose digestibility, cellulose accessibility, cellulase binding, enzyme accessibility, cellulose crytallinity	Cellulose, pretreated, cellulases	Johnson, David K. 24550868900	Natl. Renew. Ener. Lab. USA	428
18	Igarashi et al. (2011)	Enzymes, ammonia	Cellulose hydrolysis, crystalline cellulose degradation, cellulase molecules	Cellulose, cellulases	Igarashi, Kiyohiko 7402350297	VTT Tech. Res. Ctr. Finland	422
19	Henrissat et al. (1985)	Enzymes	Cellulose hydrolysis, cellobiohydrolases, endoglucanases, enzyme synergism, endo-exo cooperation	Cellulose, cellulases, degradation	Henrissat, Bernard 7005911606	King Abdulaziz Univ. S. Arabia	419
20	Fan et al. (1980)	Enzymes	Cellulose hydrolysis, cellulose crystallinity and surface area, hydrolysis rate	Cellulose, hydrolysis	Fan, Liangtseng 55743058500	Kansas State Univ. USA	416
21	Qing et al. (2010)	Enzymes	Cellulose hydrolysis, xylooligomers, hydrolysis rate, glucose yield, cellulase inhibition	Cellulose, hydrolysis	Wyman, Charles E. 7004396809	Univ. Calif. Riverside USA	409
22	Zhao et al. (2009)	ILs, enzymes	Cellulose dissolution, cellulose hydrolysis, ILs, crystallinity, accessibility, cellulase inactivation	Cellulose, hydrolysis, ILs	Zhao, Hua 7404778309	Univ. N. Carolina USA	403

*: Female; Cits.: number of citations received for each paper; Prt: biomass pretreatments.

of cellulose CI values and the complex nature of cellulase interactions with amorphous and crystalline celluloses, they cautioned against trying to correlate relatively small changes in CI with changes in cellulose digestibility.

Suganuma et al. (2008) hydrolyzed cellulose into saccharides using a range of solid catalysts in a paper with 885 citations. Crystalline pure cellulose was not hydrolyzed by conventional strong solid Brønsted acid catalysts such as niobic acid, H-mordenite, Nafion, and Amberlyst-15, whereas amorphous carbon bearing SO_3H, COOH, and OH functioned as an efficient catalyst for the reaction. The apparent activation energy for the hydrolysis of cellulose into glucose using the carbon catalyst was 110 kJ/mol, smaller than that for sulfuric acid under optimal conditions (170 kJ/mol). Further, the carbon catalyst could be readily separated from the saccharide solution after reaction for reuse in the reaction without loss of activity. They attributed the catalytic performance of the carbon catalyst to the ability of the material to adsorb $\beta\text{-}_{1,4}$ glucan, which did not adsorb to other solid acids.

Sasaki et al. (2000) dissolved and hydrolyzed microcrystalline cellulose in subcritical and supercritical water in paper with 608 citations. They performed the decomposition experiments at 25 MPa, 320°C–400°C, and 0.05–10.0 s. At 400°C, they obtained the hydrolysis products, while in 320°C–350°C water, aqueous decomposition products of glucose were the main products. Further, below 350°C, the cellulose decomposition rate was slower than the glucose and cellobiose decomposition rates, while above 350°C, the cellulose hydrolysis rate drastically increased and became higher than the glucose and cellobiose decomposition rates. However, below 280°C, cellulose particles became gradually smaller with increasing reaction time but, at high temperatures (300°C–320°C), cellulose particles disappeared with increasing transparency and much more rapidly than expected from the lower temperature results. In conclusion, cellulose hydrolysis at high temperature took place with dissolution in water due to the cleavage of intra- and intermolecular hydrogen linkages in the cellulose crystal. Thus, a homogeneous atmosphere was formed in supercritical water, and this resulted in the drastic increase of the cellulose decomposition rate above 350°C.

Sasaki et al. (1998) hydrolyzed cellulose in subcritical and supercritical water in a paper with 579 citations. They proposed a new method to hydrolyze cellulose rapidly in supercritical water to recover glucose, fructose, and oligomers (cellobiose, cellotriose, cellotetraose, etc.). They performed the decomposition experiments with a flow type reactor in the range of temperature from 290°C to 400°C at 25 MPa. They developed a high-pressure slurry feeder to feed the cellulose–water slurries. They found that the hydrolysis product yields (around 75%) in supercritical water were much higher than those in subcritical water. At a low-temperature region, the glucose or oligomer conversion rate was much faster than the hydrolysis rate of cellulose. Thus, even if the hydrolysis products, such as glucose or oligomers, were formed, their further decomposition rapidly took place and thus high yields of hydrolysis products could not be obtained. However, around the critical point, the hydrolysis rate jumped to more than an order of magnitude higher level and became faster than the glucose or oligomer decomposition rate. For this reason, they obtained a high yield of hydrolysis products in supercritical water.

Onda et al. (2008) hydrolyzed cellulose into glucose over solid acid catalysts in a paper with 519 citations. They indicated solid acid catalysis for the hydrolysis of cellulose with β-1,4-glycosidic bonds into glucose selectively higher than 90°C-%. Among the solid acid catalysts tested, such as the H-form zeolite catalysts and the sulfated and sulfonated catalysts, a sulfonated activated-carbon catalyst showed a remarkably high yield of glucose, which was due to the high hydrothermal stability and the excellent catalytic property attributed to the strong acid sites of SO_3H functional groups and the hydrophobic planes.

Dadi et al. (2006) disrupted the cellulose structure using 1-n-butyl-3-methylimidazolium chloride ([C_4mim]Cl), in a cellulose regeneration strategy which accelerated the subsequent hydrolysis reaction in a paper with 502 citations. They found that the initial enzymatic hydrolysis rates were approximately 50-fold higher for regenerated cellulose as compared to untreated cellulose (Avicel PH-101) as measured by a soluble reducing sugar assay.

Rinaldi et al. (2008) depolymerized cellulose using solid catalysts in ILs in a paper with 498 citations. They found that solid acids acted as powerful catalysts for the hydrolysis of cellulose dissolved in an IL. Cellulose underwent selective depolymerization, yielding cellulose oligomers (cellooligomers) without any substantial formation of side products. Even wood was hydrolyzed using this methodology.

Phillips et al. (2011) showed that cellobiose dehydrogenase and a copper-dependent polysaccharide monooxygenase (PMO) potentiated cellulose degradation by *Neurospora crassa* in a paper with 472 citations. These filamentous fungi used oxidative enzymes to cleave glycosidic bonds in cellulose. Deletion of *cdh-1*, the gene encoding the major cellobiose dehydrogenase of *Neurospora crassa*, reduced cellulase activity substantially, and addition of purified cellobiose dehydrogenases from *Methanosarcina thermophila* to the *Δcdh-1* strain resulted in a 1.6- to 2.0-fold stimulation in cellulase activity. Addition of cellobiose dehydrogenase to a mixture of purified cellulases showed no stimulatory effect. They showed that cellobiose dehydrogenase enhanced cellulose degradation by coupling the oxidation of cellobiose to the reductive activation of copper-dependent PMOs that catalyzed the insertion of oxygen into C–H bonds adjacent to the glycosidic linkage. Three of these PMOs had different regiospecificities resulting in oxidized products modified at either the reducing or non-reducing end of a glucan chain. They supported a direct, enzyme-catalyzed oxidation of cellulose. Cellobiose dehydrogenases and proteins related to the PMOs were found throughout both ascomycete and basidiomycete fungi, suggesting that this model for oxidative cellulose degradation might be widespread throughout the fungal kingdom. When added to mixtures of cellulases, these proteins enhanced cellulose saccharification, suggesting that they could be used to reduce the cost of biofuel production.

Yang and Wyman (2006) performed the bovine serum albumin (BSA) pretreatment to enhance enzymatic hydrolysis of cellulose in lignin containing substrates in a paper with 438 citations. They added the cellulase and BSA to Avicel cellulose and solids containing 56% cellulose and 28% lignin from dilute sulfuric acid pretreatment of corn stover. They found that little BSA was adsorbed on Avicel cellulose, while pretreated corn stover solids adsorbed considerable amounts of this protein. On the other hand, cellulase was highly adsorbed on both substrates. Adding a 1% concentration of BSA to dilute acid pretreated corn stover prior to enzyme addition at 15 FPU/g cellulose enhanced filter paper activity in solution by about a factor of 2 and β-glucosidase activity in solution by about a factor of 14. BSA treatment reduced adsorption of cellulase and particularly β-glucosidase on lignin. Similarly, BSA treatment of pretreated corn stover solids prior to enzymatic hydrolysis increased 72 h glucose yields from about 82% to about 92% at a cellulase loading of 15 FPU/g cellulose or achieved about the same yield at a loading of 7.5 FPU/g cellulose.

Hall et al. (2010) showed that cellulose crystallinity was a key predictor of the enzymatic hydrolysis rate in a paper with 430 citations. Although the crystallinity of pure cellulosic Avicel played a major role in determining the rate of hydrolysis by cellulases from *Trichoderma reesei*, they showed that it stayed constant during enzymatic conversion. They developed a convenient method for reaching intermediate degrees of crystallinity with Avicel and showed that the initial rate of the cellulase-catalyzed hydrolysis of cellulose was linearly proportional to the CI of Avicel. Despite correlation with the adsorption capacity of cellulases onto cellulose, at a given enzyme loading, the initial enzymatic rate continued to increase with a decreasing CI, even though the bound enzyme concentration stayed constant. This finding supported the determinant role of crystallinity rather than adsorption on the enzymatic rate. In the conversion of cellulose, the (021) face of the cellulose crystal was preferentially attacked by the cellobiohydrolase (Cel7A) from *T. reesei*.

Jeoh et al. (2007) showed that cellulase digestibility of pretreated biomass was limited by cellulose accessibility in a paper with 428 citations. They presented a direct method for measuring the key factors governing cellulose digestibility in a biomass sample by directly probing cellulase binding and activity using a purified cellobiohydrolase (Cel7A) from *T. reesei*. They used pretreated corn stover samples and pure cellulosic substrates to identify barriers to accessibility by this important component of cellulase preparations. They observed that cellulose conversion improved when *T.*

reesei Cel7A bound in higher concentrations, indicating that the enzyme had greater access to the substrate. Factors such as the pretreatment severity, drying after pretreatment, and cellulose crystallinity directly impacted enzyme accessibility. In conclusion, the best pretreatment schemes for rendering biomass more digestible to cellobiohydrolase enzymes were those that improved access to the cellulose in biomass cell walls, as well as those reduced the crystallinity of cell wall cellulose.

Igarashi et al. (2011) showed that the traffic jams reduced hydrolytic efficiency of cellulase on cellulose surface in a paper with 422 citations. They performed the real-time visualization of crystalline cellulose degradation by individual cellulase enzymes. They observed that the *T. reesei* cellobiohydrolase I (Cel7A) molecules slid unidirectionally along the crystalline cellulose surface but at one point exhibited collective halting analogous to a traffic jam. Further, changing the crystalline polymorphic form of cellulose by means of an ammonia treatment increased the apparent number of accessible lanes on the crystalline surface and consequently the number of moving cellulase molecules. Treatment of this bulky crystalline cellulose simultaneously or separately with *T. reesei* cellobiohydrolase II (Cel6A) resulted in a remarkable increase in the proportion of mobile enzyme molecules on the surface. Cellulose was completely degraded by the synergistic action between the two enzymes.

Henrissat et al. (1985) evaluated the action of cellobiohydrolases I and II (CBHI and CBHII) and endoglucanases I and II (EGI and EGII) purified from *T. reesei* against various substrates in a paper with 419 citations. They observed that CBHI degraded the β-D-glucan from barley in a typical endo pattern. With cellulose substrates, the synergism between CBHI and EGI or EGII depended on the structural and ultrastructural features of the substrate. This effect, unrelated to endo-exo cooperation, was found with substrates of intermediate crystallinity whereas weak or no synergism was recorded with cellulose microcrystals or the soluble carboxy-methyl cellulose derivative. They also observed the synergistic degradation of cellulose with mixtures of CBHI and CBHII. On the other hand, synergism between endoglucanases and CBHII followed the pattern expected for an endo-exo cooperation. In conclusion, there were multiple types of cooperation between the cellulolytic enzymes.

Fan et al. (1980) studied the relative effects of the crystallinity and surface area of cellulose fibers upon the enzymatic hydrolysis of cellulose and the change of the structural parameters of cellulose during the course of hydrolysis in a paper with 416 citations. They found that the hydrolysis rate was mainly dependent upon the fine structural order of cellulose which could best be represented by the crystallinity rather than the simple surface area. Further, the surface area was not a major limiting factor that slowed hydrolysis in its late stages.

Qing et al. (2010) showed that xylooligomers were strong inhibitors of cellulose hydrolysis by enzymes in a paper with 409 citations. They added xylan and various xylooligomers to Avicel hydrolysis at low enzyme loadings and found that these xylooligomers had a greater effect than adding equal amounts of xylose derived from these materials or when added separately. Furthermore, xylooligomers were inhibitorier than xylan or xylose in terms of a decreased initial hydrolysis rate and a lower final glucose yield even for a low concentration of 1.67 mg/ml. At a higher concentration of 12.5 mg/ml, xylooligomers lowered initial hydrolysis rates of Avicel by 82% and the final hydrolysis yield by 38%. Mixed DP xylooligomers showed strong inhibition on cellulase enzymes but not on β-glucosidase enzymes. Further, a large portion of the xylooligomers was hydrolyzed by Spezyme CP enzyme preparations, indicating competitive inhibition by mixed xylooligomers. Finally, xylooligomers were more powerful inhibitors than well-established glucose and cellobiose.

Zhao et al. (2009) studied a number of chloride- and acetate-based ILs for cellulose regeneration in a paper with 403 citations. They found that all regenerated celluloses were less crystalline (58%–75% lower) and more accessible to cellulase (>2 times) than untreated substrates. As a result, regenerated Avicel® cellulose, filter paper, and cotton were hydrolyzed 2–10 times faster than the respective untreated celluloses. They achieved a complete hydrolysis of Avicel® cellulose in 6 h given the *T. reesei* cellulase/substrate ratio (w/w) of 3:20 at 50°C. In addition, they observed that cellulase was more thermally stable (up to 60°C) in the presence of regenerated cellulose. The presence of various ILs during the hydrolysis induced different degrees of cellulase inactivation. Therefore, they recommended a thorough removal of IL residues after cellulose regeneration.

54.4 DISCUSSION

54.4.1 INTRODUCTION

The crude oil-based gasoline fuels have been widely used in the transportation sector since the 1920s. However, there have been great public concerns over the adverse environmental and human impact of these fuels. Hence, biomass-based bioethanol fuels have increasingly been used in blending gasoline and petrodiesel fuels, in the fuel cells, and in the biochemical production in a biorefinery context.

However, it is necessary to pretreat the biomass to enhance the yield of the bioethanol prior to the bioethanol fuel production from the feedstocks through the hydrolysis and fermentation of the biomass and hydrolysates, respectively.

One of the most-studied feedstocks for the bioethanol fuels has been the cellulose, a key constituent of the lignocellulosic feedstocks. The research in the field of the cellulose-based bioethanol fuels has intensified in this context in the key research fronts of the pretreatment and hydrolysis of the cellulose. Thus, it emerges as a distinctive research field, complementing primarily the research on the lignocellulosic biomass-based bioethanol fuels from the agricultural residues, other wastes, wood, grass, and sugar and starch feedstocks.

However, it is essential to develop efficient incentive structures for the primary stakeholders to enhance the research in this field. Although there have been a limited number of review papers for this field, there has been no review of the most-cited 25 articles in this field.

Thus, this book chapter presents a review of the most-cited 25 articles on the bioethanol fuel production and evaluation from the cellulose. Then, it discusses the key findings of these highly influential papers and comments on the future research priorities in this field.

As a first step for the search of the relevant literature, the keywords were selected using the most-cited first 200 population papers. The selected keyword list was then optimized to obtain a representative sample of papers for the searched research field. This keyword list was provided in Appendix of Konur (2023) for future replicative studies.

As a second step, a sample dataset was used for this study. The first 25 articles with at least 363 citations each were selected for the review study. Key findings from each paper were taken from the abstracts of these papers and were discussed. Additionally, a number of brief conclusions were drawn and a number of relevant recommendations were made to enhance the future research landscape.

Information about the thematic research fronts for the sample papers in the cellulose-based bioethanol fuels is given in Table 54.3. As this table shows, the most-prolific research fronts for this field are the pretreatment and hydrolysis of the cellulose with 100% and 64% of the HCPs, respectively. However, there are no HCPs for the other research fronts relating to the bioethanol production and evaluation.

TABLE 54.3
The Most-Prolific Thematic Research Fronts for the Cellulose-based Bioethanol Fuels

No.	Research fronts	N Paper (%) Review	N Paper (%) Sample	Surplus (%)
1	Biomass pretreatments	100	95	5
2	Biomass hydrolysis	64	61	3
3	Hydrolysate fermentation	0	5	−5
4	Bioethanol production	0	5	−5
5	Bioethanol fuel evaluation	0	0	0

N Paper (%) review: the number of papers in the sample of 25 reviewed papers; N paper (%) sample: the number of papers in the population sample of 197 papers.

Further, the most influential research fronts are pretreatment and hydrolysis of the food wastes with 5% and 3% surplus, respectively. Similarly, the least influential research fronts are hydrolysate fermentation, and production of bioethanol fuels with 5% deficit each.

54.4.2 Pretreatments of Cellulose

The brief information about nine most-cited papers on the pretreatments of cellulose with at least 363 citations each is given in Table 54.1. It is notable that as Table 54.3 shows, 100% of these HCPs are related to the pretreatments of the cellulose. These findings show that pretreatments of the cellulose are the fundamental processes for the bioethanol production from the cellulose.

These narrated studies highlight the importance of the pretreatment processes for the production of the bioethanol fuels from the cellulose with a high ethanol yield. These pretreatments, primarily enzymatic and chemical pretreatments, fractionate the cellulose and enhance the enzymatic digestibility of the biomass.

Cai and Zhang (2005) dissolved cellulose in LiOH/urea and NaOH/urea aqueous solutions and evaluated the dissolution behavior and solubility of cellulose and observed that cellulose could be dissolved in 7% NaOH/12% urea and 4.2% LiOH/12% urea aqueous solutions, respectively. Further, Fukaya et al. (2008) dissolved cellulose with polar ILs with different cations under mild conditions and found that especially [C$_2$mim][(MeO enabled the preparation of 10 wt% cellulose solution.

Quinlan et al. (2011) performed the oxidative degradation of cellulose by a copper metalloenzyme and showed that GH61 enzymes were a unique family of copper-dependent oxidases. Further, Remsing et al. (2006) explored the mechanism of cellulose dissolution in the [C$_4$mim]Cl on model systems and showed that the solvation of cellulose by this IL involved hydrogen bonding between the carbohydrate hydroxyl protons and the IL chloride ions in a 1:1 stoichiometry.

Heinze et al. (2005) applied [C$_4$mim]$^+$Cl$^-$, 3-methyl-N-butyl-pyridinium chloride, and benzyldimethyl(tetradecyl)ammonium chloride as solvents for cellulose and observed that these ILs had the ability to dissolve cellulose with a degree of polymerization to a very high concentration. Further, Kosan et al. (2008) dissolved cellulose with the ILs in a paper with 522 citations and used [C$_4$mim]Cl, [Emim]Cl, [Bmim]Cl, [Bmim]Ac, and [Emim]Ac.

Vitz et al. (2009) screened a wide range of potentially suitable imidazolium-based ILs for the cellulose dissolution and found that [C$_2$mim](C$_2$)$_2$OPO$_3$ was the best suitable for the dissolution of cellulose. Further, Isogai and Atalla (1998) dissolved cellulose in aqueous NaOH solutions and found that all samples prepared from microcrystalline cellulose were completely dissolved in the NaOH solution by this procedure. Finally, Barthel and Heinze (2006) studied the acylation and carbanilation of cellulose in ILs and observed that cellulose acetates with a DS in the range from 2.5 to 3.0 were accessible within 2 h at 80°C in a complete homogeneous procedure.

54.4.3 Hydrolysis of Cellulose

The brief information about 16 most-cited papers on the hydrolysis of cellulose with at least 403 citations each is given in Table 54.2. These papers also cover the research on the pretreatment of the cellulose. It is notable that as Table 54.3 shows, 64% of these HCPs are related to the hydrolysis of the cellulose. These findings show that both pretreatments and hydrolysis of the cellulose are the fundamental processes for the bioethanol production from the cellulose.

These narrated studies highlight the importance of the pretreatment and hydrolysis processes for the production of the bioethanol fuels from the cellulose with a high ethanol yield. These pretreatments, primarily enzymatic and chemical pretreatments, fractionate the cellulose and enhance the enzymatic digestibility of the biomass.

Park et al. (2010) compared four different techniques for measuring the CI using eight different cellulose preparations and explored their impact on interpreting cellulase performance and found that the simplest and most widely used method produced significantly higher crystallinity values than did the other methods for Avicel PH-101. Further, Suganuma et al. (2008) hydrolyzed cellulose into saccharides using a range of solid catalysts and found that amorphous carbon bearing SO_3H, COOH, and OH functioned as an efficient catalyst for the reaction.

Sasaki et al. (2000) dissolved and hydrolyzed microcrystalline cellulose in subcritical and super-critical water, and at 400°C, they obtained the hydrolysis products, while in 320°C–350°C water, aqueous decomposition products of glucose were the main products. Further, Sasaki et al. (1998) hydrolyzed cellulose in subcritical and supercritical water and found that the hydrolysis product yields (around 75%) in supercritical water were much higher than those in subcritical water.

Onda et al. (2008) hydrolyzed cellulose into glucose over solid acid catalysts and found that a sulfonated activated-carbon catalyst showed a remarkably high yield of glucose, which was due to the high hydrothermal stability and the excellent catalytic property. Further, Dadi et al. (2006) disrupted the cellulose structure using [C_4mim]Cl in a cellulose regeneration strategy which accel-erated the subsequent hydrolysis reaction in a paper with 502 citations.

Rinaldi et al. (2008) depolymerized cellulose using solid catalysts in ILs and found that solid acids acted as powerful catalysts for the hydrolysis of cellulose dissolved in an IL. Further, Phillips et al. (2011) showed that cellobiose dehydrogenase and a copper-dependent PMO potentiated cel-lulose degradation by *Neurospora crassa*.

Yang and Wyman (2006) performed the BSA pretreatment to enhance enzymatic hydrolysis of cellulose in lignin containing substrates and found that BSA treatment reduced adsorption of cellulase and particularly β-glucosidase on lignin. Further, Hall et al. (2010) showed that cellulose crystallinity was a key predictor of the enzymatic hydrolysis rate.

Jeoh et al. (2007) showed that cellulase digestibility of pretreated biomass was limited by cel-lulose accessibility and observed that cellulose conversion improved when *T. reesei* Cel7A bound in higher concentrations. Further, Igarashi et al. (2011) showed that the traffic jams reduced hydro-lytic efficiency of cellulase on cellulose surface and observed that the *T. reesei* cellobiohydrolase I (Cel7A) molecules slided unidirectionally along the crystalline cellulose surface but at one point exhibited collective halting analogous to a traffic jam.

Henrissat et al. (1985) evaluated the action of cellobiohydrolases I and II (CBHI and CBHII) and endoglucanases I and II (EGI and EGII) purified from *T. reesei* against various substrates and observed that CBHI degraded the β-D-glucan from barley in a typical endo pattern. Further, Fan et al. (1980) studied the relative effects of the crystallinity and surface area of cellulose fibers upon the enzymatic hydrolysis of cellulose and the change of the structural parameters of cellulose dur-ing the course of hydrolysis and found that the hydrolysis rate was mainly dependent upon the fine structural order of cellulose which could best be represented by the crystallinity rather than the simple surface area.

Qing et al. (2010) showed that xylooligomers were strong inhibitors of cellulose hydrolysis by enzymes and found that xylooligomers were inhibitorier than xylan or xylose in terms of a decreased initial hydrolysis rate and a lower final glucose yield. Further, Zhao et al. (2009) studied a number of chloride- and acetate-based ILs for cellulose regeneration and found that all regenerated celluloses were less crystalline and more accessible to cellulase than untreated substrates.

54.5 CONCLUSION AND FUTURE RESEARCH

The brief information about the key research fronts covered by the 25 most-cited papers with at least 363 citations each is given under two primary headings: the hydrolysis of the cellulose and produc-tion and evaluation of the bioethanol fuels.

The usual characteristics of these HCPs are that the pretreatments and hydrolysis of the cellulose are the primary processes for the bioethanol fuel production from cellulose to improve the ethanol yield as the celluloses are one of the most studied feedstocks at large for the bioethanol production especially for the countries with the large farmlands, forests, and crude oil deficiency.

The key findings on these research fronts should be read in light of the increasing public concerns about climate change, GHG emissions, and global warming as these concerns have been certainly behind the boom in the research on the cellulose-based bioethanol fuels as an alternative to crude oil-based gasoline and diesel fuels in the last decades. It is also a sustainable alternative to food crop -based bioethanol fuels such as corn grain-based bioethanol fuels. The recent supply shocks caused by the COVID-19 pandemics and the Russian invasion of Ukraine also highlight the importance of the production and utilization of the bioethanol fuels as an alternative to the crude oil-based gasoline and petrodiesel fuels.

The most-prolific research fronts for this field are the pretreatment and hydrolysis of the cellulose. However, there are no HCPs with the hydrolysate fermentation and bioethanol production, evaluation, and utilization.

These studies emphasize the importance of proper incentive structures for the efficient production of cellulose-based bioethanol fuels in light of North's institutional framework (North, 1991). In this context, the major producers and users of bioethanol fuels such as the USA and Brazil with vast forests and farmlands have developed strong incentive structures for the efficient cellulose-based bioethanol fuels. In light of the recent supply shocks caused primarily by the COVID-19 pandemics and Russian invasion of Ukraine, it is expected that the incentive structures such as public funding would be enhanced to increase the share of bioethanol fuels in the global fuel portfolio as a strong alternative to crude oil-based gasoline and petrodiesel fuels.

In this context, it is expected that the most-prolific researchers, institutions, countries, funding bodies, and journals in this field would have a first-mover advantage to benefit from such potential incentives. This is especially true for the US stakeholders as the USA has become the global leader in both the production and utilization of bioethanol fuels from the cellulose.

It is recommended that such review studies are performed for the primary research fronts of the cellulose-based bioethanol fuels.

ACKNOWLEDGMENTS

The contribution of the highly cited researchers in the field of the cellulose-based bioethanol fuels has been gratefully acknowledged.

REFERENCES

Alvira, P., E. Tomas-Pejo, M. Ballesteros and M. J. Negro. 2010. Pretreatment technologies for an efficient bioethanol production process based on enzymatic hydrolysis: A review. *Bioresource Technology* 101:4851–4861.

Angelici, C., B. M. Weckhuysen and P. C. A. Bruijnincx. 2013. Chemocatalytic conversion of ethanol into butadiene and other bulk chemicals. *ChemSusChem* 6:1595–1614.

Antolini, E. 2007. Catalysts for direct ethanol fuel cells. *Journal of Power Sources* 170:1–12.

Antolini, E. 2009. Palladium in fuel cell catalysis. *Energy and Environmental Science* 2:915–931.

Barthel, S. and T. Heinze. 2006. Acylation and carbanilation of cellulose in ionic liquids. *Green Chemistry* 8:301–306.

Burnham, J. F. 2006. Scopus database: A review. *Biomedical Digital Libraries* 3:1–8.

Cai, J. and L. Zhang. 2005. Rapid dissolution of cellulose in LiOH/urea and NaOH/urea aqueous solutions. *Macromolecular Bioscience* 5:539–548.

Dadi, A. P., S. Varanasi and C. A. Schall. 2006. Enhancement of cellulose saccharification kinetics using an ionic liquid pretreatment step. *Biotechnology and Bioengineering* 95:904–910.

Fan, L. T., Y. H. Lee and D. H. Beardmore. 1980. Mechanism of the enzymatic hydrolysis of cellulose: Effects of major structural features of cellulose on enzymatic hydrolysis. *Biotechnology and Bioengineering* 22:177–199.

Fernando, S., S. Adhikari, C. Chandrapal and M. Murali. 2006. Biorefineries: Current status, challenges, and future direction. *Energy & Fuels* 20:1727–1737.

Fukaya, Y., K. Hayashi, M. Wada and H. Ohno. 2008. Cellulose dissolution with polar ionic liquids under mild conditions: Required factors for anions. *Green Chemistry* 10:44–46.

Hahn-Hagerdal, B., M. Galbe, M. F. Gorwa-Grauslund, G. Liden and G. Zacchi. 2006. Bio-ethanol - The fuel of tomorrow from the residues of today. *Trends in Biotechnology* 24:549–556.

Hall, M., P. Bansal, J. H. Lee, M. J. Realff and A. S. Bommarius. 2010. Cellulose crystallinity - A key predictor of the enzymatic hydrolysis rate. *FEBS Journal* 277:1571–1582.

Heinze, T., K. Schwikal and S. Barthel. 2005. Ionic liquids as reaction medium in cellulose functionalization. *Macromolecular Bioscience* 5:520–525.

Henrissat, B., H. Driguez, C. Viet and M. Schulein. 1985. Synergism of cellulases from *Trichoderma reesei* in the degradation of cellulose. *Bio/Technology* 3:722–726.

Hill, J., E. Nelson, D. Tilman, S. Polasky and D. Tiffany. 2006. Environmental, economic, and energetic costs and benefits of biodiesel and ethanol biofuels. *Proceedings of the National Academy of Sciences of the United States of America* 103:11206–11210.

Hill, J., S. Polasky and E. Nelson, et al. 2009. Climate change and health costs of air emissions from biofuels and gasoline. *Proceedings of the National Academy of Sciences of the United States of America* 106:2077–2082.

Hsieh, W. D., R. H. Chen, T. L. Wu and T. H. Lin. 2002. Engine performance and pollutant emission of an SI engine using ethanol-gasoline blended fuels. *Atmospheric Environment* 36:403–410.

Huang, H. J., S. Ramaswamy, U. W. Tschirner and B. V. Ramarao. 2008. A review of separation technologies in current and future biorefineries. *Separation and Purification Technology* 62:1–21.

Igarashi, K., T. Uchihashi and A. Koivula, et al. 2011. Traffic jams reduce hydrolytic efficiency of cellulase on cellulose surface. *Science* 333:1279–1282.

Isogai, A. and R. H. Atalla. 1998. Dissolution of cellulose in aqueous NaOH solutions. *Cellulose* 5:309–319.

Jeoh, T., C. I. Ishizawa and M. F. Davis, et al. 2007. Cellulase digestibility of pretreated biomass is limited by cellulose accessibility. *Biotechnology and Bioengineering* 98:112–122.,

Konur, O. 2000. Creating enforceable civil rights for disabled students in higher education: An institutional theory perspective. *Disability & Society* 15:1041–1063.

Konur, O. 2002a. Access to nursing education by disabled students: Rights and duties of nursing programs. *Nurse Education Today* 22:364–374.

Konur, O. 2002b. Assessment of disabled students in higher education: Current public policy issues. *Assessment and Evaluation in Higher Education* 27:131–52.

Konur, O. 2002c. Access to employment by disabled people in the UK: Is the Disability Discrimination Act working? *International Journal of Discrimination and the Law* 5:247–279.

Konur, O. 2006a. Participation of children with dyslexia in compulsory education: Current public policy issues. *Dyslexia* 12:51–67.

Konur, O. 2006b. Teaching disabled students in higher education. *Teaching in Higher Education* 11:351–363.

Konur, O. 2007a. A judicial outcome analysis of the *Disability Discrimination Act*: A windfall for the employers? *Disability & Society* 22:187–204.

Konur, O. 2007b. Computer-assisted teaching and assessment of disabled students in higher education: The interface between academic standards and disability rights. *Journal of Computer Assisted Learning* 23:207–219.

Konur, O. 2012. The evaluation of the research on the bioethanol: A scientometric approach. *Energy Education Science and Technology Part A: Energy Science and Research* 28:1051–1064.

Konur, O. 2015. Current state of research on algal bioethanol. In *Marine Bioenergy: Trends and Developments*, Ed. S. K. Kim and C. G. Lee, pp. 217–244. Boca Raton, FL: CRC Press.

Konur, O. 2020. The scientometric analysis of the research on the bioethanol production from green macroalgae. In *Handbook of Algal Science, Technology and Medicine*, Ed. O. Konur, pp. 385–401. London: Academic Press.

Konur, O. 2023. Cellulose-based bioethanol fuels: Scientometric study. In *Feedstock-based Bioethanol Fuels. I. Non-Waste Feedstocks: Starch, Sugar, Grass, Wood, Cellulose, Algae, and Biosyngas-based Bioethanol Fuels. Handbook of Bioethanol Fuels Volume 3*, Ed. O. Konur. Boca Raton, FL: CRC Press.

Kosan, B., C. Michels and F. Meister. 2008. Dissolution and forming of cellulose with ionic liquids. *Cellulose* 15:59–66.

Lin, Y. and S. Tanaka. 2006. Ethanol fermentation from biomass resources: Current state and prospects. *Applied Microbiology and Biotechnology* 69:627–642.

Ma, X., L. Sun and C. Song. 2002. A new approach to deep desulfurization of gasoline, diesel fuel and jet fuel by selective adsorption for ultra-clean fuels and for fuel cell applications. *Catalysis Today* 77:107–116.

Morschbacker, A. 2009. Bio-ethanol based ethylene. *Polymer Reviews* 49:79–84.

Najafi, G., B. Ghobadian and T. Tavakoli, et al. 2009. Performance and exhaust emissions of a gasoline engine with ethanol blended gasoline fuels using artificial neural network. *Applied Energy* 86:630–639.

Newman, P. W. G. and J. R. Kenworthy. 1989. Gasoline consumption and cities: A comparison of U.S. cities with a global survey. *Journal of the American Planning Association* 55:24–37.

North, D. C. 1991. Institutions. *Journal of Economic Perspectives* 5:97–112.

Olsson, L. and B. Hahn-Hagerdal. 1996. Fermentation of lignocellulosic hydrolysates for ethanol production. *Enzyme and Microbial Technology* 18:312–331.

Onda, A., T. Ochi and K. Yanagisawa. 2008. Selective hydrolysis of cellulose into glucose over solid acid catalysts. *Green Chemistry* 10:1033–1037.

Park, S., J. O. Baker, M. E. Himmel, P. A. Parilla and D. K. Johnson. 2010. Cellulose crystallinity index: Measurement techniques and their impact on interpreting cellulase performance. *Biotechnology for Biofuels* 3:10.

Phillips, C. M., W. T. Beeson, J. H. Cate and M. A. Marletta. 2011. Cellobiose dehydrogenase and a copper-dependent polysaccharide monooxygenase potentiate cellulose degradation by *Neurospora crassa*. *ACS Chemical Biology* 6:1399–1406.

Pinkert, A., K. N. Marsh, S. Pang and M. P. Staiger. 2009. Ionic liquids and their interaction with cellulose. *Chemical Reviews* 109:6712–6728.

Qing, Q., B. Yang and C. E. Wyman. 2010. Xylooligomers are strong inhibitors of cellulose hydrolysis by enzymes. *Bioresource Technology* 101:9624–9630.

Quinlan, R. J., M. D. Sweeney and L. L. Leggio, et al. 2011. Insights into the oxidative degradation of cellulose by a copper metalloenzyme that exploits biomass components. *Proceedings of the National Academy of Sciences of the United States of America* 108:15079–15084.

Remsing, R. C., R. P. Swatloski, R. D. Rogers and G. Moyna. 2006. Mechanism of cellulose dissolution in the ionic liquid 1-n-butyl-3- methylimidazolium chloride: A ^{13}C and$^{35/37}$Cl NMR relaxation study on model systems. *Chemical Communications* 2006:1271–1273.

Rinaldi, R., R. Palkovits and F. Schuth. 2008. Depolymerization of cellulose using solid catalysts in ionic liquids. *Angewandte Chemie - International Edition* 47:8047–8050.

Sanchez, O. J. and C. A. Cardona. 2008. Trends in biotechnological production of fuel ethanol from different feedstocks. *Bioresource Technology* 99:5270–5295.

Sasaki, M., Z. Fang, Y. Fukushima, T. Adschiri and K. Arai. 2000. Dissolution and hydrolysis of cellulose in subcritical and supercritical water. *Industrial and Engineering Chemistry Research* 39:2883–2890.

Sasaki, M., B. Kabyemela and R. Malaluan, et al. 1998. Cellulose hydrolysis in subcritical and supercritical water. *Journal of Supercritical Fluids* 13:261–268.

Suganuma, S., K. Nakajima and M. Kitano, et al. 2008. Hydrolysis of cellulose by amorphous carbon bearing SO_3H, COOH, and OH groups. *Journal of the American Chemical Society* 130:12787–12793.

Sun, Y. and J. Cheng. 2002. Hydrolysis of lignocellulosic materials for ethanol production: A review. *Bioresource Technology* 83:1–11.

Taherzadeh, M. J. and K. Karimi. 2007. Enzyme-based hydrolysis processes for ethanol from lignocellulosic materials: A review. *Bioresources* 2:707–738.

Taherzadeh, M. J. and K. Karimi. 2008. Pretreatment of lignocellulosic wastes to improve ethanol and biogas production: A review. *International Journal of Molecular Sciences* 9:1621–1651.

Vitz, J., T. Erdmenger, C. Haensch and U. S. Schubert. 2009. Extended dissolution studies of cellulose in imidazolium based ionic liquids. *Green Chemistry* 11:417–424.

Yang, B. and C. E. Wyman. 2006. BSA treatment to enhance enzymatic hydrolysis of cellulose in lignin containing substrates. *Biotechnology and Bioengineering* 94:611–617.

Zhang, Y. H. P. and L. R. Lynd. 2004. Toward an aggregated understanding of enzymatic hydrolysis of cellulose: Noncomplexed cellulase systems. *Biotechnology and Bioengineering* 88:797–824.

Zhao, H., C. L. Jones and G. A. Baker, et al. 2009. Regenerating cellulose from ionic liquids for an accelerated enzymatic hydrolysis. *Journal of Biotechnology* 139:47–54.

Zhu, S., Y. Wu and Q. Chen, et al. 2006. Dissolution of cellulose with ionic liquids and its application: A mini-review. *Green Chemistry* 8:325–327.

Part 14

Third Generation Algal Bioethanol Fuels

55 Third Generation Algal Bioethanol Fuels
Scientometric Study

Ozcan Konur
(Formerly) Ankara Yildirim Beyazit University

55.1 INTRODUCTION

Crude oil-based gasoline fuels (Ma et al., 2002; Newman and Kenworthy, 1989) have been widely used in the transportation sector since the 1920s. However, there have been great public concerns over the adverse environmental and human impact of these fuels (Hill et al., 2006, 2009). Hence, biomass-based bioethanol fuels (Hill et al., 2006; Konur, 2012e, 2015, 2019, 2020a) have increasingly been used in blending gasoline fuels (Hsieh et al., 2002; Najafi et al., 2009), in fuel cells (Antolini, 2007, 2009), and in biochemical production (Angelici et al., 2013; Morschbacker, 2009) in a biorefinery context (Fernando et al., 2006; Huang et al., 2008).

Bioethanol fuels also play a critical role in maintaining energy security (Kruyt et al., 2009; Winzer, 2012) in supply shocks (Kilian, 2008, 2009) related to oil price shocks (Hamilton, 2003, 2009), the COVID-19 pandemic (Fauci et al., 2020; Li et al., 2020), or wars (Hamilton, 1983; Jones, 2012) in the aftermath of the Russian invasion of Ukraine (Reeves, 2014).

However, it is necessary to pretreat the biomass (Taherzadeh and Karimi, 2008; Yang and Wyman, 2008) to enhance the yield of the bioethanol (Hahn-Hagerdal et al., 2006; Sanchez and Cardona, 2008) prior to bioethanol production through hydrolysis (Sun and Cheng, 2002; Taherzadeh and Karimi, 2007) and fermentation (Lin and Tanaka, 2006; Olsson and Hahn-Hagerdal, 1996) of biomass and hydrolysates, respectively.

One of the most-studied biomass materials for bioethanol production has been algal biomass as a third generation feedstock. The research in the field of algal bioethanol fuels has intensified in this context in the key research fronts of the pretreatment of algal biomass (Choi et al., 2010; Harun and Danquah, 2011a; Harun et al., 2011), hydrolysis of algal biomass (Harun and Danquah, 2011b; Kim et al., 2014; Park et al., 2012), fermentation of algal hydrolysates (Adams et al., 2009; Horn et al., 2000; Kim et al., 2011), and algal bioethanol production in general (Deng and Coleman, 1999; Harun et al., 2010; Ho et al., 2013). Further, macroalgae (Adams et al., 2009; Horn et al., 2000; Kim et al., 2011), microalgae (Choi et al., 2010; Harun et al., 2010; Ho et al., 2013), and cyanobacteria (Deng and Coleman, 1999; Dexter and Fu, 2009; Gao et al., 2012) have been studied intensively at the expense of diatoms and dinoflagellates in this context.

However, it is essential to develop efficient incentive structures (North, 1991) for the primary stakeholders to enhance the research in this field (Konur, 2000, 2002a,b,c, 2006a,b, 2007a,b). Scientometric analysis has been used in this context to inform the primary stakeholders about the current state of the research in this research field (Garfield, 1955; Konur, 2011, 2012a,b,c,d,e,f,g,h,i, 2015, 2018b, 2019, 2020a).

As there have been no published scientometric studies in this field, this chapter presents a scientometric study of the research in algal bioethanol fuels. It examines the scientometric characteristics of both the sample and population data presenting scientometric characteristics of both these

DOI: 10.1201/9781003226451-73

datasets in the order of documents, authors, publication years, institutions, funding bodies, source titles, countries, Scopus subject categories, Scopus keywords, and research fronts.

55.2 MATERIALS AND METHODS

The search for this study was carried out using the Scopus database (Burnham, 2006) in June 2022.

As the first step for the search of the relevant literature, keywords were selected using the 200 most-cited population papers. The selected keyword list was then optimized to obtain a representative sample of papers from this research field. This keyword list was provided in the appendix for future replicative studies.

As the second step, two sets of data were used in this study. First, a population sample of 760 papers was used to examine the scientometric characteristics of the population data. Second, a sample of 100 most-cited papers, corresponding to 13% of the population papers, was used to examine the scientometric characteristics of these citation classics.

The scientometric characteristics of both these sample and population datasets were presented in the order of documents, authors, publication years, institutions, funding bodies, source titles, countries, Scopus subject categories, Scopus keywords, and research fronts.

Lastly, the key scientometric findings for both datasets were discussed to highlight the research landscape for algal bioethanol fuels. Additionally, a number of brief conclusions were drawn, and a number of relevant recommendations were made to enhance the future research landscape.

55.3 RESULTS

55.3.1 THE MOST PROLIFIC DOCUMENTS ON ALGAL BIOETHANOL FUELS

Information on the types of documents for both datasets is given in Table 55.1. Articles and conference papers, published in journals, dominate both the sample (84%) and population (89%) papers as they are under-represented in the sample papers by 5%. Further, review papers are surplus as they are over-represented in the sample papers by 11% as they constitute 16% and 5% of the sample and population papers, respectively.

It is further notable that 96% of the population papers were published in journals, while 4% and 0.1% of them were published in book series and books, respectively. On the contrary, all of the sample papers were published in the journals.

TABLE 55.1
Documents on Algal Bioethanol Fuels

Documents	Sample Dataset (%)	Population Dataset (%)	Surplus (%)
Article	80.0	86.3	−6.3
Review	16.0	5.4	10.6
Conference paper	4.0	2.2	1.8
Book chapter	0.0	5.5	−5.5
Note	0.0	0.4	−0.4
Book	0.0	0.1	−0.1
Editorial	0.0	0.1	−0.1
Letter	0.0	0.0	0.0
Short Survey	0.0	0.0	0.0
Sample size	100	760	

Population dataset, The number of papers (%) in the set of the 760 population papers; Sample dataset, The number of papers (%) in the set of 100 highly cited papers.

55.3.2 The Most Prolific Authors on Algal Bioethanol Fuels

Information about the 12 most prolific authors with at least 3% of sample papers each is given in Table 55.2. The most prolific author is Gwi-Taek Jeon of the Pukyong National University with 6% of the sample papers, followed by Jo-Shu Chang, Razif Harun, and Michael K. Danquah with 5% of the sample papers each.

Further, the most influential authors are Razif Harun and Michael K. Danquah with a 4.1% surplus each, followed by Jo-Shu Chang and Shi-Hsin Ho with a 3.3% and 3.2% surplus, respectively.

The most prolific institutions for the sample dataset are Kobe University and Pukyong National University with two authors each. In total, nine institutions house these top authors. On the other hand, the most prolific country for the sample dataset is South Korea with four authors, followed by Japan with two authors. In total, eight countries house these top authors.

There are two primary research fronts for these top authors: macroalgal and microalgal bioethanol fuels with three and nine authors, respectively. On the other hand, there is a significant gender deficit (Beaudry and Lariviere, 2016) for the sample dataset as surprisingly only one of these top researchers is female with a representation rate of 8%.

Additionally, there are other authors with a relatively low citation impact and with 0.8%–3.7% of the population papers each: Sang-Hyoun Kim, Chae Hun Ra, In Yun Sanwoo, Inn Shi Tan, Jae Hwa Lee, Sung-Mok Lee, Yong Keun Chang, Marwa M. El-Dalatony, Vanina Estrada, Kyoung Heon Kim, Mayur B. Kurade, Man Kee Lam, Keat Teong Lee, Alberto Bertucco, Astrilla Damayanti, Bintang Marhaeni, Trung Hau Nguyen, El-Sayed Salama, Pau Loke Shaw, and Carlos Eduardo de Farias Silva.

TABLE 55.2
Most Prolific Authors on Algal Bioethanol Fuels

No.	Author Name	Author Code	Sample Papers (%)	Population Papers (%)	Surplus	Institution	Country	HI	N	Res. Front
1	Jeong, Giwi-Taek	7102664797	6.0	4.6	1.4	Pukyong Natl. Univ.	South Korea	32	145	MA
2	Chang, Jo-Shu	8567368700	5.0	1.7	3.3	Tunghai Univ.	Taiwan	98	660	MI
3	Harun, Razif	35315707300	5.0	0.9	4.1	Univ. Putra Malaysia	Malaysia	21	86	MI
4	Danquah, Michael K.	12803940900	5.0	0.9	4.1	Univ. Tennessee	USA	37	213	MI
5	Ho, Shih-Hsin	36084355700	4.0	0.8	3.2	Harbin Inst. Technol.	China	57	218	MI
6	Hong, Yong-Ki	7403393284	3.0	1.6	1.4	Pukyong Natl. Univ.	South Korea	27	139	MA
7	Jeon, Byong-Hun	16052765600	3.0	1.2	1.8	Hanyang Univ.	South Korea	57	347	MI
8	Kondo, Akihiko	57203868143	3.0	0.8	2.2	Kobe Univ.	Japan	78	797	MI
9	Hasunuma, Tomohisa	23988816000	3.0	0.8	2.2	Kobe Univ.	Japan	40	189	MI
10	Abou-Shanab, Reda A.I.	9338041400	3.0	0.5	2.5	City Sci. Res. Technol. Apps.	Egypt	31	67	MI
11	Lee, Eun Yeol	14037776800	3.0	0.4	2.6	Kyung Hee Univ.	South Korea	40	210	MI
12	Bjerre, Anne B.*	6701773173	3.0	0.4	2.6	Danish Technol. Inst.	Denmark	18	34	MA

*, Female researchers; Author code, the unique code given by Scopus to the authors; MA, Macroalgae; MI, Microalgae; Population papers, the number of papers authored in the population dataset; Sample papers, the number of papers authored in the sample dataset.

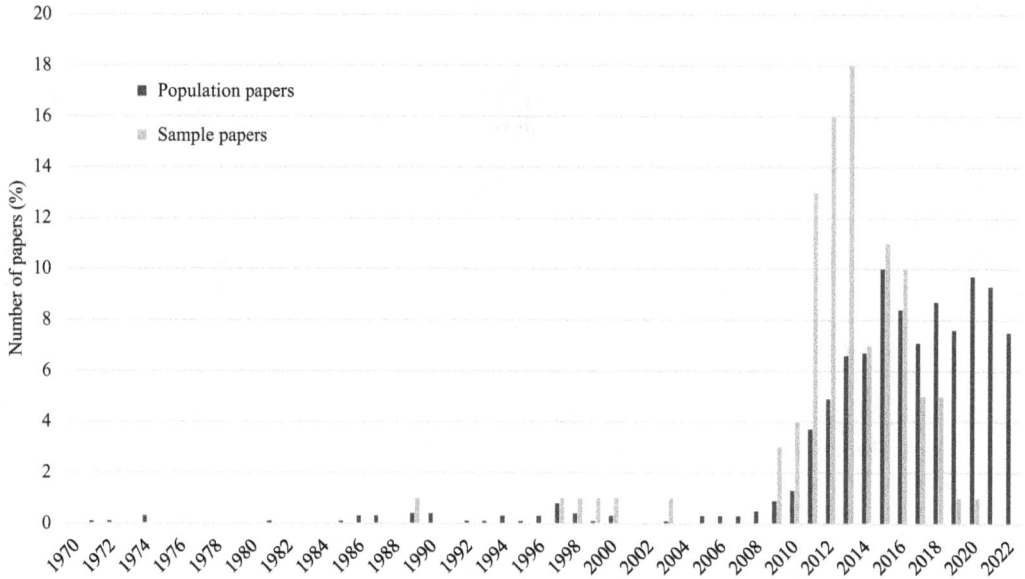

FIGURE 55.1   Research output by years regarding algal bioethanol fuels.

55.3.3   The Most Prolific Research Output by Years on Algal Bioethanol Fuels

Information about papers published between 1970 and 2022 is given in Figure 55.1. This figure clearly shows that the bulk of the research papers in the population dataset was published primarily in the 2010s and the early 2020s with 65% and 27% of the population dataset, respectively. Similarly, the publication rates for the 2000s, 1990s, 1980s, and 1970s were 3%, 3%, 1%, and 1%, respectively. Additionally, 0.2% of the population papers were published in the pre-1970s.

Similarly, the bulk of the research papers in the sample dataset was published in the 2000s and 2010s with 5% and 90% of the sample dataset, respectively. Similarly, the publication rates for the 1990s, 1980s, and 1970s were 3%, 1%, and 1% of the sample papers, respectively.

The most prolific publication years for the population dataset were 2020 and 2021 with 9.7 and 9.3% of the dataset, respectively. Further, 92% of the population papers were published between 2010 and 2022. Similarly, 92% of the sample papers were published between 2009 and 2018, while the most prolific publication year was 2013 with 18% of the sample papers. Other prolific years were 2012 and 2011 with 16% and 13% of the sample papers, respectively. Although there was a rising trend for the research output between 2009 and 2015 for the population papers, thereafter it steadied between 8% and 10% each year losing its momentum.

55.3.4   The Most Prolific Institutions on Algal Bioethanol Fuels

Information about the 18 most prolific institutions publishing papers on algal bioethanol fuels with at least 3% of the sample papers each is given in Table 55.3.

The most prolific institution is National Cheng Kung University with 8% of the sample papers, followed by Pukyong National University with 7% of the sample papers. Other prolific institutions are Korea University, Tianjin University, Monash University, and the University of Putra Malaysia with 5% of the sample papers each.

Similarly, the top country for these most prolific institutions is South Korea with six institutions, while China, Denmark, and Malaysia house two institutions each. In total, ten countries house these top institutions.

TABLE 55.3
The Most Prolific Institutions on Algal Bioethanol Fuels

No.	Institutions	Country	Sample Papers (%)	Population Papers (%)	Surplus (%)
1	National Cheng Kung Univ.	Taiwan	8.0	2.5	5.5
2	Pukyong National Univ.	South Korea	7.0	6.3	0.7
3	Korea University	South Korea	5.0	2.1	2.9
4	Tianjin Univ.	China	5.0	1.2	3.8
5	Monash Univ.	Australia	5.0	1.1	3.9
6	Univ. Putra Malaysia	Malaysia	5.0	1.1	3.9
7	Kobe Univ.	Japan	4.0	1.2	2.8
8	Dankook Univ.	South Korea	4.0	0.8	3.2
9	Kyung Hee Univ.	South Korea	4.0	0.8	3.2
10	Chinese Acad. Sci.	China	3.0	3.3	−0.3
11	Univ. Sains Malaysia	Malaysia	3.0	1.6	1.4
12	Jenderal Soedirman Univ.	Indonesia	3.0	1.4	1.6
13	Technical Univ. Denmark	Denmark	3.0	1.3	1.7
14	Central Salt Marine Chem. Res. Inst.	India	3.0	0.9	2.1
15	City Sci. Res. Technol. Apps.	Egypt	3.0	0.7	2.3
16	Danish Technol. Inst.	Denmark	3.0	0.4	2.6
17	Sogang Univ.	South Korea	3.0	0.4	2.6
18	Sungkyunkwan Univ.	South Korea	3.0	0.4	2.6

On the other hand, the institution with the highest citation impact is National Cheng Kung University with a 5.5% surplus, followed by the University of Putra Malaysia and Monash University with a 3.9% surplus each. Other influential institutions are Tianjin University, Kyung Hee University, and Dankook University with a 3.2%–3.8% surplus each. Similarly, the institution with the lowest citation impact is the Chinese Academy of Sciences with a 0.3% deficit.

Additionally, there are other institutions with a relatively low citation impact and with 0.8%–1.6% of the population papers each: Chonnam National University, Ocean University of China, CSIC, Myongji University, Silla University, the Korea Advanced Institute of Science and Technology, Kangwon National University, Hanyang University, the Academy of Scientific and Innovative Research, the Technological University of Petronas, Tanta University, National University South, Japan Science and Technology Agency, Chongqing University, University Malaya, Semarang State University, the University of Nottingham Malaysia, the University of Padua, Pennsylvania State University, the University of Sao Paulo, Curtin University Malaysia, Tunghai University, the Korea Institute of Ocean Science & Technology, the National Laboratory of Energy and Geology, and Yonsei University.

55.3.5 The Most Prolific Funding Bodies on Algal Bioethanol Fuels

Information about the 15 most prolific funding bodies funding at least 3% of the sample papers each is given in Table 55.4. Further, only 61% and 56% of the sample and population papers were funded, respectively.

The most prolific funding body is the National Research Foundation of Korea with 19% of the sample papers, followed by the Ministry of Education, Science and Technology of South Korea with 7% of the sample papers. On the other hand, the most prolific country among these top funding bodies is South Korea with seven funding bodies, followed by the USA with two funding bodies. In total, only seven countries and the EU house these top funding bodies.

The funding body with the highest citation impact is the Ministry of Education, Science and Technology of South Korea with a 4.1% surplus, while other influential funding bodies are the U.S.

TABLE 55.4

The Most Prolific Funding Bodies on Algal Bioethanol Fuels

No.	Funding Bodies	Country	Sample Paper No. (%)	Population Paper No. (%)	Surplus (%)
1	National Research Foundation of Korea	South Korea	10.0	8.6	1.4
2	Ministry of Education, Science and Technology	South Korea	7.0	2.5	4.5
3	Ministry of Science, ICT and Future Planning	South Korea	4.0	3.2	0.8
4	U.S. Department of Energy	USA	4.0	0.9	3.1
5	National Natural Science Foundation	China	3.0	6.4	−3.4
6	Ministry of Education	Taiwan	3.0	4.3	−1.3
7	National Council for Scientific and Technological Development	Brazil	3.0	3.3	−0.3
8	Ministry of Higher Education, Malaysia	Malaysia	3.0	2.5	0.5
9	Korea Institute of Energy Technology Evaluation and Planning	South Korea	3.0	1.4	1.6
10	Ministry of Land, Transport and Maritime Affairs	South Korea	3.0	1.4	1.6
11	Ministry of Knowledge Economy	South Korea	3.0	1.2	1.8
12	Council of Scientific and Industrial Research, India	India	3.0	0.8	2.2
13	Ministry for Food, Agriculture, Forestry and Fisheries	South Korea	3.0	0.8	2.2
14	Monash University	Australia	3.0	0.4	2.6
15	National Science Council	USA	3.0	0.4	2.6

Department of Energy, Monash University, and National Science Council with a 2.6%–3.1% surplus each. Similarly, the funding body with the lowest citation impact is the National Natural Science Foundation of China with a 3.4% deficit, followed by the Ministry of Education of Taiwan and the National Council for Scientific and Technological Development of Brazil with a 1.3% and 0.3% deficit, respectively.

Other funding bodies with a relatively low citation impact and with 0.8%–2.1% of the population papers each are CAPES Foundation, European Regional Development Fund, Japan Society for the Promotion of Science, the Ministry of Science and Technology of India, European Commission, the Ministry of Science and Technology of Taiwan, the Foundation for Science and Foundation for Science and Technology, the Ministry of Oceans and Fisheries, the Ministry of Science and Technology of the People's Republic of China, the National Key Research and Development Program of China, National Science Foundation, the Ministry of Education of China, the Natural Science Foundation of Shandong Province, the University of Sains Malaysia, the National Agency for the Promotion of Science and Technology, Biotechnology and Biological Sciences Research Council, Fundamental Research Funds for the Central Universities, Japan Science and Technology Agency, National High-tech Research and Development Program, and the Natural Science Foundation of Guangdong Province.

55.3.6 THE MOST PROLIFIC SOURCE TITLES ON ALGAL BIOETHANOL FUELS

Information about the 15 most prolific source titles publishing at least 2% of the sample papers each in the algal bioethanol fuels is given in Table 55.5.

The most prolific source title is Bioresource Technology with 38% of the sample papers. Other prolific titles are Renewable and Sustainable Energy Reviews, Energy and Environmental Science, Journal of Applied Phycology, Biotechnology for Biofuels, Process Biochemistry, and Applied Energy with 3%–6% of the sample papers each.

On the other hand, the source title with the highest citation impact is Bioresource Technology with a 25% surplus. Other influential titles are Renewable and Sustainable Energy Reviews, Energy

TABLE 55.5
The Most Prolific Source Titles on Algal Bioethanol Fuels

No.	Source Titles	Sample Papers (%)	Population Papers (%)	Surplus (%)
1	Bioresource Technology	38.0	13.0	25.0
2	Renewable and Sustainable Energy Reviews	6.0	1.3	4.7
3	Energy and Environmental Science	5.0	0.7	4.3
4	Journal of Applied Phycology	3.0	3.7	−0.7
5	Biotechnology for Biofuels	3.0	1.8	1.2
6	Process Biochemistry	3.0	0.9	2.1
7	Applied Energy	3.0	0.5	2.5
8	Algal Research	2.0	5.0	−3.0
9	Renewable Energy	2.0	2.6	−0.6
10	Bioprocess and Biosystems Engineering	2.0	2.0	0.0
11	Journal of Cleaner Production	2.0	1.4	0.6
12	Energy Conversion and Management	2.0	1.2	0.8
13	Energy	2.0	1.1	0.9
14	Fuel	2.0	1.1	0.9
15	Chemical Engineering Journal	2.0	0.5	1.5

and Environmental Science, Applied Energy, and Process Biochemistry with a 2.1%–4.7% surplus each. Similarly, the source title with the lowest impact is Algal Research with a 3% deficit, followed by the Journal of Applied Phycology and Renewable Energy with a 0.6%–0.7% deficit each.

Other source titles with a relatively low citation impact with 0.7%–1.6% of the population papers each are the Journal of Microbiology and Biotechnology, Bioenergy Research, Biomass Conversion and Biorefinery, Biomass and Bioenergy, Chemical Engineering Transactions, Computer Aided Chemical Engineering, Biofuels Bioproducts and Biorefining, Energies, Journal of Chemical Technology and Biotechnology, Journal of Industrial and Engineering Chemistry, Scientific Reports, Biotechnology and Bioprocess Engineering, Journal of Biotechnology, Nippon Suisan Gakkaishi (Journal of the Japanese Society of Fisheries Science), Applied Microbiology and Biotechnology, Biocatalysis and Agricultural Biotechnology, Energy Sources Part A Recovery Utilization and Environmental Effects, and Nature.

55.3.7 THE MOST PROLIFIC COUNTRIES ON ALGAL BIOETHANOL FUELS

Information about the 15 most prolific countries publishing at least 3% of sample papers each in the algal bioethanol fuels is given in Table 55.6.

The most prolific country is South Korea with 26% of the sample papers, followed by Japan, the USA, and China with 19%, 15%, and 14% of the sample papers, respectively. Malaysia, Taiwan, India, Australia, and Denmark are other prolific countries with 5%–9% of the sample papers each. Further, four European countries listed in Table 55.6 produce 14% and 8% of the sample and population papers, respectively.

On the other hand, the country with the highest citation impact is Japan with a 13.1% surplus, followed by South Korea, the USA, and Taiwan with a 6.5%, 6.1%, and 5.4% surplus, respectively. Similarly, the country with the lowest citation impact is India with a 2.9% deficit, while Indonesia, Brazil, China, and Egypt are the countries with the lowest influence with a 0.3% and 0.9% deficit each.

Additionally, there are other countries with a relatively low citation impact and with 0.7%–3.6% of the sample papers each: Spain, Italy, Thailand, Iran, Argentina, Germany, Turkey, Canada, Ireland, Colombia, France, Mexico, Sweden, Chile, Israel, Pakistan, Norway, and the Philippines.

TABLE 55.6
The Most Prolific Countries on Algal Bioethanol Fuels

No.	Countries	Sample Papers (%)	Population Papers (%)	Surplus (%)
1	South Korea	26.0	19.5	6.5
2	Japan	19.0	5.9	13.1
3	USA	15.0	8.9	6.1
4	China	14.0	14.5	−0.5
5	Malaysia	9.0	6.6	2.4
6	Taiwan	9.0	3.6	5.4
7	India	7.0	9.9	−2.9
8	Australia	5.0	2.8	2.2
9	Denmark	5.0	1.8	3.2
10	Brazil	4.0	4.7	−0.7
11	Indonesia	3.0	3.9	−0.9
12	Egypt	3.0	3.3	−0.3
13	UK	3.0	3.3	−0.3
14	Portugal	3.0	1.7	1.3
15	Netherlands	3.0	1.2	1.8

55.3.8 THE MOST PROLIFIC SCOPUS SUBJECT CATEGORIES IN ALGAL BIOETHANOL FUELS

Information about the eight most prolific Scopus subject categories indexing at least 8% of the sample papers each is given in Table 55.7.

The most prolific Scopus subject category in algal bioethanol fuels is Energy with 69% of the sample papers, followed by Environmental Science and Chemical Engineering with 60% and 54% of the sample papers, respectively. Other prolific subject categories are Biochemistry, Genetics and Molecular Biology, and Immunology and Microbiology with 25% of the sample papers each. It is notable that Social Sciences including Economics and Business accounts only for 2.1% of the population studies.

On the other hand, the Scopus subject category with the highest citation impact is Energy with a 29.8% surplus, followed by Environmental Science with a 26.6% surplus. Other influential categories are Chemical Engineering and Immunology and Microbiology with a 15% and 11% surplus, respectively. Similarly, the least influential subject category is Agricultural and Biological Sciences with a 16% deficit. Other least influential subject categories are Chemistry, Engineering, and Biochemistry, Genetics and Molecular Biology with a 1%–5% deficit each.

55.3.9 THE MOST PROLIFIC KEYWORDS IN ALGAL BIOETHANOL FUELS

Information about the Scopus keywords used with at least 6% or 2.8% of the sample or population papers, respectively, is given in Table 55.8. For this purpose, keywords related to the keyword set given in the appendix are selected from a list of the most prolific keyword set provided by the Scopus database.

These keywords are grouped under eight headings: biomass, fermentation, bacteria, hydrolysis, hydrolysates, pretreatments, other processes, and products.

The most prolific keyword related to biomass and biomass constituents is microalgae with 84% of the sample papers, followed by algae, biomass, carbohydrates, macroalgae, and seaweeds with 25%–56% of the sample papers each. Further, the most prolific keywords related to bacteria are microorganisms, yeasts, and saccharomyces with 19%–28% of the sample papers each.

TABLE 55.7

The Most Prolific Scopus Subject Categories on Algal Bioethanol Fuels

No.	Scopus Subject Categories	Sample Papers (%)	Population Papers (%)	Surplus (%)
1	Energy	69.0	39.2	29.8
2	Environmental Science	60.0	33.4	26.6
3	Chemical Engineering	54.0	38.7	15.3
4	Biochemistry, Genetics and Molecular Biology	25.0	26.1	−1.1
5	Immunology and Microbiology	25.0	13.8	11.2
6	Engineering	10.0	13.7	−3.7
7	Agricultural and Biological Sciences	9.0	25.0	−16.0
8	Chemistry	8.0	13.0	−5.0

TABLE 55.8

The Most Prolific Keywords on Algal Bioethanol Fuels

No.	Keywords	Sample Papers (%)	Population Papers (%)	Surplus (%)
1	Biomass and biomass constituents			
	Microalgae	84.0	49.6	34.4
	Algae	56.0	34.6	21.4
	Biomass	56.0	33.8	22.2
	Carbohydrates	50.0	24.7	25.3
	Macroalgae	36.0	16.1	19.9
	Seaweed	25.0	15.8	9.2
	Cyanobacteria	18.0	10.7	7.3
	Biomass production	13.0	4.3	8.7
	Feedstocks	11.0	6.3	4.7
	Red alga	10.0	5.1	4.9
	Green alga	10.0	3.3	6.7
	Cellulose	9.0	6.1	2.9
	Chlorophyta	8.0	2.9	5.1
	Laminaria	8.0	2.8	5.2
	Rhodophyta	7.0	4.7	2.3
	Gelidium	7.0	2.5	4.5
	Alginate	6.0	3.9	2.1
	Algal biomass	6.0	2.8	3.2
	Lignin	6.0	2.5	3.5
	Kappaphycus	5.0	2.9	2.1
	Ulva	5.0	2.8	2.2
	Chlorella	4.0	7.5	−3.5
	Gracilaria	4.0	2.9	1.1
	Synechocystis		3.4	−3.4
	Algal growth		2.9	−2.9
2	Fermentation			
	Fermentation	61.0	38.6	22.4

(Continued)

TABLE 55.8 (*Continued*)
The Most Prolific Keywords on Algal Bioethanol Fuels

No.	Keywords	Sample Papers (%)	Population Papers (%)	Surplus (%)
	Biomass fermentation	11.0	2.2	8.8
	Simultaneous saccharification and fermentation	9.0	3.3	5.7
	Ethanol fermentation	5.0	3.7	1.3
3	Bacteria			
	Microorganisms	28.0	15.5	12.5
	Yeast	25.0	14.3	10.7
	Saccharomyces	19.0	11.7	7.3
4	Hydrolysis			
	Hydrolysis	45.0	27.0	18.0
	Saccharification	37.0	16.8	20.2
	Enzymatic hydrolysis	19.0	11.8	7.2
	Acid hydrolysis	13.0	5.0	8.0
	Enzymatic saccharification		4.3	−4.3
5	Hydrolysates			
	Sugar	38.0	15.7	22.3
	Glucose	32.0	14.9	17.1
	Fermentable sugars	10.0	3.8	6.2
	Galactose	8.0	5.4	2.6
	Mannitol	6.0	2.2	3.8
	Reducing sugars	5.0	3.3	1.7
	Monosaccharide	4.0	3.0	1.0
6	Biomass pretreatment			
	Enzyme activity	28.0	10.0	18.0
	Sulfuric acid	19.0	7.0	12.0
	Pretreatment	19.0	5.8	13.2
	Enzymes	14.0	7.1	6.9
	Temperature	13.0	6.1	6.9
	Cellulases	11.0	4.6	6.4
	Pretreatment	10.0	6.4	3.6
	pH	9.0	7.2	1.8
7	Other processes			
	Genetic engineering	34.0	0.0	34.0
	Biotechnology	19.0	7.9	11.1
	Biorefineries	15.0	6.6	8.4
	Carbon dioxide	9.0	5.3	3.7
	Extraction	4.0	3.8	0.2
8	Products			
	Ethanol	79.0	49.6	29.4
	Biofuels	73.0	38.6	34.4
	Bioethanol	62.0	43.7	18.3
	Alcohol	41.0	19.7	21.3
	Alcohol production	28.0	11.4	16.6

(Continued)

TABLE 55.8 (*Continued*)
The Most Prolific Keywords on Algal Bioethanol Fuels

No.	Keywords	Sample Papers (%)	Population Papers (%)	Surplus (%)
	Bio-ethanol production	25.0	20.7	4.3
	Biofuel production	17.0	8.0	9.0
	Ethanol production	14.0	10.1	3.9
	Biodiesel	8.0	6.3	1.7
	Ethanol yield	6.0	0.0	6.0

The most prolific keyword related to hydrolysis is hydrolysis with 45% of the sample papers, followed by saccharification with 37% of the sample papers. Further, the most prolific keyword related to fermentation is fermentation with 61% of the sample papers.

The most prolific keyword related to hydrolysates is sugar with 38% of the sample papers, followed by glucose with 32% of the sample papers. Further, the most prolific keyword related to biomass pretreatments is enzyme activity with 28% of the sample papers, followed by sulfuric acids and pretreatments with 19% of the sample papers each.

The most prolific keyword related to other processes is genetic engineering with 34% of the sample papers, followed by biotechnology with 19% of the sample papers. Finally, the most prolific keyword related to the products is ethanol with 79% of the sample papers, followed by biofuels and bioethanol with 73% and 62% of the sample papers, respectively.

On the other hand, the most influential keywords are seaweeds, biofuel production, biomass fermentation, biomass production, biorefineries, and acid hydrolysis with an 8%–9% surplus each. Similarly, the least influential keywords are enzymatic saccharification, Chlorella, Synechocystis, and algal growth with a 3%–4% deficit each.

Similarly, the most prolific keywords across all of the research fronts are microalgae, ethanol, biofuels, bioethanol, fermentation, algae, biomass, and carbohydrates with 50%–84% of the sample papers each.

55.3.10 The Most Prolific Research Fronts in Algal Bioethanol Fuels

Information about the research fronts for sample papers in algal bioethanol fuels with regard to the algal biomass used for bioethanol production is given in Table 55.9.

As this table shows, there are three primary research fronts for this field: macroalgae, microalgae, and cyanobacteria with 51%, 32%, and 12% of the sample papers, respectively. Additionally, algae in general covers 7% of the sample papers.

Further, the most prolific research front for macroalgae is the macroalgae in general with 10% of the sample papers, followed by alginates, Gelidium, and Ulva with 6% of the sample papers each. Similarly, the most prolific research front in microalgae is microalgae in general with 12% of the sample papers, followed by Chlorella and Scenedesmus with 9% and 7% of the sample papers, respectively. Finally, Synechocystis is the most prolific cyanobacterial research front with 5% of the sample papers, followed by the research front of the cyanobacteria in general with 4% of the sample papers.

Information about the thematic research fronts for the sample papers in the algal bioethanol fuels is given in Table 55.10. As this table shows, there are four primary research fronts: algal biomass pretreatments and hydrolysis, algal hydrolysate fermentation, and algal bioethanol fuels in general with 93%, 56%, 55%, and 93% of the sample papers, respectively.

For the research front of algal biomass pretreatments, the primary research fronts are chemical, enzymatic, and microbial pretreatments with 33%, 26%, and 13% of the sample papers,

TABLE 55.9

The Most Prolific Research Fronts in Algal Bioethanol Fuels

No.	Research Fronts	N Paper (%) Sample
1	Macroalgae	51
	Macroalga in general	10
	Alginates	6
	Gelidium	6
	Ulva	6
	Gracilaria	4
	Saccharina	4
	Kappaphycus	4
	Laminaria	3
	Sargassum	2
	Chaetomorpha	2
	Undaria	2
	Other macroalgae	3
2	Microalgae	32
	Microalgae in general	12
	Chlorella	9
	Scenedesmus	7
	Chloroccum	4
	Chlamydomonas	3
	Nannochloropsis	2
	Other microalgae	2
3	Cyanobacteria	12
	Synechocystis	5
	Cyanobacteria in general	4
	Synechococcus	2
	Spirulina	2
	Other cyanobacteria	1
4	Algae in general	7

N paper (%) sample, The number of papers in the population sample of 100 papers.

respectively. Additionally, there are three further research fronts: mechanical and hydrothermal pretreatments, and pretreatments in general with 9%, 8%, and 6% of the sample papers, respectively. Further, the most prolific chemical pretreatment is the acid pretreatment with 27% of the sample papers.

Similarly, the most prolific research fronts in algal biomass hydrolysis are enzymatic and acid hydrolysis with 26% and 27% of the sample papers, respectively. The next prolific research fronts in algal hydrolysate fermentation are macroalgal and microalgal fermentation with 34% and 15% of the sample papers, respectively. Finally, the prolific research fronts in algal bioethanol fuels are macroalgal and microalgal bioethanol fuels with 44% and 30% of the sample papers, respectively.

TABLE 55.10

The Most Prolific Thematic Research Fronts in Algal Bioethanol Fuels

No.	Research Fronts	N Paper (%) Sample
1	Algal Biomass Pretreatments	93.0
	Algal biomass pretreatments in general	7.0
	Mechanical pretreatments	9.0
	Milling pretreatments	5.0
	Ultrasound pretreatments	4.0
	Hydrothermal pretreatments	8.0
	Liquid hot water pretreatment	5.0
	Hot compressed water pretreatment	1.0
	Steam explosion pretreatment	1.0
	Wet oxidation pretreatment	1.0
	Chemical pretreatments	33.0
	Acid pretreatment	27.0
	Alkaline pretreatment	3.0
	Other chemical pretreatments	2.0
	Ionic liquid pretreatment	1.0
	Enzymatic pretreatments	26.0
	Microbial pretreatments	13.0
2	Algal Biomass Hydrolysis	56.0
	Enzymatic hydrolysis	27.0
	Acid hydrolysis	26.0
	Hydrolysis in general	3.0
3	Algal Hydrolysate Fermentation	55.0
	Macroalgal fermentation	34.0
	Microalgal fermentation	15.0
	Algal fermentation	4.0
	Cyanobacterial fermentation	4.0
4	Algal Bioethanol Production in general	93.0
	Macroalgal ethanol production	44.0
	Microalgal ethanol production	30.0
	Algal ethanol production	11.0
	Cyanobacterial ethanol production	10.0

N paper (%) sample, The number of papers in the population sample of 100 papers.

55.4 DISCUSSION

55.4.1 INTRODUCTION

Crude oil-based gasoline fuels have been widely used in the transportation sector since the 1920s. However, there have been great public concerns over the adverse environmental and human impact of these fuels. Hence, biomass-based bioethanol fuels have increasingly been used in blending gasoline fuels, in fuel cells, and in biochemical production in a biorefinery context.

However, it is necessary to pretreat the biomass to enhance the yield of bioethanol prior to bio-ethanol production through hydrolysis and fermentation. One of the most studied biomass in bio-ethanol production has been algal biomass as a third generation feedstock. The research in the field of algal bioethanol fuels has intensified in this context in the key research fronts of the pretreatment and hydrolysis of algal biomass, fermentation of algal hydrolysates, and algal bioethanol fuels in general. Further, macroalgae, microalgae, and cyanobacteria have been studied intensively at the expense of diatoms and dinoflagellates in this context.

However, it is essential to develop efficient incentive structures for the primary stakeholders to enhance the research in this field. This is especially important to maintain energy security in the cases of supply shocks such as oil price shocks, war-related shocks as in the case of the Russian invasion of Ukraine, or COVID-19 shocks.

Scientometric analysis has been used in this context to inform the primary stakeholders about the current state of the research in this research field. As there has been no scientometric study in this field, this chapter presents a scientometric study of the research on algal bioethanol fuels. It examines the scientometric characteristics of both the sample and population data presenting scien-tometric characteristics of both these datasets in the order of documents, authors, publication years, institutions, funding bodies, source titles, countries, Scopus subject categories, Scopus keywords, and research fronts.

As the first step for the search of the relevant literature, keywords were selected using the 200 first most-cited papers. The selected keyword list was then optimized to obtain a representative sample of papers in this research field. A copy of this extended keyword list is provided in the appendix for future replicative studies. Further, a selected list of the keywords is presented in Table 55.8.

As the second step, two sets of data were used for this study. First, a population sample of 760 papers was used to examine the scientometric characteristics of the population data. Second, a sam-ple of 100 most-cited papers corresponding to 13% of the population data set was used to examine the scientometric characteristics of these citation classics.

The scientometric characteristics of these sample and population datasets were presented in the order of documents, authors, publication years, institutions, funding bodies, source titles, countries, Scopus subject categories, Scopus keywords, and research fronts.

Lastly, the key scientometric findings for both datasets were discussed to highlight the research landscape for algal bioethanol fuels. Additionally, a number of brief conclusions were drawn, and a number of relevant recommendations were made to enhance the future research landscape.

55.4.2 The Most Prolific Documents on Algal Bioethanol Fuels

Articles (together with conference papers) dominate both the sample (84%) and population (89%) papers (Table 55.1). Further, review papers and articles have a surplus and deficit of 11% and 5%, respectively. The representation of the reviews in the sample papers is relatively signifi-cant (16%).

Scopus differs from the Web of Science database in differentiating and showing articles (80%) and conference papers (4%) published in journals separately. However, it should be noted that these conference papers are also published in journals as articles, compared to those published only in the conference proceedings. Hence, the total number of articles and review papers in the sample dataset is 84% and 16%, respectively.

It is observed during the search process that there has been inconsistency in the classification of documents in Scopus as well as in other databases such as Web of Science. This is especially relevant for the classification of papers as reviews or articles as papers not involving a literature review may be erroneously classified as a review paper. There is also a case of review papers being classified as articles. For example, the total number of reviews in the sample data set was manually found as nearly 17% compared to 16% as indexed by Scopus, reducing the number of articles and conference papers to 83% for the sample dataset.

In this context, it would be helpful to provide a classification note for the published papers in books and journals at the first instance. It would also be helpful to use the document types listed in Table 55.1 for this purpose. Book chapters may also be classified as articles or reviews as an additional classification to differentiate review chapters from experimental chapters as it is done by the Web of Science. It would be further helpful to additionally classify conference papers as articles or review papers as well as it is done in the Web of Science database.

55.4.3 THE MOST PROLIFIC AUTHORS ON ALGAL BIOETHANOL FUELS

There have been 12 most prolific authors with at least 3% of the sample papers each as given in Table 55.2. These authors have shaped the development of research in this field.

The most prolific authors are Gwi-Taek Jeong and to a lesser extent Jo-Shu Chang, Razif Harun, and Michael K. Danquah.

It is important to note the inconsistencies in the indexing of author names in Scopus and other databases. It is especially an issue for names with more than two components such as 'Blake Sam de Hyun Jeong'. The probable outcomes are 'Jeong, B.S.D.H.', 'de Hyun Jeong, B.S.', or 'Hyun Jeong, B.S.D.'. The first choice is the gold standard of the publishing sector as the last word in the name is taken as the last name. In most of the academic databases such as PUBMED and EBSCO databases, this version is used predominantly. The second choice is a strong alternative, while the last choice is an undesired outcome as two last words are taken as the last name. It is good practice to combine the words of the last name with a hyphen: 'Hyun-Jeong, B.S.D.'. It is notable that inconsistent indexing of the author names may cause substantial inefficiencies in the search process for papers as well as allocating credit to the authors as there are different author entries for each outcome in the databases.

There are also inconsistencies in the shortening of Chinese names. For example, 'YangYing Ho' is often shortened as 'Ho, Y.', 'Ho, Y.-Y.', and 'Ho, Y.Y.' as it is done in the Web of Science database as well. However, the gold standard in this case is 'Ho, Y' where the last word is taken as the last name and the first word is taken as a single forename. In most of the academic databases such as PUBMED and EBSCO, this first version is used predominantly. However, it is helpful to use the third option to differentiate Chinese names efficiently: 'Ho, Y.Y.'. Therefore, there have been difficulties in locating papers for Chinese authors. In such cases, the use of the unique author codes provided for each author by the Scopus database has been helpful.

There is also a difficulty in allowing credit for authors, especially for authors with common names such as 'Jeong, X.' in conducting scientometric studies. These difficulties strongly influence the efficiency of scientometric studies as well as allocating credit to the authors as there are the same author entries for different authors with the same name, e.g., 'Jeong, X.' in the databases.

In this context, the coding of authors in the Scopus database is a welcome innovation compared to other databases such as the Web of Science. In this process, Scopus allocates a unique number to each author in the database (Aman, 2018). However, there might still be substantial inefficiencies in this coding system, especially for common names. For example, some of the papers for a certain author may be allocated to another researcher with a different author code. It is possible that Scopus uses a number of software programs to differentiate the author names, and the program may not be false-proof (Kim, 2018).

In this context, it does not help that author names are not given in full in some journals and books. This makes it difficult to differentiate authors with common names and makes scientometric studies further difficult in the author domain. Therefore, the author names should be given in all books and journals at the first instance. There is also a cultural issue where some authors do not use their full names in their papers. Instead, they use initials for their forenames: 'Jeong, H.J.', 'Jeong', 'Jeong, H.', or 'Jeong, J.' instead of 'Jeong, Hyun Jae'.

There are also inconsistencies in the naming of authors with more than two components by authors themselves in journal papers and book chapters. For example. 'Jeong, A.P.C.' might be given as 'Jeong, A.' or 'Jeong, A.C.' or 'Jeong, A.P.' or 'Jeong, C.' in the journals and books. This also makes scientometric studies difficult in the author domain. Hence, contributing authors should use their name consistently in their publications.

Another critical issue regarding author names is the inconsistencies in the spelling of the author names in national spellings (e.g., Göğüşçiğ, Gökçe) rather than in English spellings (e.g., Goguscig, Gokce) in Scopus database. Scopus differs from the Web of Science database and many other databases in this respect where author names are given only in English spellings. It is observed that national spellings of author names do not help much in conducting scientometric studies as well as in allocating credits to authors as sometimes there are different author entries for English and national spellings in the Scopus database.

The most prolific institutions for the sample dataset are Kobe University and Pukyong National University. Further, the most prolific countries for the sample dataset are South Korea and to a lesser extent Japan. These findings confirm the dominance of Japan and South Korea in this field. The most prolific research fronts are microalgal and macroalgal bioethanol production.

It is also notable that there is a significant gender deficit for the sample dataset surprisingly with a representation rate of 8%. This finding is the most thought-provoking with strong public policy implications. Hence, institutions, funding bodies, and policymakers should take efficient measures to reduce the gender deficit in this field as well as other scientific fields with strong gender deficits. In this context, it is worth noting the level of representation of researchers from minority groups in science on the basis of race, sexuality, age, and disability, besides gender (Blankenship, 1993; Dirth and Branscombe, 2017; Konur, 2000, 2002a,b,c, 2006a,b, 2007a,b).

55.4.4 THE MOST PROLIFIC RESEARCH OUTPUT BY YEARS ON ALGAL BIOETHANOL FUELS

The research output observed between 1970 and 2022 is illustrated in Figure 55.1. This figure clearly shows that the bulk of the research papers in the population dataset was published primarily in the 2010s and early 2020s. Similarly, the bulk of the research papers in the sample dataset was published in the past two decades. Although there was a rising trend for the research output between 2009 and 2015 for the population papers, thereafter it lost its momentum.

These findings suggest that the most prolific sample and population papers were primarily published in the 2010s. Further, 27% of the population papers were published in the early 2020s. These are thought-provoking findings as there has been a significant boom in research since 2010. In this context, the increasing public concerns about climate change (Change, 2007), greenhouse gas emissions (Carlson et al., 2017), and global warming (Kerr, 2007) have been certainly behind the boom in the research in this field in the past two decades. Furthermore, supply shocks experienced due to the COVID-19 pandemic might also be behind the research boom in this field since 2019. Indeed, there has been a sharp rise in the research output in 2020 and 2021.

Based on these findings, the size of the population papers is likely to more than double in the current decade, provided that the public concerns about climate change, greenhouse gas emissions, and global warming, as well as the supply shocks, are translated efficiently to the research funding in this field.

55.4.5 THE MOST PROLIFIC INSTITUTIONS ON ALGAL BIOETHANOL FUELS

The 18 most prolific institutions publishing papers on algal bioethanol fuels with at least 3% of the sample papers each given in Table 55.3 have shaped the development of the research in this field.

The most prolific institutions are National Cheng Kung University, Pukyong National University, and to a lesser extent Korea University, Tianjin University, Monash University, and the University of Putra Malaysia. Similarly, the top countries for these most prolific institutions are South Korea and to a lesser extent China, Denmark, and Malaysia. In total, ten countries house these top institutions.

On the other hand, the institutions with the highest impact are National Cheng Kung University and to a lesser extent the University of Putra Malaysia, Monash University, Tianjin University, Kyung Hee University, and Dankook University. These findings confirm the dominance of the South Korean, Chinese, Malaysian, Taiwanese, Danish, and Australian institutions.

55.4.6 THE MOST PROLIFIC FUNDING BODIES ON ALGAL BIOETHANOL FUELS

The information about the 15 most prolific funding bodies funding at least 3% of the sample papers each is given in Table 55.4. It is notable that only 61% and 56% of the sample and population papers were funded, respectively.

The most prolific funding bodies are the National Research Foundation of Korea and to a lesser extent the Ministry of Education, Science and Technology of South Korea. The most prolific countries for these top funding bodies are South Korea and to a lesser extent the USA. In total, only seven countries house these top funding bodies.

These findings on the funding of the research in this field suggest that the level of funding, mostly since 2010, is highly intensive, and it has been largely instrumental in enhancing the research in this field (Ebadi and Schiffauerova, 2016) in light of North's institutional framework (North, 1991). It is also notable that the funding rate in this field is superior to those in other research fronts of bioethanol fuels such as wood-based algal bioethanol fuels. Further, it is expected that this high funding rate would continue in light of the recent supply shocks.

55.4.7 THE MOST PROLIFIC SOURCE TITLES ON ALGAL BIOETHANOL FUELS

The 15 most prolific source titles publishing at least 2% of the sample papers each in the algal bioethanol fuels have shaped the development of the research in this field (Table 55.5).

The most prolific source titles are Bioresource Technology and to a lesser extent Renewable and Sustainable Energy Reviews, Energy and Environmental Science, Journal of Applied Phycology, Biotechnology for Biofuels, Process Biochemistry, and Applied Energy.

On the other hand, source titles with the highest impact are Bioresource Technology and to a lesser extent Renewable and Sustainable Energy Reviews, Energy and Environmental Science, Applied Energy, and Process Biochemistry. Similarly, source titles with the lowest impact are Algal Research and to a lesser extent the Journal of Applied Phycology and Renewable Energy.

It is notable that these top source titles are primarily related to bioresources, energy, algae, and chemical engineering. This finding suggests that Bioresource Technology and other prolific journals in these fields have significantly shaped the development of the research in this field as they focus primarily on algal bioethanol fuels with a high yield. In this context, the influence of Bioresource Technology is quite extraordinary with 38% of the sample papers.

55.4.8 THE MOST PROLIFIC COUNTRIES ON ALGAL BIOETHANOL FUELS

The 15 most prolific countries publishing at least 3% of the sample papers each have significantly shaped the development of the research in this field (Table 55.6).

The most prolific countries are South Korea and to a lesser extent the USA, China, Malaysia, Taiwan, India, Australia, and Denmark. Further, four European countries listed in Table 55.6 produce 14% and 8% of the sample and population papers, respectively.

On the other hand, countries with the highest citation impact are Japan and to a lesser extent South Korea, the USA, and Taiwan. Similarly, countries with the lowest impact are India and to a lesser extent Indonesia, Brazil, China, and Egypt.

A close examination of these findings suggests that East Pacific (South Korea, China, Japan, Taiwan, Malaysia, India, and Australia), the USA, and Europe is the major producer of research in this field. It is notable that all these top countries have large access to seas, rivers, and lakes, essential for algal research.

It is a fact that the USA has been a major player in science (Leydesdorff and Wagner, 2009). The USA has further developed a strong research infrastructure to support its corn- and grass-based bioethanol industry (Gillon, 2010).

However, China has been a rising megastar in scientific research in competition with the USA and Europe (Leydesdorff and Zhou, 2005). China is also a major player in this field as a major producer of bioethanol (Fang et al., 2010).

Next, Europe has been a persistent player in scientific research in competition with both the USA and China (Leydesdorff, 2000). Europe has also been a persistent producer of bioethanol along with the USA and Brazil (Gnansounou, 2010).

55.4.9 THE MOST PROLIFIC SCOPUS SUBJECT CATEGORIES ON ALGAL BIOETHANOL FUELS

The eight most prolific Scopus subject categories indexing at least 8% of the sample papers each, respectively, given in Table 55.7 have shaped the development of the research in this field.

The most prolific Scopus subject categories on algal bioethanol fuels are Energy and to a lesser extent Environmental Science, Chemical Engineering, Biochemistry, Genetics and Molecular Biology, and Immunology and Microbiology. On the other hand, the Scopus subject categories with the highest citation impact are Energy and to a lesser extent Environmental Science, Chemical Engineering, and Immunology and Microbiology. Similarly, the least influential subject categories are Agricultural and Biological Sciences and to a lesser extent Chemistry, Engineering, and Biochemistry, Genetics and Molecular Biology.

These findings are thought-provoking suggesting that the primary subject categories are related to energy, environmental science, and chemical engineering as the core of the research in this field concerns algal ethanol production. The other finding is that social sciences are not well represented in both the sample and population papers as in most fields in bioethanol fuels.

55.4.10 THE MOST PROLIFIC KEYWORDS ON ALGAL BIOETHANOL FUELS

A limited number of keywords have shaped the development of research in this field as shown in Table 55.8 and the appendix. These keywords are grouped under eight headings: biomass, fermentation, bacteria, hydrolysis, hydrolysates, pretreatments, other processes, and products of fermentation.

The most prolific keywords across all research fronts are microalgae, ethanol, biofuels, bioethanol, fermentation, algae, biomass, and carbohydrates. On the other hand, the most influential keywords are seaweeds, biofuel production, biomass fermentation, biomass production, biorefineries, and acid hydrolysis. Similarly, the least influential keywords are enzymatic saccharification, chlorella, synechocystis, and algal growth.

These findings suggest that it is necessary to determine the keyword set carefully to locate the relevant research in each of these research fronts. Additionally, the size of the samples for each keyword highlights the intensity of the research in the relevant research areas.

55.4.11 THE MOST PROLIFIC RESEARCH FRONTS ON ALGAL BIOETHANOL FUELS

Information about the research fronts for the sample papers on algal bioethanol fuels with regard to algal biomass used in bioethanol production is given in Table 55.9. As this table shows, there are three primary research fronts in this field: macroalgae, microalgae, and cyanobacteria. Additionally, algae in general covers 7% of the sample papers. It is notable that diatoms and dinoflagellates are not covered in this sample.

Further, the most prolific research front on macroalgae is macroalgae in general, followed by alginates, gelidium, and ulva. Similarly, the most prolific research front in microalgae is microalgae in general, followed by chlorella and scenedesmus. Finally, synechocystis is the most prolific cyanobacterial research front, followed by the research front in cyanobacteria in general.

Information about the thematic research fronts for the sample papers in algal bioethanol fuels is given in Table 55.10. As this table shows, there are four primary research fronts: algal biomass pretreatments and hydrolysis, algal hydrolysate fermentation, and algal bioethanol production in general.

For the research front of algal biomass pretreatments, the primary research fronts are chemical, enzymatic, and microbial pretreatments. Further, the most prolific chemical pretreatment is acid pretreatment.

Similarly, the most prolific research fronts on algal biomass hydrolysis are enzymatic and acid hydrolysis. Next, the prolific research fronts on algal hydrolysate fermentation are macroalgal and microalgal fermentation. Finally, the prolific research fronts on algal bioethanol fuels are macroalgal and microalgal bioethanol fuels.

These findings are thought-provoking in seeking ways to increase the algal bioethanol yield at the global scale. It is clear that all of these research fronts have public importance and merit substantial funding and other incentives. Further, it is notable that algal bioethanol fuels have become a core of bioethanol research to make it more competitive with crude oil-based gasoline and diesel fuels, especially for countries with large access to the seas and lakes, beneficial for algal research.

In the end, these most-cited papers in this field hint that the efficiency of algal bioethanol fuels could be optimized using structure, processing, and property relationships of these algal species in the fronts of algal biomass pretreatments and hydrolysis, and hydrolysate fermentation (Formela et al., 2016; Konur, 2018a, 2020b, 2021a,b,c,d; Konur and Matthews, 1989).

55.5 CONCLUSION AND FUTURE RESEARCH

Research on algal bioethanol fuels has been mapped through a scientometric study of both sample (100 papers) and population (760 papers) datasets.

The critical issue in this study has been to obtain a representative sample of the research as in any other scientometric study. Therefore, the keyword set has been carefully devised and optimized after a number of runs in the Scopus database. It is a representative sample of wider population studies. This keyword set is provided in the appendix, and the relevant keywords are presented in Table 55.8. However, it should be noted that it has been very difficult to compile a representative keyword set since this research field has been connected closely with many other fields. Therefore, it has been necessary to compile a keyword list to exclude papers concerned with the other research fields.

Another issue has been the selection of a multidisciplinary database to carry out the scientometric study of the research in this field. For this purpose, the Scopus database has been selected. The journal coverage of this database has been notably wider than that of Web of Science and other multisubject databases.

The key scientometric properties of research in this field have been determined and discussed in this chapter. It is evident that a limited number of documents, authors, institutions, publication years, institutions, funding bodies, source titles, countries, Scopus subject categories, Scopus keywords, and research fronts have shaped the development of the research in this field.

There is ample scope to increase the efficiency of scientometric studies in this field in the author and document domains by developing consistent policies and practices in both domains across all academic databases. In this respect, it seems that authors, journals, and academic databases have a lot to do. Furthermore, the significant gender deficit as in most scientific fields emerges as a public policy issue. The potential deficits on the basis of age, race, disability, and sexuality also need to be explored in this field as in other scientific fields.

The research in this field has boomed since 2010, possibly promoted by public concerns about global warming, greenhouse gas emissions, and climate change. Furthermore, the recent COVID-19 pandemic and the Russian invasion of Ukraine have resulted in global supply shocks shifting the focus of the stakeholders from crude oil-based fuels to biomass-based fuels such as bioethanol fuels. It is expected that there would be further incentives for key stakeholders to carry out research on algal bioethanol fuels to increase the ethanol yield and to make it more competitive with crude oil-based gasoline and petrodiesel fuels. This might be truer for crude oil- and foreign exchange-deficient countries to maintain energy security in the face of global supply shocks.

The relatively significant funding rate of 61% and 56% for the sample and population papers, respectively, suggests that funding in this field significantly enhanced the research in this field primarily since 2010, possibly more than doubling in the current decade. However, it is evident that there is ample room for more funding and other incentives to enhance the research in this field further.

The institutions in South Korea and to a lesser extent China, Denmark, and Malaysia have mostly shaped the research in this field. Further, East Pacific (South Korea, China, Japan, Taiwan, Malaysia, India, and Australia), the USA, and Europe have been the major producers of the research in this field as the major producers and users of bioethanol fuels from different types of biomass such as corn, sugarcane, and grass as well as other types of biomass. It is evident that these countries have well-developed research infrastructure in bioethanol fuels and their derivatives. It is also notable all of these major countries have access to seas and lakes, beneficial for algal research.

It emerges that ethanol is more popular than bioethanol as a keyword with strong implications for the search strategy. In other words, the search strategy using only bioethanol as the keyword would not be much helpful. The Scopus keywords are grouped under eight headings: biomass, fermentation, bacteria, hydrolysis, hydrolysates, pretreatments, other processes, and products of fermentation.

There are three primary research fronts for this field: macroalgae, microalgae, and cyanobacteria at the expense of diatoms and dinoflagellates. Alginates, gelidium, and ulva are the most prolific macroalgae, while chlorella and scenedesmus are the most prolific microalgae. Finally, synechocystis is the most prolific cyanobacteria.

On the other hand, there are four primary thematic research fronts: algal biomass pretreatments and hydrolysis, algal hydrolysate fermentation, and algal bioethanol production in general. Further, chemical, enzymatic, and microbial pretreatments are the most prolific pretreatments, while enzymatic and acid-algal hydrolysis are the most prolific hydrolysis. Finally, both macroalgae and microalgae dominate the research fronts of algal hydrolysate fermentation and bioethanol production at the expense of cyanobacteria, diatoms, and dinoflagellates.

These findings are thought-provoking in seeking ways to increase the bioethanol yield through algal bioethanol production at the global scale. It is clear that all of these research fronts have public importance and merit substantial funding and other incentives. Further, it is notable that algal bioethanol fuels, as a third generation biofuel, have become a core unit of bioethanol research to make it more competitive with crude oil-based gasoline and diesel fuels, especially for countries with large access to the seas, rivers, and lakes, beneficial for algal research.

Thus, scientometric analysis has a great potential to gain valuable insights into the evolution of research in this field as in other scientific fields, especially in the aftermath of significant global supply shocks such as the COVID-19 pandemic and the Russian invasion of Ukraine.

It is recommended that further scientometric studies are carried out for the primary research fronts. It is further recommended that reviews of the most-cited papers are carried out for each primary research front to complement these scientometric studies. Next, scientometric studies of hot papers in these primary fields need to be carried out.

ACKNOWLEDGMENTS

The contribution of the highly cited researchers in the field of algal bioethanol fuels has been gratefully acknowledged.

APPENDIX: THE KEYWORD SET FOR ALGAL BIOETHANOL FUELS

(((TITLE (algae OR algal OR photobioreactor* OR alga OR "photo-bioreactor*" OR chlamydomon* OR schizocytrium OR *chlorella OR navicula OR microalga* OR dunaliella* OR nannochloropsis OR scenedesmus OR *porphyridium OR botryococc* OR chlorococc* OR tetraselmis OR desmodesmus OR spirogyra OR *alginate OR agarose* OR macroalga* OR seaweed* OR porphyra OR pterocladiella OR gracilar* OR kelp* OR ulva* OR fucus OR laminaria* OR saccharina OR sargassum OR enteromorpha OR ascophyllum OR chondrus OR codium OR cladophor* OR gelidium OR kappaphycus OR macrocystis OR undaria OR "sea weed*" OR agarophyt* OR gelidiella OR wakame OR chaetomorpha OR eucheuma* OR monostroma OR padina OR palmaria* OR solieria* OR carrageenophyt* OR "sea lettuce" OR colpomenia OR mastocarpus OR acanthophora OR kombu OR diatoms OR diatom OR phaeodactylum OR cyanobact* OR *synechocystis OR *synechococcus OR chroococcus OR *microcystis OR *anabaena OR spirul* OR arthrospira) OR SRCTITLE (alga* OR phyco* OR diatom* OR algol* OR fottea* OR microalga* OR cyanobacter* OR macroalga* OR diatoms OR seaweed*) AND TITLE (saccharif* OR hydrolyze* OR *hydrolysis OR hydrolysate* OR hydrolyzate* OR ferment* OR "sulfuric acid*" OR milling OR grinding OR pretreatment* OR pretreated OR "pre-treat*" OR bioorganosolve OR steam* OR "hot water" OR "hot compressed" OR sporl OR naoh* OR "sodium hydroxide*" OR organosolv*)) AND NOT (TITLE (fatty OR pyroly* OR *diesel OR plate OR biogas OR *hydrogen OR chromat* OR protein* OR infect* OR pharm* OR leuconostoc OR anaerobic* OR biosorption OR metal* OR cartilage OR antioxidant* OR arsen* OR *saccharide OR toxi* OR drip* OR glucosidase OR *peptide OR pigment* OR oil OR gasification OR microbicide OR uptake OR lipid OR solar OR *toxin OR *methane OR cell OR liquefaction OR carot* OR *propanol OR *remediation OR anti* OR *cancer OR aman OR precursor* OR dye* OR sediment* OR diameter* OR transester* OR astaxanthin OR eco* OR *gel OR swine OR inorganic OR sausage* OR ester* OR organic* OR harvest* OR phenolic OR microcystin* OR reproductive OR entrap* OR *degradation OR docosahexaenoic OR lyase* OR periodate OR thermochem* OR osmosis OR processsing OR fermenter* OR amino OR food* OR encaps* OR *guluronic OR atp OR extract* OR nerve OR cobalt OR purif* OR *alkanoate OR fucose OR coagul* OR butanediol OR mat OR spectr* OR membrane OR phycoerythrin OR dha OR methanogen* OR lactobacillus OR cassava OR porphyrazine OR immobil* OR effluent* OR metabolite OR antiviral OR drying OR dicarboxylic OR uv OR phb OR overflow OR ash OR filtr* OR gaseous OR nano* OR actuator OR hyper OR cadmium OR ni2 OR lactic OR agar OR rat OR milk OR h2 OR anti* OR diet* OR polyamines OR ferrate OR succinic OR products OR slurry OR washing OR injury OR hplc OR processing OR *mannuronic OR linolenic OR jsc6 OR foliar OR dsm* OR fucoidin OR drum OR cost OR lovastat* OR psych OR rhodosh* OR disk OR sauce* OR naringin OR flask OR taxonomy OR *phosphates OR tlc OR mortar* OR rheol* OR films OR kraft OR porous OR urea OR sroro* OR nitrile OR fucoidan OR coal OR muscle OR nutraceut* OR dairy OR itaconic OR flash OR corn OR lutein OR fermentor OR liposome* OR straw OR lactate OR bagasse OR manure OR recycle) OR SUBJAREA (phar OR medi OR neur OR nurs OR vete OR heal OR dent OR psych) OR SRCTITLE (drug* OR toxi* OR harmful OR limo* OR poll* OR bota* OR hydrogen OR eco* OR fish* OR lipid* OR hazard* OR plant* OR aqua* OR chromat* OR geo* OR biogeo* OR *materials OR desalination OR freshw* OR limnol* OR pineal* OR polymer* OR oceean* OR chemo* OR pala* OR proto* OR phyto* OR cell OR remote OR water* OR animal OR purif* OR hydro* OR bioflux OR extrem* OR planton OR membrane OR antiox* OR dairy OR biomacro* OR macromol* OR marine OR composite* OR poultry OR biomed* OR polar OR sensors OR cereal* OR ceramic*))) OR ((((TITLE (algae OR algal OR photobioreactor* OR alga OR "photo-bioreactor*" OR chlamydomon* OR tribonema OR schizocytrium OR *chlorella OR navicula OR microalga* OR dunaliella* OR nannochloropsis OR scenedesmus OR *porphyridium OR botryococc* OR chlorococc* OR tetraselmis OR desmodesmus OR spirogyra OR *alginate OR agarose* OR macroalga* OR seaweed* OR porphyra OR pterocladiella OR gracilar* OR kelp* OR ulva* OR fucus OR laminaria* OR saccharina OR sargassum OR enteromorpha OR ascophyllum OR chondrus OR codium OR cladophor* OR gelidium OR kappaphycus OR macrocystis OR undaria OR "sea weed*" OR agarophyt* OR gelidiella OR wakame OR chaetomorpha OR eucheuma* OR monostroma OR padina OR palmaria* OR solieria* OR carrageenophyt* OR "sea lettuce" OR colpomenia OR mastocarpus OR acanthophora OR kombu OR diatoms OR diatom OR phaeodactylum OR cyanobact* OR *synechocystis OR *synechococcus OR chroococcus OR *microcystis OR *anabaena OR spirul* OR arthrospira OR cyanothece) OR SRCTITLE (alga* OR phyco* OR diatom* OR algol* OR fottea* OR microalga* OR cyanobacter* OR macroalga* OR diatoms OR seaweed*)) AND TITLE (ethanol OR bioethanol)) AND NOT (TITLE (membrane* OR liquefaction OR dehydration OR recalcitrant OR supercritical OR alterations OR docosahexaenoic OR periodate OR appraisal OR pervaporation OR lactate OR mahula OR anti* OR fucoxanthin OR jsc OR films OR diet* OR cassava OR zymomonas OR nano* OR kluy* OR carot* OR hydrogel OR liver OR transester* OR carriers OR pulsed OR admin* OR sponge OR mice OR rats OR diffusion OR cancer OR matrix OR membrane* OR chromat* OR physiol* OR vinasse OR yeast OR molass* OR cellobiose OR cost OR toxicity) OR SUBJAREA (medi OR phar OR nurs OR heal OR neur OR vete)))) AND (LIMIT-TO (SRCTYPE, "j") OR LIMIT-TO (SRCTYPE, "b") OR LIMIT-TO (SRCTYPE, "k")) AND (LIMIT-TO (LANGUAGE, "English")) AND (LIMIT-TO (DOCTYPE, "ar") OR LIMIT-TO (DOCTYPE, "cp") OR LIMIT-TO (DOCTYPE, "re") OR LIMIT-TO (DOCTYPE, "ch") OR LIMIT-TO (DOCTYPE, "no") OR LIMIT-TO (DOCTYPE, "ed") OR LIMIT-TO (DOCTYPE, "le") OR LIMIT-TO (DOCTYPE, "bk"))

REFERENCES

Adams, J. M., J. A. Gallagher and I. S. Donnison. 2009. Fermentation study on *Saccharina latissima* for bioethanol production considering variable pre-treatments. *Journal of Applied Phycology* 21:569–574.

Aman, V. 2018. Does the Scopus author ID suffice to track scientific international mobility? A case study based on Leibniz laureates. *Scientometrics* 117:705–720.

Angelici, C., B. M. Weckhuysen and P. C. A. Bruijnincx. 2013. Chemocatalytic conversion of ethanol into butadiene and other bulk chemicals. *ChemSusChem* 6:1595–1614.

Antolini, E. 2007. Catalysts for direct ethanol fuel cells. *Journal of Power Sources* 170:1–12.

Antolini, E. 2009. Palladium in fuel cell catalysis. *Energy and Environmental Science* 2:915–931.

Beaudry, C. and V. Lariviere. 2016. Which gender gap? Factors affecting researchers' scientific impact in science and medicine. *Research Policy* 45:1790–1817.

Blankenship, K. M. 1993. Bringing gender and race in: US employment discrimination policy. *Gender & Society* 7:204–226.

Burnham, J. F. 2006. Scopus database: A review. *Biomedical Digital Libraries* 3:1–8.

Carlson, K. M., J. S. Gerber and D. Mueller, et al. 2017. Greenhouse gas emissions intensity of global croplands. *Nature Climate Change* 7:63–68.

Change, C. 2007. Climate change impacts, adaptation and vulnerability. *Science of the Total Environment* 326:95–112.

Choi, S. P., M. T. Nguyen and S. J. Sim. 2010. Enzymatic pretreatment of *Chlamydomonas reinhardtii* biomass for ethanol production. *Bioresource Technology* 101:5330–5336.

Deng, M. D. and J. R. Coleman. 1999. Ethanol synthesis by genetic engineering in cyanobacteria. *Applied and Environmental Microbiology* 65:523–528.

Dexter, J. and P. Fu. 2009. Metabolic engineering of cyanobacteria for ethanol production. *Energy and Environmental Science* 2:857–864.

Dirth, T. P. and N. R. Branscombe. 2017. Disability models affect disability policy support through awareness of structural discrimination. *Journal of Social Issues* 73:413–442.

Ebadi, A. and A. Schiffauerova. 2016. How to boost scientific production? A statistical analysis of research funding and other influencing factors. *Scientometrics* 106:1093–1116.

Fang, X., Y. Shen, J. Zhao, X. Bao and Y. Qu. 2010. Status and prospect of lignocellulosic bioethanol production in China. *Bioresource Technology* 101:4814–4819.

Fauci, A. S., H. C. Lane and R. R. Redfield. 2020. Covid-19-navigating the uncharted. *New England Journal of Medicine* 382:1268–1269.

Fernando, S., S. Adhikari, C. Chandrapal and M. Murali. 2006. Biorefineries: Current status, challenges, and future direction. *Energy & Fuels* 20:1727–1737.

Formela, K., A. Hejna, L. Piszczyk, M. R. Saeb and X. Colom. 2016. Processing and structure-property relationships of natural rubber/wheat bran biocomposites. *Cellulose* 23:3157–3175.

Gao, Z., H. Zhao, Z. Li, X. Tan and X. Lu. 2012. Photosynthetic production of ethanol from carbon dioxide in genetically engineered cyanobacteria. *Energy and Environmental Science* 5:9857–9865.

Garfield, E. 1955. Citation indexes for science. *Science* 122:108–111.

Gillon, S. 2010. Fields of dreams: Negotiating an ethanol agenda in the Midwest United States. *Journal of Peasant Studies* 37:723–748.

Gnansounou, E. 2010. Production and use of lignocellulosic bioethanol in Europe: Current situation and perspectives. *Bioresource Technology* 101:4842–4850.

Hahn-Hagerdal, B., M. Galbe, M. F. Gorwa-Grauslund, G. Liden and G. Zacchi. 2006. Bio-ethanol - The fuel of tomorrow from the residues of today. *Trends in Biotechnology* 24:549–556.

Hamilton, J. D. 1983. Oil and the macroeconomy since World War II. *Journal of Political Economy* 91:228–248.

Hamilton, J. D. 2003. What is an oil shock? *Journal of Econometrics* 113:363–398.

Hamilton, J. D. 2009. Causes and consequences of the oil shock of 2007-08. *Brookings Papers on Economic Activity* 2009:215–261.

Harun, R. and M. K. Danquah. 2011a. Influence of acid pre-treatment on microalgal biomass for bioethanol production. *Process Biochemistry* 46:304–309

Harun, R. and M. K. Danquah. 2011b. Enzymatic hydrolysis of microalgal biomass for bioethanol production. *Chemical Engineering Journal* 168:1079–1084.

Harun, R., M. K. Danquah and G. M. Forde. 2010. Microalgal biomass as a fermentation feedstock for bioethanol production. *Journal of Chemical Technology and Biotechnology* 85:199–203.

Harun, R., W. S. Y. Jason, T. Cherrington and M. K. Danquah. 2011. Exploring alkaline pre-treatment of microalgal biomass for bioethanol production. *Applied Energy* 88:3464–3467.

Hill, J., E. Nelson, D. Tilman, S. Polasky and D. Tiffany. 2006. Environmental, economic, and energetic costs and benefits of biodiesel and ethanol biofuels. *Proceedings of the National Academy of Sciences of the United States of America* 103:11206–11210.

Hill, J., S. Polasky and E. Nelson, et al. 2009. Climate change and health costs of air emissions from biofuels and gasoline. *Proceedings of the National Academy of Sciences of the United States of America* 106:2077–2082.

Ho, S. H., S. W. Huang and C. Y. Chen, et al. 2013. Bioethanol production using carbohydrate-rich microalgae biomass as feedstock. *Bioresource Technology* 135:191–198.

Horn, S. J., I. M. Aasen and K. Ostgaard. 2000. Ethanol production from seaweed extract. *Journal of Industrial Microbiology and Biotechnology* 25:249–254.

Hsieh, W. D., R. H. Chen, T. L. Wu and T. H. Lin. 2002. Engine performance and pollutant emission of an SI engine using ethanol-gasoline blended fuels. *Atmospheric Environment* 36:403–410.

Huang, H. J., S. Ramaswamy, U. W. Tschirner and B. V. Ramarao. 2008. A review of separation technologies in current and future biorefineries. *Separation and Purification Technology* 62:1–21.

Jones, T. C. 2012. America, oil, and war in the Middle East. *Journal of American History* 99:208–218.

Kerr, R. A. 2007. Global warming is changing the world. *Science* 316:188–190.

Kilian, L. 2008. Exogenous oil supply shocks: How big are they and how much do they matter for the US economy? *Review of Economics and Statistics* 90:216–240.

Kilian, L. 2009. Not all oil price shocks are alike: Disentangling demand and supply shocks in the crude oil market. *American Economic Review*, 99:1053–1069.

Kim, J. 2018. Evaluating author name disambiguation for digital libraries: A case of DBLP. *Scientometrics* 116:1867–1886.

Kim, K. H., I. S. Choi, H. M. Kim, S. G. Wi and H. J. Bae. 2014. Bioethanol production from the nutrient stress-induced microalga *Chlorella vulgaris* by enzymatic hydrolysis and immobilized yeast fermentation. *Bioresource Technology* 153:47–54.

Kim, N. J., H. Li, K. Jung, H. N. Chang and P. C. Lee. 2011. Ethanol production from marine algal hydrolysates using *Escherichia coli* KO11. *Bioresource Technology* 102:7466–7469.

Konur, O. 2000. Creating enforceable civil rights for disabled students in higher education: An institutional theory perspective. *Disability & Society* 15:1041–1063.

Konur, O. 2002a. Access to nursing education by disabled students: Rights and duties of nursing programs. *Nurse Education Today* 22:364–374.

Konur, O. 2002b. Assessment of disabled students in higher education: Current public policy issues. *Assessment and Evaluation in Higher Education* 27:131–152.

Konur, O. 2002c. Access to employment by disabled people in the UK: Is the Disability Discrimination Act working? *International Journal of Discrimination and the Law* 5:247–279.

Konur, O. 2006a. Participation of children with dyslexia in compulsory education: Current public policy issues. *Dyslexia* 12:51–67.

Konur, O. 2006b. Teaching disabled students in higher education. *Teaching in Higher Education* 11:351–363.

Konur, O. 2007a. A judicial outcome analysis of the *Disability Discrimination Act*: A windfall for the employers? *Disability & Society* 22:187–204.

Konur, O. 2007b. Computer-assisted teaching and assessment of disabled students in higher education: The interface between academic standards and disability rights. *Journal of Computer Assisted Learning* 23:207–219.

Konur, O. 2011. The scientometric evaluation of the research on the algae and bio-energy. *Applied Energy* 88:3532–3540.

Konur, O. 2012a. The evaluation of the biogas research: A scientometric approach. *Energy Education Science and Technology Part A: Energy Science and Research* 29:1277–1292.

Konur, O. 2012b. The evaluation of the educational research: A scientometric approach. *Energy Education Science and Technology Part B: Social and Educational Studies* 4:1935–1948.

Konur, O. 2012c. The evaluation of the global energy and fuels research: A scientometric approach. *Energy Education Science and Technology Part A: Energy Science and Research* 30:613–628.

Konur, O. 2012d. The evaluation of the research on the biodiesel: A scientometric approach. *Energy Education Science and Technology Part A: Energy Science and Research* 28:1003–1014.

Konur, O. 2012e. The evaluation of the research on the bioethanol: A scientometric approach. *Energy Education Science and Technology Part A: Energy Science and Research* 28:1051–1064.

Konur, O. 2012f. The evaluation of the research on the biofuels: A scientometric approach. *Energy Education Science and Technology Part A: Energy Science and Research* 28:903–916.

Konur, O. 2012g. The evaluation of the research on the biohydrogen: A scientometric approach. *Energy Education Science and Technology Part A: Energy Science and Research* 29:323–338.

Konur, O. 2012h. The evaluation of the research on the microbial fuel cells: A scientometric approach. *Energy Education Science and Technology Part A: Energy Science and Research* 29:309–322.

Konur, O. 2012i. The scientometric evaluation of the research on the production of bioenergy from biomass. *Biomass and Bioenergy* 47:504–515.

Konur, O. 2015. Current state of research on algal bioethanol. In *Marine Bioenergy: Trends and Developments*, Ed. S. K. Kim and C. G. Lee, pp. 217–244. Boca Raton, FL: CRC Press.

Konur, O., Ed. 2018a. *Bioenergy and Biofuels*. Boca Raton, FL: CRC Press.

Konur, O. 2018b. Bioenergy and biofuels science and technology: Scientometric overview and citation classics. In *Bioenergy and Biofuels*, Ed. O. Konur, pp. 3–63. Boca Raton: CRC Press.

Konur, O. 2019. Cyanobacterial bioenergy and biofuels science and technology: A scientometric overview. In *Cyanobacteria: From Basic Science to Applications*, Ed. A. K. Mishra, D. N. Tiwari and A. N. Rai, pp. 419–442. Amsterdam: Elsevier.

Konur, O. 2020a. The scientometric analysis of the research on the bioethanol production from green macroalgae. In *Handbook of Algal Science, Technology and Medicine*, Ed. O. Konur, pp. 385–401. London: Academic Press.

Konur, O., Ed. 2020b. *Handbook of Algal Science, Technology and Medicine*. London: Academic Press.

Konur, O., Ed. 2021a. *Handbook of Biodiesel and Petrodiesel Fuels: Science, Technology, Health, and Environment*. Boca Raton, FL: CRC Press.

Konur, O., Ed. 2021b. *Handbook of Biodiesel and Petrodiesel Fuels: Science, Technology, Health, and Environment. Volume 1. Biodiesel Fuels: Science, Technology, Health, and Environment*. Boca Raton, FL: CRC Press.

Konur, O., Ed. 2021c. *Handbook of Biodiesel and Petrodiesel Fuels: Science, Technology, Health, and Environment. Volume 2. Biodiesel Fuels based on the Edible and Nonedible Feedstocks, Wastes, and Algae: Science, Technology, Health, and Environment*. Boca Raton, FL: CRC Press.

Konur, O., Ed. 2021d. *Handbook of Biodiesel and Petrodiesel Fuels: Science, Technology, Health, and Environment. Volume 3. Petrodiesel Fuels: Science, Technology, Health, and Environment*. Boca Raton, FL: CRC Press.

Konur, O. and F. L. Matthews. 1989. Effect of the properties of the constituents on the fatigue performance of composites: A review. *Composites* 20:317–328.

Kruyt, B., D. P. van Vuuren, H. J. de Vries and H. Groenenberg. 2009. Indicators for energy security. *Energy Policy* 37:2166–2181.

Leydesdorff, L. 2000. Is the European Union becoming a single publication system? *Scientometrics* 47:265–280.

Leydesdorff, L. and C. Wagner. 2009. Is the United States losing ground in science? A global perspective on the world science system. *Scientometrics* 78:23–36.

Leydesdorff, L. and P. Zhou. 2005. Are the contributions of China and Korea upsetting the world system of science? *Scientometrics* 63:617–630.

Li, H., S. M. Liu, X. H. Yu, S. L. Tang and C. K. Tang. 2020. Coronavirus disease 2019 (COVID-19): Current status and future perspectives. *International Journal of Antimicrobial Agents* 55:105951.

Lin, Y. and S. Tanaka. 2006. Ethanol fermentation from biomass resources: Current state and prospects. *Applied Microbiology and Biotechnology* 69:627–642.

Ma, X., L. Sun and C. Song. 2002. A new approach to deep desulfurization of gasoline, diesel fuel and jet fuel by selective adsorption for ultra-clean fuels and for fuel cell applications. *Catalysis Today* 77:107–116.

Morschbacker, A. 2009. Bio-ethanol based ethylene. *Polymer Reviews* 49:79–84.

Najafi, G., B. Ghobadian and T. Tavakoli, et al. 2009. Performance and exhaust emissions of a gasoline engine with ethanol blended gasoline fuels using artificial neural network. *Applied Energy* 86:630–639.

Newman, P. W. G. and J. R. Kenworthy. 1989. Gasoline consumption and cities: A comparison of U.S. cities with a global survey. *Journal of the American Planning Association* 55:24–37.

North, D. C. 1991. Institutions. *Journal of Economic Perspectives* 5:97–112.

Olsson, L. and B. Hahn-Hagerdal. 1996. Fermentation of lignocellulosic hydrolysates for ethanol production. *Enzyme and Microbial Technology* 18:312–331.

Park, J. H., J. Y. Hong and H. C. Jang, et al. 2012. Use of *Gelidium amansii* as a promising resource for bioethanol: A practical approach for continuous dilute-acid hydrolysis and fermentation. *Bioresource Technology* 108:83–88.

Reeves, S. 2014. To Russia with love: How moral arguments for a humanitarian intervention in Syria opened the door for an invasion of the Ukraine. *Michigan State University International Law Review* 23:199.

Sanchez, O. J. and C. A. Cardona. 2008. Trends in biotechnological production of fuel ethanol from different feedstocks. *Bioresource Technology* 99:5270–5295.

Sun, Y. and J. Cheng. 2002. Hydrolysis of lignocellulosic materials for ethanol production: A review. *Bioresource Technology* 83:1–11.

Taherzadeh, M. J. and K. Karimi. 2007. Enzyme-based hydrolysis processes for ethanol from lignocellulosic materials: A review. *Bioresources* 2:707–738.

Taherzadeh, M. J. and K. Karimi. 2008. Pretreatment of lignocellulosic wastes to improve ethanol and biogas production: A review. *International Journal of Molecular Sciences* 9:1621–1651.

Winzer, C. 2012. Conceptualizing energy security. *Energy Policy* 46:36–48.

Yang, B. and C. E. Wyman. 2008. Pretreatment: The key to unlocking low-cost cellulosic ethanol. *Biofuels, Bioproducts and Biorefining* 2:26–40.

56 Third Generation Algal Bioethanol Fuels

Review

Bahareh Nowruzi
Islamic Azad University

Ozcan Konur
(Formerly) Ankara Yildirim Beyazit University

56.1 INTRODUCTION

Crude oil-based gasoline fuels (Ma et al., 2002; Newman and Kenworthy, 1989) have been widely used in the transportation sector since the 1920s. However, there have been great public concerns over the adverse environmental and human impact of these fuels (Hill et al., 2006, 2009). Hence, biomass-based bioethanol fuels (Hill et al., 2006; Konur, 2012, 2015, 2019, 2020) have increasingly been used in blending gasoline fuels (Hsieh et al., 2002; Najafi et al., 2009), in fuel cells (Antolini, 2007, 2009), and in biochemical production (Angelici et al., 2013; Morschbacker, 2009) in a biorefinery context (Fernando et al., 2006; Huang et al., 2008).

However, it is necessary to pretreat the biomass (Alvira et al., 2010; Taherzadeh and Karimi, 2008) to enhance the yield of bioethanol (Hahn-Hagerdal et al., 2006; Sanchez and Cardona, 2008) prior to algal bioethanol production through hydrolysis (Sun and Cheng, 2002; Taherzadeh and Karimi, 2007) and fermentation (Lin and Tanaka, 2006; Olsson and Hahn-Hagerdal, 1996) of biomass and hydrolysates, respectively.

One of the most studied biomass materials for bioethanol fuels has been algal biomass. Research in the field of algal bioethanol fuels has intensified in this context in the key research fronts of the pretreatment of algal biomass (Choi et al., 2010; Harun and Danquah, 2011a; Harun et al., 2011), hydrolysis of algal biomass (Harun and Danquah, 2011b; Kim et al., 2014; Park et al., 2012), fermentation of algal hydrolysates (Adams et al., 2009; Horn et al., 2000; Kim et al., 2011), and algal bioethanol fuels in general (Deng and Coleman, 1999; Harun et al., 2010; Ho et al., 2013). Further, macroalgae (Adams et al., 2009; Horn et al., 2000; Kim et al., 2011), microalgae (Choi et al., 2010; Harun et al., 2010; Ho et al., 2013), and cyanobacteria (Deng and Coleman, 1999; Dexter and Fu, 2009; Gao et al., 2012) have been studied intensively at the expense of diatoms and dinoflagellates in this context.

However, it is essential to develop efficient incentive structures (North, 1991) for the primary stakeholders to enhance research in this field (Konur, 2000, 2002a,b,c, 2006a,b, 2007a,b). Although there are a number of review papers on algal bioethanol fuels (de Farias Silva and Bertucco, 2016; John et al., 2011; Yanagisawa et al., 2013), there has been no review of the 25 most-cited papers in this field.

Thus, this chapter presents a review of the 25 most-cited articles in the field of algal bioethanol fuels. Then, it discusses the key findings of these highly influential papers and comments on future research priorities in this field.

 DOI: 10.1201/9781003226451-74

56.2 MATERIALS AND METHODS

The search for this study was carried out using the Scopus database (Burnham, 2006) in June 2022.

As the first step for the search of the relevant literature, keywords were selected using the 200 most-cited sample papers. The selected keyword list was then optimized to obtain a representative sample of papers for this research field. This final keyword set was provided in the Appendix of Konur (2023) for future replication studies.

As the second step, a sample dataset was used in this study. The first 25 articles with at least 139 citations each were selected for review. Key findings from each paper were taken from the abstracts of these papers and were discussed. Additionally, a number of brief conclusions were drawn, and a number of relevant recommendations were made to enhance the future research landscape.

56.3 RESULTS

The brief information about the 25 most-cited papers with at least 139 citations each on algal bioethanol fuels is given below. The primary research fronts are the pretreatment of algal biomass, fermentation of algal hydrolysates, and algal bioethanol production in general with 7, 4, and 14 highly cited papers (HCPs), respectively.

56.3.1 ALGAL BIOMASS PRETREATMENTS OF ALGAL BIOETHANOL FUELS

There are seven HCPs for algal biomass pretreatments of algal bioethanol production (Table 56.1). The most prolific pretreatments are chemical and enzymatic pretreatments of algal biomass with five and three HCPs, respectively. Further, the most prolific chemical pretreatment is the acid pretreatment of algal biomass with four HCPs. Another chemical pretreatment is alkaline pretreatment with only one HCP. The study dealing with both acid and enzymatic hydrolysis is presented separately.

56.3.1.1 Algal Biomass Pretreatments in General

Hernandez et al. (2015) performed physical, chemical, and enzymatic pretreatments on three microalgal species to disrupt and break down complex carbohydrates into simple sugars, as a preliminary stage to produce bioethanol in a paper with 139 citations. They performed these pretreatments alone and combined with each other. They obtained the highest concentration of monosaccharides per gram of microalgae dry weight by the combination of acid and enzymatic pretreatments for *Chlorella sorokiniana* and *Nannochloropsis gaditana* at 128 and 129 mg/g, respectively. In the case of *Scenedesmus almeriensis,* they obtained the highest monosaccharide concentration (88 mg/g) after acid hydrolysis with sulfuric acid (H_2SO_4) for 60 min at 121°C.

56.3.1.2 Algal Biomass Chemical Pretreatments

The most prolific chemical pretreatment is the acid pretreatment of algal biomass with four HCPs. Another chemical pretreatment is the alkaline pretreatment of algal biomass with one HCP only.

56.3.1.2.1 Acid Pretreatments

Harun and Danquah (2011a) studied the effect of H_2SO_4 pretreatment on the ethanol yield during bioethanol production in a paper with 266 citations. They investigated acid concentration, temperature, microalgae loading, and pretreatment time. They obtained the highest bioethanol concentration using 7.20 g/L when the pretreatment step was performed with 15 g/L of microalgae at 140°C using 1% (v/v) of H_2SO_4 for 30 min. In terms of the ethanol yield, they obtained a maximum ethanol yield of 52 wt% (g ethanol/g microalgae) using 10 g/L of microalgae and 3% (v/v) of H_2SO_4 under 160°C for 15 min. The temperature was the most critical factor during acid pretreatment of microalgae for bioethanol production.

TABLE 56.1

Algal Biomass Pretreatments in Algal Bioethanol Production

No.	Papers	Biomass/Hydrolysate	Prts.	Yeasts	Parameters	Keywords	Lead Author	Affil.	Cits
1	Harun and Danquah (2011a)	Microalgae	H_2SO_4	Na	Bioethanol production, acid hydrolysis, ethanol yield, temperature	Microalgal, bioethanol, pre-treatment	Harun, Razif 35315707300	Univ. Putra Malaysia Malaysia	266
2	Choi et al. (2010)	*Chlamydomonas reinhardtii* UTEX 90	Enzymes	Na	Bioethanol production, enzymatic hydrolysis, ethanol yield, SHF	Chlamydomonas, ethanol, pretreatment	Choi, Seung P. 26433223800	Korea Univ. S. Korea	261
3	Harun et al. (2011)	Chlorococcum infusionum	NaOH	Na	Bioethanol production, NaOH pretreatment, sugar and ethanol yield	Microalgal, bioethanol, pre-treatment	Harun, Razif 35315707300	Univ. Putra Malaysia Malaysia	195
4	Miranda et al. (2012)	Scenedesmus obliquus	H_2SO_4	Na	Scenedesmus pretreatment optimization, acid pretreatment, sugar yield	Scenedesmus, bioethanol, pre-treatment	Gouveia, Luisa* 7004135834	Univ. Algarve Portugal	185
5	Nguyen et al. (2009)	*Chlamydomonas reinhardtii* UTEX 90	H_2SO_4	S. cerevisiae S288C	Bioethanol production, acid pretreatment, sugar and ethanol yield	Chlamydomonas, ethanol, pretreatment	Sim, Sang J. 55665584200	Seoul Jun Res. Ctr. S. Korea	168
6	Harun and Danquah (2011b)	*Chlorococcum* sp.	Cellulases	Na	Enzymatic pretreatment, glucose and cellobiose production	Microalgal, bioethanol, hydrolysis	Harun, Razif 35315707300	Univ. Putra Malaysia Malaysia	145
7	Hernandez et al. (2015)	*Chlorella sorokiniana, Nannochloropsis gaditana, Scenedesmus almeriensis*	H_2SO_4, enzymes	Na	Acid and enzymatic pretreatments, sugar yield	Microalgal, bioethanol, saccharification, pre-treatment	Garcia-Gonzalez, M. C. 25649355100	Agric. Technol. Inst. Spain	139

*, Female; Cits., Number of citations received for each paper; Na, non-available; Prt, Biomass pretreatments.

Miranda et al. (2012) optimized the pretreatment optimization of *Scenedesmus obliquus* for bioethanol production in a paper with 185 citations. They obtained the best results with acid hydrolysis by H_2SO_4 at 120°C for 30min using dried biomass. Further, the sugar extraction efficiency level achieved was 95.6% when compared to the harsh quantitative acid hydrolysis. They also studied the effect of other parameters such as biomass loading and the number of extraction cycles. The latter case showed that a unique hydrolysis step was sufficient.

Nguyen et al. (2009) studied the acid pretreatment of *Chlamydomonas reinhardtii* UTEX 90 for bioethanol production in a paper with 168 citations. With dry cells of 5% (w/v), they pretreated microalgal biomass with H_2SO_4 (1%–5%) under temperatures from 100°C to 120°C, for 15 to 120min. As a result, they observed that the glucose release from this biomass was maximum at 58% (w/w) after pretreatment with 3% H_2SO_4 at 110°C for 30min. This method enabled the hydrolysis of starch and other oligosaccharides in algal cells with high efficiency. They then fermented the pretreated slurry by *Saccharomyces cerevisiae* S288C, resulting in an ethanol yield of 29.2%.

56.3.1.2.2 Alkaline Pretreatments

Harun et al. (2011) explored the alkaline (NaOH) pretreatment of *Chlorococcum infusionum* for bioethanol production in a paper with 196 citations. They investigated the concentration of NaOH, temperature, pretreatment time, bioethanol concentration, glucose concentration, and cell size to determine the effectiveness of the pretreatment process. They found that the highest glucose yield was 350 mg/g, and the maximum bioethanol yield obtained was 0.26 g ethanol/g algae using 0.75% (w/v) of NaOH and 120°C for 30min. They concluded that the NaOH pretreatment was a promising option to pretreat microalgal biomass for bioethanol production.

56.3.1.3 Algal Biomass Enzymatic Pretreatments

There are three HCPs for the enzymatic pretreatment of algal biomass.

Choi et al. (2010) studied the enzymatic pretreatment of *C. reinhardtii* UTEX 90 by two commercial hydrolytic enzymes for bioethanol production in a paper with 261 citations. They observed that almost all starch was released and converted into glucose. They investigated enzyme concentration, pH, temperature, and residence time to obtain an optimum combination. They produced 235 mg of ethanol from 1.0 g of algal biomass by a separate hydrolysis and fermentation (SHF) process. This process had advantages such as the low cost of chemicals, short residence time, and simple equipment system.

Harun and Danquah (2011b) produced bioethanol from *Chloroccum* sp. by the enzymatic hydrolysis of biomass in a paper with 145 citations. They used cellulase obtained from *Trichoderma reesei*, ATCC 26921. They performed hydrolysis under varying conditions of temperature, pH, and substrate concentration, with constant enzyme dosage. They obtained the highest glucose yield of 64.2% (w/w) at a temperature of 40°C, pH 4.8, and a substrate concentration of 10 g/L of microalgal biomass. They observed twice as fast glucose production than cellulobiose. Further, the value of $K_{m,app}$ was higher for the hydrolysis of cellobiose (15.18 g/L) compared to that of the substrate (1.48 g/L).

56.3.2 ALGAL HYDROLYSATE FERMENTATION OF ALGAL BIOETHANOL FUELS

There are two research fronts for the fermentation of algal hydrolysates for algal bioethanol production: fermentation of macroalgal and microalgal hydrolysates with three and one HCPs, respectively (Table 56.2).

56.3.2.1 Fermentation of Macroalgal Hydrolysates

Adams et al. (2009) produced bioethanol from *Saccharina latissima* without thermal acid pretreatments in a paper with 250 citations. They showed that the thermal (65°C) and acid (pH2) pretreatments were not required for the fermentation of *S. latissima* and obtained higher ethanol yields in untreated fermentation compared to those with thermal acid pretreatments. They used *S. cerevisiae* and 1 U/kg laminarinase.

TABLE 56.2

Algal Hydrolysate Fermentation in Algal Bioethanol Production

No.	Papers	Biomass/Hydrolysate	Prts.	Yeasts	Parameters	Keywords	Lead Author	Affil.	Cits
1	Adams et al. (2009)	*Saccharina latissima*	Laminarinase	*S. cerevisiae*	Bioethanol production, fermentation, no thermal acid pretreatments, ethanol yield	Saccharina, ethanol, fermentation	Adams, John M. 56876080600	Univ. Exeter UK	250
2	Horn et al. (2000)	*Laminaria hyperborea*	Na	*Pichia angophorae*	Bioethanol production, mannitol and laminaran fermentation, ethanol yield	Seaweed, ethanol	Horn, Swein J. 7103259928	Norwegian Univ. Life Sci. Norway	236
3	Kim et al. (2011)	*Laminaria japonica*	Acid, enzymes	*Escherichia coli* KO11	Bioethanol production, glucose and laminaran fermentation, ethanol yield, SSF	Algal, ethanol, hydrolysates	Chang, Ho N. 24431318500	KAIST S. Korea	224
4	Hirano et al. (1997)	*Chlorella vulgaris* IAM C-534, *Chlamydomonas reinhardtii* UTEX 2247, Sak-1	Na	Yeasts	Bioethanol production, biomass fermentation, dark fermentation, ethanol yield	Microalgal, ethanol	Hirano, Atsushi 57209632970	Tokyo Elect. Power Inc. Japan	188

Cits., Number of citations received for each paper; Na, non-available; Prt, Biomass pretreatments.

Horn et al. (2000) produced bioethanol from *Laminaria hyperborea* in a paper with 236 citations. They tested four microorganisms to carry out the fermentation of this macroalgal biomass, one bacterium and three yeasts. They found that only *Pichia angophorae* utilized both laminaran and mannitol for bioethanol production. Further, laminaran and mannitol were consumed simultaneously but with different relative rates. However, in batch fermentations, mannitol was the preferred substrate. Its proportion of the total laminaran and mannitol consumption rate increased with the oxygen transfer rate (OTR) and pH. In continuous fermentations, laminaran was the preferred substrate at a low OTR, whereas at higher OTRs, laminaran and mannitol were consumed at similar rates. Optimization of the ethanol yield required a low OTR, and the best yield of 0.43 g ethanol/(g substrate) was achieved in batch culture at pH 4.5 and 5.8 mmol O_2/L/h.

Kim et al. (2011) produced bioethanol from macroalgal hydrolysates using *Escherichia coli* KO11 in a paper with 224 citations. They treated *Ulva lactuca*, *Gelidium amansii*, *Laminaria japonica*, and *Sargassum fulvellum*, with acid and commercially available hydrolytic enzymes. Their hydrolysates contained glucose, mannose, galactose, and mannitol, among other sugars, at different ratios. The *L. japonica* hydrolysate contained up to 30.5% mannitol and 6.98% glucose in the hydrolysate solids. *E. coli* KO11 utilized both mannitol and glucose and produced 0.4 g ethanol per g of carbohydrate when cultured in *L. japonica* hydrolysate supplemented with Luria-Bertani broth and hydrolytic enzymes. They concluded that acid hydrolysis followed by simultaneous enzymatic pretreatment and inoculation with *E. coli* KO11 could be a viable strategy for bioethanol production from macroalgae.

56.3.2.2 Fermentation of Microalgal Hydrolysates

Hirano et al. (1997) produced ethanol from microalgae by conventional SHF and intracellular anaerobic fermentation in a paper with 188 citations. They isolated more than 250 microalgal strains to examine ethanol productivity. Some strains had a high growth rate of 20–30 g dry biomass/m²/day and a high starch content of more than 20% (dry base). They found that the *Chlorella vulgaris* IAM C-534 strain had a high starch content of 37%. Starch was extracted from the cells of *Chlorella*, saccharified, and fermented with yeasts; 65% of the ethanol conversion rate was obtained as compared to the theoretical rate from starch. They also examined intracellular starch fermentation under dark and anaerobic conditions. All of the tested strains showed intracellular starch degradation and ethanol production, but the levels of ethanol production were significantly different from each other. They obtained a higher ethanol yield with *C. reinhardtii* UTEX 2247 and Sak-1 isolated from seawater. These showed a maximum ethanol concentration of 1% (w/w). They concluded that intracellular ethanol production was simpler and less energy-intensive than the conventional SHF process.

56.3.3 ALGAL BIOETHANOL FUELS IN GENERAL

There are 14 HCPs for algal bioethanol production in general with eight, three, and three HCPS for macroalgal, cyanobacterial, and microalgal bioethanol production, respectively (Table 56.3).

56.3.3.1 Microalgal Bioethanol Production

Ho et al. (2013) produced bioethanol from *C. vulgaris* FSP-E in a paper with 413 citations. They observed that the enzymatic hydrolysis of the microalgal biomass containing 51% carbohydrate per dry weight gave a glucose yield of 90.4% (or 0.461 g/(gbiomass)). Further, the SHF and simultaneous saccharification and fermentation (SSF) processes converted the enzymatic microalgal hydrolysate into ethanol with a 79.9% and 92.3% theoretical yield, respectively. On the other hand, dilute acid hydrolysis with 1% H_2SO_4 was also very effective in saccharifying *C. vulgaris* FSP-E biomass, achieving a glucose yield of nearly 93.6% from microalgal carbohydrates at a starting biomass concentration of 50 g/L. Further, using the acidic hydrolysate of *C. vulgaris* FSP-E biomass, the SHF process produced ethanol at a concentration of 11.7 g/L and an 87.6% theoretical yield.

TABLE 56.3

Algal Bioethanol Production in General

No.	Papers	Biomass/ Hydrolysate	Prts.	Yeasts	Parameters	Keywords	Lead Author	Affil.	Cits
1	Ho et al. (2013)	Chlorella vulgaris FSP-E	Acids, enzymes	Na	Bioethanol production, acid and enzymatic hydrolysis, SSF, SHF, sugar and ethanol yield	Microalgae, bioethanol	Chang, Jo-Shu 8567368700	Tunghai Univ. Taiwan	412
2	Harun et al. (2010)	Chlorococcum sp. residues	Na	S. bayanus	Bioethanol production, fermentation, ethanol yield	Microalgal, bioethanol	Harun, Razif 35315707300	Univ. Putra Malaysia Malaysia	399
3	Deng and Coleman (1999)	Synechococcus sp. PCC 7942	Pdc, adh	Na	Bioethanol production, engineering of cyanobacterium	Cyanobacteria, ethanol	Coleman, John R. 7402803364	Univ. Toronto, Canada	340
4	Gao et al. (2012)	Synechocystis sp. PCC6803	Pdc, adh	Na	Bioethanol production, engineering of cyanobacterium, ethanol productivity	Cyanobacteria, ethanol	Lu, Xuefeng 55619293343	Chinese Acad. Sci. China	259
5	Dexter and Fu (2009)	Synechocystis sp. PCC 6803	Pdc, adh	Na	Bioethanol production, engineering of cyanobacterium, bioethanol productivity	Cyanobacteria, ethanol	Fu, Pengcheng 7202037416	Hainan Univ. China	249
6	Enquist-Newman et al. (2014)	Alginate	Na	S. cerevisiae	Bioethanol production, engineering of alginate, bioethanol productivity and yield, mannitol, DEH	Macroalgae, ethanol	Yoshikuni, Yasuo 7102890080	Joint Genome Inst. USA	239
7	Kumar et al. (2013)	Gracilaria verrucosa	Enzymes	S. cerevisiae	Bioethanol production, sugar and ethanol yield	Gracilaria, bioethanol	Sahoo, Dinabandhu 16176224600	Univ. Delhi India	188
8	Van der Wal et al. (2013)	Ulva lactuca	LHW, cellulases	Clostridium acetobutylicum, Clostridium beijerinckii.	ABE production, yeast type, sugar and ABE yield	Seaweed, ulva, ethanol	Lopez-Contreras, Ana M.* 6602239834	Wageningen Univ. Res. Netherlands	184
9	Park et al. (2012)	Gelidium amansii	Acids	Na	Bioethanol production, sugar and ethanol yield, fermentation inhibitors	Gelidium, bioethanol, hydrolysis, fermentation	Kim, Yong J. 36065720100	Korea Inst. Machinery Mater. S. Korea	173

(Continued)

TABLE 56.3 (*Continued*)
Algal Bioethanol Production in General

No.	Papers	Biomass/ Hydrolysate	Prts.	Yeasts	Parameters	Keywords	Lead Author	Affil.	Cits
10	Borines et al. (2013)	*Sargassum* spp.	H₂SO₄, cellulases, β-glucosidase	*S. cerevisiae*	Bioethanol production, sugar and ethanol yield	Sargassum, bioethanol, macroalgae	Borines, Myra G.* 6507143659	Univ. Philippines Philippines	155
11	Jang et al. (2012)	*Saccharina japonica*	Milling, Termamyl 120L, H₂SO₄, *Bacillus* sp. JS-1	*Pichia angophorae* KCTC 17574	Bioethanol production, SSF, optimal hydrolysis, sugar and ethanol yield	Saccharina, seaweed, ethanol, saccharification, fermentation	Kim, Sung-Koo 57195386876	Pukyong Natl. Univ. S. Korea	155
12	Ellis et al. (2012)	Algae	Acid, alkali, glucose, xylanase, cellulase	Clostridium saccha-roperbutylacetonicum N1-4	ABE production, pretreatment type, fermentation, ABE yield	Algae, ethanol	Miller, Charles D. 57213751887	Utah State Univ. USA	152
13	Ge et al. (2011)	Alginate residues	H₂SO₄, cellulase and cellobiase	*S. cerevisiae*	Bioethanol production, acid and enzymatic hydrolysis, fermentation, sugar and ethanol yield	Seaweed, ethanol, saccharification	Mou, Haijin 6604026403	Ocean Univ. China China	152
14	Kim et al. (2014)	*Chlorella vulgaris*	Bead beating, pectinases	Yeasts	Bioethanol production, bead beating and enzymatic pretreatments, sugar and ethanol yields	Chlorella, microalga, ethanol, hydrolysis, fermentation	Bae, Hyeun-Jong 24280549300	Chonnam Natl. Univ. S. Korea	149

*, Female; Cits., Number of citations received for each paper; Na, non-available; Prt, Biomass pretreatments.

Harun et al. (2010) produced bioethanol from *Chlorococcum* sp. residues using *Saccharomyces bayanus* in a paper with 299 citations. They obtained a maximum ethanol concentration of 3.83 g/L obtained from 10 g/L of lipid-extracted microalgae residues with a productivity level of 38% w/w.

Kim et al. (2014) produced bioethanol from nutrient stress-induced *C. vulgaris* by enzymatic hydrolysis and immobilized yeast fermentation in a paper with 149 citations. They employed nutrient stress cultivation to enhance the carbohydrate content of *C. vulgaris*. Nitrogen limitation increased the carbohydrate content to 22.4% from the normal content of 16.0% on a dry weight basis. They found that the bead-beating pretreatment increased hydrolysis by 25% compared with the processes lacking pretreatment. In the enzymatic hydrolysis process, the pectinase enzyme group was superior for releasing fermentable sugars from carbohydrates in microalgae. In particular, pectinase from *Aspergillus aculeatus* displayed a 79% saccharification yield after 72 h at 50°C. Using continuous immobilized yeast fermentation, they converted microalgal hydrolysate into ethanol at a yield of 89%.

56.3.3.2 Cyanobacterial Bioethanol Production

Deng and Coleman (1999) produced bioethanol from *Synechococcus* sp. strain PCC 7942 in a paper with 341 citations. They introduced new genes into this cyanobacterium in order to create a novel pathway for fixed carbon utilization resulting in the synthesis of ethanol. For this purpose, they cloned the coding sequences of pyruvate decarboxylase (*pdc*) and alcohol dehydrogenase II (*adh*) from *Zymomonas mobilis* into the shuttle vector pCB4 and then used them to transform *S.* sp. strain PCC 7942. Under control of the promoter from the *rbcLS* operon encoding cyanobacterial ribulose-1,5- bisphosphate carboxylase-oxygenase (RuBisCo), they observed that the *pdc* and *adh* genes were expressed at high levels. The transformed cyanobacterium synthesized ethanol, which diffused from the cells into the culture medium.

Gao et al. (2012) studied the photosynthetic production of bioethanol from genetically engineered *Synechocystis* sp. PCC6803 in a paper with 259 citations. They integrated photosynthetic biomass production and microbial conversion producing ethanol together into *S.* sp. PCC6803, which could directly convert carbon dioxide to ethanol in one single biological system. They constructed a *S.* sp. PCC6803 mutant strain with ethanol-producing efficiency (5.50 g/L, 212 mg/L/day) by genetically introducing pyruvate decarboxylase (*pdc*) from *Z. mobilis* and overexpressing endogenous alcohol dehydrogenase (*adh*) through homologous recombination at two different sites of the chromosome and disrupting the biosynthetic pathway of poly-β-hydroxybutyrate (PHB). In total, they cloned nine alcohol dehydrogenases from different cyanobacterial strains and expressed them in *E. coli* to test the ethanol-producing efficiency.

Dexter and Fu (2009) studied the metabolic engineering of *Synechocystis* sp. PCC 6803 for bioethanol production in a paper with 249 citations. They constructed a *S.* sp. PCC 6803 strain that could photoautotrophically convert CO_2 to bioethanol using a double homologous recombination system to integrate the pyruvate decarboxylase (*pdc*) and alcohol dehydrogenase II (*adh*) genes from *Z. mobilis* into the *S.* sp. PCC 6803 chromosome under the control of the psbAII promoter. They established a computerized photobioreactor system for the experimental design and data acquisition for the analysis of the cyanobacterial cell cultures and bioethanol production. They found that this system had an average ethanol yield of 5.2 mmol OD 730/unit/L/day.

56.3.3.3 Macroalgal Bioethanol Production

Enquist-Newman et al. (2014) produced bioethanol from brown macroalgal sugars by a synthetic yeast platform in a paper with 239 citations. They reengineered the alginate and mannitol catabolic pathways in *S. cerevisiae* and discovered an alginate monomer, 4-deoxy-l-erythro-5-hexoseulose uronate (DEH), a transporter from the *Asteromyces cruciatus*. They found that the genomic integration and overexpression of the gene encoding this transporter conferred the ability of a *S. cerevisiae* strain to efficiently metabolize DEH and mannitol. When this platform was further adapted to grow on mannitol and DEH under anaerobic conditions, they observed that it was capable of ethanol fermentation from mannitol and DEH, achieving titers of 4.6% (v/v) (36.2 g/L) and yields up to 83% of the maximum theoretical yield from consumed sugars.

Kumar et al. (2013) produced bioethanol from *Gracilaria verrucosa* in a paper with 188 citations. This macroalgae harvested at various time durations resulted in the extraction of ~27%–33% agar. The leftover pulp contained 62%–68% holocellulose, which on enzymatic hydrolysis yielded 0.87 g sugars/g cellulose. Further, the enzymatic hydrolysate on fermentation with *S. cerevisiae* produced ethanol with an ethanol yield of 0.43 g/g sugars.

Van der Wal et al. (2013) produced acetone, butanol, and ethanol (ABE) from *Ulva lactuca* in a paper with 184 citations. They obtained solubilization of over 90% of sugars by liquid hot water (LHW) pretreatment followed by enzymatic hydrolysis using commercial cellulases. They used a hydrolysate for the production of ABE by *Clostridium acetobutylicum* and *Clostridium beijerinckii*. They found that *C. beijerinckii* utilized all sugars in the hydrolysate and produced ABE at high yields (0.35. g ABE/g sugar consumed), while *C. acetobutylicum* produced mostly organic acids (acetic and butyric acids).

Park et al. (2012) produced bioethanol from *Gelidium amansii* with continuous dilute acid hydrolysis and fermentation processes in a paper with 173 citations. In the hydrolysis step, they compared the hydrolysates obtained from a batch reactor and a continuous reactor based on fermentable sugar yield and inhibitor formation. They found that there were many advantages to the continuous hydrolysis process. For example, the low melting point of the agar component in *G. amansii* facilitated improved raw material fluidity in the continuous reactor. In addition, the hydrolysate obtained from the continuous process delivered a high sugar and low inhibitor concentration, thereby leading to both high yield and high final ethanol titer in the fermentation process.

Borines et al. (2013) produced bioethanol from *Sargassum* spp. in a paper with 155 citations. They focused on the pretreatment, enzymatic saccharification, and fermentation of *S.* spp. They achieved optimal acid pretreatment conditions in terms of glucose and reducing sugar yields at 3.4%–4.6% (w/v) H_2SO_4 concentration, 115°C, and 1.50 h. They hydrolyzed the pretreated biomass with cellulase enzymes supplemented with β-glucosidase. After fermentation by *S. cerevisiae* at 40°C, pH of 4.5 for 48 h, they found that the ethanol conversion rate of the enzyme hydrolysate reached 89%, which was markedly higher than the theoretical yield of 51% based on glucose as substrate.

Jang et al. (2012) produced bioethanol from *Saccharina japonica* by the SSF process in a paper with 155 citations. They dried the seaweed with hot air, ground it with a hammer mill, and filtered it with a 200-mesh sieve prior to pretreatment. They hydrolyzed the pretreated seaweed by thermal acid hydrolysis with H_2SO_4 and the industrial enzyme, Termamyl 120 L. They then used the isolated *Bacillus* sp. JS-1 to increase the yield of saccharification. The optimal saccharification conditions were 10% (w/v) seaweed slurry, 40 mM H_2SO_4, and 1 g dcw/L isolated *Bacillus* sp. JS-1. Using this saccharification procedure, the reducing sugar concentration was 45.6 g/L, and the total yield of saccharification under optimal conditions and *S. japonica* was 69.1%. They finally carried out the SSF process for ethanol production. They obtained the highest ethanol concentration, 7.7 g/L (9.8 mL/L) with a theoretical yield of 33.3%, by SSF with 0.39 g dcw/L *Bacillus* sp. JS-1 and 0.45 g dcw/L of *P. angophorae* KCTC 17574.

Ellis et al. (2012) produced acetone, butanol, and ethanol (ABE) from wastewater algae by *Clostridium saccharoperbutylacetonicum* N1–4 in a paper with 152 citations. They performed batch fermentations with 10% algae as the feedstock. They found that the fermentation of acid/base-pretreated algae produced 2.74 g/L of total ABE, as compared with 7.27 g/L from pretreated algae supplemented with 1% glucose. Additionally, 9.74 g/L of total ABE was produced when xylanase and cellulase enzymes were supplemented with the pretreated algae media. The 1% glucose supplement increased the total ABE production by approximately 160%, while supplementing with enzymes resulted in a 250% increase in the total ABE production when compared to production from pretreated algae with no supplementation of glucose and enzymes. Additionally, enzyme supplementation resulted in the highest total ABE production yield of 0.311 g/g and volumetric productivity of 0.102 g/L h. The use of non-pretreated algae produced 0.73 g/L of total ABE.

Ge et al. (2011) produced bioethanol from alginate wastes by the dilute H_2SO_4 pretreatment and further enzymatic hydrolysis in a paper with 152 citations. They performed the dilute H_2SO_4 pretreatment using H_2SO_4 at concentrations of 0%, 0.1%, 0.2%, 0.5%, and 1.0% (w/v) for 0.5, 1.0, and 1.5 h, respectively, at 121°C. They then used cellulase and cellobiase. The residues had a high cellulose content (30.0%) and little hemicellulose (2.2%). They observed that the acid pretreatment improved the hydrolysis efficiency of cellulase and cellobiase by increasing the reaction surface area of residues and enhanced the final yield of glucose for fermentation. The maximum yield of glucose reached 277.5 mg/g residues under optimal conditions of the dilute H_2SO_4 pretreatment (0.1% w/v, 121°C, 1.0 h) followed by enzymatic hydrolysis (50°C, pH 4.8, 48 h). After fermentation by *S. cerevisiae* at 30°C for 36 h, the ethanol conversion rate of the concentrated hydrolysates reached 41.2%, which corresponds to 80.8% of the theoretical yield.

56.4 DISCUSSION

56.4.1 INTRODUCTION

Crude oil-based gasoline fuels have been widely used in the transportation sector since the 1920s. However, there have been great public concerns over the adverse environmental and human impact of these fuels. Hence, biomass-based bioethanol fuels have increasingly been used in blending gasoline and petrodiesel fuels, in fuel cells, and in biochemical production in a biorefinery context.

However, it is necessary to pretreat the biomass to enhance the yield of bioethanol prior to algal bioethanol production through hydrolysis and fermentation of biomass and hydrolysates, respectively. One of the most studied biomass materials for bioethanol fuels has been algal biomass. Research in the field of algal bioethanol fuels has intensified in this context in the key research fronts of the pretreatment and hydrolysis of algal biomass, fermentation of algal hydrolysates, and algal bioethanol fuels in general. Further, macroalgae, microalgae, and cyanobacteria have been studied intensively at the expense of diatoms and dinoflagellates in this context.

However, it is essential to develop efficient incentive structures for the primary stakeholders to enhance research in this field. Although there are a number of review papers in this field, there has been no review of the 25 most-cited articles in this field.

Thus, this chapter presents a review of the 25 most-cited articles on algal bioethanol fuels. Then, it discusses the key findings of these highly influential papers and comments on future research priorities in this field.

As the first step for the search of the relevant literature, keywords were selected using the 200 most-cited population papers. The selected keyword list was then optimized to obtain a representative sample of papers in this research field. This keyword list was provided in the Appendix of Konur (2023) for future replicative studies.

As the second step, a sample dataset was used in this study. The first 25 articles with at least 139 citations each were selected for review. Key findings from each paper were taken from the abstracts of these papers and were discussed. Additionally, a number of brief conclusions were drawn, and a number of relevant recommendations were made to enhance the future research landscape.

Information about the research fronts for the sample papers on algal bioethanol fuels with regard to algal biomass used in these processes is given in Table 56.4. As this table shows, there are three primary research fronts in this field: macroalgal, microalgal, and cyanobacterial bioethanol fuels with 40%, 44%, and 12% of these HCPs, respectively. Another research front is algal bioethanol fuels with 4% of these HCPs.

Further, on an individual basis, the most prolific algal biomass materials are chlamydomonas chlorella and chlorococcum with 12% of these HCPs each. Other prolific algal biomass materials are alginates and synechocystis with 8% of these HCPs each.

On the other hand, microalgae is the most influential research front with a 12% surplus, while macroalgae is the least influential research front with an 11% deficit.

Further, on an individual basis, chlamydomonas, laminaria, chloroccum, chlorella, and synechocystis are the most influential research fronts with a 3%–9% surplus each. Similarly, macroalga in general, microalgae in general, cyanobacteria in general, and kappaphycus are the least influential research fronts with a 4%–10% deficit each.

TABLE 56.4
The Most Prolific Research Fronts on Algal Bioethanol Fuels

No.	Research Fronts	N Paper (%) Review	N Paper (%) Sample	Surplus (%)
1	Macroalgae	40	51	−11
	Alginates	8	6	2
	Laminaria	8	3	5
	Gelidium	4	6	−2
	Ulva	4	6	−2
	Gracilaria	4	4	0
	Saccharina	4	4	0
	Sargassum	4	2	2
	Macroalga in general	0	10	−10
	Kappaphycus	0	4	−4
	Other macroalgae	0	3	−3
	Chaetomorpha	0	2	−2
	Undaria	0	2	−2
2	Microalgae	44	32	12
	Chlorella	12	9	3
	Chloroccum	12	4	8
	Chlamydomonas	12	3	9
	Microalgae in general	4	12	−8
	Scenedesmus	4	7	−3
	Nannochloropsis	4	2	2
	Other microalgae	0	2	−2
3	Cyanobacteria	12	12	0
	Synechocystis	8	5	3
	Synechococcus	4	2	2
	Cyanobacteria in general	0	4	−4
	Spirulina	0	2	−2
	Other cyanobacteria	0	1	−1
4	Algae in general	4	7	−3

N Paper (%) review, The number of papers in the sample of 25 reviewed papers; N paper (%) sample, The number of papers in the population sample of 100 papers.

Information about the thematic research fronts for the sample papers on algal bioethanol fuels is given in Table 56.5. As this table shows, there are four primary research fronts in this field: pretreatment and hydrolysis of algal biomass, fermentation of algal hydrolysates, and algal bioethanol production in general with 84%, 52%, 40%, and 84% of these HCPs, respectively.

Further, the most prolific pretreatments are chemical and enzymatic pretreatments with 52% of these HCPs each. Another research front is the mechanical pretreatment of algal biomass with 4% of the HCPs. Similarly, the most prolific hydrolysis is enzymatic and acid hydrolysis of algal biomass with 32 and 24% of these HCPs, respectively.

The most prolific research fronts for the fermentation of algal hydrolysates are fermentation of macroalgal and microalgal hydrolysates with 24% and 12% of these HCPs, respectively. Another research front is the fermentation of algal hydrolysates in general with 4% of these HCPs. Finally, the key research fronts for algal bioethanol fuels in general are macroalgal, microalgal, and

TABLE 56.5

The Most Prolific Thematic Research Fronts on Algal Bioethanol Fuels

No.	Research Fronts	N Paper (%) Review	N Paper (%) Sample	Surplus (%)
1	Algal Biomass Pretreatments	84	93	−9
	Algal biomass pretreatments in general	0	7	−7
	Mechanical pretreatments	4	9	−5
	Milling pretreatments	4	5	−1
	Ultrasound pretreatments	0	4	−4
	Hydrothermal pretreatments	0	8	−8
	Steam explosion pretreatment	0	1	−1
	Liquid hot water pretreatment	0	5	−5
	Hot compressed water pretreatment	0	1	−1
	Wet oxidation pretreatment	0	1	−1
	Chemical pretreatments	52	33	19
	Acid pretreatment	44	27	17
	Alkaline pretreatment	8	3	5
	Ionic liquid pretreatment	0	1	−1
	Other chemical pretreatments	0	2	−2
	Enzymatic pretreatments	52	26	26
	Microbial pretreatments	16	13	3
2	Algal Biomass Hydrolysis	52	56	−4
	Enzymatic hydrolysis	32	27	5
	Acid hydrolysis	24	26	−2
	Hydrolysis in general	0	3	−3
3	Algal Hydrolysate Fermentation	40	55	−15
	Macroalgal fermentation	24	34	−10
	Microalgal fermentation	12	15	−3
	Cyanobacterial fermentation	0	4	−4
	Algal fermentation	4	4	0
4	Algal Bioethanol Production in General	84	93	−9
	Macroalgal ethanol production	40	44	−4
	Microalgal ethanol production	32	30	2
	Cyanobacterial ethanol production	12	10	2
	Algal ethanol production	4	11	−7

N Paper (%) review, The number of papers in the sample of 25 reviewed papers; N paper (%) sample, The number of papers in the population sample of 100 papers

cyanobacterial bioethanol fuels with 40%, 32%, and 12% of these HCPs, respectively. Another research front is algal bioethanol fuels in general with 4% of these HCPs.

Further, algal hydrolysate fermentation is the least influential front with a 15% deficit, followed by algal biomass pretreatment, algal bioethanol production, and algal biomass hydrolysis with a 9%, 9%, and 4% deficit, respectively. On the other hand, on an individual basis, enzymatic pretreatments, chemical pretreatments, and acid pretreatment are the most prolific research fronts with 26%, 19%, and 17% surplus, respectively. Similarly, macroalgal fermentation, hydrothermal pretreatments, algal ethanol production, and algal biomass pretreatments in general are the least influential research fronts with a 10%, 8%, 7%, and 7% deficit, respectively.

56.4.2 Algal Biomass Pretreatments of Algal Bioethanol Fuels

There are seven HCPs for algal biomass pretreatments of algal bioethanol fuels (Table 56.1). The most prolific pretreatments are chemical and enzymatic pretreatments of algal biomass with five and three HCPs, respectively. Further, the most prolific chemical pretreatment is the acid pretreatment of algal biomass with four HCPs. Another chemical pretreatment is alkaline pretreatment with only one HCP. The study dealing with both acid and enzymatic hydrolysis is presented separately.

These HCPs show a sample of research on the pretreatment of algal biomass for algal bioethanol fuels. These studies hint that the chemical and enzymatic pretreatments enhance both the sugar and ethanol yield during the production of algal bioethanol fuels. In other words, the pretreatment stage is one of the most important phases of algal bioethanol fuels (Alvira et al., 2010; Taherzadeh and Karimi, 2008).

It is notable that the most prolific chemical pretreatment is the acid pretreatment of algal biomass. Further, it is notable that there is no dedicated HCP in the research fronts of hydrothermal and mechanical pretreatments of algal biomass such as ultrasound and steam explosion pretreatments. It is interesting to note that all the algal biomass studies for these pretreatments are related to microalgal biomass such as *Chlamydomonas reinhardtii*.

56.4.2.1 Biomass Pretreatments in General

Hernandez et al. (2015) performed physical, chemical, and enzymatic pretreatments on three microalgal species to disrupt and break down complex carbohydrates into simple sugars as a preliminary stage to produce bioethanol and obtained the highest concentration of monosaccharides per gram of microalgae dry weight by the combination of acid and enzymatic pretreatments for *Chlorella sorokiniana* and *Nannochloropsis gaditana*, while in the case of *Scenedesmus almeriensis*, the highest monosaccharide concentration was obtained after acid hydrolysis.

56.4.2.2 Chemical Pretreatments

The most prolific chemical pretreatment is the acid pretreatment of algal biomass with four HCPs. Another chemical pretreatment is the alkaline pretreatment of algal biomass with one HCP only.

56.4.2.2.1 Acid Pretreatments

Harun and Danquah (2011a) studied the effect of the H_2SO_4 pretreatment on the ethanol yield during bioethanol production and found that temperature was the most critical factor in the acid pretreatment of microalgae for bioethanol production. Further, Miranda et al. (2012) optimized the pretreatment optimization of *Scenedesmus obliquus* for bioethanol production and obtained the best results with acid hydrolysis by H_2SO_4 at 120°C for 30 min using dried biomass. Finally, Nguyen et al. (2009) studied the acid pretreatment of *Chlamydomonas reinhardtii* UTEX 90 for bioethanol production and observed that the glucose release from this biomass was maximum at 58% (w/w) after pretreatment with 3% H_2SO_4 at 110°C for 30 min.

56.4.2.2.2 Alkali Pretreatments

Harun et al. (2011) explored the NaOH pretreatment of *Chlorococcum infusionum* for bioethanol production and found that the highest glucose yield was 350 mg/g, and the maximum bioethanol yield obtained was 0.26 g ethanol/g algae using 0.75% (w/v) of NaOH and 120°C for 30 min.

56.4.2.3   Enzymatic Pretreatments

There are three HCPs for the enzymatic pretreatment of algal biomass. Choi et al. (2010) studied the enzymatic pretreatment of *C. reinhardtii* UTEX 90 by two commercial hydrolytic enzymes for bioethanol production and observed that almost all starch was released and converted into glucose. Further, Harun and Danquah (2011b) produced bioethanol from *Chloroccum* sp. by the enzymatic hydrolysis of biomass and obtained the highest glucose yield of 64.2% (w/w) at a temperature of 40°C, pH 4.8, and a substrate concentration of 10 g/L of microalgal biomass.

56.4.3   Algal Hydrolysate Fermentation of Algal Bioethanol Fuels

There are two research fronts for the fermentation of algal hydrolysates for algal bioethanol production: fermentation of the macroalgal and microalgal hydrolysates with three and one HCPs, respectively (Table 56.2).

These studies hint that the fermentation of algal hydrolysates enhances the ethanol yield during the algal bioethanol production process. In other words, the fermentation stage is one of the most critical phases of algal bioethanol production.

56.4.3.1   Fermentation of Macroalgal Hydrolysates

Adams et al. (2009) produced bioethanol from *Saccharina latissima* without thermal acid pretreatments. They showed that thermal (65°C) and acid (pH2) pretreatments were not required for the fermentation of *S. latissima* and obtained higher ethanol yields in untreated fermentation compared to those with thermal acid pretreatments. Further, Horn et al. (2000) produced bioethanol from *Laminaria hyperborea* and found that only *Pichia angophorae* utilized both laminaran and mannitol for bioethanol production. Finally, Kim et al. (2011) produced bioethanol macroalgal hydrolysates using *E. coli* KO11 and found that *E. coli* KO11 utilized both mannitol and glucose and produced 0.4 g ethanol per gram of carbohydrate when cultured in *L. japonica* hydrolysate supplemented with Luria-Bertani broth and hydrolytic enzymes.

56.4.3.2   Fermentation of Microalgal Hydrolysates

Hirano et al. (1997) produced ethanol from microalgae by the conventional SHF and intracellular anaerobic fermentation and found that starch extracted from the cells of *Chlorella* had 65% of the ethanol conversion rate as compared to the theoretical rate from starch.

56.4.4   Algal Bioethanol Fuels in General

There are 14 HCPs for algal bioethanol production in general with eight, three, and three HCPS for macroalgal, cyanobacterial, and microalgal bioethanol production, respectively (Table 56.3).

These HCPs show a sample of research on algal bioethanol production. These studies hint that this research front covers both the conventional hydrolysis–fermentation processes, mostly for microalgae and macroalgae, and the direct conversion of algal biomass to bioethanol by genetic engineering of algal biomass, mostly for cyanobacteria. It is a general trend that conventional studies cover the pretreatment, hydrolysis, and fermentation stages of the bioethanol production process.

It is notable there is no dedicated HCP for bioethanol production from diatoms and dinoflagellates.

56.4.4.1   Microalgal Bioethanol Production

Ho et al. (2013) produced bioethanol from *C. vulgaris* FSP-E and observed that the enzymatic hydrolysis of microalgal biomass gave a glucose yield of 90.4% (or 0.461 g/(gbiomass)). Further, Harun et al. (2010) produced bioethanol from *Chlorococcum* sp. residues using *Saccharomyces bayanus* and obtained a maximum ethanol concentration with a productivity level of 38% w/w. Finally, Kim et al.

(2014) produced bioethanol from nutrient stress-induced *C. vulgaris* by enzymatic hydrolysis and immobilized yeast fermentation and found that the bead-beating pretreatment increased hydrolysis by 25% compared with the processes lacking pretreatment.

56.4.4.2 Cyanobacterial Bioethanol Production

Deng and Coleman (1999) produced bioethanol from the *Synechococcus* sp. strain PCC 7942 and observed that the *pdc* and *adh* genes were expressed at high levels and the transformed cyanobacterium synthesized ethanol, which diffused from the cells into the culture medium. Further, Gao et al. (2012) studied bioethanol production from genetically engineered *Synechocystis* sp. PCC6803, cloned nine alcohol dehydrogenases from different cyanobacterial strains, and expressed them in *E. coli* to test the ethanol-producing efficiency. Finally, Dexter and Fu (2009) studied the metabolic engineering of *Synechocystis* sp. PCC 6803 for bioethanol production and found that this system had an average ethanol yield of 5.2 mmol OD 730/unit/L/day.

56.4.4.3 Macroalgal Bioethanol Production

Enquist-Newman et al. (2014) produced bioethanol from brown macroalgal sugars by a synthetic yeast platform and found that the genomic integration and overexpression of the gene encoding this transporter conferred the ability of a *S. cerevisiae* strain to efficiently metabolize DEH and mannitol. Further, Kumar et al. (2013) produced bioethanol from *Gracilaria verrucosa* and found that the enzymatic hydrolysate on fermentation with *S. cerevisiae* produced ethanol with a yield of 0.43 g/g sugars.

Van der Wal et al. (2013) produced acetone, butanol, and ethanol (ABE) from *Ulva lactuca* and found that *C. beijerinckii* utilized all sugars in the hydrolysate and produced ABE at high yields (0.35 g ABE/g sugar consumed), while *C. acetobutylicum* produced mostly organic acids (acetic and butyric acids). Further, Park et al. (2012) produced bioethanol from *Gelidium amansii* with continuous dilute acid hydrolysis and fermentation processes and found that the hydrolysate obtained from the continuous process delivered a high sugar and low inhibitor concentration, thereby leading to both high yield and high final ethanol titer in the fermentation process.

Borines et al. (2013) produced bioethanol from *Sargassum* spp. and after fermentation by *S. cerevisiae* and found that the ethanol conversion rate of the enzymatic hydrolysate reached 89%, which was markedly higher than the theoretical yield of 51% based on glucose as substrate. Further, Jang et al. (2012) produced bioethanol from *Saccharina japonica* by the SSF process and obtained the highest ethanol concentration with a theoretical yield of 33.3%, by SSF.

Ellis et al. (2012) produced acetone, butanol, and ethanol (ABE) from wastewater algae by *Clostridium saccharoperbutylacetonicum* N1–4 and found that the fermentation of acid/base-pretreated algae produced 2.74. g/L of the total ABE. Further, Ge et al. (2011) produced bioethanol from alginate wastes by the dilute H_2SO_4 pretreatment and further enzymatic hydrolysis and after fermentation by *S. cerevisiae* and found that the ethanol conversion rate of concentrated hydrolysates reached 41.2%, which corresponds to 80.8% of the theoretical yield.

56.5 CONCLUSION AND FUTURE RESEARCH

The brief information about the key research fronts covered by the 25 most-cited papers with at least 139 citations each is given under four primary headings: pretreatment and hydrolysis of algal biomass, fermentation of algal hydrolysates, and algal bioethanol production in general.

The usual characteristics of these HCPs are that the pretreatment and hydrolysis of algal biomass and fermentation of algal hydrolysates are the primary processes for algal bioethanol fuels to improve the ethanol yield as algal biomass is one of the most studied feedstocks for bioethanol fuels, especially for countries with the large access to seas and lakes such as the USA and the East Pacific countries.

The key findings on these research fronts should be read in light of the increasing public concerns about climate change, GHG emissions, and global warming as these concerns have been certainly behind the research boom in algal bioethanol fuels as an alternative to crude oil-based gasoline and diesel fuels in the past decades. It is also a sustainable alternative to food crop- and waste-based bioethanol fuels. The recent supply shocks caused by the COVID-19 pandemic and the Russian invasion of Ukraine also highlight the importance of the production and utilization of bioethanol fuels as an alternative to crude oil-based gasoline and petrodiesel fuels.

There are three primary research fronts in this field with respect to algal biomass: macroalgal, microalgal, and cyanobacterial bioethanol fuels at the expense of diatoms and dinoflagellates. Further, on an individual basis, the most prolific algal biomass materials are chlamydomonas, chlorella, and chlorococcum, while microalgae is the most influential research front and macroalgae is the least influential research front.

Similarly, there are four primary thematic research fronts in this field: pretreatment and hydrolysis of algal biomass, fermentation of algal hydrolysates, and algal bioethanol fuels in general. Further, the most prolific pretreatments are chemical and enzymatic pretreatments, while the most prolific hydrolysis is enzymatic and acid hydrolysis of algal biomass. Finally, fermentations of macroalgal and microalgal hydrolysates are the most prolific fermentations, while macroalgal, microalgal, and cyanobacterial bioethanol fuels are the most prolific research fronts.

These studies emphasize the importance of proper incentive structures for efficient algal bioethanol fuels in light of North's institutional framework (North, 1991). In this context, the major producers and users of bioethanol fuels such as the USA and far East countries such as South Korea, Malaysia, and Japan with large access to seas and lakes have developed strong incentive structures for efficient algal bioethanol production. In light of the supply shocks caused primarily by the COVID-19 pandemic and the Russian invasion of Ukraine, it is expected that the incentive structures such as public funding would be enhanced to increase the proportion of bioethanol fuels in the global fuel portfolio as a strong alternative to crude oil-based gasoline and diesel fuels. In this context, it is expected that the most prolific researchers, institutions countries, funding bodies, and journals in this field would have a first-mover advantage to benefit from such potential incentives.

It is recommended that such review studies are performed for the primary research fronts of algal bioethanol fuels.

ACKNOWLEDGMENTS

The contribution of the highly cited researchers in the field of algal bioethanol fuels has been gratefully acknowledged.

REFERENCES

Adams, J. M., J. A. Gallagher and I. S. Donnison. 2009. Fermentation study on *Saccharina latissima* for bioethanol production considering variable pre-treatments. *Journal of Applied Phycology* 21:569–574.

Alvira, P., E. Tomas-Pejo, M. Ballesteros and M. J. Negro. 2010. Pretreatment technologies for an efficient bioethanol production process based on enzymatic hydrolysis: A review. *Bioresource Technology* 101:4851–4861.

Angelici, C., B. M. Weckhuysen and P. C. A. Bruijnincx. 2013. Chemocatalytic conversion of ethanol into butadiene and other bulk chemicals. *ChemSusChem* 6:1595–1614.

Antolini, E. 2007. Catalysts for direct ethanol fuel cells. *Journal of Power Sources* 170:1–12.

Antolini, E. 2009. Palladium in fuel cell catalysis. *Energy and Environmental Science* 2:915–931.

Borines, M. G., R. L. de Leon and J. L. Cuello. 2013. Bioethanol production from the macroalgae *Sargassum* spp. *Bioresource Technology* 138:22–29.

Burnham, J. F. 2006. Scopus database: A review. *Biomedical Digital Libraries* 3:1–8.

Choi, S. P., M. T. Nguyen and S. J. Sim. 2010. Enzymatic pretreatment of *Chlamydomonas reinhardtii* biomass for ethanol production. *Bioresource Technology* 101:5330–5336.

De Farias Silva, C. E. and A. Bertucco. 2016. Bioethanol from microalgae and cyanobacteria: A review and technological outlook. *Process Biochemistry* 51:1833–1842.

Deng, M. D. and J. R. Coleman. 1999. Ethanol synthesis by genetic engineering in cyanobacteria. *Applied and Environmental Microbiology* 65:523–528.

Dexter, J. and P. Fu. 2009. Metabolic engineering of cyanobacteria for ethanol production. *Energy and Environmental Science* 2:857–864.

Ellis, J. T., N. N. Hengge, R. C. Sims and C. D. Miller 2012. Acetone, butanol, and ethanol production from wastewater algae. *Bioresource Technology* 111:491–495.

Enquist-Newman, M., A. M. E. Faust and D. D. Bravo, et al. 2014. Efficient ethanol production from brown macroalgae sugars by a synthetic yeast platform. *Nature* 505:239–243.

Fernando, S., S. Adhikari, C. Chandrapal and M. Murali. 2006. Biorefineries: Current status, challenges, and future direction. *Energy & Fuels* 20:1727–1737.

Gao, Z., H. Zhao, Z. Li, X. Tan and X. Lu. 2012. Photosynthetic production of ethanol from carbon dioxide in genetically engineered cyanobacteria. *Energy and Environmental Science* 5:9857–9865.

Ge, L., P. Wang and H. Mou. 2011. Study on saccharification techniques of seaweed wastes for the transformation of ethanol. *Renewable Energy* 36:84–89.

Hahn-Hagerdal, B., M. Galbe, M. F. Gorwa-Grauslund, G. Liden and G. Zacchi. 2006. Bio-ethanol - The fuel of tomorrow from the residues of today. *Trends in Biotechnology* 24:549–556.

Harun, R. and M. K. Danquah. 2011a. Influence of acid pre-treatment on microalgal biomass for bioethanol production. *Process Biochemistry* 46:304–309.

Harun, R. and M. K. Danquah. 2011b. Enzymatic hydrolysis of microalgal biomass for bioethanol production. *Chemical Engineering Journal* 168:1079–1084.

Harun, R., M. K. Danquah and G. M. Forde. 2010. Microalgal biomass as a fermentation feedstock for bioethanol production. *Journal of Chemical Technology and Biotechnology* 85:199–203.

Harun, R., W. S. Y. Jason, T. Cherrington and M. K. Danquah. 2011. Exploring alkaline pre-treatment of microalgal biomass for bioethanol production. *Applied Energy* 88:3464–3467.

Hernandez, D., B. Riano, M. Coca and M. C. Garcia-Gonzalez. 2015. Saccharification of carbohydrates in microalgal biomass by physical, chemical and enzymatic pre-treatments as a previous step for bioethanol production. *Chemical Engineering Journal* 262:939–945.

Hill, J., E. Nelson, D. Tilman, S. Polasky and D. Tiffany. 2006. Environmental, economic, and energetic costs and benefits of biodiesel and ethanol biofuels. *Proceedings of the National Academy of Sciences of the United States of America* 103:11206–11210.

Hill, J., S. Polasky and E. Nelson, et al. 2009. Climate change and health costs of air emissions from biofuels and gasoline. *Proceedings of the National Academy of Sciences of the United States of America* 106:2077–2082.

Hirano, A., R. Ueda, S. Hirayama and Y. Ogushi. 1997. CO_2 fixation and ethanol production with microalgal photosynthesis and intracellular anaerobic fermentation. *Energy* 22:137–142.

Ho, S. H., S. W. Huang and C. Y. Chen, et al. 2013. Bioethanol production using carbohydrate-rich microalgae biomass as feedstock. *Bioresource Technology* 135:191–198.

Horn, S. J., I. M. Aasen and K. Ostgaard. 2000. Ethanol production from seaweed extract. *Journal of Industrial Microbiology and Biotechnology* 25:249–254.

Hsieh, W. D., R. H. Chen, T. L. Wu and T. H. Lin. 2002. Engine performance and pollutant emission of an SI engine using ethanol-gasoline blended fuels. *Atmospheric Environment* 36:403–410.

Huang, H. J., S. Ramaswamy, U. W. Tschirner and B. V. Ramarao. 2008. A review of separation technologies in current and future biorefineries. *Separation and Purification Technology* 62:1–21.

Jang, J. S., Y. K. Cho, G.T. Jeong and S. K. Kim. 2012. Optimization of saccharification and ethanol production by simultaneous saccharification and fermentation (SSF) from seaweed, *Saccharina japonica*. *Bioprocess and Biosystems Engineering* 35:11–18.

John, R. P., G. S. Anisha, K. M. Nampoothiri and A. Pandey. 2011. Micro and macroalgal biomass: A renewable source for bioethanol. *Bioresource Technology* 102:186–193.

Kim, K. H., I. S. Choi, H. M. Kim, S. G. Wi and H. J. Bae. 2014. Bioethanol production from the nutrient stress-induced microalga *Chlorella vulgaris* by enzymatic hydrolysis and immobilized yeast fermentation. *Bioresource Technology* 153:47–54.

Kim, N. J., H. Li, K. Jung, H. N. Chang and P. C. Lee. 2011. Ethanol production from marine algal hydrolysates using *Escherichia coli* KO11. *Bioresource Technology* 102:7466–7469.

Konur, O. 2000. Creating enforceable civil rights for disabled students in higher education: An institutional theory perspective. *Disability & Society* 15:1041–1063.

Konur, O. 2002a. Access to nursing education by disabled students: Rights and duties of nursing programs. *Nurse Education Today* 22:364–374.

Konur, O. 2002b. Assessment of disabled students in higher education: Current public policy issues. *Assessment and Evaluation in Higher Education* 27:131–52.

Konur, O. 2002c. Access to employment by disabled people in the UK: Is the Disability Discrimination Act working? *International Journal of Discrimination and the Law* 5:247–279.

Konur, O. 2006a. Participation of children with dyslexia in compulsory education: Current public policy issues. *Dyslexia* 12:51–67.

Konur, O. 2006b. Teaching disabled students in higher education. *Teaching in Higher Education* 11:351–363.

Konur, O. 2007a. A judicial outcome analysis of the *Disability Discrimination Act*: A windfall for the employers? *Disability & Society* 22:187–204.

Konur, O. 2007b. Computer-assisted teaching and assessment of disabled students in higher education: The interface between academic standards and disability rights. *Journal of Computer Assisted Learning* 23:207–219.

Konur, O. 2012. The evaluation of the research on the bioethanol: A scientometric approach. *Energy Education Science and Technology Part A: Energy Science and Research* 28:1051–1064.

Konur, O. 2015. Current state of research on algal bioethanol. In *Marine Bioenergy: Trends and Developments*, Ed. S. K. Kim and C. G. Lee, pp. 217–244. Boca Raton, FL: CRC Press.

Konur, O. 2019. Cyanobacterial bioenergy and biofuels science and technology: A scientometric overview. In *Cyanobacteria: From Basic Science to Applications*, Ed. A. K. Mishra, D. N. Tiwari and A. N. Rai, pp. 419–442. Amsterdam: Elsevier.

Konur, O. 2020. The scientometric analysis of the research on the bioethanol production from green macroalgae. In *Handbook of Algal Science, Technology and Medicine*, Ed. O. Konur, pp. 385–401. London: Academic Press.

Konur, O. 2023. Third generation algal bioethanol fuels: Scientometric study. In *Feedstock-based Bioethanol Fuels. I. Non-Waste Feedstocks: Starch, Sugar, Grass, Wood, Cellulose, Algae, and Biosyngas-based Bioethanol Fuels. Handbook of Bioethanol Fuels Volume 3*, Ed. O. Konur. Boca Raton, FL: CRC Press.

Kumar, S., R. Gupta, G. Kumar, D. Sahoo and R. C. Kuhad. 2013. Bioethanol production from *Gracilaria verrucosa*, a red alga, in a biorefinery approach. *Bioresource Technology* 135:150–156.

Lin, Y. and S. Tanaka. 2006. Ethanol fermentation from biomass resources: Current state and prospects. *Applied Microbiology and Biotechnology* 69:627–642.

Ma, X., L. Sun and C. Song. 2002. A new approach to deep desulfurization of gasoline, diesel fuel and jet fuel by selective adsorption for ultra-clean fuels and for fuel cell applications. *Catalysis Today* 77:107–116.

Miranda, J. R., P. C. Passarinho and L. Gouveia. 2012. Pre-treatment optimization of *Scenedesmus obliquus* microalga for bioethanol production. *Bioresource Technology* 104:342–348.

Morschbacker, A. 2009. Bio-ethanol based ethylene. *Polymer Reviews* 49:79–84.

Najafi, G., B. Ghobadian and T. Tavakoli, et al. 2009. Performance and exhaust emissions of a gasoline engine with ethanol blended gasoline fuels using artificial neural network. *Applied Energy* 86:630–639.

Newman, P. W. G. and J. R. Kenworthy. 1989. Gasoline consumption and cities: A comparison of U.S. cities with a global survey. *Journal of the American Planning Association* 55:24–37.

Nguyen, M. T., S. P. Choi, J. Lee, J. H. Lee and S. J. Sim. 2009. Hydrothermal acid pretreatment of *Chlamydomonas reinhardtii* biomass for ethanol production. *Journal of Microbiology and Biotechnology* 19:161–166.

North, D. C. 1991. Institutions. *Journal of Economic Perspectives* 5:97–112.

Olsson, L. and B. Hahn-Hagerdal. 1996. Fermentation of lignocellulosic hydrolysates for ethanol production. *Enzyme and Microbial Technology* 18:312–331.

Park, J. H., J. Y. Hong and H. C. Jang, et al. 2012. Use of *Gelidium amansii* as a promising resource for bioethanol: A practical approach for continuous dilute-acid hydrolysis and fermentation. *Bioresource Technology* 108:83–88.

Sanchez, O. J. and C. A. Cardona. 2008. Trends in biotechnological production of fuel ethanol from different feedstocks. *Bioresource Technology* 99:5270–5295.

Sun, Y. and J. Cheng. 2002. Hydrolysis of lignocellulosic materials for ethanol production: A review. *Bioresource Technology* 83:1–11.

Taherzadeh, M. J. and K. Karimi. 2007. Enzyme-based hydrolysis processes for ethanol from lignocellulosic materials: A review. *Bioresources* 2:707–738.

Taherzadeh, M. J. and K. Karimi. 2008. Pretreatment of lignocellulosic wastes to improve ethanol and biogas production: A review. *International Journal of Molecular Sciences* 9:1621–1651.

Van der Wal, H., B. L. H. M. Sperber, B. Houweling-Tan, et al. 2013. Production of acetone, butanol, and ethanol from biomass of the green seaweed *Ulva lactuca*. *Bioresource Technology* 128:431–437.

Yanagisawa, M., S. Kawai and K. Murata. 2013. Strategies for the production of high concentrations of bioethanol from seaweeds. *Bioengineered* 4:224–235.

Part 15

Biosyngas-based Bioethanol Fuels

57 Biosyngas-based Bioethanol Fuels
Scientometric Study

Ozcan Konur
(Formerly) Ankara Yildirim Beyazit University

57.1 INTRODUCTION

Crude oil-based gasoline fuels (Ma et al., 2002; Newman and Kenworthy, 1989) have been widely used in the transportation sector since the 1920s. However, there have been great public concerns over the adverse environmental and human impact of these fuels (Hill et al., 2006, 2009). Hence, biomass-based bioethanol fuels (Hill et al., 2006; Konur, 2012e, 2015, 2019, 2020a) have increasingly been used in blending gasoline fuels (Hsieh et al., 2002; Najafi et al., 2009), in fuel cells (Antolini, 2007, 2009), and in biochemical production (Angelici et al., 2013; Morschbacker, 2009) in a biorefinery context (Fernando et al., 2006; Huang et al., 2008).

Bioethanol fuels also play a critical role in maintaining energy security (Kruyt et al., 2009; Winzer, 2012) in supply shocks (Kilian, 2008, 2009) related to oil price shocks (Hamilton, 2003, 2009), the COVID-19 pandemic (Fauci et al., 2020; Li et al., 2020), or wars (Hamilton, 1983; Jones, 2012) in the aftermath of the Russian invasion of Ukraine (Reeves, 2014).

One of the most studied feedstocks for bioethanol fuels has been synthesis biogas (biosyngas) produced from biomass with constituents such as carbon dioxide (CO_2), carbon monoxide (CO), and hydrogen (H_2). Research in the field of biosyngas-based bioethanol fuels has intensified in this context in the key research fronts of the electrochemical reduction of biosyngas (Hoang et al., 2018; Ren et al., 2015, 2016), catalytic conversion of biosyngas (Choi and Liu, 2009; Gong et al., 2012; Pan et al., 2007), and production of bioethanol fuels by other processes such as the photosynthetic conversion of biosyngas (Gao et al., 2012; Hirano et al., 1997), microbial fermentation of biosyngas (Datar et al., 2004; Younesi et al., 2005), and photocatalytic conversion of biosyngas (Liu et al., 2009). Further, CO_2 (Hoang et al., 2018; Ren et al., 2015, 2016), biosyngas (Choi and Liu, 2009; Datar et al., 2004; Gong et al., 2012), and to a lesser extent CO (Abubackar et al., 2015; Bertheussen et al., 2016; Rajagopalan et al., 2016) have been studied intensively in this context.

However, it is essential to develop efficient incentive structures (North, 1991) for the primary stakeholders to enhance the research in this field (Konur, 2000, 2002a,b,c, 2006a,b, 2007a,b). Scientometric analysis has been used in this context to inform the primary stakeholders about the current state of research in this research field (Garfield, 1955; Konur, 2011, 2012a,b,c,d,e,f,g,h,i, 2015, 2018b, 2019, 2020a).

As there have been no published scientometric studies in this field, this chapter presents a scientometric study of research on biosyngas-based bioethanol fuels. It examines the scientometric characteristics of both the sample and population data and presents them in the order of documents, authors, publication years, institutions, funding bodies, source titles, countries, Scopus subject categories, Scopus keywords, and research fronts.

DOI: 10.1201/9781003226451-76

57.2 MATERIALS AND METHODS

The search for this study was carried out using the Scopus database (Burnham, 2006) in September 2022.

As the first step for the search of the relevant literature, keywords were selected using the 100 most-cited population papers. The selected keyword list was then optimized to obtain a representative sample of papers in this research field. This keyword list was provided in the Appendix for future replicative studies.

As the second step, two sets of data were used in this study. First, a population sample of 603 papers was used to examine the scientometric characteristics of the population data. Secondly, a sample of 100 most-cited papers, corresponding to 17% of the population papers, was used to examine the scientometric characteristics of these citation classics.

The scientometric characteristics of both these sample and population datasets were presented in the order of documents, authors, publication years, institutions, funding bodies, source titles, countries, Scopus subject categories, Scopus keywords, and research fronts.

Lastly, the key scientometric findings for both datasets were discussed to highlight the research landscape of biosyngas-based bioethanol fuels. Additionally, a number of brief conclusions were drawn, and a number of relevant recommendations were made to enhance the future research landscape.

57.3 RESULTS

57.3.1 THE MOST PROLIFIC DOCUMENTS ON BIOSYNGAS-BASED BIOETHANOL FUELS

Information on the types of documents for both datasets is given in Table 57.1. Articles and conference papers, published in journals, dominate both the sample (91%) and population (94%) papers with a 3% deficit. Further, review papers and short surveys have a 6% surplus as they are over-represented in the sample papers, constituting 9% and 3% of the sample and population papers, respectively.

It is further notable that 97% of the population papers were published in journals, while 2% and 1% of them were published in book series and books, respectively. Similarly, 99% of the sample papers were published in journals, while 1% were published in book series.

TABLE 57.1
Documents on Biosyngas-based Bioethanol Fuels

Documents	Sample Dataset (%)	Population Dataset (%)	Surplus (%)
Article	87.0	89.7	−2.7
Review	9.0	3.2	5.8
Conference paper	4.0	3.6	0.4
Book chapter	0.0	1.7	−1.7
Letter	0.0	0.8	−0.8
Note	0.0	0.7	−0.7
Editorial	0.0	0.2	−0.2
Short Survey	0.0	0.2	−0.2
Book	0.0	0.0	0.0
Sample size	100	603	

Population dataset, The number of papers (%) in the set of the 603 population papers; Sample dataset, The number of papers (%) in the set of 100 highly cited papers.

57.3.2 The Most Prolific Authors on Biosyngas-based Bioethanol Fuels

Information about the 25 most prolific authors with at least 4% of the sample papers each is given in Table 57.2. The most prolific authors are Hasan K. Atiyeh, Christian Kennes, Maria C. Veiga, and Raymond L. Huhnke with 6% of the sample papers each. Other prolific authors are Haris N. Abubackar and Boon Siang Yeo with 5% of the sample papers each. On the other hand, the most influential authors are Haris N. Abubackar and Boon Siang Yeo with a 4.2% surplus each, followed by Christian Kennes and Maria C. Veiga with a 3.8% surplus each.

The most prolific institution for the sample dataset is the University of Toronto with seven authors, followed by the National University of Singapore and the University of Coruna with three authors each. Other prolific institutions are Cornell University, Oklahoma State University, and the University of Oklahoma with two authors each.

On the other hand, the most prolific country for the sample dataset is Canada with eight authors, followed by the USA with seven authors. Other prolific countries are Singapore and Spain with three authors each. In total, only eight countries house these top authors.

There are three research fronts for these top authors: bioethanol fuels based on CO_2, biosyngas, and CO with 14, 11, and 3 authors, respectively. On the other hand, there is a significant gender deficit (Beaudry and Lariviere, 2016) for the sample dataset as surprisingly only one of these top researchers is female with a representation rate of 4%.

Additionally, there are other authors with a relatively low citation impact and with 1.0%–2.5% of the population papers each: Yuan Liu, Guohui Yang, Wei Huang, Noritatsu Tsubaki, Yisheng Tan, Jiaming Wang, Kang An, Hironori Arakawa, Randy S. Lewis, Kiyomi Okabe, Huixian Zhong, Alberto C. Badino, Hitoshi Kusama, Ghasem Najafpour, Kazuhiro Sayama, Yuhan Sun, Baojun Wang, Habibollah Younesi, and Riguang Zhang.

57.3.3 The Most Prolific Research Output by Years on Biosyngas-based Bioethanol Fuels

Information about papers published between 1970 and 2022 is given in Figure 57.1. This figure clearly shows that the bulk of the research papers in the population dataset was published primarily in the 2010s and the early 2020s with 51% and 27% of the population dataset, respectively. Similarly, the publication rates for the 2000s, 1990s, 1980s, and 1970s were 9%, 7%, 4%, and 1% respectively. Additionally, 1% of the population papers were published in the pre-1970s.

Similarly, the bulk of the research papers in the sample dataset was published in the 2010s and 2000s with 68% and 10% of the sample dataset, respectively. Similarly, the publication rates for the 1990s, 1980s, and 1970s were 8%, 6%, and 0% of the sample papers, respectively.

The most prolific publication year for the population dataset was 2022 with 9.8% of the dataset, followed by 2020 and 2021 with 8.6% and 8.3% of the population, respectively. Further, 78% of the population papers were published between 2010 and 2022. Similarly, 79% of the sample papers were published between 2009 and 2020, while the most prolific publication year was 2016 with 13% of the sample papers. Other prolific publication years were 2011, 2020, 2014, 2017, and 2018 with 10%, 8%, 7%, 7%, and 7% of the sample dataset, respectively.

57.3.4 The Most Prolific Institutions on Biosyngas-based Bioethanol Fuels

Information about the 15 most prolific institutions publishing papers on biosyngas-based bioethanol fuels with at least 3.0% of the sample papers each is given in Table 57.3.

The most prolific institution is Oklahoma State University with 9% of the sample papers, followed by the Chinese Academy of Sciences, Tianjin University, the University of Coruna, and the National University of Singapore with 7%, 6%, 6%, and 5% of the sample papers, respectively.

TABLE 57.2
The Most Prolific Authors on Biosyngas-based Bioethanol Fuels

No.	Author Name	Author Code	Sample Papers (%)	Population Papers (%)	Surplus	Institution	Country	HI	N	Res. Front
1	Atiyeh, Hasan K.	36452802100	6.0	2.5	3.5	Oklahoma State Univ.	USA	76	27	S
2	Kennes, Christian	8627862800	6.0	2.2	3.8	Univ. Coruna	Spain	183	45	D, S
3	Veiga, Maria C.*	35550232800	6.0	2.2	3.8	Univ. Coruna	Spain	179	44	D, S
4	Huhnke, Raymond L.	6603911746	6.0	1.8	4.2	Oklahoma State Univ.	USA	126	39	S
5	Abubackar, Haris N.	36815751200	5.0	2.0	3.0	Univ. Coruna	Spain	29	17	D, S
6	Yeo, Boon Siang	35099678500	5.0	0.8	4.2	Natl. Univ. Singapore	Singapore	79	42	C
7	Tanner, Ralph S.	7202659981	4.0	1.2	2.8	Univ. Oklahoma	USA	73	36	S
8	Wilkins, Mark R.	56492323200	4.0	1.2	2.8	Univ. Oklahoma	USA	97	28	S
9	Sargent, Edward H.	7102619565	4.0	1.0	3.0	Univ. Toronto	Canada	850	150	C
10	Sinton, David	6603799389	4.0	1.0	3.0	Univ. Toronto	Canada	344	72	C
11	Gong, Jinlong	55218059900	4.0	0.8	3.2	Natl. Univ. Singapore	Singapore	372	95	S
12	Spivey, James J.	7005741953	4.0	0.8	3.2	Louisiana State Univ.	USA	249	46	S
13	Wang, Ying	56011497000	4.0	0.8	3.2	Univ. Toronto	Canada	47	26	C
14	Angenent, Largus T.	6602535088	4.0	0.7	3.3	Cornell Univ.	USA	196	65	S
15	Chen, Bin	57207850002	4.0	0.7	3.3	Univ. Elect. Sci. technol.	China	63	30	C
16	Dinh, Cao T.	35109117700	4.0	0.7	3.3	Queen's Univ.	Canada	95	55	C
17	Li, Fengwang	55061301900	4.0	0.7	3.3	Univ. Sydney	Australia	77	42	C
18	Li, Jun	57276063200	4.0	0.7	3.3	Univ. Toronto	Canada	73	37	C
19	Li, Yuguang C	56826233400	4.0	0.7	3.3	Univ. Toronto	Canada	40	27	C
20	Lum, Yanwei	56685212800	4.0	0.7	3.3	Natl. Univ. Singapore	Singapore	37	29	C
21	Luo, Mingchuan	57191430744	4.0	0.7	3.3	Univ. Toronto	Canada	74	41	C
22	Nam, Dae Hyun	12760238700	4.0	0.7	3.3	Daegu Gyeongbuk Inst. Sci. Technol.	South Korea	48	26	C
23	Richter, Hanno	55552042300	4.0	0.7	3.3	Cornell Univ.	USA	26	21	S
24	Wang, Ziyun	54895729100	4.0	0.7	3.3	Univ. Auckland	N. Zealand	65	34	C
25	Wicks, Joshua	57209207515	4.0	0.7	3.3	Univ. Toronto	Canada	31	30	C

*, Female; Author code, the unique code given by Scopus to the authors; C, CO_2; D, CO; Population papers, the number of papers authored in the population dataset; S, Biosyngas; Sample papers, the number of papers authored in the sample dataset.

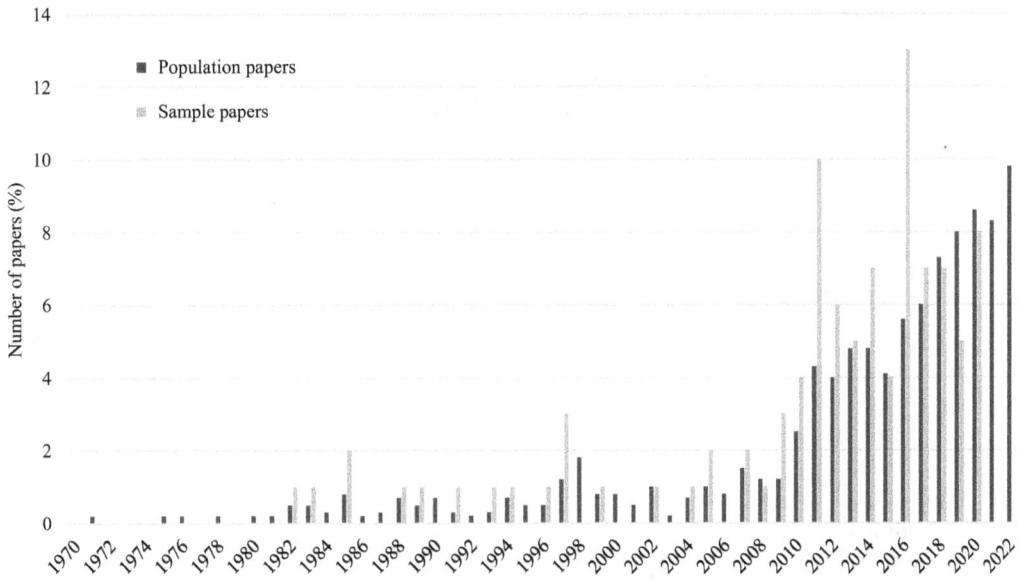

FIGURE 57.1 Research output by years regarding biosyngas-ased bioethanol fuels.

TABLE 57.3
The Most Prolific Institutions on Biosyngas-based Bioethanol Fuels

No.	Institutions	Country	Sample Papers (%)	Population Papers (%)	Surplus (%)
1	Oklahoma State Univ.	USA	9.0	3.2	5.8
2	Chinese Acad. Sci.	China	7.0	10.3	−3.3
3	Tianjin Univ.	China	6.0	5.0	1.0
4	Univ. Coruna	Spain	6.0	2.3	3.7
5	Natl. Univ. Singapore	Singapore	5.0	1.3	3.7
6	State Key Lab. Phys. Chem. Sol. Surf.	China	4.0	1.5	2.5
7	Univ. Oklahoma	USA	4.0	1.2	2.8
8	Univ. Toronto	Canada	4.0	1.0	3.0
9	Louisiana State Univ.	USA	4.0	0.8	3.2
10	Cornell Univ.	USA	4.0	0.7	3.3
11	Xiamen Univ.	China	4.0	2.3	1.7
12	Argonne Natl. lab.	USA	3.0	1.0	2.0
13	Oak Ridge Natl. Lab.	USA	3.0	1.0	2.0
14	Iowa State Univ.	USA	3.0	0.7	2.3
15	Beijing Univ. Chem. Technol.	China	3.0	2.3	0.7

Similarly, the top country for these most prolific institutions is the USA with seven institutions, followed by China with five institutions. In total, only five countries house these top institutions.

On the other hand, the institution with the highest citation impact is Oklahoma State University with a 5.8% surplus, followed by the University of Coruna and the National University of Singapore with 3.7% of the sample papers each. Other influential institutions are Cornell University, Louisiana State University, and the University of Toronto with a 3%–3.3% surplus each.

Additionally, there are other institutions with a relatively low citation impact and with 0.8%–3.6% of the population papers each: Taiyuan University of Technology, University of Toyama, Dalian University of Technology, CNRS, ShanghaiTech University, Russian Academy of Sciences,

National Institute of Materials and Chemical Research, Zhejiang University, Brigham Young University, University of Sao Paulo, Fudan University, Federal University of Sao Carlos, Japan Science and Technology Agency, State University of Campinas, Tokyo Institute of Technology, Babol Noshirvani University of Technology, Delft University of Technology, National Research Council Canada, Kyushu University, Technical University of Denmark, Peking University, National Cheng Kung University, and Tarbiat Modares University.

57.3.5 The Most Prolific Funding Bodies on Biosyngas-based Bioethanol Fuels

Information about the 16 most prolific funding bodies funding at least 3% and 1% of the sample and population papers, separately, is given in Table 57.4. Further, only 58% and 60% of the sample and population papers were funded, respectively.

The most prolific funding body is the National Natural Science Foundation of China with 16% of the sample papers, followed by the U.S. Department of Energy, Office of Science, and Argonne National Laboratory with 10%, 6%, and 6% of the sample papers, respectively. Other prolific funding bodies are the Ministry of Science and Technology of China, European Regional Development Fund, National Science Foundation, and Oklahoma State University with 5% of the sample papers each.

On the other hand, the most prolific countries for these top funding bodies are the USA and China with six and four funding bodies, respectively. Another prolific country is Canada with two funding bodies. In total, only six countries and the EU house these top funding bodies.

The funding body with the highest citation impact is the U.S. Department of Energy with a 6.2% surplus, followed by Argonne National Laboratory, Office of Science, and Oklahoma State University with 4.8%, 4.2%, and 3.7% surplus, respectively. Further, the funding body with the lowest citation impact is the National Natural Science Foundation of China with an 8.5% deficit. Another least influential body is the National Key Research and Development Program of China with a 1.1% deficit.

TABLE 57.4
The Most Prolific Funding Bodies on Biosyngas-based Bioethanol Fuels

No.	Funding bodies	Country	Sample Paper No. (%)	Population Paper No. (%)	Surplus (%)
1	National Natural Science Foundation of China	China	16.0	24.5	−8.5
2	U.S. Department of Energy	USA	10.0	3.8	6.2
3	Office of Science	USA	6.0	1.8	4.2
4	Argonne National Laboratory	USA	6.0	1.2	4.8
5	Ministry of Science and Technology of China	China	5.0	2.8	2.2
6	European Regional Development Fund	EU	5.0	1.7	3.3
7	National Science Foundation	USA	5.0	1.7	3.3
8	Oklahoma State University	USA	5.0	1.3	3.7
9	National Basic Research Program of China (973 Program)	China	4.0	2.2	1.8
10	Ministry of Economy and Competitiveness	Spain	4.0	2.0	2.0
11	Natural Sciences and Engineering Research Council	Canada	4.0	1.8	2.2
12	Suncor Energy Inc.	USA	4.0	0.8	3.2
13	Ministry of Education - Singapore	Singapore	4.0	0.7	3.3
14	National Key Research and Development Program of China	China	3.0	4.1	−1.1
15	Japan Society for the Promotion of Science	Japan	3.0	1.2	1.8
16	Canada Foundation for Innovation	Canada	3.0	1.0	2.0

Other funding bodies with a relatively low citation impact and with 1.0%–2.3% of the population papers each are the Chinese Academy of Sciences, Fundamental Research Funds for the Central Universities, National Council for Scientific and Technological Development, Natural Science Foundation of Shanxi Province, Higher Education Personnel Improvement Coordination, European Commission, Ministry of Education of China, National Research Foundation of Korea, China Postdoctoral Science Foundation, Research Support Foundation of the State of Sao Paulo, Ministry of Education, Culture, Sports, Science and Technology, Ministry of Science, Technology, and Innovation, Science and Technology Commission of Shanghai Municipality, Basic Energy Sciences, China Scholarship Council, German Research Foundation, and Natural Science Foundation of Zhejiang Province.

57.3.6 THE MOST PROLIFIC SOURCE TITLES ON BIOSYNGAS-BASED BIOETHANOL FUELS

Information about the 16 most prolific source titles publishing at least 2% and 0.7% of the sample and population papers, respectively, on biosyngas-based bioethanol fuels is given in Table 57.5.

The most prolific source title is Bioresource Technology with 13% of the sample papers, followed by the Journal of Catalysis and Angewandte Chemie International Edition with 7% of the sample papers, each. Other prolific titles are the Journal of the American Chemical Society, ACS Catalysis, and Catalysis Today with 6%, 5%, and 4% of the sample papers, respectively.

On the other hand, the source title with the highest citation impact is Bioresource Technology with a 7.7% surplus. Other influential titles are Angewandte Chemie International Edition, Journal of Catalysis, and Journal of the American Chemical Society with a 4.7%–5.0% surplus each.

Other source titles with a relatively low citation impact with 0.8%–2.3% of the population papers each are Fuel, Industrial and Engineering Chemistry Research, Chemcatchem, Catalysis Letters, Energy and Fuels, Applied Catalysis A General, Applied Surface Science, Catalysts, Chemical Communications, Chemical Engineering Journal, Journal of CO_2 Utilization, Journal of Power Sources, Catalysis Communications, Catalysis Science and Technology, Chemical Engineering Science, International Journal of Hydrogen Energy, Journal of Chemical Technology and Biotechnology, Physical Chemistry Chemical Physics, Small, and Studies in Surface Science and Catalysis.

TABLE 57.5
The Most Prolific Source Titles on Biosyngas-based Bioethanol Fuels

No.	Source Titles	Sample Papers (%)	Population Papers (%)	Surplus (%)
1	Bioresource Technology	13.0	5.3	7.7
2	Journal of Catalysis	7.0	2.3	4.7
3	Angewandte Chemie International Edition	7.0	2.0	5.0
4	Journal of the American Chemical Society	6.0	1.3	4.7
5	ACS Catalysis	5.0	1.3	3.7
6	Catalysis Today	4.0	2.0	2.0
7	Energy	3.0	0.8	2.2
8	Journal of Chemical Technology and Biotechnology	3.0	0.8	2.2
9	Nature Catalysis	3.0	0.5	2.5
10	Applied Catalysis B Environmental	2.0	1.2	0.8
11	Applied Energy	2.0	1.2	0.8
12	Biochemical Engineering Journal	2.0	0.8	1.2
13	Chemsuschem	2.0	0.8	1.2
14	Biotechnology and Bioengineering	2.0	0.7	1.3
15	Energy and Environmental Science	2.0	0.7	1.3
16	Nature Communications	2.0	0.7	1.3

57.3.7 THE MOST PROLIFIC COUNTRIES ON BIOSYNGAS-BASED BIOETHANOL FUELS

Information about the 11 most prolific countries publishing at least 2% of sample papers each on biosyngas-based bioethanol fuels is given in Table 57.6.

The most prolific country is the USA with 48% of the sample papers, followed by China and Japan with 24% and 12% of the sample papers, respectively. Other prolific countries are Spain, Singapore, Canada, and South Korea with 4%–8% of the sample papers each. Further, the two European countries listed in Table 57.6 produce 11% and 6% of the sample and population papers, respectively.

On the other hand, the country with the highest citation impact is the USA with a 26.8% surplus. Other influential countries are Singapore and Spain with 3.8% and 3.7% surplus, respectively. Similarly, the country with the lowest citation impact is China with a 12.2% deficit.

Additionally, there are other countries with a relatively low citation impact and with 0.5%–5% of the sample papers each: Brazil, Germany, India, France, Russia, the UK, Australia, Denmark, Saudi Arabia, Sweden, Switzerland, Italy, Poland, Portugal, Bangladesh, Thailand, Belgium, Indonesia, Mexico, Pakistan, Romania, and Ukraine.

57.3.8 THE MOST PROLIFIC SCOPUS SUBJECT CATEGORIES ON BIOSYNGAS-BASED BIOETHANOL FUELS

Information about the eight most prolific Scopus subject categories indexing at least 7% of the sample papers each is given in Table 57.7.

The most prolific Scopus subject category on biosyngas-based bioethanol fuels is Chemical Engineering with 69% of the sample papers, followed by Chemistry with 48% of the sample papers. Other prolific subject categories are Energy, Environmental Science, and Biochemistry, Genetics, and Molecular Biology with 31%–36% of the sample papers each. It is notable that Social Sciences including Economics and Business accounts for 0.0% and 1.5% of the sample and population studies, respectively.

On the other hand, the Scopus subject category with the highest citation impact is Biochemistry, Genetics and Molecular Biology with a 15% surplus, followed by Environmental Science, Chemical Engineering, and Energy with a 14%, 13%, and 11% surplus, respectively. Similarly, the least influential subject category is Materials Science with a 7% deficit, followed by Chemistry and Engineering with a 4% and 3% deficit, respectively.

TABLE 57.6
The Most Prolific Countries on Biosyngas-based Bioethanol Fuels

No.	Countries	Sample Papers (%)	Population Papers (%)	Surplus (%)
1	USA	48.0	21.2	26.8
2	China	24.0	36.2	−12.2
3	Japan	12.0	10.8	1.2
4	Spain	8.0	4.3	3.7
5	Singapore	5.0	1.2	3.8
6	Canada	4.0	4.0	0.0
7	South Korea	4.0	2.7	1.3
8	Netherlands	3.0	1.8	1.2
9	Iran	2.0	2.2	−0.2
10	Malaysia	2.0	1.5	0.5
11	Taiwan	2.0	1.5	0.5

TABLE 57.7
The Most Prolific Scopus Subject Categories on Biosyngas-based Bioethanol Fuels

No.	Scopus Subject Categories	Sample Papers (%)	Population Papers (%)	Surplus (%)
1	Chemical Engineering	69.0	55.6	13.4
2	Chemistry	48.0	52.2	−4.2
3	Energy	36.0	24.9	11.1
4	Environmental Science	34.0	20.4	13.6
5	Biochemistry, Genetics and Molecular Biology	31.0	16.3	14.7
6	Engineering	13.0	16.3	−3.3
7	Immunology and Microbiology	11.0	5.3	5.7
8	Materials Science	7.0	13.8	−6.8

57.3.9 THE MOST PROLIFIC KEYWORDS ON BIOSYNGAS-BASED BIOETHANOL FUELS

Information about the Scopus keywords used with at least 6% or 3% of the sample or population papers, respectively, is given in Table 57.8. For this purpose, keywords related to the keyword set given in the Appendix are selected from a list of the most prolific keyword set provided by the Scopus database.

These keywords are grouped under four headings: feedstocks, catalysts, processes, and products. The most prolific keywords related to the feedstocks are carbon dioxide, synthesis gas, carbon monoxide, syn-gas, hydrogen, and syngas with 14%–48% of the sample papers each. Further, the prolific keywords related to catalysts are catalysts, copper, catalyst selectivity, carbon, and catalysis with 15%–21% of the sample papers each.

Further, the most prolific keywords related to the processes are clostridium, fermentation, syngas fermentations, electrolytic reduction, reduction, hydrogenation, and Faradaic efficiencies with 12%–35% of the sample papers each. Finally, the most prolific keywords related to products are ethanol, ethanol production, ethylene, and bioethanol with 14%–85% of the sample papers each.

On the other hand, the most influential keywords are reduction, Faradaic efficiencies, synthesis gas, carbon, catalysis, hydrogen, ethanol production, syngas fermentations, ethanol, electrochemical reductions, electro reduction, and bioethanol with 6%–10% surplus each. Similarly, the most prolific keywords across all research fronts are ethanol, carbon dioxide, synthesis gas, clostridium, carbon monoxide, fermentation, syn-gas, catalysts, syngas fermentations, electrolytic reduction, and copper with 18%–85% of the sample papers each.

57.3.10 THE MOST PROLIFIC RESEARCH FRONTS ON BIOSYNGAS-BASED BIOETHANOL FUELS

Information about the research fronts for sample papers on biosyngas-based bioethanol fuels with regard to the feedstocks used in bioethanol production is given in Table 57.9. As this table shows, the most prolific starch feedstock is biosyngas with 47% of the sample papers, followed by CO_2 and CO with 40% and 13% of the sample papers, respectively.

Further, information about the thematic research fronts for the sample papers on biosyngas-based bioethanol fuels is given in Table 57.10. As this table shows, there are five primary research fronts: fermentation, catalytic reduction, electrochemical reduction, photosynthetic production, and photocatalytic reduction of biosyngas with 36%, 31%, 24%, 5%, and 3% of the sample papers, respectively.

TABLE 57.8
The Most Prolific Keywords on Biosyngas-based Bioethanol Fuels

No.	Keywords	Sample Papers (%)	Population Papers (%)	Surplus (%)
1	Feedstocks			
	Carbon dioxide	48.0	49.4	−1.4
	Synthesis gas	35.0	25.9	9.1
	Carbon monoxide	31.0	11.6	19.4
	Syn-gas	25.0	12.1	12.9
	Hydrogen	15.0	7.1	7.9
	Syngas	14.0	12.1	1.9
	Methane	7.0	2.7	4.3
2	Catalysts			
	Catalysts	21.0	15.8	5.2
	Copper	18.0	7.1	10.9
	Catalyst selectivity	17.0	14.1	2.9
	Carbon	17.0	8.3	8.7
	Catalysis	15.0	7.0	8.0
	Rhodium	7.0	4.8	2.2
	Electrocatalysts	6.0	4.1	1.9
	Electrocatalysis	5.0	3.0	2.0
	Catalyst activity	3.0	7.0	−4.0
	Catalytic performance	3.0	3.8	−0.8
	Copper oxides	4.0	3.6	0.4
	Copper compounds	4.0	3.3	0.7
	Cobalt	4.0	3.2	0.8
	Copper alloys	4.0	3.0	1.0
	Binary alloys	3.0	3.6	−0.6
3	Processes			
	Clostridium	35.0	12.4	22.6
	Fermentation	30.0	17.4	12.6
	Biosyngas fermentation	19.0	12.4	6.6
	Electrolytic reduction	18.0	7.0	11.0
	Reduction	15.0	5.3	9.7
	Hydrogenation	14.0	11.3	2.7
	Faradaic efficiencies	14.0	4.3	9.7
	Synthesis	12.0	7.3	4.7
	CO_2 reduction	11.0	7.5	3.5
	Electro reduction	10.0	4.1	5.9
	Electrochemical reduction	9.0	3.0	6.0
	Co hydrogenation	7.0	4.6	2.4
	Fermentation medium	7.0	1.7	5.3
	Bioconversion	6.0	2.3	3.7
	Biotransformation	6.0	2.2	3.8
4	Products			
	Ethanol	85.0	78.4	6.6

(Continued)

TABLE 57.8 (*Continued*)
The Most Prolific Keywords on Biosyngas-based Bioethanol Fuels

No.	Keywords	Sample Papers (%)	Population Papers (%)	Surplus (%)
	Ethanol production	17.0	10.3	6.7
	Ethylene	17.0	6.8	10.2
	Bioethanol	14.0	8.1	5.9
	Butanol	8.0	2.3	5.7
	Methanol	7.0	2.7	4.3
	Bio-ethanol production	6.0	3.8	2.2
	Ethanol synthesis		4.8	−4.8
	Ethanol fuels		3.8	−3.8

TABLE 57.9
The Most Prolific Research Fronts on Biosyngas-based Bioethanol Fuels

No.	Research Fronts	N Paper (%)
1	Biosyngas	47.0
2	CO_2	40.0
3	CO	13.0

Paper (%) sample, The number of papers in the population sample of 100 papers.

TABLE 57.10
The Most Prolific Thematic Research Fronts on Biosyngas-based Bioethanol Fuels

No.	Research Fronts	N Paper (%)
1	Fermentation	36.0
2	Catalytic reduction	31.0
3	Electrochemical reduction	24.0
4	Photosynthetic production	5.0
5	Photocatalytic reduction	3.0

57.4 DISCUSSION

57.4.1 INTRODUCTION

Crude oil-based gasoline fuels have been widely used in the transportation sector since the 1920s. However, there have been great public concerns over the adverse environmental and human impact of these fuels. Hence, biomass-based bioethanol fuels have increasingly been used in blending gasoline fuels, in fuel cells, and in biochemical production in a biorefinery context.

One of the most studied feedstocks for bioethanol fuels has been synthesis biogas (biosyngas) produced from biomass with constituents such as CO_2, carbon monoxide, and H_2. Research in the field of biosyngas-based bioethanol fuels has intensified in this context in the key research fronts of

the electrochemical reduction and catalytic conversion of biosyngas, and the production of bioethanol fuels by other processes such as photosynthetic and photocatalytic conversion, and microbial
fermentation of biosyngas. Further, CO_2, biosyngas, and to a lesser extent CO have been studied
intensively in this context.

However, it is essential to develop efficient incentive structures for the primary stakeholders to
enhance the research in this field. This is especially important to maintain energy security in the
cases of supply shocks such as oil price shocks, war-related shocks as in the case of the Russian
invasion of Ukraine, or COVID-19 shocks.

Scientometric analysis has been used in this context to inform the primary stakeholders about the
current state of research in this research field. As there has been no scientometric study in this field,
this chapter presents a scientometric study of the research on biosyngas-based bioethanol fuels. It
examines the scientometric characteristics of both the sample and population data and presents
them in the order of documents, authors, publication years, institutions, funding bodies, source
titles, countries, Scopus subject categories, Scopus keywords, and research fronts.

As the first step for the search of the relevant literature, keywords were selected using the 100
most-cited papers. The selected keyword list was then optimized to obtain a representative sample
of papers for this research field. A copy of this extended keyword list was provided in the Appendix
for future replicative studies. Further, a selected list of the keywords is presented in Table 57.8.

As the second step, two sets of data were used in this study. First, a population sample of 603
papers was used to examine the scientometric characteristics of the population data. Secondly, a
sample of 203 most-cited papers, corresponding to 17% of the population dataset was used to examine the scientometric characteristics of these citation classics.

The scientometric characteristics of these sample and population datasets were presented in the
order of documents, authors, publication years, institutions, funding bodies, source titles, countries,
Scopus subject categories, Scopus keywords, and research fronts.

Lastly, the key scientometric findings for both datasets were discussed to highlight the research
landscape for biosyngas-based bioethanol fuels. Additionally, a number of brief conclusions were
drawn, and a number of relevant recommendations were made to enhance the future research
landscape.

57.4.2 THE MOST PROLIFIC DOCUMENTS ON BIOSYNGAS-BASED BIOETHANOL FUELS

Articles (together with conference papers) dominate both the sample (91%) and population (94%)
papers with a 3% deficit (Table 57.1). Further, review papers have a surplus (6%). The representation
of reviews in the sample papers is relatively substantial (9%).

Scopus differs from the Web of Science database in differentiating and showing articles (87%)
and conference papers (4%) published in journals separately. However, it should be noted that these
conference papers are also published in journals as articles, compared to those published only in
conference proceedings. Hence, the total number of articles and review papers in the sample dataset
is 91% and 9%, respectively.

It is observed during the search process that there has been inconsistency in the classification
of documents in Scopus as well as in other databases such as Web of Science. This is especially
relevant for the classification of papers as reviews or articles as papers not involving a literature
review may be erroneously classified as a review paper. There is also a case of review papers being
classified as articles. For example, the total number of reviews in the sample dataset was manually
found as nearly 11% compared to 9% as indexed by Scopus, decreasing the number of articles and
conference papers to 89% for the sample dataset.

In this context, it would be helpful to provide a classification note for published papers in books
and journals at the first instance following the good practice employed by some journals. It would
also be helpful to use document types listed in Table 57.1 for this purpose. Book chapters may also
be classified as articles or reviews as an additional classification to differentiate review chapters

from experimental chapters as it is done by Web of Science. It would be further helpful to additionally classify conference papers as articles or review papers as it is done in the Web of Science database.

57.4.3 THE MOST PROLIFIC AUTHORS ON BIOSYNGAS-BASED BIOETHANOL FUELS

There have been 25 most-prolific authors with at least 4% of the sample papers each as given in Table 57.2. These authors have shaped the development of research in this field.

The most prolific authors are Hasan K. Atiyeh, Christian Kennes, Maria C. Veiga, Raymond L. Huhnke, Haris N. Abubackar, and Boon Siang Yeo. On the other hand, the most influential authors are Haris N. Abubackar, Boon Siang Yeo, Christian Kennes, Maria C. Veiga.

It is important to note the inconsistencies in the indexing of author names in Scopus and other databases. It is especially an issue for names with more than two components such as 'Blake Sam de Hyun Kennes'. The probable outcomes are 'Kennes, B.S.D.H.', 'de Hyun Kennes, B.S.', or 'Hyun Kennes, B.S.D.'. The first choice is the gold standard of the publishing sector as the last word in the name is taken as the last name. In most of the academic databases such as PUBMED and EBSCO databases, this version is used predominantly. The second choice is a strong alternative, while the last choice is an undesired outcome as two last words are taken as the last name. It is good practice to combine the words of the last name with a hyphen: 'Hyun- Kennes, B.S.D.'. It is notable that inconsistent indexing of author names may cause substantial inefficiencies in the search process for papers as well as allocating credit to the authors as there are different author entries for each outcome in the databases.

There are also inconsistencies in the shortening of Chinese names. For example. 'YangYing Wang' is often shortened as 'Wang, Y.', 'Wang, Y.-Y.', and 'Wang, Y.Y.' as it is done in the Web of Science database as well. However, the gold standard in this case is 'Wang, Y' where the last word is taken as the last name and the first word is taken as a single forename. However, it would be helpful to use the third option to differentiate Chinese names efficiently: 'Wang, Y.Y.'. In most of the academic databases such as PUBMED and EBSCO, this first version is used predominantly. Therefore, there have been difficulties in locating papers for Chinese authors. In such cases, the use of unique author codes provided for each author by the Scopus database has been helpful.

There is also a difficulty in allocating credit for authors, especially to those with common names such as 'Wang, Y.' in conducting scientometric studies. These difficulties strongly influence the efficiency of scientometric studies as well as allocating credit to authors as there are the same author entries for different authors with the same name, e.g., 'Wang, Y.' in the databases.

In this context, the coding of authors in the Scopus database is a welcome innovation compared to other databases such as Web of Science. In this process, Scopus allocates a unique number to each author in the database (Aman, 2018). However, there might still be substantial inefficiencies in this coding system, especially for common names. For example, some of the papers for a certain author may be allocated to another researcher with a different author code. It is possible that Scopus uses a number of software programs to differentiate the author names and the program may not be false-proof (Kim, 2018).

In this context, it does not help that author names are not given in full in some journals and books. This makes it difficult to differentiate authors with common names and makes scientometric studies further difficult in the author domain. Therefore, author names should be given in all books and journals at the first instance. There is also a cultural issue where some authors do not use their full names in their papers. Instead, they use initials for their forenames: 'Kennes, H.J.', 'Kennes, H.', or 'Kennes, J.' instead of 'Kennes, Hyun Jae'.

There are also inconsistencies in the naming of authors with more than two components by authors themselves in journal papers and book chapters. For example. 'Kennes, A.P.C.' might be given as 'Kennes, A.' or 'Kennes, A.C.' or 'Kennes, A.P.' or 'Kennes, C' in journals and books. This also makes scientometric studies difficult in the author domain. Hence, contributing authors should use their name consistently in their publications.

Another critical issue regarding author names is inconsistencies in the spelling of author names in the national spellings (e.g., Şöğütçığsevil, Gökçe) rather than in the English spellings (e.g., Sogutcigsevil, Gokce) in the Scopus database. Scopus differs from the Web of Science database and many other databases in this respect where author names are given only in English spellings. It is observed that national spellings of author names do not help much in conducting scientometric studies as well as in allocating credits to authors as sometimes there are different author entries for English and National spellings in the Scopus database.

The most prolific institutions for the sample dataset are the University of Toronto and to a lesser extent the National University of Singapore, the University of Coruna, Cornell University, Oklahoma State University, and the University of Oklahoma. Further, the most prolific country for the sample dataset is Canada, the USA, and to a lesser extent Singapore and Spain. These findings confirm the dominance of the USA and Canada and to a lesser extent Singapore and Spain in this field. On the other hand, the primary research fronts are bioethanol fuels based on CO_2, biosyngas, and CO.

It is also notable that there is a significant gender deficit for the sample dataset surprisingly with a representation rate of 4%. This finding is the most thought-provoking with strong public policy implications. Hence, institutions, funding bodies, and policymakers should take efficient measures to reduce the gender deficit in this field as well as other scientific fields with strong gender deficits. In this context, it is worth noting the level of representation of researchers from minority groups in science on the basis of race, sexuality, age, and disability, besides gender (Blankenship, 1993; Dirth and Branscombe, 2017; Konur, 2000, 2002a,b,c, 2006a,b, 2007a,b).

57.4.4 THE MOST PROLIFIC RESEARCH OUTPUT ON BIOSYNGAS-BASED BIOETHANOL FUELS

The research output observed between 1970 and 2022 is illustrated in Figure 57.1. This figure clearly shows that the bulk of the research papers in the population dataset was published primarily in the 2010s and early 2020s. Similarly, the bulk of the research papers in the sample dataset was published in the 2010s and to a lesser extent in the 2000s. These findings suggest that the most prolific sample and population papers were primarily published in the 2010s.

These are the thought-provoking findings as there has been a significant research boom since 2010. In this context, the increasing public concerns about climate change (Change, 2007), greenhouse gas emissions (Carlson et al., 2017), and global warming (Kerr, 2007) have been certainly behind the research boom in this field since 2010. Furthermore, the recent supply shock experiences due to the COVID-19 pandemic might also be behind the research boom in this field since 2019.

Based on these findings, the size of the population papers is likely to more than double in the current decade considering the rising trends for both sample and population datasets, provided that public concerns about climate change, greenhouse gas emissions, and global warming, as well as the supply shocks, are translated efficiently to research funding in this field.

57.4.5 THE MOST PROLIFIC INSTITUTIONS ON BIOSYNGAS-BASED BIOETHANOL FUELS

The 15 most prolific institutions publishing papers on biosyngas-based bioethanol fuels with at least 3.0% of the sample papers given in Table 57.3 have shaped the development of the research in this field.

The most prolific institutions are Oklahoma State University, the Chinese Academy of Sciences, and to a lesser extent Tianjin University, the University of Coruna, and the National University of Singapore. Similarly, the top country for these most prolific institutions is the USA, closely followed by China. In total, only five countries house these top institutions.

On the other hand, institutions with the highest impact are Oklahoma State University and to a lesser extent the University of Coruna, the National University of Singapore, Cornell University,

Louisiana State University, and the University of Toronto. These findings confirm the dominance of the USA and to a lesser extent institutions from Singapore, Spain, and Canada for these HCPs.

57.4.6 THE MOST PROLIFIC FUNDING BODIES ON BIOSYNGAS-BASED BIOETHANOL FUELS

The 16 most prolific funding bodies funding at least 3% and 1% of the sample and population papers, respectively, are given in Table 57.4. It is notable that 58% and 60% of the sample and population papers were funded, respectively.

The most prolific funding bodies are the National Natural Science Foundation of China, U.S. Department of Energy, and to a lesser extent Office of Science, Argonne National Laboratory, Ministry of Science and Technology of China, European Regional Development Fund, National Science Foundation, and Oklahoma State University. The most prolific countries for these top funding bodies are the USA, China, and a lesser extent Canada.

On the other hand, the most influential funding bodies are the U.S. Department of Energy, Argonne National Laboratory, and to a lesser extent Office of Science, and Oklahoma State University. Further, the funding body with the lowest citation impact is the National Natural Science Foundation of China.

These findings on the funding of research in this field suggest that the level of funding, mostly since 2010, is highly intensive, and it has been largely instrumental in enhancing the research in this field (Ebadi and Schiffauerova, 2016) in light of North's institutional framework (North, 1991). It is also notable that the funding rate in this field is relatively significant compared to those in other research fronts of bioethanol fuels such as cellulose-based bioethanol fuels. Further, it is expected that this high funding rate would continue in light of the recent supply shocks. Further, it emerges that the USA has heavily funded the research on corn grain- and corn residue-based bioethanol fuels.

57.4.7 THE MOST PROLIFIC SOURCE TITLES ON BIOSYNGAS-BASED BIOETHANOL FUELS

The 16 most prolific source titles publishing at least 2% and 0.7% of the sample and population papers, respectively, in biosyngas-based bioethanol fuels have shaped the development of research in this field (Table 57.5).

The most prolific source titles are Bioresource Technology and to a lesser extent the Journal of Catalysis, Angewandte Chemie International Edition, American Chemical Society, ACS Catalysis, and Catalysis Today. On the other hand, source titles with the highest impact are Bioresource Technology and to a lesser extent Angewandte Chemie International Edition, Journal of Catalysis, and Journal of the American Chemical Society.

It is notable that these top source titles are primarily related to catalysis, chemistry, and to a lesser extent bioresources. This finding suggests that Bioresource Technology and other prolific journals in these fields have significantly shaped the development of research in this field as they focus primarily on biosyngas-based bioethanol fuel production with high ethanol selectivity and productivity. In this context, the influence of these top journals is quite extraordinary.

57.4.8 THE MOST PROLIFIC COUNTRIES ON BIOSYNGAS-BASED BIOETHANOL FUELS

The 11 most prolific countries publishing at least 2% of the sample papers each have significantly shaped the development of research in this field (Table 57.6).

The most prolific countries are the USA, China, Japan, and to a lesser extent Spain, Singapore, Canada, and South Korea. On the other hand, countries with the highest citation impact are the USA and to a lesser extent Singapore and Spain. Similarly, the country with the lowest impact is China.

A close examination of these findings suggests that the USA, China, Europe, Japan, and to a lesser extent, Singapore, Canada, and South Korea are the major producers of research in this field. It is a fact that the USA has been a major player in science (Leydesdorff and Wagner, 2009). The

USA has further developed a strong research infrastructure to support its corn- and grass-based bioethanol industry (Gillon, 2010).

However, China has been a rising megastar in scientific research in competition with the USA and Europe (Leydesdorff and Zhou, 2005). China is also a major player in this field as a major producer of bioethanol (Fang et al., 2010).

Next, Europe has been a persistent player in scientific research in competition with both the USA and China (Leydesdorff, 2000). Europe has also been a persistent producer of bioethanol along with the USA and Brazil (Gnansounou, 2010).

57.4.9 THE MOST PROLIFIC SCOPUS SUBJECT CATEGORIES ON BIOSYNGAS-BASED BIOETHANOL FUELS

The eight most prolific Scopus subject categories indexing at least 7% of the sample papers each, respectively, given in Table 57.7 have shaped the development of research in this field.

The most prolific Scopus subject categories on biosyngas-based bioethanol fuels are Chemical Engineering and to a lesser extent Chemistry, Energy, Environmental Science, and Biochemistry, Genetics and Molecular Biology. It is also notable that Social Sciences including Economics and Business has a minimal presence in both sample and population studies.

On the other hand, Scopus subject categories with the highest citation impact are Biochemistry, Genetics and Molecular Biology, and to a lesser extent Environmental Science, Chemical Engineering, and Energy. Similarly, the least influential subject categories are Materials Science and to a lesser extent Chemistry and Engineering.

These findings are thought-provoking, suggesting that the primary subject categories are related to chemical engineering, chemistry, energy, environmental science, and biochemistry as the core of the research in this field concerns the production of biosyngas-based bioethanol fuels. Another finding is that social sciences is not well represented in both the sample and population papers in line with most of the fields in bioethanol fuels. Social, environmental, and economics studies account for the field of social sciences.

57.4.10 THE MOST PROLIFIC KEYWORDS ON BIOSYNGAS-BASED BIOETHANOL FUELS

A limited number of keywords have shaped the development of research in this field as shown in Table 57.8 and the Appendix. These keywords are grouped under four headings: feedstocks, catalysts, processes, and products.

The most prolific keywords across all research fronts are ethanol, carbon dioxide, synthesis gas, clostridium, carbon monoxide, fermentation, syn-gas, catalysts, biosyngas fermentations, electrolytic reduction, and copper. Similarly, the most influential keywords are reduction, Faradaic efficiencies, synthesis gas, carbon, catalysis, hydrogen, ethanol production, biosyngas fermentations, ethanol, electrochemical reductions, electro reduction, and bioethanol.

These findings suggest that it is necessary to determine the keyword set carefully to locate the relevant research in each of these research fronts. Additionally, the size of the samples for each keyword highlights the intensity of research in the relevant research areas.

57.4.11 THE MOST PROLIFIC RESEARCH FRONTS ON BIOSYNGAS-BASED BIOETHANOL FUELS

Information about the research fronts for the sample papers on biosyngas-based bioethanol fuels with regard to feedstocks used for bioethanol production is given in Table 57.9. As this table shows, the most prolific feedstocks are biosyngas, CO_2, and to a lesser extent CO.

Information about the thematic research fronts for the sample papers in biosyngas-based bioethanol fuels is given in Table 57.10. As this table shows, there are five primary research fronts:

fermentation, catalytic reduction, electrochemical reduction, photosynthetic production, and photocatalytic reduction. The first three research fronts dominate the research in this field compared to photosynthetic production and photocatalytic reduction.

These findings are thought-provoking in seeking ways to increase biosyngas-based bioethanol productivity and selectivity at the global scale. It is clear that all these research fronts have public importance and merit substantial funding and other incentives. Further, it is notable that biosyngas-based bioethanol fuels have become a core unit of bioethanol research to make it more competitive with crude oil-based gasoline and diesel fuels, especially for the USA, Europe, and China.

It is also notable that the evaluation of biosyngas-based bioethanol fuels such as technoeconomics, life cycle, economics, social science, and environmental impact-related studies emerges as a neglected research field. In this context, the USA has been the global leader in the production and use of biomass-based bioethanol fuels since the 1970s in the aftermath of the global crude oil crisis in the early 1970s.

Finally, it is important to research biosyngas-based bioethanol fuel production in the wider research field of biosyngas transformation to fuels and value-added products with over 35,000 papers and book chapters (Hu et al., 2016; Kopke et al., 2010; Li et al., 2016). It is also closely related to the research field on CO_2 utilization with over 80,000 papers and book chapters (Habisreutinger et al., 2013; Sakakura et al., 2007; Wang et al., 2011). Although the conventional copper (Cu) and rhodium (Rh)-based catalysts (Gao et al., 2016; Li and Kanan, 2012) are used in these wider studies, the recent emphasis has been on the development and application of the nanocatalysts (Chen et al., 2012; Fu et al., 2012) as in this study.

In this context, catalytic hydrogenation (Wang et al., 2011), photocatalytic reduction (Habisreutinger et al., 2013), photoelectrocatalytic reduction (Inoue et al., 1979), electrochemical reduction (Peterson et al., 2010), catalytic reduction (Porosoff et al., 2016), microbial fermentation (Kopke et al., 2010), and photosynthetic conversion (Tang et al., 2011) emerge as the key processes for the production of bioethanol fuels from biosyngas and its constituent, CO_2.

In the end, these most-cited papers in this field hint that the production of biosyngas-based bioethanol fuels could be optimized using the structure, processing, and property relationships of these feedstocks primarily in the fronts of the fermentation, catalytic reduction, and electrochemical reduction of biosyngas (Formela et al., 2016; Konur, 2018a, 2020b, 2021a,b,c,d; Konur and Matthews, 1989).

57.5 CONCLUSION AND FUTURE RESEARCH

Research on biosyngas-based bioethanol fuels has been mapped through a scientometric study of both sample (100 papers) and population (603 papers) datasets.

The critical issue in this study has been to obtain a representative sample of research as in any other scientometric study. Therefore, the keyword set has been carefully devised and optimized after a number of runs in the Scopus database. It is a representative sample of wider population studies. This keyword set was provided in the Appendix, and the relevant keywords are presented in Table 57.8. However, it should be noted that it has been very difficult to compile a representative keyword set since this research field has been connected closely with many other fields. Therefore, it has been necessary to compile a keyword list to exclude papers concerned with the other research fields.

It is notable in this context that research on the production of bioethanol fuels from biosyngas and its constituents, CO_2 and CO, is highly different from the research on bioethanol production from food crops, lignocellulosic feedstocks, and wastes as the emphasis has been on the reduction of these feedstocks using a wide range of catalysts. It is more related to research on the production of fuels and value-added compounds from syngas and CO_2 utilization.

Another issue has been the selection of a multidisciplinary database to carry out the scientometric study of research in this field. For this purpose, the Scopus database has been selected. The journal coverage of this database has been notably wider than that of the Web of Science and other multisubject databases.

The key scientometric properties of the research in this field have been determined and discussed in this chapter. It is evident that a limited number of documents, authors, institutions, publication years, institutions, funding bodies, source titles, countries, Scopus subject categories, Scopus keywords, and research fronts have shaped the development of the research in this field.

There is ample scope to increase the efficiency of scientometric studies in this field in the author and document domains by developing consistent policies and practices in both domains across all academic databases. In this respect, it seems that authors, journals, and academic databases have a lot to do. Furthermore, the significant gender deficit as in most scientific fields emerges as a public policy issue. The potential deficits on the basis of age, race, disability, and sexuality need also to be explored in this field as in other scientific fields.

Research in this field has boomed since 2010, possibly promoted by public concerns about global warming, greenhouse gas emissions, and climate change. Furthermore, the recent COVID-19 pandemic and the Russian invasion of Ukraine have resulted in a global supply shock, shifting the focus of the stakeholders from crude oil-based fuels to biomass-based fuels such as bioethanol fuels. It is expected that there would be further incentives for the key stakeholders to carry out research on biosyngas-based bioethanol fuels from biosyngas to increase ethanol productivity and selectivity and to make it more competitive with crude oil-based gasoline and petrodiesel fuels. This might be true for crude oil- and foreign exchange-deficient countries to maintain energy and food security in the face of global supply shocks.

The relatively significant funding rate of 58% and 60% for the sample and population papers, respectively, suggests that funding in this field significantly enhanced the research in this field primarily since 2010, possibly more than doubling in the current decade. However, it is evident that there is ample room for more funding and other incentives to enhance the research in this field further.

Institutions from the USA and to a lesser extent China have mostly shaped the research in this field. Further, the USA, China, and to a lesser extent, Japan, Europe, Singapore, Canada, and South Korea have been the major producers of research in this field as they are the major producers and users of bioethanol fuels from different types of biomass such as corn, wheat, and rice as well as other types of biomass. It is evident that these countries have well-developed research infrastructure in bioethanol fuels and their derivatives. It is also notable these major countries mostly have access to large farmlands.

It emerges that ethanol is more popular than bioethanol as a keyword with strong implications for the search strategy. In other words, the search strategy using only bioethanol as the keyword would not be much helpful. Scopus keywords are grouped under five headings: feedstocks, fermentation, hydrolysis and hydrolysates, products, and evaluation.

The most prolific feedstocks are biosyngas, CO_2, and to a lesser extent CO used in these studies. On the other hand, there are five primary research fronts: fermentation, catalytic reduction, electrochemical reduction, and to a lesser extent photosynthetic production and photocatalytic reduction of biosyngas. The first three research fronts dominate the research in this field to a higher extent compared to the last two fronts.

These findings are thought-provoking in seeking ways to increase bioethanol selectivity and productivity through biosyngas-based bioethanol fuel production at the global scale. It is clear that all these research fronts have public importance and merit substantial funding and other incentives. Further, it is notable that biosyngas-based bioethanol fuels have become a core unit of bioethanol research to make it more competitive with crude oil-based gasoline and petrodiesel fuels, especially for countries with large access to farmlands.

Thus, scientometric analysis has a great potential to gain valuable insights into the evolution of research in this field as in other scientific fields, especially in the aftermath of significant global supply shocks such as the COVID-19 pandemic and the Russian invasion of Ukraine.

It is recommended that further scientometric studies are carried out for the primary research fronts. It is further recommended that reviews of the most-cited papers are carried out for each primary research front to complement these scientometric studies. Next, scientometric studies of hot papers in these primary fields need to be carried out.

ACKNOWLEDGMENTS

The contribution of the highly cited researchers in the field of biosyngas-based bioethanol fuels has been gratefully acknowledged.

APPENDIX: THE KEYWORD SET FOR THE BIOSYNGAS-BASED BIOETHANOL FUELS

((TITLE (ethanol OR bioethanol OR c2h5oh) AND TITLE (*syngas OR "carbon dioxide" OR "catalytic particles" OR "synthesis gas" OR co2 OR "carbon monoxide" OR "producer gas" OR "co h-2" OR "co h2" OR "waste gas" OR "synthetic gas" OR "h2 co" OR "h 2 co")) AND NOT (TITLE (oxidizing OR balance OR dissolution OR carbonate* OR oxidation OR supercritical OR solubility OR phase OR reforming OR emission* OR solutions OR *solvents OR cao OR gibbs OR elongation OR capture OR splitting OR extraction OR holdup OR rhb OR sago OR ions OR promoter OR caprylate OR methanol OR "adding value" OR {fermentation CO2} Or sequest* OR upgrading) OR SRCTITLE (data OR phase OR fluid* OR atmosph* OR hort* OR hepat*) OR SUBJAREA (medi OR phar OR neur))) AND (LIMIT-TO (SRCTYPE, "j") OR LIMIT-TO (SRCTYPE, "k") OR LIMIT-TO (SRCTYPE, "b")) AND (LIMIT-TO (DOCTYPE, "ar") OR LIMIT-TO (DOCTYPE, "cp") OR LIMIT-TO (DOCTYPE, "re") OR LIMIT-TO (DOCTYPE, "ch") OR LIMIT-TO (DOCTYPE, "no") OR LIMIT-TO (DOCTYPE, "le") OR LIMIT-TO (DOCTYPE, "ed") OR LIMIT-TO (DOCTYPE, "sh")) AND (LIMIT-TO (LANGUAGE, "English")))

REFERENCES

Abubackar, H. N., M. C. Veiga and C. Kennes. 2015. Carbon monoxide fermentation to ethanol by *Clostridium autoethanogenum* in a bioreactor with no accumulation of acetic acid. *Bioresource Technology* 186:122–127.

Aman, V. 2018. Does the Scopus author ID suffice to track scientific international mobility? A case study based on Leibniz laureates. *Scientometrics* 117:705–720.

Angelici, C., B. M. Weckhuysen and P. C. A. Bruijnincx. 2013. Chemocatalytic conversion of ethanol into butadiene and other bulk chemicals. *ChemSusChem* 6:1595–1614.

Antolini, E. 2007. Catalysts for direct ethanol fuel cells. *Journal of Power Sources* 170:1–12.

Antolini, E. 2009. Palladium in fuel cell catalysis. *Energy and Environmental Science* 2:915–931.

Beaudry, C. and V. Lariviere. 2016. Which gender gap? Factors affecting researchers' scientific impact in science and medicine. *Research Policy* 45:1790–1817.

Bertheussen, E., A. Verdaguer-Casadevall and D. Ravasio, et al. 2016. Acetaldehyde as an intermediate in the electroreduction of carbon monoxide to ethanol on oxide-derived copper. *Angewandte Chemie - International Edition* 55:1450–1454.

Blankenship, K. M. 1993. Bringing gender and race in: US employment discrimination policy. *Gender & Society* 7:204–226.

Burnham, J. F. 2006. Scopus database: A review. *Biomedical Digital Libraries* 3:1–8.

Carlson, K. M., J. S. Gerber and D. Mueller, et al. 2017. Greenhouse gas emissions intensity of global croplands. *Nature Climate Change* 7:63–68.

Change, C. 2007. Climate change impacts, adaptation and vulnerability. *Science of the Total Environment* 326:95–112.

Chen, Y., C. W. Li and M. W. Kanan. 2012. Aqueous CO_2 reduction at very low overpotential on oxide-derived Au nanoparticles. *Journal of the American Chemical Society* 134:19969–19972.

Choi, Y. and P. Liu. 2009. Mechanism of ethanol synthesis from syngas on Rh(111). *Journal of the American Chemical Society* 131:13054–13061.

Datar, R. P., R. M. Shenkman, B. G. Cateni, R. I. Huhnke and R.S. Lewis. 2004. Fermentation of biomass-generated producer gas to ethanol. *Biotechnology and Bioengineering* 86:587–594.

Dirth, T. P. and N. R. Branscombe. 2017. Disability models affect disability policy support through awareness of structural discrimination. *Journal of Social Issues* 73:413–442.

Ebadi, A. and A. Schiffauerova. 2016. How to boost scientific production? A statistical analysis of research funding and other influencing factors. *Scientometrics* 106:1093–1116.

Fang, X., Y. Shen, J. Zhao, X. Bao and Y. Qu. 2010. Status and prospect of lignocellulosic bioethanol production in China. *Bioresource Technology* 101:4814–4819.

Fauci, A. S., H. C. Lane and R. R. Redfield. 2020. Covid-19-navigating the uncharted. *New England Journal of Medicine* 382:1268–1269.

Fernando, S., S. Adhikari, C. Chandrapal and M. Murali. 2006. Biorefineries: Current status, challenges, and future direction. *Energy & Fuels* 20:1727–1737.

Formela, K., A. Hejna, L. Piszczyk, M. R. Saeb and X. Colom. 2016. Processing and structure-property relationships of natural rubber/wheat bran biocomposites. *Cellulose* 23:3157–3175.

Fu, Y., D. Sun and Y. Chen, et al. 2012. An amine-functionalized titanium metal-organic framework photocatalyst with visible-light-induced activity for CO_2 reduction. *Angewandte Chemie - International Edition* 51:3364–3367.

Gao, S., Y. Lin and X. Jiao, et al. 2016. Partially oxidized atomic cobalt layers for carbon dioxide electroreduction to liquid fuel. *Nature* 529:68–71.

Gao, Z., H. Zhao, Z. Li, X. Tan and X. Lu. 2012. Photosynthetic production of ethanol from carbon dioxide in genetically engineered cyanobacteria. *Energy and Environmental Science* 5:9857–9865.

Garfield, E. 1955. Citation indexes for science. *Science* 122:108–111.

Gillon, S. 2010. Fields of dreams: Negotiating an ethanol agenda in the Midwest United States. *Journal of Peasant Studies* 37:723–748.

Gnansounou, E. 2010. Production and use of lignocellulosic bioethanol in Europe: Current situation and perspectives. *Bioresource Technology* 101:4842–4850.

Gong, J., H. Yue and Y. Zhao, et al. 2012. Synthesis of ethanol via syngas on Cu/SiO_2 catalysts with balanced Cu^0-Cu^+ sites. *Journal of the American Chemical Society* 134:13922–13925.

Habisreutinger, S. N., L. Schmidt-Mende and J. K. Stolarczyk. 2013. Photocatalytic reduction of CO_2 on TiO_2 and other semiconductors. *Angewandte Chemie - International Edition* 52:7372–7408.

Hamilton, J. D. 1983. Oil and the macroeconomy since World War II. *Journal of Political Economy* 91:228–248.

Hamilton, J. D. 2003. What is an oil shock? *Journal of Econometrics* 113:363–398.

Hamilton, J. D. 2009. Causes and consequences of the oil shock of 2007-08. *Brookings Papers on Economic Activity* 2009:215–261.

Hill, J., E. Nelson, D. Tilman, S. Polasky and D. Tiffany. 2006. Environmental, economic, and energetic costs and benefits of biodiesel and ethanol biofuels. *Proceedings of the National Academy of Sciences of the United States of America* 103:11206–11210.

Hill, J., S. Polasky and E. Nelson, et al. 2009. Climate change and health costs of air emissions from biofuels and gasoline. *Proceedings of the National Academy of Sciences of the United States of America* 106:2077–2082.

Hirano, A., R. Ueda, S. Hirayama and Y. Ogushi. 1997. CO_2 fixation and ethanol production with microalgal photosynthesis and intracellular anaerobic fermentation. *Energy* 22:137–142.

Hoang, T. T. H., S. Verma and S. Ma, et al. 2018. Nanoporous copper-silver alloys by additive-controlled electrodeposition for the selective electroreduction of CO_2 to ethylene and ethanol. *Journal of the American Chemical Society* 140:5791–5797.

Hsieh, W. D., R. H. Chen, T. L. Wu and T. H. Lin. 2002. Engine performance and pollutant emission of an SI engine using ethanol-gasoline blended fuels. *Atmospheric Environment* 36:403–410.

Hu, J., S. Jin and Q. Shen, et al. 2016. Cobalt carbide nanoprisms for direct production of lower olefins from syngas. *Nature* 538:84–87.

Huang, H. J., S. Ramaswamy, U. W. Tschirner and B. V. Ramarao. 2008. A review of separation technologies in current and future biorefineries. *Separation and Purification Technology* 62:1–21.

Inoue, T., A. Fujishima, S. Konishi and K. Honda. 1979. Photoelectrocatalytic reduction of carbon dioxide in aqueous suspensions of semiconductor powders. *Nature* 277:637–638.

Jones, T. C. 2012. America, oil, and war in the Middle East. *Journal of American History* 99:208–218.

Kerr, R. A. 2007. Global warming is changing the world. *Science* 316:188–190.

Kilian, L. 2008. Exogenous oil supply shocks: How big are they and how much do they matter for the US economy? *Review of Economics and Statistics* 90:216–240.

Kilian, L. 2009. Not all oil price shocks are alike: Disentangling demand and supply shocks in the crude oil market. *American Economic Review*, 99:1053–1069.

Kim, J. 2018. Evaluating author name disambiguation for digital libraries: A case of DBLP. *Scientometrics* 116:1867–1886.

Konur, O. 2000. Creating enforceable civil rights for disabled students in higher education: An institutional theory perspective. *Disability & Society* 15:1041–1063.

Konur, O. 2002a. Access to nursing education by disabled students: Rights and duties of nursing programs. *Nurse Education Today* 22:364–374.

Konur, O. 2002b. Assessment of disabled students in higher education: Current public policy issues. *Assessment and Evaluation in Higher Education* 27:131–152.

Konur, O. 2002c. Access to employment by disabled people in the UK: Is the Disability Discrimination Act working? *International Journal of Discrimination and the Law* 5:247–279.

Konur, O. 2006a. Participation of children with dyslexia in compulsory education: Current public policy issues. *Dyslexia* 12:51–67.

Konur, O. 2006b. Teaching disabled students in higher education. *Teaching in Higher Education* 11:351–363.

Konur, O. 2007a. A judicial outcome analysis of the *Disability Discrimination Act*: A windfall for the employers? *Disability & Society* 22:187–204.

Konur, O. 2007b. Computer-assisted teaching and assessment of disabled students in higher education: The interface between academic standards and disability rights. *Journal of Computer Assisted Learning* 23:207–219.

Konur, O. 2011. The scientometric evaluation of the research on the algae and bio-energy. *Applied Energy* 88:3532–3540.

Konur, O. 2012a. The evaluation of the biogas research: A scientometric approach. *Energy Education Science and Technology Part A: Energy Science and Research* 29:1277–1292.

Konur, O. 2012b. The evaluation of the educational research: A scientometric approach. *Energy Education Science and Technology Part B: Social and Educational Studies* 4:1935–1948.

Konur, O. 2012c. The evaluation of the global energy and fuels research: A scientometric approach. *Energy Education Science and Technology Part A: Energy Science and Research* 30:613–628.

Konur, O. 2012d. The evaluation of the research on the biodiesel: A scientometric approach. *Energy Education Science and Technology Part A: Energy Science and Research* 28:1003–1014.

Konur, O. 2012e. The evaluation of the research on the bioethanol: A scientometric approach. *Energy Education Science and Technology Part A: Energy Science and Research* 28:1051–1064.

Konur, O. 2012f. The evaluation of the research on the biofuels: A scientometric approach. *Energy Education Science and Technology Part A: Energy Science and Research* 28:903–916.

Konur, O. 2012g. The evaluation of the research on the biohydrogen: A scientometric approach. *Energy Education Science and Technology Part A: Energy Science and Research* 29:323–338.

Konur, O. 2012h. The evaluation of the research on the microbial fuel cells: A scientometric approach. *Energy Education Science and Technology Part A: Energy Science and Research* 29:309–322.

Konur, O. 2012i. The scientometric evaluation of the research on the production of bioenergy from biomass. *Biomass and Bioenergy* 47:504–515.

Konur, O. 2015. Current state of research on algal bioethanol. In *Marine Bioenergy: Trends and Developments*, Ed. S. K. Kim and C. G. Lee, pp. 217–244. Boca Raton, FL: CRC Press.

Konur, O., Ed. 2018a. *Bioenergy and Biofuels*. Boca Raton, FL: CRC Press.

Konur, O. 2018b. Bioenergy and biofuels science and technology: Scientometric overview and citation classics. In *Bioenergy and Biofuels*, Ed. O. Konur, pp. 3–63. Boca Raton: CRC Press.

Konur, O. 2019. Cyanobacterial bioenergy and biofuels science and technology: A scientometric overview. In *Cyanobacteria: From Basic Science to Applications*, Ed. A. K. Mishra, D. N. Tiwari and A. N. Rai, pp. 419–442. Amsterdam: Elsevier.

Konur, O. 2020a. The scientometric analysis of the research on the bioethanol production from green macroalgae. In *Handbook of Algal Science, Technology and Medicine*, Ed. O. Konur, pp. 385–401. London: Academic Press.

Konur, O., Ed. 2020b. *Handbook of Algal Science, Technology and Medicine*. London: Academic Press.

Konur, O., Ed. 2021a. *Handbook of Biodiesel and Petrodiesel Fuels: Science, Technology, Health, and Environment*. Boca Raton, FL: CRC Press.

Konur, O., Ed. 2021b. *Handbook of Biodiesel and Petrodiesel Fuels: Science, Technology, Health, and Environment. Volume 1. Biodiesel Fuels: Science, Technology, Health, and Environment*. Boca Raton, FL: CRC Press.

Konur, O., Ed. 2021c. *Handbook of Biodiesel and Petrodiesel Fuels: Science, Technology, Health, and Environment. Volume 2. Biodiesel Fuels based on the Edible and Nonedible Feedstocks, Wastes, and Algae: Science, Technology, Health, and Environment*. Boca Raton, FL: CRC Press.

Konur, O., Ed. 2021d. *Handbook of Biodiesel and Petrodiesel Fuels: Science, Technology, Health, and Environment. Volume 3. Petrodiesel Fuels: Science, Technology, Health, and Environment*. Boca Raton, FL: CRC Press.

Konur, O. and F. L. Matthews. 1989. Effect of the properties of the constituents on the fatigue performance of composites: A review. *Composites* 20:317–328.

Kopke, M., C. Held and S. Hujer, et al. 2010. *Clostridium ljungdahlii* represents a microbial production platform based on syngas. Proceedings of the National Academy of Sciences of the United States of America 107:13087–13092.

Kruyt, B., D. P. van Vuuren, H. J. de Vries and H. Groenenberg. 2009. Indicators for energy security. *Energy Policy* 37:2166–2181.

Leydesdorff, L. 2000. Is the European Union becoming a single publication system? *Scientometrics* 47:265–280.

Leydesdorff, L. and C. Wagner. 2009. Is the United States losing ground in science? A global perspective on the world science system. *Scientometrics* 78:23–36.

Leydesdorff, L. and P. Zhou. 2005. Are the contributions of China and Korea upsetting the world system of science? *Scientometrics* 63:617–630.

Li, C. W. and M. W. Kanan. 2012. CO_2 reduction at low overpotential on Cu electrodes resulting from the reduction of thick Cu_2O films. *Journal of the American Chemical Society* 134:7231–7234.

Li, H., S. M. Liu, X. H. Yu, S. L. Tang and C. K. Tang. 2020. Coronavirus disease 2019 (COVID-19): Current status and future perspectives. *International Journal of Antimicrobial Agents* 55:105951.

Li, J., Y. Zhu and D. Xiao, et al. 2016. Selective conversion of syngas to light olefins. *Science* 351:1065–1068.

Liu, Y., B. Huang and Y. Dai, et al. 2009. Selective ethanol formation from photocatalytic reduction of carbon dioxide in water with $BiVO_4$ photocatalyst. *Catalysis Communications* 11:210–213.

Ma, X., L. Sun and C. Song. 2002. A new approach to deep desulfurization of gasoline, diesel fuel and jet fuel by selective adsorption for ultra-clean fuels and for fuel cell applications. *Catalysis Today* 77:107–116.

Morschbacker, A. 2009. Bio-ethanol based ethylene. *Polymer Reviews* 49:79–84.

Najafi, G., B. Ghobadian and T. Tavakoli, et al. 2009. Performance and exhaust emissions of a gasoline engine with ethanol blended gasoline fuels using artificial neural network. *Applied Energy* 86:630–639.

Newman, P. W. G. and J. R. Kenworthy. 1989. Gasoline consumption and cities: A comparison of U.S. cities with a global survey. *Journal of the American Planning Association* 55:24–37.

North, D. C. 1991. Institutions. *Journal of Economic Perspectives* 5:97–112.

Pan, X., Z. Fan and W. Chen, et al. 2007. Enhanced ethanol production inside carbon-nanotube reactors containing catalytic particles. *Nature Materials* 6:507–511.

Peterson, A. A., F. Abild-Pedersen, F. Studt, J. Rossmeisl and J. K. Norskov. 2010. How copper catalyzes the electroreduction of carbon dioxide into hydrocarbon fuels. *Energy and Environmental Science* 3:1311–1315.

Porosoff, M. D., B. Yan and J. G. Chen. 2016. Catalytic reduction of CO_2 by H_2 for synthesis of CO, methanol and hydrocarbons: Challenges and opportunities. *Energy and Environmental Science* 9:62–73.

Rajagopalan, S., R. P. Datar and R. S. Lewis. 2002. Formation of ethanol from carbon monoxide via a new microbial catalyst. *Biomass and Bioenergy* 23:487–493.

Reeves, S. 2014. To Russia with love: How moral arguments for a humanitarian intervention in Syria opened the door for an invasion of the Ukraine. *Michigan State University International Law Review* 23:199.

Ren, D., B. S. H. Ang and B. S. Yeo. 2016. Tuning the selectivity of carbon dioxide electroreduction toward ethanol on oxide-derived Cu_xZn catalysts. *ACS Catalysis* 6:8239–8247.

Ren, D., Y. Deng and A. D. Handoko, et al. 2015. Selective electrochemical reduction of carbon dioxide to ethylene and ethanol on copper(I) oxide catalysts. *ACS Catalysis* 5:2814–2821.

Sakakura, T., J. C. Choi and H. Yasuda. 2007. Transformation of carbon dioxide. *Chemical Reviews* 107:2365–2387.

Tang, D., W. Han, P. Li, X. Miao and J. Zhong. 2011. CO_2 biofixation and fatty acid composition of *Scenedesmus obliquus* and *Chlorella pyrenoidosa* in response to different CO_2 levels. *Bioresource Technology* 102:3071–3076.

Wang, W., S. Wang, X. Ma and J. Gong. 2011. Recent advances in catalytic hydrogenation of carbon dioxide. *Chemical Society Reviews* 40:3703–3727.

Winzer, C. 2012. Conceptualizing energy security. *Energy Policy* 46:36–48.

Younesi, H., G. Najafpour and A. R. Mohamed. 2005. Ethanol and acetate production from synthesis gas via fermentation processes using anaerobic bacterium, *Clostridium ljungdahlii*. *Biochemical Engineering Journal* 27:110–119.

58 Biosyngas-based Bioethanol Fuels

Review

Ozcan Konur
(Formerly) Ankara Yildirim Beyazit University

58.1 INTRODUCTION

Crude oil-based gasoline fuels (Ma et al., 2002; Newman and Kenworthy, 1989) have been widely used in the transportation sector since the 1920s. However, there have been great public concerns over the adverse environmental and human impact of these fuels (Hill et al., 2006, 2009). Hence, biomass-based bioethanol fuels (Hill et al., 2006; Konur, 2012, 2015, 2019, 2020) have increasingly been used in blending gasoline fuels (Hsieh et al., 2002; Najafi et al., 2009), in fuel cells (Antolini, 2007, 2009), and in biochemical production (Angelici et al., 2013; Morschbacker, 2009) in a biorefinery context (Fernando et al., 2006; Huang et al., 2008).

One of the most studied feedstocks for bioethanol fuels has been synthesis biogas (biosyngas) and its constituents such as carbon dioxide (CO_2), carbon monoxide (CO), and hydrogen (H_2). Research in the field of biosyngas-based bioethanol fuels has intensified in this context in the key research fronts of the electrochemical reduction of biosyngas (Hoang et al., 2018; Ren et al., 2015, 2016), catalytic conversion of biosyngas (Choi and Liu, 2009; Gong et al., 2012; Pan et al., 2007), and the production of bioethanol fuels by other processes such as photosynthetic conversion of biosyngas (Gao et al., 2012; Hirano et al., 1997), microbial fermentation of biosyngas (Datar et al., 2004; Younesi et al., 2005), and photocatalytic conversion of biosyngas (Liu et al., 2009). Further, CO_2 (Hoang et al., 2018; Ren et al., 2015, 2016), biosyngas (Choi and Liu, 2009; Datar et al., 2004; Gong et al., 2012), and to a lesser extent CO (Abubackar et al., 2015; Bertheussen et al., 2016; Rajagopalan et al., 2002) have been studied intensively in this context.

However, it is essential to develop efficient incentive structures (North, 1991) for the primary stakeholders to enhance the research in this field (Konur, 2000, 2002a,b,c, 2006a,b, 2007a,b). Although there have been a number of review papers on biosyngas-based bioethanol fuels (Abubackar et al., 2011; Gupta et al., 2011; Subramani and Gangwal, 2008), there has been no review of the most cited 25 papers in this field.

Thus, this book chapter presents a review of the most cited 25 articles in the field of biosyngas-based bioethanol fuels. Then, it discusses the key findings of these highly influential papers and comments on future research priorities in this field.

58.2 MATERIALS AND METHODS

The search for this study was carried out using Scopus database (Burnham, 2006) in September 2022.

As a first step for the search of the relevant literature, the keywords were selected using the most cited first 100 population papers. The selected keyword list was then optimized to obtain a representative sample of papers for the searched research field. This final keyword set was provided in the appendix of Konur (2023) for future replication studies.

DOI: 10.1201/9781003226451-77

As a second step, a sample dataset was used for this study. The first 25 articles with at least 163 citations each were selected for the review study. Key findings from each paper were taken from the abstracts of these papers and were discussed. Additionally, a number of brief conclusions were drawn and a number of relevant recommendations were made to enhance the future research landscape.

58.3 RESULTS

The brief information about 25 most cited papers with at least 163 citations each on biosyngas-based bioethanol fuels is given below. The primary research fronts are the electrochemical reduction and catalytic conversion of biosyngas and production of bioethanol fuels by the other processes with 13, 7, and 5 highly cited papers (HCPs), respectively. Further, there are two, two, and one HCPs for the photosynthetic production of bioethanol fuels and photocatalytic reduction and microbial fermentation of biosyngas, respectively, for the last research front.

58.3.1 The Electrochemical Reduction of Biosyngas to Bioethanol Fuels

The brief information about 13 most cited papers on the electrochemical reduction of biosyngas to bioethanol fuels with at least 165 citations each is given in Table 58.1.

Ren et al. (2015) carried out the selective electrochemical reduction of CO_2 to ethylene and ethanol on copper(I) oxide (Cu_2O) catalysts at various electrochemical potentials in a paper with 568 citations. They used aqueous 0.1 M potassium bicarbonate ($KHCO_3$) as electrolyte. They observed that the Faradaic yields of ethylene and ethanol could be systematically tuned by changing the thickness of the deposited over-layers. Films 1.7–3.6 μm thick exhibited the best selectivity for these C_2 compounds at −0.99 V vs reversible hydrogen electrode (RHE), with Faradaic efficiencies (FE) of 34%–39% and 9%–16% for ethylene and ethanol, respectively. Less than 1% methane was formed while a high ethylene/methane products' ratio of up to ~100 could be achieved. The Cu_2O films reduced rapidly and remained as metallic Cu^0 particles during the CO_2 reduction. An increase in local pH at the surface of the electrode was not the only factor in enhancing the formation of C_2 products. An optimized surface population of edges and steps on the catalyst was also necessary to facilitate the dissociation of CO_2 and the dimerization of the pertinent CH_xO intermediates to ethylene and ethanol.

Hoang et al. (2018) developed nanoporous copper–silver (CuAg) alloys by additive-controlled electrodeposition for the selective electroreduction of CO_2 to ethylene and ethanol in a paper with 402 citations. They observed that the electrodeposition of CuAg alloy films from plating baths containing 3,5-diamino-1,2,4-triazole (DAT) as an inhibitor yielded high-surface-area catalysts for the active and selective electroreduction of CO_2 to multi-carbon hydrocarbons and oxygenates. The co-deposited alloy film was homogeneously mixed. The alloy film containing 6% Ag had the best CO_2 electroreduction performance, with the Faradaic efficiency for ethylene and ethanol production reaching nearly 60% and 25%, respectively, at a cathode potential of just −0.7 V vs RHE and a total current density of ~ −300 mA/cm². The origin of the high selectivity toward C_2 products was a combined effect of the enhanced stabilization of the Cu_2O over-layer and the optimal availability of the CO intermediate due to the Ag incorporated in the alloy.

Ren et al. (2016) tuned the selectivity of CO_2 electrochemical reduction toward ethanol on oxide-derived Cu–zinc (Cu_xZn) catalysts in a paper with 367 citations. They showed that the selectivity of CO_2 reduction toward ethanol could be tuned by introducing a co-catalyst (Zn) to generate an *in situ* source of mobile CO reactant. They prepared Cu-based oxides with different amounts of Zn dopants (Cu, $Cu_{10}Zn$, Cu_4Zn, and Cu_2Zn) and used them as catalysts under ambient pressure in aqueous 0.1 M $KHCO_3$ electrolyte. By varying the amount of Zn in the bimetallic catalysts, they found that the selectivity of ethanol versus ethylene production, defined by the ratio of their Faradaic efficiencies ($FE_{ethanol}/FE_{ethylene}$), could be tuned by a factor of up to ~12.5. Ethanol formation was

TABLE 58.1
The Electrochemical Reduction of Biosyngas to Bioethanol Fuels

No.	Papers	Feedstock	Catalysts	Parameters	Keywords	Lead Author	Affil.	Cits
1	Ren et al. (2015)	CO_2	Cu_2O	Electrochemical reduction, CO_2, Cu_2O catalysts	Ethanol, carbon dioxide, electrochemical	Yeo, Boon Siang 35099678500	Natl. Univ. Singapore, Singapore	568
2	Hoang et al. (2018)	CO_2	CuAg	Catalytic conversion, CO_2, CuAg catalysts, ethanol production efficiency and selectivity	Ethanol, electroreduction	Gewirth, Andrew A. 7006865533	Univ. Ill. U. C. USA	402
3	Ren et al. (2016)	CO_2	Cu_xZn	Electrochemical reduction, CO_2, Cu_xZn catalysts, ethanol selectivity	Ethanol, carbon dioxide, electroreduction	Yeo, Boon Siang 35099678500	Natl. Univ. Singapore, Singapore	367
4	Ma et al. (2020)	CO_2	F modified Cu	Electrochemical reduction, CO_2, F-modified Cu catalysts, ethanol selectivity	Ethanol, CO_2, electrocatalytic reduction	Zhang, Qinghong 57204367863	Xiamen Univ. China	306
5	Ma et al. (2016)	CO_2	Cu NP	Electrochemical reduction, CO_2, Cu NP catalysts, ethanol selectivity	Ethanol, CO_2, electrosynthesis	Yamauchi, Miho* 7201591670	Tohoku Univ. Japan	301
6	Song et al. (2017)	CO_2	N-doped mesoporous C	Ethanol, electrochemical reduction, CO_2, N-doped mesoporous C, ethanol selectivity	Ethanol, CO_2, electroreduction	Chen, Wei 57198582365	Chinese Acad. Sci. China	225
7	Song et al. (2016)	CO_2	Cu NP-N-doped graphene	Electrochemical reduction, Cu NP-N-doped graphene catalysts, ethanol selectivity	Ethanol, CO_2, electrochemical Conversion	Rondinone, Adam J. 6603096980	Oak Ridge Natl. Lab. USA	225
8	Lee et al. (2017)	CO_2	$Ag-Cu_2O$	Electrochemical reduction, CO_2, $Ag-Cu_2O$ catalysts, ethanol selectivity	Ethanol, electrochemical reduction, CO_2	Lee, Jaeyoung 57195131278	Gwangju Inst. Sci. Technol. S. Korea	193
9	Li et al. (2019)	CO_2	Ag-Cu	Electrochemical reduction, CO_2, Ag-Cu catalysts, ethanol selectivity	Ethanol, CO_2, electroreduction	Sargent, Edward H. 7102619565	Univ. Toronto Canada	191
10	Karapinar et al. (2019)	CO_2	Cu-N-doped C	Electrochemical reduction, CO_2, Cu-N-doped C catalysts, ethanol selectivity	Ethanol, CO_2, electroreduction	Mougel, Victor 35102991200	Swiss Fed. Inst. Technol. Zurich Switzerland	185
11	Li et al. (2020)	CO_2	Cu	Electrochemical reduction, CO_2, Cu catalysts, ethanol selectivity	Ethanol, conversion, CO_2	Sargent, Edward H. 7102619565	Univ. Toronto Canada	177
12	Wang et al. (2020)	CO_2	Cu-N-doped C	Electrochemical reduction, CO_2, Cu-N-doped C catalysts, ethanol selectivity, deoxygenation	Ethanol, deoxygenation, CO_2	Sargent, Edward H. 7102619565	Univ. Toronto Canada	171
13	Liu et al. (2017)	CO_2	BND	Electrochemical reduction, CO_2, BND catalysts, ethanol selectivity	Electrochemical reduction, carbon dioxide, ethanol	Quan, Xie 55533618700	Dalian Univ. Technol. China	165

*, Female; Cits., Number of citations received for each paper; Na: not available.

maximized on Cu_4Zn at $-1.05\,V$ vs RHE, with a remarkable FE and current density of 29.1% and $-8.2\,mA/cm^2$, respectively. The Cu_4Zn catalyst was also catalytically stable for the production of ethanol for at least 5 h. They emphasized the importance of Zn as a CO-producing site by performing CO_2 reduction on Cu–Ni and Cu–Ag bimetallic catalysts. The as-deposited Cu-based oxide films were reduced to the metallic state during CO_2 reduction, after which only signals belonging to CO adsorbed on Cu sites were recorded. This showed that the reduction of CO_2 probably occurred on metallic sites rather than on metal oxides. Finally, they proposed a two-site mechanism to rationalize the selective reduction of CO_2 to ethanol.

Ma et al. (2020) carried out the one-step electrosynthesis of ethylene and ethanol from CO_2 in an alkaline electrolyzer in a paper with 306 citations. They used a fluorine (F)-modified Cu catalyst that exhibited an ultrahigh current density of $1.6\,A/cm^2$ with a C_{2+} (mainly ethylene and ethanol) Faradaic efficiency of 80% for electrocatalytic CO_2 reduction in a flow cell. They found that the C_{2-4} selectivity reached 85.8% at a single-pass yield of 16.5%. They showed a hydrogen-assisted C–C coupling mechanism between adsorbed cyclohexene oxide (CHO) intermediates for C_{2+} formation. F enhanced water activation, CO adsorption, and hydrogenation of adsorbed CO to CHO intermediate that could readily undergo coupling. In conclusion, F-modified Cu was highly active and selective CO_2 electroreduction catalyst.

Ma et al. (2016) carried out the one-step electrosynthesis of ethylene and ethanol from CO_2 in an alkaline electrolyzer in a paper with 301 citations. They synthesized four Cu nanoparticle (NP) catalysts of different morphologies and compositions (amount of surface oxide). They observed that the use of catalysts with large surface roughness resulted in a combined Faradaic efficiency (46%) for the electroreduction of CO_2 to ethylene and ethanol in combination with current densities of $\sim200\,mA/cm^2$. They attributed the high production levels of ethylene and ethanol mainly to the use of alkaline electrolyte to improve kinetics and the suppressed evolution of H_2, as well as the application of gas diffusion electrodes covered with active and rough Cu NPs in the electrolyzer.

Song et al. (2017) synthesized ethanol by CO_2 electrochemical reduction over a nitrogen (N)-doped ordered cylindrical mesoporous carbon (C) in a paper with 225 citations. They obtained nearly 100% selectivity toward ethanol with a high Faradaic efficiency of 77% at $-0.56\,V$ vs. RHE. They attributed this remarkable performance to the synergetic effect of the N heteroatoms and the cylindrical channel configurations as they both facilitate the dimerization of key CO* intermediates and the subsequent proton–electron transfers, resulting in superior electrocatalytic performance for synthesizing ethanol from CO_2.

Song et al. (2016) explored the high-selectivity electrochemical conversion of CO_2 to ethanol using a Cu NP/N-doped graphene electrode in a paper with 225 citations. They developed this nanostructured catalyst for the direct electrochemical conversion of CO_2 to ethanol with high Faradaic efficiency (63% at $-1.2\,V$ vs RHE) and high selectivity (84%) that operated in water and at ambient temperature and pressure. This catalyst was comprised of Cu NPs on a highly textured, N-doped graphene film. CO_2 was electrochemically reduced to CO, which dimerized while the electrochemical reduction of the dimer yielded ethanol with an overall Faradaic efficiency of 63%. The active sites on the Cu NPs and the graphene films worked in tandem to control the electrochemical reduction of the CO dimer to alcohol.

Lee et al. (2017) explored the Ag–Cu biphasic boundaries for selective electrochemical reduction of CO_2 to ethanol in a paper with 193 citations. They developed a Ag-incorporated cuprous oxide ($Ag-Cu_2O$) electrode. They observed that the incorporation of Ag into Cu_2O led to the suppression of hydrogen (H_2) evolution. Furthermore, by varying the elemental arrangement (phase-separated and phase-blended) of Ag and Cu, they further found that ethanol selectivity could be controlled. Consequently, the Faradaic efficiency for ethanol on phase-blended $Ag-Cu_2O$ ($Ag-Cu_2OPB$) was three times higher than that of the Cu_2O without the Ag dopant. The electrochemical reaction behavior was not solely associated with a role of the Ag dopant, CO leading to an ethanol formation pathway over ethylene, but also the doping pattern related population of Ag–Cu biphasic boundaries relatively suppressed the H_2 evolution reaction and encouraged the reaction of mobile CO generated on Ag to a residual intermediate on a Cu site.

Li et al. (2019) developed a catalyst with a binding site diversity to promote CO_2 electroreduction to ethanol in a paper with 191 citations. They introduced diverse binding sites to a Cu catalyst to destabilize the ethylene reaction intermediates and thereby promote ethanol production. They then developed a bimetallic Ag–Cu catalyst. They obtained a record Faradaic efficiency of 41% toward ethanol at $250\,mA/cm^2$ and $-0.67\,V$ vs RHE, leading to a cathodic-side (half-cell) energy efficiency of 24.7%. They confirmed the diversity of binding configurations. This physical picture, involving multisite binding, accounted for the enhanced ethanol production for bimetallic catalysts.

Karapinar et al. (2019) explored the electroreduction of CO_2 on single-site Cu–N-doped C material for the selective formation of ethanol in a paper with 185 citations. They prepared this Cu–N–C material via a simple pyrolytic route that exclusively featured single Cu atoms with a CuN_4 coordination environment, atomically dispersed in a N-doped conductive C matrix. They observed that this catalyst achieved aqueous CO_2 electroreduction to ethanol at a Faradaic yield of 55% under optimized conditions (electrolyte: $0.1\,m\ CsHCO_3$, potential: $-1.2\,V$ vs. RHE and gas-phase recycling setup), as well as CO electroreduction to ethanol and ethylene with a Faradaic yield of 80%. During electrolysis, the isolated sites transiently converted into metallic Cu NPs, which were likely the catalytically active species. Remarkably, this process was reversible and the initial material was recovered intact after electrolysis. In conclusion, this catalyst was highly selective for CO_2 electroreduction to ethanol, and the active species during electrolysis were transient small Cu NPs.

Li et al. (2020) explored the cooperative CO_2-to-ethanol conversion via enriched intermediates at molecule–metal catalyst interfaces in a paper with 177 citations. They presented a cooperative catalyst design of molecule–metal catalyst interfaces with the goal of producing a reaction-intermediate-rich local environment, which improved the electrosynthesis of ethanol from CO_2 and H_2O. They implemented this strategy by functionalizing the Cu surface with a family of porphyrin-based metallic complexes that catalyzed CO_2 to CO. They found that the high concentration of local CO facilitated C–C coupling and steered the reaction pathway toward ethanol. They had a CO_2-to-ethanol Faradaic efficiency of 41% and a partial current density of $124\,mA/cm^2$ at $-0.82\,V$ versus the RHE. They finally integrated this catalyst into a membrane electrode assembly-based system and achieved an overall energy efficiency of 13%.

Wang et al. (2020) produced ethanol from CO_2 via suppression of deoxygenation in a paper with 171 citations. They developed a class of catalysts that achieved an ethanol Faradaic efficiency of 52% and an ethanol cathodic energy efficiency of 31%. For this purpose, they exploited the fact that suppression of the deoxygenation of the intermediate HOCCH* to ethylene promoted ethanol production and hence that confinement using capping layers having strong electron-donating ability on active catalysts promoted C–C coupling and increased the reaction energy of HOCCH* deoxygenation. Thus, they developed an electrocatalyst with confined reaction volume by coating Cu catalysts with N-doped C. The strong electron-donating ability and confinement of the nitrogen-doped C layers led to the observed pronounced selectivity toward ethanol.

Liu et al. (2017) explored the selective electrochemical reduction of CO_2 to ethanol on a boron (B)- and N-co-doped nanodiamond (BND) in a paper with 171 citations. They obtained good ethanol selectivity on the BND with high Faradaic efficiency of 93.2% ($-1.0\,V$ vs. RHE), which overcame the limitation of low selectivity for multi-C or high-heating-value fuels. Its superior performance mainly originated from the synergistic effect of B and N co-doping, high N content, and overpotential for hydrogen evolution. The possible pathway for CO_2 reduction was CO2→ *COOH→ *CO→ *COCO→ *COCH$_2$OH→ *CH$_2$OCH2OH→ CH$_3$CH$_2$OH. In conclusion, they obtained efficient and selective electrochemical reduction of CO_2 to ethanol on the BND electrocatalyst as they used the synergistic effect of co-doping, N content, and H_2 evolution potential as key factors for tailoring ethanol selectivity.

58.3.2 The Catalytic Conversion of Biosyngas to Bioethanol Fuels

The brief information about seven most cited papers on the catalytic conversion of biosyngas to bioethanol fuels with at least 163 citations each is given in Table 58.2.

Pan et al. (2007) produced bioethanol inside carbon nanotube (CNT) reactors containing catalytic rhodium (Rh) particles in a paper with 814 citations. They enhanced the catalytic activity of Rh particles confined inside CNTs for the conversion of CO and H_2 (biosyngas) to bioethanol. They observed that the overall formation rate of ethanol (30.0 mol/mol Rh/h) inside the nanotubes

TABLE 58.2
The Catalytic Conversion of Biosyngas to Bioethanol Fuels

No.	Papers	Feedstock	Catalysts	Parameters	Keywords	Lead Author	Affil.	Cits
1	Pan et al. (2007)	CO, H_2	Rh	Catalytic conversion, CO, H_2, CNT reactors, ethanol formation rate, Rh catalysts	Ethanol, catalytic	Pan, Xiulian* 7401930966	Chinese Acad. Sci. China	814
2	Gong et al. (2012)	Biosyngas	Cu/SiO_2	Catalytic conversion, Cu/SiO_2 catalysts, catalytic activity, CU site densities, ethanol selectivity	Ethanol, biosyngas	Ma, Xinbin 7404550196	Tianjin Univ. China	523
3	Choi and Liu (2009)	Biosyngas	Rh(III)	Catalytic conversion, biosyngas, Rh(III) catalysts, ethanol selectivity mechanisms, methane formation, C–C bond formation	Ethanol, biosyngas, synthesis	Liu, Ping* 57220877686	Brookhaven Natl. Lab. USA	291
4	Bai et al. (2017)	CO_2	Pd-Cu NPs	Catalytic conversion, CO_2, hydrogenation,	Ethanol, CO_2, hydrogenation	Huang, Xiaoqing 56515570400	Xiamen Univ. China	232
5	Haider et al. (2009)	Biosyngas	Rh/SiO_2-TiO_2	Catalytic conversion, biosyngas, Rh catalyst, support effect, Fe promoter	Ethanol, biosyngas, conversion	Davis, Robert J. 55482764100	Univ. Minnesota USA	199
6	Prieto et al. (2014)	Biosyngas	Cu-Co NP	Catalytic conversion, Cu-Co NP catalysts, ethanol selectivity	Ethanol synthesis gas, conversion	De Jong, Petra H.* 6603809695	Utrecht Univ. Netherlands	189
7	Mei et al. (2010)	Biosyngas	Rh-Mn/SiO_2	Catalytic conversion, Rh-Mn/SiO_2 catalysts, ethanol selectivity, CO hydrogenation	Ethanol synthesis, biosyngas	Mei, Donghai 7005968630	Tiangong Univ. China	163

*, Female; Cits., number of citations received for each paper.

exceeded that on the outside of the nanotubes by more than an order of magnitude, although the latter was much more accessible.

Gong et al. (2012) synthesized ethanol from biosyngas on Cu/silicon dioxide (SiO_2) catalysts with balanced Cu^0-Cu^+ sites with 83% yield in a paper with 523 citations. They attributed the remarkable stability and efficiency of this catalyst to the unique lamellar structure and the cooperative effect between surface Cu^0 and Cu^+ obtained by an ammonia evaporation hydrothermal method. They observed that the Cu^0 and Cu^+ were formed during the reduction process, originating from well-dispersed copper oxide (CuO) and copper phyllosilicate, respectively. As there was a correlation between the catalytic activity and the Cu^0 and Cu^+ site densities, Cu^0 could be the sole active site and primarily responsible for the activity of the catalyst. Further, they showed that the selectivity for ethanol or ethylene glycol could be tuned simply by regulating the reaction temperature.

Choi and Liu (2009) explored the mechanism of ethanol synthesis from biosyngas on Rh(111) catalysts in a paper with 291 citations. They performed calculations based on density functional theory (DFT) to investigate the complex ethanol synthesis on Rh(111). They found that ethanol synthesis on Rh(111) started with formyl formation from CO hydrogenation, followed by subsequent hydrogenation reactions and CO insertion. Three major products were involved in this process: methane, methanol, and ethanol, where the ethanol productivity was low and Rh(111) was highly selective to methane rather than ethanol or methanol. The rate-limiting step of the overall conversion was the hydrogenation of CO to formyl species, while the selectivity to ethanol was controlled by methane formation and C–C bond formation between methyl species and CO. The strong Rh–CO interaction impeded the CO hydrogenation and therefore slowed down the overall reaction; however, its high affinity to methyl, oxygen, and acetyl species indeed helped the C–O bond breaking of methoxy species and therefore the direct ethanol synthesis via CO insertion. In conclusion, to achieve high productivity and selectivity for ethanol, Rh has to get help from the promoters, which should be able to suppress methane formation and/or boost C–C bond formation.

Bai et al. (2017) carried out the highly active and selective hydrogenation of CO_2 to ethanol by highly ordered palladium (Pd) –Cu NPs in a paper with 232 citations. By tuning the composition of the Pd–Cu NPs and catalyst supports, they optimized the efficiency of CO_2 hydrogenation to ethanol with Pd_2Cu NPs/P_{25} exhibiting high selectivity to ethanol of up to 92.0% and the highest turnover frequency of 359.0 h^{-1}. They attributed the high ethanol production and selectivity of this catalyst boosting *CO (adsorption CO) hydrogenation to *HCO, the rate-determining step for the CO_2 hydrogenation to ethanol.

Haider et al. (2009) explored the effect of support (SiO_2 or titania (TiO_2)) and loading of iron (Fe) promoter on the activity and selectivity of Rh-based catalysts for the direct synthesis of ethanol from biosyngas in a paper with 199 citations. They performed this reaction in a fixed-bed reactor system typically operating at 543 K, 20 atm, weight hourly space velocity (WHSV) of 8,000 cm^3/g_{cat}/h and H_2:CO ratio of 1:1. The Rh was very highly dispersed on the supports and was in direct contact with the Fe promoter. Although little ethanol was produced over 2 wt% Rh on silica, a similar loading of Rh on titania was active for this reaction. Promotion of 2 wt% Rh/SiO_2 by 1 wt% Fe produced a catalyst that exhibited a 22% selectivity to ethanol, with methane being the primary side product. Addition of Fe to 2 wt% Rh/titania also improved the selectivity to ethanol with the highest selectivity being 37% for a sample with 5 wt% Fe. They also explored the effects of temperature, pressure, and H_2:CO ratio on the performance of 2 wt% Rh/TiO_2 and 2 wt% Rh–2.5 wt% Fe/TiO_2. Although the influence of pressure and H_2:CO ratio was moderate, higher temperatures clearly increased methane production at the expense of ethanol and methanol. They then studied the adsorption and thermal desorption of CO in argon (Ar) or H_2 on 2 wt% Rh/TiO_2 and 2 wt% Rh–2.5 wt% Fe/TiO_2. The gem-dicarbonyl species that was the primary species on these catalysts at room temperature after exposure to CO was more thermally stable on the Fe-promoted catalyst.

Prieto et al. (2014) synthesized Cu–cobalt (Cu-Co) NP/molybdate (MoO_x) catalysts for the selective conversion of biosyngas (CO+H_2) to ethanol and higher alcohols in a paper with 189 citations. They identified mixed Cu–Co alloy sites, at Co-enriched surfaces, as ideal for the selective production of long-chain alcohols. Accordingly, they developed a versatile synthesis route based on metal NP exsolution from a molybdate precursor compound whose crystalline structure isomorphically accommodated Cu^{2+} and Co^{2+} cations in a wide range of compositions. Superior mixing of Cu and Co species promoted formation of CuCo alloy nanocrystals after activation, leading to two orders of magnitude higher yield to high alcohols than a benchmark CuCoCr catalyst. The yield to high alcohols was maximized in parallel to the CuCo alloy contribution, for Co-rich surface compositions, for which Cu phase segregation was prevented. In conclusion, maximizing the contribution from mixed Cu–Co sites, while preventing Cu phase segregation, resulted in superior yields to high alcohols.

Mei et al. (2010) explored the ethanol synthesis from biosyngas over Rh-based/SiO_2 catalysts in a paper with 163 citations. They focused the reaction kinetics of ethanol synthesis from CO hydrogenation over SiO_2-supported Rh/manganese (Mn) NP alloy catalysts. They found that a Mn promoter could exist in a binary alloy with Rh and play a critical role in lowering the CO insertion reaction barriers, thus improving the selectivity toward ethanol. The binary Rh/Mn alloy was thermodynamically more stable than the mixed metal/metal oxides under the reducing reaction condition. Finally, they explored the effects of various promoters (M=Ir, Ga, V, Ti, Sc, Ca, and Li) on the CO insertion reaction over Rh/M alloy NPs. They found that alloying the promoters with the electronegativity difference, $\Delta\chi$, between the promoter (M) and Rh being 0.7 was most effective in lowering the barriers of CO insertion reaction, which led to higher selectivity to ethanol. In conclusion, the electronegativity difference criterion was very useful in improving the catalytic performance using transition metal-based catalysts for ethanol synthesis from CO hydrogenation.

58.3.3   THE PRODUCTION OF BIOSYNGAS-BASED BIOETHANOL FUELS
BY THE OTHER PROCESSES

The brief information about five most cited papers on the production of bioethanol fuels by the other processes with at least 172 citations each is given in Table 58.3.

58.3.3.1   The Photosynthetic Production of Bioethanol Fuels from Biosyngas

Gao et al. (2012) carried out the photosynthetic production of ethanol from CO_2 in genetically engineered cyanobacteria in a paper with 262 citations. They applied a consolidated bioprocessing strategy to integrate photosynthetic biomass production and microbial conversion producing ethanol together into *Synechocystis* sp. PCC6803, which could directly convert CO_2 to ethanol in one single biological system. They first constructed a *Synechocystis* sp. PCC6803 mutant strain with significantly high ethanol-producing efficiency (5.50 g/L, 212 mg/L/day) by genetically introducing pyruvate decarboxylase from *Zymomonas mobilis* and overexpressing endogenous alcohol dehydrogenase (ADH) through homologous recombination at two different sites of the chromosome, and disrupting the biosynthetic pathway of poly-β-hydroxybutyrate. In total, they cloned nine ADHs from different cyanobacterial strains and expressed in *Escherichia coli* to test ethanol-producing efficiency.

Hirano et al. (1997) explored the CO_2 fixation and ethanol production with microalgal photosynthesis and intracellular anaerobic fermentation in a paper with 189 citations. They extracted starch from *Chlorella vulgaris* (IAM C-534), saccharified, and fermented with yeasts with 65% of the ethanol conversion rate as compared to the theoretical rate from starch. They then obtained higher ethanol productions with *Chlamydomonas reinhardtii* (UTEX2247) and Sak-1 isolated from seawater under dark and anaerobic conditions. These showed a maximum ethanol concentration of 1 (w/w)%. In conclusion, intracellular ethanol production was simpler and less energy intensive than the conventional ethanol fermentation process.

58.3.3.2   The Microbial Fermentation of Biosyngas to Bioethanol Fuels

Datar et al. (2004) fermented biosyngas (primarily CO, CO_2, CH_4, H_2, and N_2) to ethanol in a paper with 182 citations. They generated the gas from switchgrass via gasification with a composition of

TABLE 58.3
The Production of Biosyngas-based Bioethanol Fuels by the Other Processes

No.	Papers	Feedstock	Res. Fronts	Catalysts	Parameters	Keywords	Lead Author	Affil.	Cits
1	Gao et al. (2012)	CO_2	Photosynthetic conversion	Synechocystis sp.	Photosynthetic conversion, CO_2, Synechocystis sp. genetic engineering	Ethanol, carbon dioxide, photosynthetic production, cyanobacteria	Lu, Xuefeng 55619293343	Chinese Acad. Sci. China	262
2	Liu et al. (2009)	CO_2	Photocatalytic reduction	$BiVO_4$	Photocatalytic reduction, CO_2, $BiVO_4$ catalysts	Ethanol, photocatalytic reduction, carbon dioxide	Huang, Baibiao 7403682200	Shandong Univ. China	237
3	Hirano et al. (1997)	CO_2	Photosynthetic conversion	C. vulgaris, C. reinhardtii	Photosynthetic conversion, CO_2, microalgae, ethanol productivity	Ethanol, photosynthesis, CO_2, fermentation	Hirano, Atsushi 57209632970	Tokyo Elect. Power Co. Japan	189
4	Datar et al. (2004)	Biosyngas	Fermentation	Na	Fermentation, biomass based biosyngas, ethanol production	Ethanol, producer gas, fermentation	Lewis, Randy S. 35262249500	Brigham Young Univ. USA	182
5	Younesi et al. (2005)	Biosyngas	Fermentation	Na	Fermentation, biomass based biosyngas, ethanol production, pressure effect	Ethanol, synthesis gas, fermentation	Najafpour, Ghasem 55485049700	Babol Noshirvani Univ. Technol. Iran	172

Cits., number of citations received for each paper; Na, not available.

56.8% N_2, 14.7% CO, 16.5% CO_2, 4.4% H_2, and 4.2% CH_4. They used this gas in a 4-L bioreactor to generate ethanol and other products via fermentation using a novel clostridial bacterium. The cells stopped growing but were still viable while ethanol was primarily produced once the cells stopped growing (ethanol was nongrowth associated). Further, H_2 utilization stopped while cells began growing again if 'clean' bottled gases were introduced following exposure to the producer gas.

Younesi et al. (2005) produced ethanol and acetate from biosyngas via fermentation processes using *Clostridium ljungdahlii* in a paper with 172 citations. They performed the experiments with various initial total pressures of biosyngas at 0.8–1.8 atm with 0.2 intervals. They found that formation of acetate was almost the same for all initial pressures as well as cell concentrations while ethanol concentration was promoted by H_2 and CO_2 in the culture media. They obtained the maximum acetate production (1.3 g/L) at biosyngas total pressures of 1.4 atm while they obtained maximum ethanol concentration of 0.6 g/L with the biosyngas total pressures of 1.6 and 1.8 atm. The cell and product yields were 0.3 g cell/g CO and 0.41 g products/g CO, respectively. The product ratio of ethanol and acetate was 0.54 g ethanol/g acetate at biosyngas total pressures of 1.6 and 1.8 atm. The maximum cell dry weight (xm), inhibition constant (k), and maximum specific growth rate were 1.2 g/L, 0.003 h^{-1}, and 0.07 h^{-1}, respectively. Finally, the inhibition constant was obtained at 2 mmol CO/L.

58.3.3.3   The Photocatalytic Reduction of Biosyngas to Bioethanol Fuels

Liu et al. (2009) carried out the selective ethanol formation from photocatalytic reduction of CO_2 in water with bismuth vanadate ($BiVO_4$) photocatalyst in a paper with 237 citations. They observed the selective formation of ethanol under the condition of high-intensity visible-light irradiation where the intense light irradiation generated a large number of C1 intermediate species anchored on the surface of $BiVO_4$, which dimerized to form ethanol. Further, for the photocatalytic reduction of CO_2 into ethanol, monoclinic $BiVO_4$ was more efficient than tetragonal $BiVO_4$.

58.4   DISCUSSION

58.4.1   INTRODUCTION

Crude oil-based gasoline fuels have been widely used in the transportation sector since the 1920s. However, there have been great public concerns over the adverse environmental and human impact of these fuels. Hence, biomass-based bioethanol fuels have increasingly been used in blending gasoline and petrodiesel fuels, in fuel cells, and in biochemical production in a biorefinery context.

One of the most studied feedstocks for bioethanol fuels has been biosyngas produced from biomass and its constituents such as CO_2, CO, and H_2. Research in the field of biosyngas-based bioethanol fuels has intensified in this context in the key research fronts of electrochemical reduction and catalytic conversion of biosyngas and the production of the bioethanol fuels by the other processes such as the photosynthetic and photocatalytic conversion and microbial fermentation of biosyngas. Further, CO_2, biosyngas, and CO have been studied intensively as feedstocks in this context.

However, it is essential to develop efficient incentive structures for the primary stakeholders to enhance the research in this field. Although there have been a number of review papers for this field, there has been no review of the most cited 25 articles in this field.

Thus, this book chapter presents a review of the most cited 25 articles on bioethanol fuel production and evaluation from biosyngas and its constituents. Then, it discusses the key findings of these highly influential papers and comments on future research priorities in this field.

As a first step for the search of the relevant literature, the keywords were selected using the most cited first 100 population papers. The selected keyword list was then optimized to obtain a representative sample of papers for the searched research field. This keyword list was provided in the appendix of Konur (2023) for future replicative studies.

As a second step, a sample data set was used for this study. The first 25 articles with at least 163 citations each were selected for the review study. Key findings from each paper were taken from the abstracts of these papers and were discussed. Additionally, a number of brief conclusions were drawn and a number of relevant recommendations were made to enhance the future research landscape.

Information about the research fronts for the sample papers in biosyngas-based bioethanol fuels with regard to feedstocks used in these processes is given in Table 58.4. As this table shows, there are three primary research fronts for this field: CO_2, biosyngas, and CO with 68%, 32%, and 0% of the HCPs, respectively. Further, CO_2 is the most influential feedstock with 28% surplus while biosyngas and CO are the least influential ones with 15% and 13% deficits.

Information about the thematic research fronts for the sample papers in biosyngas-based bioethanol fuels is given in Table 58.5. As this table shows, there are five research fronts for this field: electrochemical reduction, catalytic reduction, fermentation, photosynthetic production, and photocatalytic reduction with 52%, 28%, 8%, 8%, and 4% of the HCPs, respectively.

Further, the most influential research front is the electrochemical reduction of the biosyngas with 28% surplus. Similarly, fermentation of the biosyngas is the least influential research front with 28% deficit.

Finally, it is important to put the research on biosyngas-based bioethanol fuel production in the wider research field of the biosyngas transformation to fuels and value-added products with over 35,000 papers and book chapters (Hu et al., 2016; Kopke et al., 2010; Li et al., 2016). It is also closely related to the research field on the CO_2 utilization with over 80,000 papers and book chapters (Habisreutinger et al., 2013; Sakakura et al., 2007; Wang et al., 2011). Although the Cu- and Rh-based conventional catalysts (Gao et al., 2016; Li and Kanan, 2012) are used in these wider studies, the recent emphasis has been on the development and application of the nanocatalysts (Chen et al., 2012; Fu et al., 2012) as in this study.

In this context, catalytic hydrogenation (Wang et al., 2011), photocatalytic reduction (Habisreutinger et al., 2013), photoelectrocatalytic reduction (Inoue et al., 1979), electrochemical

TABLE 58.4
The Most Prolific Research Fronts for the Biosyngas-based Bioethanol Fuels

No.	Research Fronts	N Paper (%) Review	N Paper (%) Sample	Surplus (%)
1	CO_2	68.0	40.0	28.0
2	Biosyngas	32.0	47.0	−15.0
3	CO	0.0	13.0	−13.0

N Paper (%) review, the number of papers in the sample of 25 reviewed papers; N paper (%) sample, the number of papers in the population sample of 100 papers.

TABLE 58.5
The Most Prolific Thematic Research Fronts for the Residual Starch Feedstock-based Bioethanol Fuels

No.	Research Fronts	N Paper (%) Review	N Paper (%) Sample	Surplus (%)
1	Electrochemical reduction	52.0	24.0	28.0
2	Catalytic reduction	28.0	31.0	-3.0
3	Fermentation	8.0	36	-28.0
4	Photosynthetic production	8.0	5.0	3.0
5	Photocatalytic reduction	4.0	3.0	1.0

N Paper (%) review, the number of papers in the sample of 25 reviewed papers; N paper (%) sample, the number of papers in the population sample of 100 papers.

reduction (Peterson et al., 2010), catalytic reduction (Porosoff et al., 2016), microbial fermentation (Kopke et al., 2010), and photosynthetic conversion of the biosyngas (Tang et al., 2011) emerge as the key processes for the production of bioethanol fuels from biosyngas and its constituent, CO_2.

58.4.2 THE ELECTROCHEMICAL REDUCTION OF BIOSYNGAS TO BIOETHANOL FUELS

The brief information about 13 most cited papers on electrochemical reduction of biosyngas to bioethanol fuels with at least 165 citations each is given in Table 58.1.

These HCPs show a sample of the research on the production of bioethanol fuels from biosyngas and its constituent, CO_2. The development and application of electrocatalysts with high ethanol selectivity are the main focus of these papers.

Further, as bioethanol fuels have been used in the vehicles, blending them with gasoline or diesel fuels, the studies on the evaluation of these bioethanol fuels are a significant research front together with the research front of bioethanol fuel production. However, there are no such papers for this sample in this context.

Ren et al. (2015) carried out selective electrochemical reduction of CO_2 to ethylene and ethanol on Cu_2O catalysts at various electrochemical potentials and observed that the Faradaic yields of ethylene and ethanol could be systematically tuned by changing the thickness of the deposited over-layers. Further, Hoang et al. (2018) developed nanoporous CuAg alloys by additive-controlled electrodeposition for the selective electroreduction of CO_2 to ethylene and ethanol and observed that the electrodeposition of CuAg alloy films as an inhibitor yielded high-surface-area catalysts for the active and selective electroreduction of CO_2 to multi-C hydrocarbons and oxygenates.

Ren et al. (2016) tuned the selectivity of CO_2 electrochemical reduction toward ethanol on oxide-derived Cu_xZn catalysts and showed that the selectivity of CO_2 reduction toward ethanol could be tuned by introducing a co-catalyst (Zn) to generate an *in situ* source of a mobile CO reactant. Further, Ma et al. (2020) carried out the one-step electrosynthesis of ethylene and ethanol from CO_2 in an alkaline electrolyzer and observed an ethylene and ethanol Faradaic efficiency of 80% for electrocatalytic CO_2 reduction in a flow cell.

Ma et al. (2016) carried out the one-step electrosynthesis of ethylene and ethanol from CO_2 in an alkaline electrolyzer and observed that the use of catalysts with large surface roughness resulted in a combined Faradaic efficiency (46%) for the electroreduction of CO_2 to ethylene and ethanol. Further, Song et al. (2017) synthesized ethanol by CO_2 electrochemical reduction over a N-doped ordered cylindrical mesoporous C and obtained nearly 100% selectivity toward ethanol with a high Faradaic efficiency of 77% at −0.56 V vs. RHE.

Song et al. (2016) explored the high-selectivity electrochemical conversion of CO_2 to ethanol using a Cu NP/N-doped graphene electrode and observed high Faradaic efficiency and high selectivity (84%) that operated in water and at ambient temperature and pressure. Further, Lee et al. (2017) explored the Ag–Cu biphasic boundaries for selective electrochemical reduction of CO_2 to ethanol and observed that the incorporation of Ag into Cu_2O led to the suppression of H_2 evolution.

Li et al. (2019) developed a catalyst with a binding site diversity to promote CO_2 electroreduction to ethanol and obtained a record Faradaic efficiency of 41% toward ethanol, leading to a cathodic-side (half-cell) energy efficiency of 24.7%. Further, Karapinar et al. (2019) explored the electroreduction of CO_2 on single-site Cu-N-doped C catalyst for the selective formation of ethanol and observed that this catalyst achieved aqueous CO_2 electroreduction to ethanol at a Faradaic yield of 55% under optimized conditions.

Li et al. (2020) explored the cooperative CO_2-to-ethanol conversion via enriched intermediates at molecule–metal catalyst interfaces and had a CO_2-to-ethanol Faradaic efficiency of 41%. Further, Wang et al. (2020) produced ethanol from CO_2 via suppression of deoxygenation and achieved an ethanol Faradaic efficiency of 52% and an ethanol cathodic energy efficiency of 31%. Finally,

Liu et al. (2017) explored the selective electrochemical reduction of CO_2 to ethanol on a BND and obtained good ethanol selectivity on the BND with high Faradaic efficiency of 93.2 %.

58.4.3 The Catalytic Conversion of Biosyngas to Bioethanol Fuels

The brief information about seven most cited papers on catalytic conversion of biosyngas to bioethanol fuels with at least 163 citations each is given in Table 58.2.

These HCPs show a sample of the research on the production of bioethanol fuels from biosyngas and its constituent, CO_2. The development and application of electrocatalysts with high ethanol selectivity are the main focus of these papers.

Further, as bioethanol fuels have been used in the vehicles, blending them with gasoline or diesel fuels, the studies on the evaluation of these bioethanol fuels are a significant research front together with the research front of bioethanol fuel production. However, there are no such papers for this sample in this context.

Pan et al. (2007) produced bioethanol inside CNT reactors containing catalytic Rh particles and observed that the overall formation rate of ethanol inside the CNTs exceeded that on the outside of the CNTs by more than an order of magnitude, although the latter was much more accessible. Further, Gong et al. (2012) synthesized ethanol from biosyngas on Cu/SiO_2 catalysts with balanced Cu^0-Cu^+ sites with 83% yield.

Choi and Liu (2009) explored the mechanism of ethanol synthesis from biosyngas on Rh(111) catalysts and found that ethanol synthesis on Rh(111) started with formyl formation from CO hydrogenation, followed by subsequent hydrogenation reactions and CO insertion. Further, Bai et al. (2017) carried out highly active and selective hydrogenation of CO_2 to ethanol by highly ordered Pd-Cu NPs and observed high selectivity to ethanol of up to 92.0% and the highest turnover frequency.

Haider et al. (2009) explored the effect of support (SiO_2 or TiO_2) and loading of Fe promoter on the activity and selectivity of Rh-based catalysts for the direct synthesis of ethanol from biosyngas and observed that although little ethanol was produced over 2 wt% Rh on silica, a similar loading of Rh on TiO_2 was active for this reaction. Further, Prieto et al. (2014) synthesized Cu-Co NP/MoO_x catalysts for the selective conversion of biosyngas to ethanol and higher alcohols and observed two orders of magnitude higher yield to high alcohols than a benchmark CuCoCr catalyst. Finally, Mei et al. (2010) explored the ethanol synthesis from biomass-based biosyngas over Rh/SiO_2 catalysts and found that a Mn promoter could exist in a binary alloy with Rh and play a critical role in lowering the CO insertion reaction barriers, thus improving the selectivity toward ethanol.

58.4.4 The Production of Biosyngas-based Bioethanol Fuels by the Other Processes

The brief information about five most cited papers on the production of bioethanol fuels by the other processes with at least 172 citations each is given in Table 58.3. There are two, two, and one HCPs for the photosynthetic production of bioethanol fuels, microbial fermentation, and photocatalytic reduction of biosyngas, respectively.

These HCPs show a sample of research on the production of bioethanol fuels from biosyngas and its constituent, CO_2. The development and application of electrocatalysts, yeasts, and algae with high ethanol selectivity and productivity are the main focus of these papers.

Further, as bioethanol fuels have been used in the vehicles, blending them with gasoline or diesel fuels, the studies on the evaluation of these bioethanol fuels are a significant research front together with the research front of bioethanol fuel production. However, there are no such papers for this sample in this context.

58.4.4.1 The Photosynthetic Production of Bioethanol Fuels from Biosyngas

Gao et al. (2012) carried out photosynthetic production of ethanol from CO_2 in genetically engineered cyanobacteria and cloned nine ADHs from different cyanobacterial strains and expressed in *E. coli* to test ethanol-producing efficiency. Further, Hirano et al. (1997) explored the CO_2 fixation and ethanol production with microalgal photosynthesis and intracellular anaerobic fermentation and found that intracellular ethanol production was simpler and less energy intensive than the conventional ethanol fermentation process.

58.4.4.2 The Microbial Fermentation of Biosyngas to Bioethanol Fuels

Datar et al. (2004) fermented biosyngas to ethanol and found that the cells stopped growing but were still viable while ethanol was primarily produced once the cells stopped growing. Further, Younesi et al. (2005) produced ethanol and acetate from biosyngas via fermentation processes using *C. ljungdahlii* and obtained maximum ethanol concentration.

58.4.4.3 The Photocatalytic Reduction of Biosyngas to Bioethanol Fuels

Liu et al. (2009) carried out the selective ethanol formation from photocatalytic reduction of CO2 in water with BiVO4 photocatalyst and observed selective formation of ethanol under the condition of high-intensity visible-light irradiation.

58.5 CONCLUSION AND FUTURE RESEARCH

The brief information about the key research fronts covered by the 25 most cited papers with at least 163 citations each was given above under three primary headings: The electrochemical and catalytic reduction of the biosyngas and the production of biosyngas-based bioethanol fuels by other processes. The emphasis is on the CO_2-based bioethanol fuel production compared to the biosyngas-based bioethanol fuel production.

Similarly, the most prolific catalysts are the Cu and its alloys and to a lesser extent Rh, C, and Ag. Thus, the key issues for this sample is the development and application of catalysts for the CO_2- and biosyngas-based bioethanol fuel production to ensure high ethanol selectivity and productivity during the production processes. With the limited application of nanomaterials for this sample, there is ample room for the development of nanocatalysts with a remarkable ethanol sensitivity and productivity.

The key findings on these research fronts should be read in light of increasing public concerns about climate change, GHG emissions, and global warming as these concerns have been certainly behind the boom in the research on the biomass-based biosyngas-based bioethanol fuels as an alternative to crude oil-based gasoline and petrodiesel fuels in the last decades. It is also a sustainable alternative to food crop-based bioethanol fuels such as corn grain-based bioethanol fuels. The recent supply shocks caused by the COVID-19 pandemic and the Russian invasion of Ukraine also highlight the importance of the production and utilization of bioethanol fuels from biosyngas as an alternative to crude oil-based gasoline and petrodiesel fuels.

There are three research fronts for the feedstocks used for this field: CO_2 and biosyngas. It is notable that CO_2 is the most influential feedstock while biosyngas and CO are the least influential ones.

Similarly, there are five research fronts for this field: electrochemical reduction and catalytic reduction and to a lesser extent fermentation, photosynthetic production, and photocatalytic reduction of biosyngas. It is notable that electrochemical reduction of biosyngas has surplus while fermentation of the biosyngas has deficit.

These studies emphasize the importance of proper incentive structures for the efficient production of biosyngas-based bioethanol fuels in light of North's institutional framework (North, 1991). In this context, the major producers and users of bioethanol fuels such as the USA and Canada with vast farmlands have developed strong incentive structures for the efficient biosyngas-based bioethanol fuels. In light of the recent supply shocks caused primarily by the COVID-19 pandemic and Russian invasion of Ukraine, it is expected that the incentive structures such as public funding would be enhanced to increase the share of bioethanol fuels in the global fuel portfolio as a strong alternative to crude oil-based gasoline and petrodiesel fuels. In this context, it is expected that the most prolific researchers, institutions, countries, funding bodies, and journals in this field would have a first-mover advantage to benefit from such potential incentives. This is especially true for the US, Chinese, and European stakeholders as they have become the global leaders in both the production and utilization of bioethanol fuels from the biomass-based biosyngas and its constituent, CO_2.

It is recommended that such review studies are performed for the primary research fronts of biosyngas-based bioethanol fuels.

ACKNOWLEDGMENTS

The contribution of the highly cited researchers in the field of biosyngas-based bioethanol fuels has been gratefully acknowledged.

REFERENCES

Abubackar, H. N., M. C. Veiga and C. Kennes. 2011. Biological conversion of carbon monoxide: Rich syngas or waste gases to bioethanol. *Biofuels, Bioproducts and Biorefining* 5:93–114.

Abubackar, H. N., M. C. Veiga and C. Kennes. 2015. Carbon monoxide fermentation to ethanol by *Clostridium autoethanogenum* in a bioreactor with no accumulation of acetic acid. *Bioresource Technology* 186:122–127.

Angelici, C., B. M. Weckhuysen and P. C. A. Bruijnincx. 2013. Chemocatalytic conversion of ethanol into butadiene and other bulk chemicals. *ChemSusChem* 6:1595–1614.

Antolini, E. 2007. Catalysts for direct ethanol fuel cells. *Journal of Power Sources* 170:1–12.

Antolini, E. 2009. Palladium in fuel cell catalysis. *Energy and Environmental Science* 2:915–931.

Bai, S., Q. Shao and P. Wang, et al. 2017. Highly active and selective hydrogenation of CO_2 to ethanol by ordered Pd-Cu nanoparticles. *Journal of the American Chemical Society* 139:6827–6830.

Bertheussen, E., A. Verdaguer-Casadevall and D. Ravasio, et al. 2016. Acetaldehyde as an intermediate in the electroreduction of carbon monoxide to ethanol on oxide-derived copper. *Angewandte Chemie - International Edition* 55:1450–1454.

Burnham, J. F. 2006. Scopus database: A review. *Biomedical Digital Libraries* 3:1–8.

Chen, Y., C. W. Li and M. W. Kanan. 2012. Aqueous CO_2 reduction at very low overpotential on oxide-derived Au nanoparticles. *Journal of the American Chemical Society* 134:19969–19972.

Choi, Y. and P. Liu. 2009. Mechanism of ethanol synthesis from syngas on Rh(111). *Journal of the American Chemical Society* 131:13054–13061.

Datar, R. P., R. M. Shenkman, B. G. Cateni, R. I. Huhnke and R.S. Lewis. 2004. Fermentation of biomass-generated producer gas to ethanol. *Biotechnology and Bioengineering* 86:587–594.

Fernando, S., S. Adhikari, C. Chandrapal and M. Murali. 2006. Biorefineries: Current status, challenges, and future direction. *Energy & Fuels* 20:1727–1737.

Fu, Y., D. Sun and Y. Chen, et al. 2012. An amine-functionalized titanium metal-organic framework photocatalyst with visible-light-induced activity for CO_2 reduction. *Angewandte Chemie - International Edition* 51:3364–3367.

Gao, S., Y. Lin and X. Jiao, et al. 2016. Partially oxidized atomic cobalt layers for carbon dioxide electroreduction to liquid fuel. *Nature* 529:68–71.

Gao, Z., H. Zhao, Z. Li, X. Tan and X. Lu. 2012. Photosynthetic production of ethanol from carbon dioxide in genetically engineered cyanobacteria. *Energy and Environmental Science* 5:9857–9865.

Gong, J., H. Yue and Y. Zhao, et al. 2012. Synthesis of ethanol via syngas on Cu/SiO_2 catalysts with balanced Cu^0-Cu^+ sites. *Journal of the American Chemical Society* 134:13922–13925.

Gupta, M., M. L. Smith and J. J. Spivey. 2011. Heterogeneous catalytic conversion of dry syngas to ethanol and higher alcohols on Cu-based catalysts. *ACS Catalysis,* 1:641–656.

Habisreutinger, S. N., L. Schmidt-Mende and J. K. Stolarczyk. 2013. Photocatalytic reduction of CO_2 on TiO_2 and other semiconductors. *Angewandte Chemie - International Edition* 52:7372–7408.

Haider, M. A., M. R. Gogate and R. J. Davis. 2009. Fe-promotion of supported Rh catalysts for direct conversion of syngas to ethanol. *Journal of Catalysis* 261:9–16.

Hill, J., E. Nelson, D. Tilman, S. Polasky and D. Tiffany. 2006. Environmental, economic, and energetic costs and benefits of biodiesel and ethanol biofuels. *Proceedings of the National Academy of Sciences of the United States of America* 103:11206–11210.

Hill, J., S. Polasky and E. Nelson, et al. 2009. Climate change and health costs of air emissions from biofuels and gasoline. *Proceedings of the National Academy of Sciences of the United States of America* 106:2077–2082.

Hirano, A., R. Ueda, S. Hirayama and Y. Ogushi. 1997. CO_2 fixation and ethanol production with microalgal photosynthesis and intracellular anaerobic fermentation. *Energy* 22:137–142.

Hoang, T. T. H., S. Verma and S. Ma, et al. 2018. Nanoporous copper-silver alloys by additive-controlled electrodeposition for the selective electroreduction of CO_2 to ethylene and ethanol. *Journal of the American Chemical Society* 140:5791–5797.

Hsieh, W. D., R. H. Chen, T. L. Wu and T. H. Lin. 2002. Engine performance and pollutant emission of an SI engine using ethanol-gasoline blended fuels. *Atmospheric Environment* 36:403–410.

Hu, J., S. Jin and Q. Shen, et al. 2016. Cobalt carbide nanoprisms for direct production of lower olefins from syngas. *Nature* 538:84–87.

Huang, H. J., S. Ramaswamy, U. W. Tschirner and B. V. Ramarao. 2008. A review of separation technologies in current and future biorefineries. *Separation and Purification Technology* 62:1–21.

Inoue, T., A. Fujishima, S. Konishi and K. Honda. 1979. Photoelectrocatalytic reduction of carbon dioxide in aqueous suspensions of semiconductor powders. *Nature* 277:637–638.

Karapinar, D., N. T. Huan and N. R. Sahraie, et al. 2019. Electroreduction of CO_2 on single-site copper-nitrogen-doped carbon material: Selective formation of ethanol and reversible restructuring of the metal sites. *Angewandte Chemie - International Edition* 58:15098–15103.

Konur, O. 2000. Creating enforceable civil rights for disabled students in higher education: An institutional theory perspective. *Disability & Society* 15:1041–1063.

Konur, O. 2002a. Access to nursing education by disabled students: Rights and duties of nursing programs. *Nurse Education Today* 22:364–374.

Konur, O. 2002b. Assessment of disabled students in higher education: Current public policy issues. *Assessment and Evaluation in Higher Education* 27:131–52.

Konur, O. 2002c. Access to employment by disabled people in the UK: Is the Disability Discrimination Act working? *International Journal of Discrimination and the Law* 5:247–279.

Konur, O. 2006a. Participation of children with dyslexia in compulsory education: Current public policy issues. *Dyslexia* 12:51–67.

Konur, O. 2006b. Teaching disabled students in higher education. *Teaching in Higher Education* 11:351–363.

Konur, O. 2007a. A judicial outcome analysis of the *Disability Discrimination Act*: A windfall for the employers? *Disability & Society* 22:187–204.

Konur, O. 2007b. Computer-assisted teaching and assessment of disabled students in higher education: The interface between academic standards and disability rights. *Journal of Computer Assisted Learning* 23:207–219.

Konur, O. 2012. The evaluation of the research on the bioethanol: A scientometric approach. *Energy Education Science and Technology Part A: Energy Science and Research* 28:1051–1064.

Konur, O. 2015. Current state of research on algal bioethanol. In *Marine Bioenergy: Trends and Developments*, Ed. S. K. Kim and C. G. Lee, pp. 217–244. Boca Raton, FL: CRC Press.

Konur, O. 2019. Cyanobacterial bioenergy and biofuels science and technology: A scientometric overview. In *Cyanobacteria: From Basic Science to Applications*, Ed. A. K. Mishra, D. N. Tiwari and A. N. Rai, pp. 419–442. Amsterdam: Elsevier.

Konur, O. 2020. The scientometric analysis of the research on the bioethanol production from green macroalgae. In *Handbook of Algal Science, Technology and Medicine*, Ed. O. Konur, pp. 385–401. London: Academic Press.

Konur, O. 2023. Biosyngas-based bioethanol fuels: Scientometric study. In *Feedstock-based Bioethanol Fuels. I. Non-Waste Feedstocks: Starch, Sugar, Grass, Wood, Cellulose, Algae, and Biosyngas-based Bioethanol Fuels. Handbook of Bioethanol Fuels Volume 3*, Ed. O. Konur. Boca Raton, FL: CRC Press.

Kopke, M., C. Held and S. Hujer, et al. 2010. *Clostridium ljungdahlii* represents a microbial production platform based on syngas. *Proceedings of the National Academy of Sciences of the United States of America* 107:13087–13092.

Lee, S., G. Park and J. Lee. 2017. Importance of Ag-Cu biphasic boundaries for selective electrochemical reduction of CO_2 to ethanol. *ACS Catalysis* 7:8594–8604.

Li, C. W. and M. W. Kanan. 2012. CO_2 reduction at low overpotential on Cu electrodes resulting from the reduction of thick Cu_2O films. *Journal of the American Chemical Society* 134:7231–7234.

Li, F., Y. C. Li and Z. Wang, et al. 2020. Cooperative CO_2-to-ethanol conversion via enriched intermediates at molecule-metal catalyst interfaces. *Nature Catalysis* 3:75–82.

Li, J., Y. Zhu and D. Xiao, et al. 2016. Selective conversion of syngas to light olefins. *Science* 351:1065–1068.

Li, Y. C., Z. Wang and T. Yuan, et al. 2019. Binding site diversity promotes CO_2 electroreduction to ethanol. *Journal of the American Chemical Society* 141:8584–8591.

Liu, Y., B. Huang and Y. Dai, et al. 2009. Selective ethanol formation from photocatalytic reduction of carbon dioxide in water with $BiVO_4$ photocatalyst. *Catalysis Communications* 11:210–213.

Liu, Y., Y. Zhang and K. Cheng, et al. 2017. Selective electrochemical reduction of carbon dioxide to ethanol on a boron- and nitrogen-co-doped nanodiamond. *Angewandte Chemie - International Edition* 56:15607–15611.

Ma, S., M. Sadakiyo and R. Luo, et al. 2016. One-step electrosynthesis of ethylene and ethanol from CO_2 in an alkaline electrolyzer. *Journal of Power Sources* 301:219–228.

Ma, W., S. Xie and T. Liu, et al. 2020. Electrocatalytic reduction of CO_2 to ethylene and ethanol through hydrogen-assisted C-C coupling over fluorine-modified copper. *Nature Catalysis* 3:478–487.

Ma, X., L. Sun and C. Song. 2002. A new approach to deep desulfurization of gasoline, diesel fuel and jet fuel by selective adsorption for ultra-clean fuels and for fuel cell applications. *Catalysis Today* 77:107–116.

Mei, D., R. Rousseau and S. M. Kathmann, et al. 2010. Ethanol synthesis from syngas over Rh-based/SiO_2 catalysts: A combined experimental and theoretical modeling study. *Journal of Catalysis* 271:325–342.

Morschbacker, A. 2009. Bio-ethanol based ethylene. *Polymer Reviews* 49:79–84.

Mosier, N., R. Hendrickson, N. Ho, M. Sedlak and M. R. Ladisch. 2005. Optimization of pH controlled liquid hot water pretreatment of corn stover. *Bioresource Technology* 96:1986–1993.

Najafi, G., B. Ghobadian and T. Tavakoli, et al. 2009. Performance and exhaust emissions of a gasoline engine with ethanol blended gasoline fuels using artificial neural network. *Applied Energy* 86:630–639.

Newman, P. W. G. and J. R. Kenworthy. 1989. Gasoline consumption and cities: A comparison of U.S. cities with a global survey. *Journal of the American Planning Association* 55:24–37.

North, D. C. 1991. Institutions. *Journal of Economic Perspectives* 5:97–112.

Pan, X., Z. Fan and W. Chen, et al. 2007. Enhanced ethanol production inside carbon-nanotube reactors containing catalytic particles. *Nature Materials* 6:507–511.

Peterson, A. A., F. Abild-Pedersen, F. Studt, J. Rossmeisl and J. K. Norskov. 2010. How copper catalyzes the electroreduction of carbon dioxide into hydrocarbon fuels. *Energy and Environmental Science* 3:1311–1315.

Porosoff, M. D., B. Yan and J. G. Chen. 2016. Catalytic reduction of CO_2 by H_2 for synthesis of CO, methanol and hydrocarbons: Challenges and opportunities. *Energy and Environmental Science* 9:62–73.

Prieto, G., S. Beijer and M. I. Smith, et al. 2014. Design and synthesis of copper-cobalt catalysts for the selective conversion of synthesis gas to ethanol and higher alcohols. *Angewandte Chemie - International Edition* 53:6397–6401.

Rajagopalan, S., R. P. Datar and R. S. Lewis. 2002. Formation of ethanol from carbon monoxide via a new microbial catalyst. *Biomass and Bioenergy* 23:487–493.

Ren, D., B. S. H. Ang and B. S. Yeo. 2016. Tuning the selectivity of carbon dioxide electroreduction toward ethanol on oxide-derived Cu_xZn catalysts. *ACS Catalysis* 6:8239–8247.

Ren, D., Y. Deng and A. D. Handoko, et al. 2015. Selective electrochemical reduction of carbon dioxide to ethylene and ethanol on copper(I) oxide catalysts. *ACS Catalysis* 5:2814–2821.

Sakakura, T., J. C. Choi and H. Yasuda. 2007. Transformation of carbon dioxide. *Chemical Reviews* 107:2365–2387.

Song, Y., R. Peng and D. K. Hensley, et al. 2016. High-selectivity electrochemical conversion of CO_2 to ethanol using a copper nanoparticle/N-doped graphene electrode. *ChemistrySelect* 1:6055–6061.

Song, Y., W. Chen and C. Zhao, et al. 2017. Metal-free nitrogen-doped mesoporous carbon for electroreduction of CO_2 to ethanol. *Angewandte Chemie - International Edition* 56:10840–10844.

Subramani, V. and S. K. Gangwal. 2008. A review of recent literature to search for an efficient catalytic process for the conversion of syngas to ethanol. *Energy and Fuels* 22:814–839.

Tang, D., W. Han, P. Li, X. Miao and J. Zhong. 2011. CO_2 biofixation and fatty acid composition of *Scenedesmus obliquus* and *Chlorella pyrenoidosa* in response to different CO_2 levels. *Bioresource Technology* 102:3071–3076.

Wang, W., S. Wang, X. Ma and J. Gong. 2011. Recent advances in catalytic hydrogenation of carbon dioxide. *Chemical Society Reviews* 40:3703–3727.

Wang, X., Z. Wang and F. P. G. de Arquer, et al. 2020. Efficient electrically powered CO_2-to-ethanol via suppression of deoxygenation. *Nature Energy* 5:478–486.

Younesi, H., G. Najafpour and A. R. Mohamed. 2005. Ethanol and acetate production from synthesis gas via fermentation processes using anaerobic bacterium, *Clostridium ljungdahlii. Biochemical Engineering Journal* 27:110–119.

Index

For Product Safety Concerns and Information please contact our EU
representative GPSR@taylorandfrancis.com
Taylor & Francis Verlag GmbH, Kaufingerstraße 24, 80331 München, Germany

www.ingramcontent.com/pod-product-compliance
Lightning Source LLC
Chambersburg PA
CBHW080134220326
41598CB00032B/5060